Perspectives in Concurrency Theory

A Festschrift for P S Thiagarajan

Editors

Kamal Lodaya
Madhavan Mukund*
R Ramanujam

Institute of Mathematical Sciences, Chennai
* Chennai Mathematical Institute

Universities Press

CRC Press
Taylor & Francis Group
Boca Raton London New York

CRC is an imprint of the Taylor & Francis Group,
an informa business

Universities Press (India) Private Limited

Registered Office
3-6-747/1/A & 3-6-754/1, Himayatnagar, Hyderabad 500 029 (A.P.) India
email: info@universitiespress.com

Distributed in India, China, Pakistan, Bangladesh, Sri Lanka, Nepal, Bhutan, Indonesia, Malaysia, Singapore and Hong Kong by
Orient Blackswan Private Limited
Registered Office
3-6-752 Himayatnagar, Hyderabad 500 029 (A.P.) India

Other Offices
Bangalore / Bhopal / Bhubaneshwar / Chennai
Ernakulam / Guwahati / Hyderabad / Jaipur / Kolkata
Lucknow / Mumbai / New Delhi / Patna

Distributed in the rest of the world by
CRC Press LLC, Taylor and Francis Group,
6000 Broken Sound Parkway, NW, Suite 300, Boca Raton, FL 33487, USA

First published in India by
Universities Press (India) Private Limited 2009

ISBN (13) 978 1 4398 0943 3
ISBN (10) 1 4398 0943 7

Set in Garamond 10/12

Printed in India by
Graphica Printers, Hyderabad

Published by
Universities Press (India) Private Limited
3-6-747/1/A & 3-6-754/1, Himayatnagar, Hyderabad 500 029 (A.P.) India

Contents

Preface

This volume is a collection of articles constituting a Perspective in Concurrency Theory especially from the viewpoint of automata and logics. There are many anthologies of work on concurrency theory, but this one is special, for these are articles in honour of **P.S. Thiagarajan**, on the occasion of his 60th birthday in November 2008.

P.S. Thiagarajan, or *Thiagu* as he is known in the concurrency community, is one of the principal researchers in the subject. He has contributed extensively to the study of models and logics for concurrency, and the range of themes spanned by articles in this volume by leading researchers, many of whom have collaborated with Thiagu, offers some idea of his breadth.

We thank all the authors for their contributions, and the staff of Universities Press, especially Sreelatha Menon and Madhu Reddy, for their help in bringing this volume in time for the Festschrift. We thank S.P. Suresh of CMI for his critical help with creating the LaTeX class file used for the volume.

Thiagarajan's professional career spans three continents and five countries (as yet). After his undergraduate degree from Indian Institute of Technology (IIT) Madras in 1970, within a short span of time, he got his Ph.D. from Rice University, Texas, USA, in 1973. After a brief stint as a Research Associate at Project MAC (the predecessor of the current Laboratory for Computer Science), Massachusetts Institute of Technology (MIT) (1973–1975), he spent 8 years at Gesellschaft für Mathematik und Datenverarbeitung (GMD), Bonn, in Germany. By then, he was one of the leading contributors to the theory of Petri Nets. (Thiagu himself has always labelled his work "Petri's Net Theory", identifying his commitment to the underlying ideas of net theory, rather than the specific class of structures defined by Petri nets.)

With his shifting to Aarhus University in Denmark, Thiagu's work too shifted focus, from working within concurrency models, to studying relationships between models and exploring logics underlying models. He returned to India in 1986, where he set up the Theoretical Computer Science group at the Institute of Mathematical Sciences, Chennai. In 1989, he moved and helped set up the Chennai Mathematical Institute. During this period, the range of his work expanded to include automata theory, timed systems, control synthesis and hybrid systems. Since 2001, Thiagarajan has been with the National University of Singapore, and his current interests include system-level design methods for embedded systems, real time and hybrid systems and computational systems biology.

Thiagu's forté is conceptual clarity and technical thoroughness. His presentations are carefully structured, and theorems seem to arrive inevitably at destined times, like *pallavi* in a good Carnatic music concert. Marked by Thiagu's characteristic flashes of wit and his ready sense of humour, discussions with him are always lively.

This volume is a birthday salute to P.S. Thiagarajan from all his colleagues in the computer science community, with a round of applause, and in anticipation of his inevitable joke at any possibly solemn occasion. We wish him good health, and a long and happy research career ahead.

Kamal Lodaya, Madhavan Mukund, R. Ramanujam
Chennai, November, 2008

P.S. Thiagarajan

Then ...

... and now

Modular Synthesis of Petri Nets from Regular Languages

Eric Badouel[1], Philippe Darondeau[1], Laure Petrucci[2]
[1]*IRISA, campus de Beaulieu, F-35042 Rennes Cedex*
{Eric.Badouel,Philippe.Darondeau}@irisa.fr
[2]*LIPN, 99, avenue Jean-Baptiste Clément, F-93430 Villetaneuse*
Laure.Petrucci@lipn.univ-paris13.fr

Abstract

We propose a framework in which the synthesis of Petri nets from products of regular languages may be dealt with in a modular way, without evaluating any global language. For this purpose, we focus on distributed Petri nets, made of subnets residing in different sites of a communication network. The behaviour of each component is specified by a regular language on the union of the alphabets of this component and the components immediately upstream.

Keywords: Petri nets, distributed synthesis, communication networks

1 Introduction

The basic Petri Net Synthesis problem consists in finding whether an automaton or a labelled graph may be realized up to an isomorphism by the reachable state graph of a Petri net with injectively labelled transitions. The problem was first examined for elementary nets, and it was decided using the key concept of *regions* of a graph [12, 13, 4, 10, 5]. The similar problem for P/T-nets was decided later on in [1] using the extended concept of regions defined in [15] and in [3].

Another Petri Net Synthesis problem consists in finding whether a prefix-closed language may be realized by the set of firing sequences of a Petri net with injectively labelled transitions. This problem was solved abstractly in [14] using the concept of regions of a language. For prefix-closed regular languages, the problem was decided later on in [1].

A drawback of the synthesis procedures constructed so far in both contexts is the lack of modularity: the graph or language to be realized is given by a monolithic specification. This limitation restricts the range of the applications of Petri Net Synthesis. In this paper, we will show that the synthesis of nets from regular languages is not incompatible with modular specifications. Under some conditions which we believe reasonable, one can in fact synthesize a bounded P/T-net $\mathcal{N} = (P,T,F,M_0)$ from a finite family of regular languages $L_s \subseteq \widehat{T}_s^*$, where $\widehat{T}_s \subseteq T$ and $T = \cup_s \widehat{T}_s$, without computing their product $\otimes_s L_s$ nor any automaton accepting this product.

If $T = \cup_s \widehat{T}_s$ but no other assumption is made on the alphabets $\widehat{T}_s$ of the languages L_s, we do not know any modular solution to the net synthesis problem. Therefore, we assume that languages L_s bijectively correspond to the sites $s \in S$ of a communication network $G = (S, \rightarrow)$, where $s' \rightarrow s$ represents a channel from s' to s. We assume moreover a partition $T = \uplus_s T_s$ such that for all s, $\widehat{T}_s$ is the union of T_s and all sets $T_{s'}$ such that $s' \rightarrow s$ in G. The idea is that the synthesized P/T-nets $\mathcal{N}$ should be distributed, and a transition $t' \in T_{s'}$ cannot produce tokens consumed by $t \in T_s$ unless there is a channel from s' to s in G. According to this interpretation, $\widehat{T}_s$ is the set of all transitions $t \in T_s$ that can take place at site s plus all remote transitions that can influence the firing of the transitions $t \in T_s$, by sending tokens on channels from s' to s in G.

In this framework, if a specification $\{L_s \mid s \in S\}$ is coherent, that is the languages L_s are the respective projections of some language $L \subseteq T^*$, one can decide in a modular way whether there exists a bounded and distributed net implementing them. Even though coherence can in general not be checked in a modular way, it is fortunately the case when G is minimally connected, which excludes e.g. ring architectures but still covers a lot of practical cases.

The rest of the paper is organized as follows. In Section 2, we recall the concept of Distributed Petri Nets and the way to convert them into communicating finite state machines with bounded channels. In Section 3, we state the Distributed Net Synthesis Problem to be solved, and we show that this problem may be decomposed over sites, thus leading to the Open Net Synthesis Problem. The Open Net Synthesis Problem is addressed in Section 4, using a suitable adaptation of the concept of regions of a language. A brief conclusion completes the paper.

2 Distributed Petri Nets

In this section, we refine the concept of Distributed Petri Nets introduced in [2] and we recall their relationship to communicating finite state machines. To begin with, we recall the definition of Place/Transition nets.

Definition 1 *[P/T-net] A* P/T-net *is a bi-partite graph* $N=(P,T,F)$, *where* P *and* T *are disjoint sets of vertices, called* places *and* transitions, *respectively, and* $F:(P\times T)\cup(T\times P)\to\mathbb{N}$ *is a set of directed edges with non-negative integer weights. A* marking *of* N *is a map* $M:P\to\mathbb{N}$. *The* state graph *of* N *is a labelled graph, with markings as vertices, where there is an edge from* M *to* M' *with label* $t\in T$ *(notation:* $M[t\rangle M'$*) if and only if, for every place* $p\in P$, $M(p)\geq F(p,t)$ *and* $M'(p)=M(p)-F(p,t)+F(t,p)$. *The* reachability graph *of an* initialized *P/T-net* $\mathcal{N}=(P,T,F,M_0)$ *with the* initial marking M_0 *is the induced restriction of its state graph on the set of markings that may be reached from* M_0. *The net* $\mathcal{N}$ *is* finite *if* P *and* T *are finite. The net* $\mathcal{N}$ *is* bounded *if its reachability graph is finite. The* language $L(\mathcal{N})$ *of the net* $\mathcal{N}$ *is the set of words* $t_1t_2\dots t_l\in T^*$ *that label firing sequences* $M_0[t_1\rangle M_1[t_2\rangle\dots M_{l-1}[t_l\rangle M_l$. *We use* $M_0[t_1\dots t_l\rangle M_l$ *as a shorter notation for such a firing sequence.*

It follows from the net firing rule that whenever $M[w\rangle M'$ for some word $w\in T^*$, $M'(p)=M(p)+\sum_{t\in T}\psi(w)(t)\times(F(t,p)-F(p,t))$ for every place $p\in P$, where $\psi(w)$ is the firing count vector of w, also called the Parikh image of w.

Definition 2 *The* Parikh image *of a word* w *of* $\widehat{T}^*$, *where* $\widehat{T}=\{t_1,\dots,t_k\}$, *is the map* $\psi(w):\widehat{T}\to\mathbb{N}$ *such that* $\psi(w)(t_h)$ *counts the occurrences of* t_h *in* w. *By a slight abuse of notation, we sometimes use the alternative map* $\psi(w):\{1,\dots,k\}\to\mathbb{N}$ *such that* $\psi(w)(h)$ *counts the occurrences of* t_h *in* w.

Distributed Petri Nets are P/T-nets in which the weighted flow relation $F:(P\times T)\cup(T\times P)\to\mathbb{N}$ includes predefined constraints $F(p,t)=0$ or $F(t,p)=0$. These constraints reflect the architecture of a communication network, and they ensure that the net may be realized with a communicating finite state machine mapped on this architecture.

Definition 3 *[Distributed Net Architecture] Given a set of transitions* T, *a* distributed net architecture *is defined by a* location map $\lambda:T\to S$ *and a* communication graph $G=(S,\to)$ *where* S *is a finite set of* sites. *For* $s\in S$, *we note* $T_s=\lambda^{-1}\{s\}$ *and* $\widehat{T}_s=T_s\cup\lambda^{-1}\{s'\in S\,|\,s'\to s\}$.

T_s is the set of all transitions that can occur at site s. $\widehat{T}_s$ is the set of all transitions that can exert direct influence on the firability of transitions in T_s. In the sequel, $\pi_s:T^*\to T_s^*$ is the unique monoid morphism such that $\pi_s(t) = t$ for $t \in T_s$ and $\pi_s(t) = \epsilon$ (the empty word) otherwise. Similarly, $\widehat{\pi}_s : T^* \rightarrow \widehat{T}_s^*$ is the unique monoid morphism such that $\widehat{\pi}_s(t) = t$ for $t \in \widehat{T}_s$ and $\widehat{\pi}_s(t) = \epsilon$ otherwise.

Definition 4 *[Distributed P/T-net] Given a location map $\lambda : T \rightarrow S$ and a* connected *communication graph $G = (S, \rightarrow)$, a* distributed P/T-net *is a P/T-net in which the following requirements are satisfied:*
- $(\forall p \in P)(\exists t \in T)\, F(p,t) \neq 0$
- $(\forall p \in P)(\forall t, t' \in T)\, F(p,t) \neq 0 \wedge F(p,t') \neq 0 \Rightarrow \lambda(t) = \lambda(t')$
- $(\forall p \in P)(\forall t, t' \in T)\, F(p,t) \neq 0 \wedge F(t',p) \neq 0 \Rightarrow \lambda(t) = \lambda(t') \vee \lambda(t') \rightarrow \lambda(t)$

In view of the first two requirements in Def. 4, the location map $\lambda : T \rightarrow S$ extends in a unique way to a map $\lambda : T \cup P \rightarrow S$ such that $F(p,t) \neq 0 \Rightarrow \lambda(p) = \lambda(t)$. So, the places of a distributed net are located in sites. The first two requirements stipulate that a transition t located in site $\lambda(t)$ cannot consume tokens from a place p located in a different site $\lambda(p) \neq \lambda(t)$. The third condition stipulates that a transition t' located in site $\lambda(t')$ can produce and send tokens to a place p only if $\lambda(p) = \lambda(t')$ or there is an edge from $\lambda(t')$ to $\lambda(p)$ in graph G, figuring a channel from $\lambda(t')$ to $\lambda(p)$.

Example 1 Let us consider a communication system, sketched in figure 1, with four machines PC_1, PC_2, PC_3 and PC_4 connected by a network comprising a *router*. Machines PC_1 and PC_3 send messages, while machines PC_2 and PC_4 wait for incoming messages. Machines PC_1 and PC_3 send messages to machine PC_2 via the router. The router uses a *store and forward* strategy, i.e. each incoming message is stored and then forwarded to its destination. Copies of all messages from PC_3 to PC_2 are sent directly to PC_4.

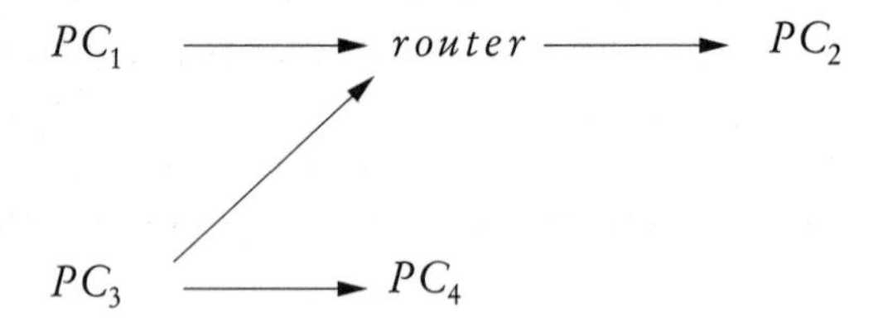

Figure 1: *Example of a routing system*

Figure 2 shows a distributed Petri net modelling this system. The distribution is such that: $S = \{PC_1, PC_2, PC_3, PC_4, router\}$ with the communication graph as in figure 1 and

$\lambda(P_{11}) = \lambda(P_{12}) = \lambda(T_{11}) = \lambda(T_{12}) = PC_1$
$\lambda(B_{R2}) = \lambda(P_{21}) = \lambda(P_{22}) = \lambda(T_{21}) = \lambda(T_{22}) = PC_2$
$\lambda(P_{31}) = \lambda(P_{32}) = \lambda(T_{31}) = \lambda(T_{32}) = PC_3$
$\lambda(B_{34}) = \lambda(P_{41}) = \lambda(P_{42}) = \lambda(T_{41}) = \lambda(T_{42}) = PC_4$
$\lambda(R_1) = \lambda(R_2) = \lambda(P_{R1}) = \lambda(P_{R2}) = \lambda(T_{1R}) = \lambda(T_{3R}) = \lambda(T_{R2}) = router$

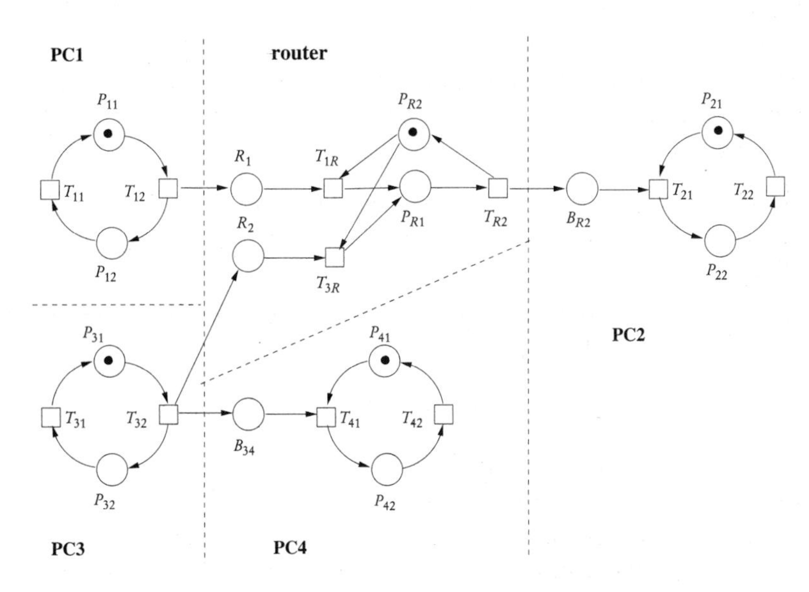

Figure 2: *Distributed Petri net modelling the routing system*

The sets T_s and $\widehat{T}_s$ are thus defined by:

$T_{PC_1} = \{T_{11}, T_{12}\}$, $\widehat{T}_{PC_1} = \{T_{11}, T_{12}\}$
$T_{PC_2} = \{T_{21}, T_{22}\}$, $\widehat{T}_{PC_2} = \{T_{21}, T_{22}, T_{1R}, T_{3R}, T_{R2}\}$
$T_{PC_3} = \{T_{31}, T_{32}\}$, $\widehat{T}_{PC_3} = \{T_{31}, T_{32}\}$
$T_{PC_4} = \{T_{41}, T_{42}\}$, $\widehat{T}_{PC_4} = \{T_{41}, T_{42}, T_{31}, T_{32}\}$
$T_{router} = \{T_{1R}, T_{3R}, T_{R2}\}$, $\widehat{T}_{router} = \{T_{1R}, T_{3R}, T_{R2}, T_{11}, T_{12}, T_{31}, T_{32}\}$

From now on, $\mathcal{N} = (P, T, F, M_0)$ is a finite initialized P/T-net, bounded and distributed w.r.t. $\lambda : T \cup P \to S$ and $G = (S, \to)$. The distributed architecture (λ, G) induces a decomposition of $\mathcal{N}$ into an indexed family of components $\mathcal{N}_s$ for s ranging over S, as follows. Each subnet $\mathcal{N}_s$ is the induced restriction of $\mathcal{N}$ on the subset of places $P_s = P \cap \lambda^{-1}\{s\}$ and on the subset of transitions $\widehat{T}_s$, i.e. $\mathcal{N}_s = (P_s, \widehat{T}_s, F_s, M_{0s})$ where M_{0s} is the restriction of M_0 on P_s and

F_s is the restriction of F on $(P_s \times \widehat{T}_s) \cup (\widehat{T}_s \times P_s)$. Thus $\mathcal{N} = \oplus_s \mathcal{N}_s$ according to the definition below.

Definition 5 *The sum of a family of place disjoint nets* $\mathcal{N}_s = (P_s, \widehat{T}_s, F_s, M_{0s})$, $s \in S$, *is the net* $\oplus_s \mathcal{N}_s = (P, T, F, M_0)$ *defined with* $P = \uplus_s P_s$, $T = \cup_s \widehat{T}_s$, $F(p,t) = F_s(p,t)$ *if* $p \in P_s$ *and* $t \in \widehat{T}_s$ *for some* $s \in S$ *else* 0, $F(t,p) = F_s(t,p)$ *if* $p \in P_s$ *and* $t \in \widehat{T}_s$ *for some* $s \in S$ *else* 0, *and* $M_0(p) = M_{0s}(p)$ *for* $p \in P_s$.

In order to explain the relationship between the language $L(\mathcal{N})$ of the net and the languages of the net components $L(\mathcal{N}_s)$, we recall the following definition.

Definition 6 *[mixed product of languages, adapted from [11]] Given* $L_s \subseteq \widehat{T}_s^*$ *for* s *ranging over* S, *let* $\otimes_s L_s \subseteq T^*$ *be defined as* $\{w \mid \forall s : \widehat{\pi}_s(w) \in L_s\}$.

The following propositions follow immediately from the definition of distributed P/T-nets.

Proposition 7 $L(\mathcal{N}_s) \cdot (\widehat{T}_s \setminus T_s) \subseteq L(\mathcal{N}_s)$.

Proposition 8 $L(\mathcal{N}_s) \supseteq \widehat{\pi}_s(L(\mathcal{N}))$.

Proposition 9 $L(\mathcal{N}) = L(\oplus_s \mathcal{N}_s) = \otimes_s L(\mathcal{N}_s)$.

We state now two other useful properties of the component nets $\mathcal{N}_s$ induced by the decomposition of a bounded and distributed P/T-net.

Proposition 10 *The following two properties hold for all* $s \in S$ *and* $t \in T_s$:
i) $(\forall w \in L(\mathcal{N}))\ wt \notin L(\mathcal{N}) \Rightarrow \widehat{\pi}_s(wt) \notin L(\mathcal{N}_s)$,
ii) $(\forall w' \in \widehat{\pi}_s(L(\mathcal{N})))\ w't \notin \widehat{\pi}_s(L(\mathcal{N})) \Rightarrow w't \notin L(\mathcal{N}_s)$.

Proof We first prove (i). Let $w \in L(\mathcal{N})$ and $t \in T_s$ be such that $wt \notin L(\mathcal{N})$. In view of the net firing rule, $0 \leq M_0(p) + \sum_{t' \in T} \psi(w)(t') \times (F(t',p) - F(p,t')) < F(p,t)$ for some place $p \in P$, and $F(p,t) > 0$ entails that $\lambda(p) = \lambda(t) = s$. From Def. 4, $F(p,t') = 0$ and $F(t',p) = 0$ for every transition t' outside $\widehat{T}_s$, while by Def. 5, $F(p,t') = F_s(p,t')$ and $F(t',p) = F_s(t',p)$ for every transition t' in $\widehat{T}_s$ (in particular for t). So, if we set $w' = \widehat{\pi}_s(w)$, then $M_{0s}(p) + \sum_{t' \in \widehat{T}_s} \psi(w')(t') \times (F_s(t',p) - F_s(p,t')) < F_s(p,t)$, showing that $w't = \widehat{\pi}_s(wt) \notin L(\mathcal{N}_s)$.

We now show that (i) entails (ii). Let $w' \in \widehat{\pi}_s(L(\mathcal{N}))$ and $t \in T_s$ such that $w't \notin \widehat{\pi}_s(L(\mathcal{N}))$. Since $w' \in \widehat{\pi}_s(L(\mathcal{N}))$, $w' = \widehat{\pi}_s(w)$ for some $w \in L(\mathcal{N})$. From $w' = \widehat{\pi}_s(w)$ and $w't \notin \widehat{\pi}_s(L(\mathcal{N}))$, necessarily $wt \notin L(\mathcal{N})$ hence by (i), $\widehat{\pi}_s(wt) \notin L(\mathcal{N}_s)$, i.e. $w't \notin L(\mathcal{N}_s)$. ⊣

Definition 11 *[restricted boundedness] Given $s \in S$ and $L_s \subseteq \widehat{T}_s^*$, the net $\mathcal{N}_s = (P_s, \widehat{T}_s, F_s, M_{0s})$ is said to be* bounded in restriction to L_s *if there exists some finite bound B such that $(\forall p \in P_s)(\forall w' \in L_s \cap L(\mathcal{N}_s))$ $M_{0s}[w'\rangle M' \Rightarrow M'(p) \leq B$.*

Proposition 12 *Let $\mathcal{N}$ be a bounded and distributed P/T-net. Then for all $s \in S$, the component subnet $\mathcal{N}_s$ of $\mathcal{N}$ is bounded in restriction to $\widehat{\pi}_s(L(\mathcal{N}))$.*

Proof Let $M_0[w\rangle M$ in $\mathcal{N}$ and $M_{0s}[w'\rangle M'$ in $\mathcal{N}_s$, with $w' = \widehat{\pi}_s(w)$. As $\mathcal{N}$ is a bounded net, for any place $p \in P$, $M(p) \leq B$ for some finite bound B independent of M and p. From Def. 4, for any place $p \in P_s$, $M(p) = M_{0s}(p) + \sum_{t' \in \widehat{T}_s} \psi(w')(t') \times (F_s(t', p) - F_s(p, t')) = M'(p)$, hence $M'(p) \leq B$. ⊣

In the end of this section, we sketch a simple translation from bounded and distributed P/T-nets to communicating finite state machines with an equivalent behaviour. Let $\mathcal{N}$ be a bounded and distributed P/T-net over (λ, G). For any place $p \in P$, there exists a finite bound $B(p)$ such that $M(p) \leq B(p)$ for every reachable marking M of $\mathcal{N}$. Note that, for any place $p \in P_s$, $B(p)$ may be determined locally from $\widehat{\pi}_s(L(\mathcal{N}))$ if this language is known, without computing the global state graph of $\mathcal{N}$. This will be the case for nets $\mathcal{N}$ synthesized from a specification $\{L_s \mid s \in S\}$ since $\widehat{\pi}_s(L(\mathcal{N})) = L_s$ for such nets.

Given a bounded and distributed P/T-net $\mathcal{N}$, we shall simulate the behaviour of this net with a communicating machine $(\mathcal{A}, \mathcal{C})$ where $\mathcal{A} = \{A_s \mid s \in S\}$ is a set of finite automata and $\mathcal{C} = \{(s, s') \mid s \rightarrow s'$ in $G\}$ is a set of channels. The set of messages that may be sent to or received from a channel (s, s') is the set of names of places p' such that $p' \in P_{s'}$ and $F(t, p') > 0$ for some transition $t \in T_s$. For each $s \in S$, the automaton A_s has an extended alphabet $T_s \cup \tau_s$, where T_s is a set of internal actions representing homonymic transitions of $\mathcal{N}$, and τ_s is a set of communication actions. Each alphabet τ_s comprises two types of actions on channels: an action $!s'(p')$ means sending message p' on channel (s, s') in order to simulate the emission of a token (intended for place p'); an action $?s(p')$ means receiving message p' from channel (s, s') in order to simulate the delayed arrival of a token in place p'. Channels are unordered, i.e. messages sent on channel (s, s') may be received in a different order. The state of a channel may therefore be seen as a vector of counters (one for each p'). The state of the communicating machine is defined as the set of states of the component automata A_s plus the set of states of the channels. An action $!s'(p')$ is always enabled when it is enabled in some automaton

A_s such that $(s,s') \in \mathscr{C}$. An action $?s(p')$ is enabled if it is enabled in some automaton A_s such that $(s,s') \in \mathscr{C}$ and this channel contains at least one occurrence of the message p'.

We now define, for each $s \in S$, the set of states Q_s and the transition function δ_s of the automaton A_s. First, we let Q_s be the set of all maps $M' : P_s \to \mathbb{N}$ such that $(\forall p \in P_s) M'(p) \leq B(p)$. Second, we let $q_{0s} = M_{0s}$ and $A_s = (Q_s, T_s \cup \tau_s, \delta_s, q_{0s})$ where $\delta_s : Q_s \times (T_s \cup \tau_s)^* \to Q_s$ is the partial transition function defined as follows:

- for each $t \in T_s$ and $q, q' \in Q_s$ such that $q[t\rangle q'$ is a firing step of $\mathcal{N}_s$ (although q is not necessarily reachable from M_{0s}), let $\delta_s(q, t \cdot send(t)) = q'$ for an arbitrary linearisation $send(t) \in \tau_s^*$ of the multiset $\{!s'(p') \times F(t,p') \mid s' \neq s \wedge p' \in P_{s'}\}$,
- for each $s' \neq s$ and $p \in P_s$ such that $F(t',p) \neq 0$ for some $t' \in T_{s'}$, provided that $q(p) < B(p)$, let $\delta_s(q, ?s'(p)) = q'$ where $q'(p) = q(p) + 1$ and $q'(p') = q(p')$ for $p' \neq p$.

Note that when a transition produces tokens for one or several distant places, these tokens are sent on the communication network in one batch while they are received one at a time. In order to restore the symmetry between send actions and receive actions, one may expand the automata A_s to automata with larger sets of states Q_s' and partial transition functions $\delta_s : Q_s' \times (T_s \cup \tau_s) \to Q_s'$. We claim that the resulting family of finite state machines $\{A_s \mid s \in S\}$ communicating through the channels (s', s) or $s' \to s$ defined in G, is channel bounded and implements $L(\mathcal{N})$. By adapting the proofs given in [2], one may further show that $\mathcal{N}$ and the considered communicating finite state machines have branching bisimilar behaviours [17] when all communication actions $?s(p)$ and $!s(p)$ are dealt with as silent actions τ.

Remarks: Following the indications given in section 5 of [2], one may define an alternative translation from distributed nets to communicating systems $(\{\mathcal{N}_s' \mid s \in S\}, \mathscr{C})$ where each component $\mathcal{N}_s'$ is a Petri net with possible concurrency.

3 The Distributed Net Synthesis Problem

The problem we want to address may be stated as follows.

Definition 13 *[Net Specification] Given a distributed net architecture* (λ, G) *with* $\lambda : T \to S$ *and* $G = (S, \to)$, *a* net specification $\mathscr{L}$ *is an* *indexed family of prefix-closed and regular languages* $L_s \subseteq \widehat{T}_s^*$ *where* s *ranges over* S.

Problem 1 (Net Synthesis Problem) *Given a distributed net architecture* (λ, G) *and a specification* $\{L_s \mid s \in S\}$, *decide whether there exists and construct a bounded and distributed P/T-net* $\mathcal{N}$ *such that* $\widehat{\pi}_s(L(\mathcal{N})) = L_s$ *for all* $s \in S$.

We moreover want to solve this problem in a *modular* way, without ever computing the mixed product $\otimes_s L_s$ nor the global language of any net over (λ, G). In particular, we forbid ourselves to compute from the specification some global net $\mathcal{N}$ and then check whether $\widehat{\pi}_s(L(\mathcal{N})) = L_s$ for all $s \in S$ as required. It is worth noting that a specification $\mathcal{L} = \{L_s \mid s \in S\}$ does generally not define unambiguously any global language L over T, as the following example shows.

Example 2 Let $S = \{1,2,3\}$ with $T_1 = \{a\}$, $T_2 = \{b\}$, $T_3 = \{c\}$, and $G = \{1 \to 2, 1 \to 3\}$. Consider the net specification given by $L_1 = pref\{a\}$, $L_2 = pref\{ab\}$, and $L_3 = pref\{ac\}$ where $pref\,E$ means the set of all left factors of words in E. Then there exists several languages $L \subseteq T^*$ such that $L_i = \widehat{\pi}_i(L)$ for every i in $\{1,2,3\}$, for instance $L = pref\,\{abc, acb\}$, $L = pref\,\{abc\}$, $L = pref\,\{acb\}$, or $L = pref\,\{ab, ac\}$. In fact, only $L = pref\,\{abc, acb\}$ coincides with the language of a solution $\mathcal{N}$ to the distributed net synthesis problem from $\mathcal{L} = \{L_1, L_2, L_3\}$.

It seems however desirable that a specification $\mathcal{L} = \{L_s \mid s \in S\}$ over (λ, G) defines at least one global language L over T, hence the following definition.

Definition 14 *[Coherency] Given a distributed net architecture* (λ, G) *with* $\lambda : T \to S$ *and* $G = (S, \to)$, *a specification* $\{L_s \mid s \in S\}$ *is* coherent *if there exists some language* L *over* T *such that* $\widehat{\pi}_s(L) = L_s$ *for all* $s \in S$.

Proposition 15 *A specification* $\mathcal{L} = \{L_s \mid s \in S\}$ *is coherent if and only if* $(\forall s')\, L_{s'} = \widehat{\pi}_{s'}(\otimes_s L_s)$.

Proof Suppose $\widehat{\pi}_s(L) = L_s$ for all $s \in S$ then clearly $L \subseteq \otimes_s L_s$ and for all s', $L_{s'} = \widehat{\pi}_{s'}(\otimes_s L_s)$ because $\widehat{\pi}_{s'}(L) \subseteq \widehat{\pi}_{s'}(\otimes_s L_s) \subseteq L_{s'}$ and $\widehat{\pi}_{s'}(L) = L_{s'}$. The converse implication is immediate. ⊣

Proposition 16 *Let* $\mathcal{L} = \{L_s \mid s \in S\}$ *be a coherent specification and let* $\mathcal{N} = \oplus_s \mathcal{N}_s$ *be a distributed net. Then* $L(\mathcal{N}) = \otimes_s L_s$ *if and only if* $\widehat{\pi}_s(L(\mathcal{N})) = L_s$ *for all* $s \in S$.

Proof Suppose that $\widehat{\pi}_s(L(\mathcal{N})) = L_s$ for all $s \in S$. Then $L(\mathcal{N}) \subseteq \otimes_s L_s$. Moreover, for all $s \in S$, $\widehat{\pi}_s(L(\mathcal{N})) \subseteq L(\mathcal{N}_s)$ because

$L(\mathcal{N}) = \otimes_s L(\mathcal{N}_s)$. Therefore $\otimes_s L_s \subseteq \otimes_s L(\mathcal{N}_s) = L(\mathcal{N})$. Altogether, $L(\mathcal{N}) = \otimes_s L_s$. The converse implication follows from Prop.15. ⊣

In view of Prop. 16, for coherent specifications, Problem 1 is equivalent to the following.

Problem 2 *Given a net architecture* (λ, G) *and a specification* $\{L_s \mid s \in S\}$*, decide whether there exists and construct a bounded and distributed P/T-net* $\mathcal{N}$ *over* (λ, G) *such that* $L(\mathcal{N}) = \otimes_s L_s$.

In view of Def. 14, the coherency of a specification $\mathcal{L} = \{L_s \mid s \in S\}$ is necessary to the existence of solutions to Problem 1, but Problem 2 may have solutions for incoherent specifications. Indeed $L(\mathcal{N}) = \otimes_s L(\mathcal{N}_s)$ for any distributed net $\mathcal{N}$, but in most cases $L(\mathcal{N}_s)$ is a strict superset of $\widehat{\pi}_s L(\mathcal{N})$ and therefore $\{L(\mathcal{N}_s) \mid s \in S\}$ is not a coherent specification.

Coherency is a decidable property since it expresses as $(\forall s \in S)$ $L_s = \widehat{\pi}_s(\otimes_s L_s)$ and any mixed product or projection of regular languages is a regular language. However one can generally not check coherency without computing the mixed product $\otimes_s L_s$. Computing from $\mathcal{L}$, without a preliminary check of coherency, some net $\mathcal{N}$ candidate as a solution to Problem 1 does not help overcoming the difficulty because, unless coherency is assumed, it is generally not possible to check that $\widehat{\pi}_s(L(\mathcal{N})) = L_s$ for all $s \in S$ without computing $L(\mathcal{N})$. The modular synthesis of P/T-nets can therefore not be envisaged without imposing constraints on distributed net architectures, hence the following definition.

Definition 17 *[Minimally Connected] A communication graph* $G = (S, \rightarrow)$ *is* minimally connected *if the underlying undirected multigraph is a graph and this graph is a tree (with arbitrary root vertex).*

Example 3 The communication graph of example 1 (figure 1) is minimally connected.

Lemma 18 *Let* (λ, G) *be a net architecture with a minimally connected graph* $G = (S, \rightarrow)$. *Then* $\widehat{T}_{s'} \cap \widehat{T}_s = T_{s'}$ *for every edge* $s' \rightarrow s$ *of* G.

Proof By definition, $\widehat{T}_{s'}$ is the union of $T_{s'}$ and all subsets $T_{s''}$ such that $s'' \rightarrow s'$. Similarly, $\widehat{T}_s$ is the union of T_s and all subsets $T_{s''}$ such that $s'' \rightarrow s$, thus including $T_{s'}$. By minimal connectedness of G, $s'' \rightarrow s'$ entails $s'' \neq s$, and there exists no vertex s'' such that $s'' \rightarrow s$ and $s'' \rightarrow s'$. As subsets $T_{s''}$ are pairwise disjoint, the lemma follows. ⊣

Lemma 19 *Let (λ, G) be a net architecture with a minimally connected graph $G = (S, \rightarrow)$. Then, for any two distinct vertices s' and s, $\widehat{T}_{s'} \cap \widehat{T}_s = \emptyset$ unless $s' \rightarrow s$ or $s \rightarrow s'$ or $s'' \rightarrow s'$ and $s'' \rightarrow s$ for some (necessarily unique) s''. Moreover, in the latter case, $\widehat{T}_{s'} \cap \widehat{T}_s = T_{s''}$.*
Proof Left to the reader. ⊣

Proposition 20 *Let $\mathcal{L} = \{L_s \mid s \in S\}$ be a specification over (λ, G) where G is minimally connected. Then $\mathcal{L}$ is coherent if and only if $\pi_{s'}(L_{s'}) = \pi_{s'}(L_s)$ for every edge $s' \rightarrow s$ in G.*
Proof By Prop. 15, $\{L_s \mid s \in S\}$ is coherent if and only if, for any $s' \in S$ and for any $w \in L_{s'}$, there exists an indexed family of words $\{w_s \mid s \in S \wedge w_s \in L_s\}$ such that $w = w_{s'}$ and $w \in \widehat{\pi}_{s'}(\otimes_s \{w_s\})$.

By construction of the alphabets T_s and $\widehat{T}_s$, $s \in S$, $T_{s'} = \widehat{T}_{s'} \cap \widehat{T}_s$ whenever $s' \rightarrow s$, and then $\pi_{s'} \circ \widehat{\pi}_{s'} = \pi_{s'} \circ \widehat{\pi}_s$. Thus if the specification $\mathcal{L}$ is coherent, $\pi_{s'}(L_{s'}) = \pi_{s'} \circ \widehat{\pi}_{s'}(\otimes_s L_s) = \pi_{s'} \circ \widehat{\pi}_s(\otimes_s L_s) = \pi_{s'}(L_s)$.

Suppose that $\pi_{s'}(L_{s'}) = \pi_{s'}(L_s)$ for every edge $s' \rightarrow s$ in G. We show that $\{L_s \mid s \in S\}$ is coherent. Let s' be an arbitrary vertex of G, and let $w_{s'} \in L_{s'}$. For all $n \geq 0$, let S_n be the subset of vertices at a distance at most n from s', thus $S_0 = \{s'\}$ and $S_m = S_{m+1}$ for some m. We proceed by induction on $n < m$.

Assume that for each $s \in S_n$, w_s has been chosen from L_s such that, for all vertices $s'', s \in S_n$, if $s'' \rightarrow s$ is an edge of G then $\pi_{s''}(w_{s''}) = \pi_{s''}(w_s)$. As G is minimally connected, any vertex $s'' \in S_{n+1} \setminus S_n$ is connected by an edge $s'' \rightarrow s$ or $s \rightarrow s''$ to *exactly one* vertex $s \in S_n$, and there is no edge between two distinct vertices in $S_{n+1} \setminus S_n$. As $(s'' \rightarrow s) \Rightarrow (\pi_{s''}(L_{s''}) = \pi_{s''}(L_s))$ and $(s \rightarrow s'') \Rightarrow (\pi_s(L_s) = \pi_s(L_{s''}))$, one can choose independently for all $s'' \in S_{n+1} \setminus S_n$ some $w_{s''}$ from $L_{s''}$ such that $\pi_{s''}(w_{s''}) = \pi_{s''}(w_s)$ if $s'' \rightarrow s$ or $\pi_s(w_{s''}) = \pi_s(w_s)$ if $s \rightarrow s''$.

Now let $n = m$, thus w_s has been defined for all $s \in S$, and $\pi_{s''}(w_{s''}) = \pi_{s''}(w_s)$ for every edge $s'' \rightarrow s$ in G. At this stage, Lemmas 18 and 19 show that $\otimes_s w_s$ is not the empty set, hence $w \in \widehat{\pi}_{s'}(\otimes_s \{w_s\})$. ⊣

Prop. 20 shows that when the communication graph G is minimally connected, the coherency of specifications over (λ, G) can be checked in a modular way. For the purpose of modular net synthesis, we impose on specifications a slightly stronger requirement of coherency as follows.

Definition 21 *[Strong Coherency] A specification $\mathcal{L} = \{L_s \mid s \in S\}$ over (λ, G) is* strongly coherent *if it is coherent and the following condition is satisfied for every edge $s' \rightarrow s$ in G:*
$(\forall w \in L_s)(\forall t \in T_{s'})\, \pi_{s'}(wt) \in \pi_{s'}(L_{s'}) \Rightarrow wt \in L_s$.

Example 4 We now consider example 1 again. Let PC_1, PC_2, PC_3, PC_4 and *router* be the subnets shown on figure 2, and let *router'* be the net obtained by removing places $R1$ and $R2$ from *router*. For any two subnets $\mathcal{N}'$ and $\mathcal{N}''$, let $\mathcal{N}'+\mathcal{N}''$ be the net containing the places and transitions of both nets plus the connecting arcs drawn on figure 2. Define $L_{PC_1} = L(PC_1)$, $L_{PC_2} = L(PC_2 + router')$, $L_{PC_3} = L(PC_3)$, $L_{PC_4} = L(PC_4 + PC_3)$, and $L_{router} = L(router + PC_1 + PC_3)$. Then $\mathcal{L} = \{L_s \mid s \in S\}$ is a strongly coherent specification.

For coherent specifications, in view of Prop. 20, the above condition may be rewritten to $(\forall w \in L_s)(\forall t \in T_{s'})\, \pi_{s'}(wt) \in \pi_{s'}(L_s) \Rightarrow wt \in L_s$, hence strong coherency may be checked in a modular way.

Proposition 22 *Let $\mathcal{N}$ be a distributed P/T-net over (λ, G). If G is minimally connected, then $\mathcal{L} = \{\widehat{\pi}_s(L(\mathcal{N})) \mid s \in S\}$ is strongly coherent.*
Proof The coherency of $\mathcal{L}$ follows clearly from its definition. It remains to show that $\mathcal{L}$ is strongly coherent. Let $s, s' \in S$ such that $s' \to s$ in G. Let $w \in \widehat{\pi}_s(L(\mathcal{N}))$ and $t \in T_{s'}$ such that $\pi_{s'}(wt) \in \pi_{s'} \circ \widehat{\pi}_{s'}(L(\mathcal{N}))$. We show that $wt \in \widehat{\pi}_s(L(\mathcal{N}))$. First, it is possible to choose a word $w' \in \widehat{\pi}_{s'}(L(\mathcal{N}))$ such that $\pi_{s'}(w) = \pi_{s'}(w')$ and $w't \in \widehat{\pi}_{s'}(L(\mathcal{N}))$. Second, proceeding as in the proof of Prop. 20, it is possible to extend this choice to a full family of words $w_\sigma \in \widehat{\pi}_\sigma(L(\mathcal{N}))$, $\sigma \in S$, such that $w_s = w$, $w_{s'} = w'$, and $w_{s''} \in \widehat{\pi}_{s''}(\otimes_\sigma w_\sigma)$ for all s''. By Prop. 8, $w_{s''} \in L(\mathcal{N}_{s''})$ for all s''. Now consider the second family of words u_σ, $\sigma \in S$, defined with $u_\sigma = w_\sigma t$ if $t \in \widehat{T}_\sigma$ and $u_\sigma = w_\sigma$ otherwise. Then, $u_\sigma \in L(\mathcal{N}_\sigma)$ for all $\sigma \in S$, and $u_{s''} \in \widehat{\pi}_{s''}(\otimes_\sigma u_\sigma)$ for all s''. By Prop. 9, $u_\sigma \in \widehat{\pi}_\sigma(L(\mathcal{N}))$ for all $\sigma \in S$. In particular, $u_s = wt \in \widehat{\pi}_s(L(\mathcal{N}))$. Therefore, $\mathcal{L}$ is strongly coherent.
$\dashv$

Imposing minimally connected architectures is restrictive, since this excludes in particular ring architectures. We postpone comments on this point. On the contrary, in view of Prop. 22 and the statement of Problem 1, strong coherency is a necessary condition for a specification to have some distributed net realization. It is therefore not restrictive to assume strong coherency, which can be checked in a modular way for minimally connected architectures. We feel that modular synthesis of nets is not possible without imposing this restriction on architectures. In order to solve Problem 1 for a specification $\mathcal{L} = \{L_s \mid s \in S\}$ over an architecture (λ, G) which is not minimally connected, we can only suggest to choose if possible a non-trivial equivalence relation $\equiv$ on S such that $G/\equiv$ is minimally connected, and to solve Problem 1 for the aggregated specification $\mathscr{L}' = \{L'_x \mid x \in X\}$, where $X = (S/\equiv)$ and $L'_x = \otimes_{s \in X} L_s$.

In the sequel, we consider always strongly coherent specifications over minimally connected architectures. In view of Prop. 16, Problem 1 is then equivalent to the following.

Problem 3 *Given a net architecture* (λ, G) *in which* G *is minimally connected, and given a strongly coherent specification* $\{L_s \mid s \in S\}$*, decide whether there exists and construct a bounded and distributed P/T-net* $\mathcal{N}$ *over* (λ, G) *such that* $L(\mathcal{N}) = \otimes_s L_s$.

We will solve Problem 3 without ever computing the mixed product $\otimes_s L_s$ nor the global language of any net over (λ, G). The keys to the modular synthesis of bounded and distributed P/T-nets are given by the following two propositions.

Proposition 23 *Let* $\{L_s \mid s \in S\}$ *be a strongly coherent specification, and let* $\{L'_s \mid s \in S\}$ *be an arbitrary set of prefix-closed languages* $L'_s \subseteq \widehat{T}^*_s$ *such that* $L'_s \cdot (\widehat{T}_s \setminus T_s) \subseteq L'_s$ *for all* s*. Then* $\otimes_s L_s = \otimes_s L'_s$ *if and only if, for all* $s \in S$*:* $L_s \subseteq L'_s \subseteq \complement((L_s \cdot T_s \setminus L_s) \cdot \widehat{T}^*_s)$ *where* $\complement$ *means complementation w.r.t.* $\widehat{T}^*_s$.

Proof **If part** Suppose that the stated conditions are satisfied. Clearly, $\otimes_s L_s \subseteq \otimes_s L'_s$. In order to show the converse inclusion, we proceed by contradiction, and assume the existence of some minimal word $wt \in (\otimes_s L'_s) \setminus (\otimes_s L_s)$ with $t \in T$. From the minimality assumption, $w \in (\otimes_s L'_s) \cap (\otimes_s L_s)$, and therefore $w \in \otimes_s \{w_s\}$ for some family of words $w_s \in L_s$ such that $w_s = \widehat{\pi}_s(w) \in L_s \subseteq L'_s$ for all s. Let $t \in T_{s'}$. Define a second family of words w'_s as follows.

- for $s = s'$ or $s' \to s$ let $w'_s = w_s t$,
- in any other case let $w'_s = w_s$.

Clearly, $wt \in \otimes_s \{w'_s\}$. As $wt \in (\otimes_s L'_s)$, the word $w'_{s'} = \widehat{\pi}_{s'}(wt) = \widehat{\pi}_{s'}(w)t$ belongs both to $L'_{s'}$ and to $L_{s'} \cdot T_{s'}$. Therefore, by the second condition of inclusion, $w'_{s'}$ belongs to $L_{s'}$. In order that $wt \notin (\otimes_s L_s)$, it must be the case that $w'_s = w_s t \notin L_s$ for some s such that $s' \to s$. But this is excluded by the condition of strong coherency, since $w_s \in L_s$, $\pi_{s'}(w_s t) = \pi_{s'}(w'_s) = \pi_{s'}(w'_{s'})$ and $w'_{s'}$ belongs to $L_{s'}$. It follows from this contradiction that $\otimes_s L_s = \otimes_s L'_s$.

Only If part Suppose that $\otimes_s L_s = \otimes_s L'_s$. By the condition of coherency, $L_s = \widehat{\pi}_s(\otimes_s L_s)$, hence $L_s = \widehat{\pi}_s(\otimes_s L'_s) \subseteq L'_s$, and the first inclusion holds. Now let $w_{s'} \in L_{s'}$ and $t \in T_{s'}$ such that $w_{s'}t \notin L_{s'}$, and assume for a contradiction that $w_{s'}t \in L'_{s'}$. As the specification $\{L_s \mid s \in S\}$ is coherent, there exists a family of words $w_s \in L_s$, $s \neq s'$, and a word $w \in \otimes_s w_s$ such that $\widehat{\pi}_s(w) = w_s$ for all $s \in S$. Define a second family of words w'_s as follows.

- for $s = s'$ or $s' \to s$ let $w'_s = w_s t$,
- in any other case let $w'_s = w_s$.

Then clearly, $wt \in \otimes_s w'_s$. By assumption, $w'_{s'} = w_{s'}t \in L'_{s'}$. Moreover $s' \to s$ entails that $w'_s = w_s t \in L'_s$ since $w_s \in L_s \subseteq L'_s$ and $L'_s \cdot T_{s'} \subseteq L'_s$. Therefore, $wt \in \otimes_s L'_s = \otimes_s L_s$. It follows from the coherency of the specification that $\widehat{\pi}_{s'}(wt) = w_{s'}t$ belongs to $L_{s'}$, in contradiction with the assumptions. This establishes the second inclusion, hence the proposition. ⊣

Proposition 24 *Let $\mathcal{L} = \{L_s \mid s \in S\}$ be a strongly coherent specification over (λ, G) where $G = (S, \to)$ is minimally connected. Then $\otimes_s L_s = L(\mathcal{N})$ for some distributed and bounded P/T-net $\mathcal{N} = \oplus \mathcal{N}_s$ where $\mathcal{N}_s = (P_s, \widehat{T}_s, F_s, M_{0s})$ if and only if the following conditions are satisfied:*

1. $p \in P_s \wedge t \in \widehat{T}_s \setminus T_s \Rightarrow F_s(p,t) = 0$,
2. $p \in P_s \Rightarrow F_s(p,t) > 0$ for some $t \in T_s$,
3. $w \in L_s \wedge t \in T_s \wedge wt \in L_s \Rightarrow M_{0s}[w\rangle M'$ with $F_s(p,t) \leq M'(p)$ for $p \in P_s$,
4. $w \in L_s \wedge t \in T_s \wedge wt \notin L_s \Rightarrow M_{0s}[w\rangle M'$ with $F_s(p,t) > M'(p)$ for some p,
5. the net $\mathcal{N}_s$ is bounded in restriction to L_s.

Proof **If part** From conditions 1 and 2, $\mathcal{N}$ is a distributed P/T-net. Therefore, $L(\mathcal{N}) = \otimes_s L(\mathcal{N}_s)$ and $L(\mathcal{N}_s) \cdot (\widehat{T}_s \setminus T_s) \subseteq L(\mathcal{N}_s)$. From conditions 3 and 4, $L_s \subseteq L(\mathcal{N}_s) \subseteq \complement((L_s \cdot T_s \setminus L_s) \cdot \widehat{T}_s^*)$. As $\mathcal{L} = \{L_s \mid s \in S\}$ is a strongly coherent specification, $\otimes_s L_s = L(\mathcal{N})$ follows by Prop. 23. It remains to show that $\mathcal{N}$ is bounded. We proceed by contradiction. Let $s \in S$ and $p \in P_s$ and suppose that for any n, there exists some firing sequence $M_0[w\rangle M$ of the net $\mathcal{N}$ with $M(p) > n$. As $\mathcal{N} = \oplus_s \mathcal{N}_s$, there exists some corresponding firing sequence $M_{0s}[\widehat{\pi}_s(w)\rangle M'$ of the net $\mathcal{N}_s$ with $M'(p) = M(p)$. As

$w \in L(\mathcal{N}) = \otimes_s L_s$, $\widehat{\pi}_s(w) \in L_s$. Therefore, the net $\mathcal{N}_s$ is not bounded in restriction to L_s, contradicting condition 5.

Only If part In view of Def. 4 and Propositions 8, 10 and 12, all conditions stated in Prop. 24 are necessary to the existence of a solution to the bounded and distributed net synthesis problem from $\mathcal{L}$. ⊣

By Prop. 24, the bounded and distributed net synthesis problem from a strongly coherent specification $\{L_s \mid s \in S\}$ decomposes modularly to $|S|$ independent instances of the following.

Problem 4 (Open Net Synthesis Problem)
Given finite alphabets T and $\widehat{T}$, with $T \subseteq \widehat{T}$, and a non-empty regular and prefix-closed language $L \subseteq \widehat{T}^*$, decide whether there exists and construct a finite P/T-net $\mathcal{N} = (P, \widehat{T}, F, M_0)$ such that:

1. $p \in P \wedge t \in \widehat{T} \setminus T \Rightarrow F(p,t) = 0$,
2. $p \in P \Rightarrow F(p,t) > 0$ *for some* $t \in T$,
3. $w \in L \wedge t \in T \wedge wt \in L \Rightarrow M_0[w\rangle M$ *with* $F(p,t) \leq M(p)$ *for all* $p \in P$,
4. $w \in L \wedge t \in T \wedge wt \notin L \Rightarrow M_0[w\rangle M$ *with* $F(p,t) > M(p)$ *for some* p,
5. $\mathcal{N}$ *is bounded in restriction to* L.

For fixed alphabets $\widehat{T}$ and $T \subseteq \widehat{T}$, a net $\mathcal{N} = (P, \widehat{T}, F, M_0)$ satisfying condition 1 w.r.t. T is called an *open net* in the sequel. Clearly, $L(\mathcal{N}) \cdot (\widehat{T} \setminus T) \subseteq L(\mathcal{N})$ for any such open net.

4 Open Net Synthesis using Regions

In this section, we solve Problem 4 using the concept of regions of a language, originally defined in [14]. The presentation of regions given below draws inspiration from [1] and [8], with minor adaptations reflecting the slightly different statement of the synthesis problem.

Definition 25 *[regions] Let $L \subseteq \widehat{T}^*$ be a prefix closed language over $\widehat{T} = \{t_1, \ldots, t_k\}$. A* region *of L is a non-negative integer vector*

$$r = \langle r_{init}, r^\circ t_1, t_1{}^\circ r, \ldots, r^\circ t_k, t_k{}^\circ r \rangle$$

such that for all $t \in \widehat{T}$, $wt \in L \Rightarrow r/w \geq r^\circ t$ where r/w is defined inductively on the words of $\widehat{T}^*$ with $r/\epsilon = r_{init}$ and $r/wt =$

$r/w - r°t + t°r$. The region r is a bounded *region of L if there exists $B \in \mathbb{N}$ such that $r/w \leq B$ for all $w \in L$.*

According to Def. 25, the regions r of L correspond bijectively with the *one-place* nets $\mathcal{N}_r = (\{p_r\}, \widehat{T}, F, M_0)$ with language larger than L, viz. $M_0(p_r) = r_{init}$, $F(p_r, t_h) = r°t_h$ and $F(t_h, p_r) = t_h°r$ for all $t_h \in \widehat{T}$. In order to reflect condition 1 in Problem 4, we now introduce open regions as follows.

Definition 26 *[Open Regions] Given two alphabets $T \subseteq \widehat{T}$ and a non-empty prefix-closed and regular language $L \subseteq \widehat{T}^*$, an open region of L is any bounded region r of L such that $r°t_h = 0$ for all letters $t_h \in \widehat{T} \setminus T$.*

The suitability of the concept of regions for solving Problem 4 is shown by the following proposition, where $\complement$ means the complementation in $\widehat{T}^*$.

Proposition 27 *Problem 4 has a solution $\mathcal{N}$ if and only if for all $w \in L$ and $t \in T$, $wt \notin L \Rightarrow r/w < r°t$ for some open region r of L (r* disables *t after w).*
Let R be any finite and minimal *set of open regions of L such that, for any $t \in T$ and for any minimal word $wt \notin L$, some region in R disables t after w. Then $L \subseteq L(\mathcal{N}) \subseteq \complement((LT \setminus L) \cdot \widehat{T}^*)$ for the open net $\mathcal{N} = (P, \widehat{T}, F, M_0)$ as follows:*
- P is a set of places p_r in bijective correspondence with the regions $r \in R$,
- $M_0(p_r) = r_{init}$ and for any $t \in \widehat{T}$, $F(p_r, t) = r°t$ and $F(t, p_r) = t°r$.
Moreover, $\mathcal{N}$ is bounded in restriction to L, hence it is a solution to Problem 4.

Proof Let $\mathcal{N} = (P, \widehat{T}, F, M_0)$ be a solution to Problem 4. Let $w \in L$ and $t \in T$ such that $wt \notin L$. In view of the net firing rule, there must exist some place $p \in P$ such that $F(p,t) > M_0(p) + \sum_h \psi(w)(t_h) \times (F(t_h, p) - F(p, t_h))$. Let $r = \langle r_{init}, r°t_1, t_1°r, \ldots, r°t_k, t_k°r \rangle$ be the associated vector defined with $r_{init} = M_0(p)$, $r°t_h = F(p, t_h)$ and $t_h°r = F(t_h, p)$ for all $t_h \in \widehat{T}$. Then r is an open region of L and r disables t after w.

Conversely let R be a finite set of open regions of L as described in the proposition, and let $\mathcal{N}$ be the associated net defined *ibidem*. In view of the definition of open regions, the minimality of R, and the correspondence between regions in R and places of $\mathcal{N}$, conditions 1, 2 and 3 of Problem 4 are clearly satisfied. Condition 4 follows from the hypothesis that for any minimal word $wt \notin L$, some region in R disables t after w. Condition 5 follows from the fact that any open region of L is by definition a bounded region of L. Finally, $L \subseteq L(\mathcal{N})$

and $L(\mathcal{N}) \subseteq \complement((LT \setminus L) \cdot \widehat{T}^*)$ are mere restatements of the conditions 3 and 4. ⊣

The next proposition helps computing the open regions of a regular language.

Proposition 28 *Given an integral vector $r = \langle r_{init}, r^\circ t_1, t_1{}^\circ r, \ldots, r^\circ t_k, t_k{}^\circ r \rangle$ with non-negative entries, let $r \times w$ be defined for $w \in \widehat{T}^*$ inductively with $r \times \epsilon = 0$ and $r \times wt = r \times w - r^\circ t + t^\circ r$. Then for any prefix-closed regular language L over $\widehat{T}$, r is an open region of L if and only if the following conditions hold:*

i) $r \times w = 0$ *for every word w such that $vw^* \subseteq L$,*
ii) $r / w \geq r^\circ t$ *for every $w \in L$ such that $wt \in L$,*
iii) $r^\circ t_h = 0$ *for every letter $t_h \in \widehat{T} \setminus T$.*

Proof Condition (i) is necessary because $r / vw^n = r / v + n(r \times w)$ for all n. Indeed, if $r \times w < 0$ then it is not possible that $r / vw^n \geq 0$ for all n, as required by the definition of regions, and if $r \times w > 0$ then it is not possible that $r / vw^n < B$ for some bound B for all n, as required by the definition of bounded regions. The remaining two conditions are obviously necessary.

Conversely, suppose that conditions (i,ii,iii) hold. In order to show that r is an open region of L, it suffices to prove that there exists a finite bound $B \in \mathbb{N}$ such that $r / w \leq B$ for all $w \in L$. As L is regular and prefix-closed, $L = L(A)$ for some finite deterministic automaton $A = (Q, \widehat{T}, \delta, q_0)$ where $\delta : Q \times \widehat{T} \rightharpoonup Q$ is a partial map, all states in Q can be reached from q_0, and they are all accepting. Define $L' = \{w' \in L \,|\, w' = uvv' \Rightarrow \delta(q_0, u) \neq \delta(q_0, uv) \vee v = \varepsilon\}$. Clearly, L' is a finite set of words. We claim that $B = max\{r / w' \,|\, w' \in L'\}$ is an adequate bound. To establish this claim, it suffices to show that for any $w \in L \setminus L'$, $r / w = r / w'$ for some $w' \in L'$. We prove this property by induction on the length of words. Let $w \in L \setminus L'$, hence $w = uvv'$ and $\delta(q_0, u) = \delta(q_0, uv)$ for some $v \neq \varepsilon$. As $uv^* \subseteq L$, $r \times v = 0$ by condition (i). Therefore, $r / w = r_{init} + r \times w = r_{init} + r \times uv' = r / uv'$. Now $uv' \in L$ and it is strictly shorter than w. The validity of the claim follows by induction. ⊣

Propositions 27 and 28 provide a basis for deciding Problem 4 and hence for synthesizing open nets from regular languages. The presentation of the decision and synthesis procedure given in the end of the section is cut in two parts. In the first part, we show that the open regions of a regular language may be characterized by a finite linear system. In the second part, we show that deciding Problem 4 amounts to deciding for a *finite* set of minimal words $wt \notin L$ whether some open region disables t after w. The net $\mathcal{N}$ solution to Problem 4 is synthesized from this finite set of disabling regions as indicated in Prop. 27.

4.1 A finite linear characterization of open regions

Let $L \subseteq \widehat{T}^*$ be a prefix closed and regular language over $\widehat{T} = \{t_1, \ldots, t_k\}$. Then $L = L(A)$ for some finite deterministic automaton $A = (Q, \widehat{T}, \delta, q_0)$ where $\delta : Q \times \widehat{T} \rightarrow Q$ is a partial map, all states in Q can be reached from q_0, and they are all accepting. We construct from A a finite linear system such that a non-negative integer vector $r = \langle r_{init}, r\,^\circ t_1, t_1{}^\circ r, \ldots, r\,^\circ t_k, t_k{}^\circ r\rangle$ is an open region of L if and only if it is a solution of this system.

The construction of the linear system is based on a partial unfolding of A into a finite automaton $\mathcal{U}A = (Q', \widehat{T}, \delta', q_0')$ with components as follows. Q' is a subset of words of $L(A)$, constructed inductively from the single element $q_0' = \epsilon$ according to the completion rule stated hereafter. Let $w \in Q'$, then for any $t \in \widehat{T}$, $wt \in Q'$ and $\delta'(w, t) = wt$ if and only if $\delta(q_0, wt)$ is defined and differs from $\delta(q_0, v)$ for every prefix v of w. This yields a finite *spanning* tree. A finite number of *chords* are then added by setting $\delta'(vu, t) = v$ whenever $vu, v \in Q'$ and $\delta(q_0, vut) = \delta(q_0, v)$.

Proposition 29 *Let* $r = \langle r_{init}, r\,^\circ t_1, t_1{}^\circ r, \ldots, r\,^\circ t_k, t_k{}^\circ r\rangle$ *be a non-negative integer vector. Then* r *is an open region of* $L(A)$ *if and only if:*

i) $r \times ut = 0$ *for every chord* $vu \xrightarrow{t} v$ *in* $\mathcal{U}A$,
ii) $r/w \geq r\,^\circ t$ *for every edge* $w \xrightarrow{t} w'$ *in* $\mathcal{U}A$,
iii) $r\,^\circ t_h = 0$ *for every letter* $t_h \in \widehat{T} \setminus T$.

Proof In view of Prop. 28, the stated conditions are necessary. Conversely, if (i,ii,iii) hold, then r is an open region of $L(A)$ because $L(A) = L(\mathcal{U}A)$. One may indeed reproduce the reasoning followed in the proof of Prop. 28 to show that r is a bounded region of $L(\mathcal{U}A)$. ⊣

Each condition $r \times ut = 0$ in Prop. 29 is equivalent to the linear homogeneous equation:

$$\Sigma_{h=1}^{k} \vec{X}(h) \times (t_h{}^{\circ}r - r^{\circ}t_h) = 0 \tag{1}$$

where $\vec{X} = \psi(ut)$ is the Parikh image of the cycle ut in $\mathcal{U}A$. Similarly, each condition $r/w \geq r^{\circ}t$ in Prop. 29 is equivalent to the linear homogeneous inequality:

$$r_{init} + \Sigma_{h=1}^{k} \vec{Y}(h) \times (t_h{}^{\circ}r - r^{\circ}t_h) - (r^{\circ}t) \geq 0 \tag{2}$$

where $\vec{Y} = \psi(w)$ is the Parikh image of the path w in $\mathcal{U}A$.

Let $\mathcal{REG}$ denote the finite system of linear equations (1) and inequalities (2) in the $2k+1$ integer variables r_{init}, $t_h{}^{\circ}r$ and $r^{\circ}t_h$ derived from $\mathcal{U}A$, augmented with $r_{init} \geq 0$, $t_h{}^{\circ}r \geq 0$ and $r^{\circ}t_h \geq 0$ for all h, and $r^{\circ}t_h = 0$ for every letter $t_h \in \widehat{T} \setminus T$.

Proposition 30 *An integer vector r is an open region of $L = L(A)$ if and only if all linear constraints in $\mathcal{REG}$ are satisfied.*

Proof This is an immediate consequence of Prop. 29. ⊣

4.2 The decision and synthesis procedure

From propositions 27 and 30, $L(A) = L(\mathcal{N})$ for some open P/T-net $\mathcal{N}$ if and only if, for each letter $t \in T$ and for each state w of $\mathcal{U}A$, either $w \xrightarrow{t} w'$ for some w' or $r/w < r^{\circ}t$ for some open region r. Now $r/w < r^{\circ}t$ is equivalent to the linear homogeneous inequality:

$$r_{init} + \Sigma_{h=1}^{k} \vec{Y}(h) \times (t_h{}^{\circ}r - r^{\circ}t_h) - (r^{\circ}t) < 0 \tag{3}$$

where $\vec{Y}$ is the Parikh image of w. Inequality (3) holds for some open region r of $L(A)$ if and only if the linear system $\mathcal{REG}$ extended with the inequality:

$$r_{init} + \Sigma_{h=1}^{k} \vec{Y}(h) \times (t_h{}^{\circ}r - r^{\circ}t_h) - (r^{\circ}t) \leq -1 \tag{4}$$

is feasible in $\mathbb{Q}^{2k+1}$, which can be decided in polynomial time. Indeed, rational solutions always induce integer solutions. Let R be any minimal set of open regions of $L = L(A)$ large enough for disabling t after w whenever $t \in T$, w is a state of $\mathcal{U}A$, and $wt \notin L(A)$. Then $L = L(\mathcal{N})$ where $\mathcal{N}$ is constructed as indicated in proposition 27.

5 Conclusion

Let us first summarize the paper. We have refined the model of distributed Petri nets studied in [2], by considering net architectures (λ, G) where $\lambda : T \to S$ partitions the set of transitions over sites and $G = (S, \to)$ specifies the possible communications between sites. We have stated conditions under which one can check in a modular way the coherency of a product specification $\{L_s \mid s \in S\}$, where each L_s is a regular language over $\cup\{\lambda^{-1}(s') \mid s' = s \vee s' \to s\}$. We have shown that the distributed P/T-net synthesis problem may then be decomposed into $|S|$ independent "open" net synthesis problems, which may be solved by a simple adaptation of already known techniques.

We now compare our approach to the synthesis of distributed systems with the earlier approach taken in [6]. In that work, the authors start from global specifications, namely a transition system TS over a global alphabet T. The goal is to produce an implementation by a product of transition systems $\{TS_s \mid s \in S\}$ over some distributed alphabet $\{T_s \mid s \in S\}$ with $T = \cup_s T$. The drawback is the impossibility to deal with large specifications, since these are monolithic. We tried in [9] to adopt this framework for the synthesis of P/T-nets from modular specifications. For that purpose, we replaced the global specification TS with a Modular Transition System [16], as follows.

A modular transition system comprises a set of *local modules* TS'_s over *disjoint* alphabets $T'_s = T_s \setminus \cup\{T_{s'} \mid s' \neq s\}$ plus a *synchronizing module* $\mathcal{S}$. The synchronizing module defines all remaining transitions in $\cup\{T_s \cap T_{s'} \mid s \neq s'\}$ as jumps between vectors of states of the local modules. The synthesis of P/T-nets from modular specifications iş unfortunately more symbolic than modular: one can avoid computing a global transition system, but one cannot cut the synthesis problem to smaller independent synthesis problems. This is the reason why we chosed here to start from a product specification $\{L_s \mid s \in S\}$, which would rather look as implementations in the spirit of [6].

Our goals differ also significantly from the goals pursued in [6] in that we want to synthesize distributed P/T-nets, and this introduces a lower level of implementation. This led us to consider that the communication network on which the implementation is mapped has some importance, hence the particular form of our product specifications $\{L_s \mid s \in S\}$ where each L_s is a regular language over $\cup\{\lambda^{-1}(s') \mid s' = s \vee s' \to s\}$.

Bibliography

[1] Badouel, E., Bernardinello, L., Darondeau, Ph.: Polynomial Algorithms for the Synthesis of Bounded Nets. Proc. CAAP, LNCS **915** (1995) 647-679.

[2] Badouel, E., Caillaud, B., Darondeau, Ph.: Distributing finite automata through Petri net synthesis. *Formal Aspects of Computing* **13** (2002) 447-470.

[3] Bernardinello, L., De Michelis, G., Petruni, K., Vigna, S.: On the Synchronic Sructure of Transition Systems. In J. Desel, editor, *Structures in Concurrency Theory*, Workshops on Computing, Springer Verlag (1996) 11-31.

[4] Bernardinello, L.: Synthesis of Net Systems. Proc. ATPN, LNCS **691** (1993) 11-31.

[5] Cortadella, J., Kishinevsky, M., Lavagno, L., Yakovlev, A.: Deriving Petri Nets from Finite Transition Systems. *IEEE Trans. on Computers* **47**,8 (1998) 859-882.

[6] Castellani, I., Mukund, M., Thiagarajan, P.S.: Synthesizing Distributed Transition Systems from Global Specifications. Proc. FSTTCS, LNCS **1738** (1999) 219-231.

[7] Christensen, S., Petrucci, L.: Modular State Space Analysis of Coloured Petri nets. Proc. ATPN, LNCS **935** (1995) 201-217.

[8] Darondeau, P.: Region Based Synthesis of P/T-nets and its Potential Applications. Proc. ATPN, LNCS **1825** (2000) 16-23.

[9] Darondeau, P., Petrucci, L.: Modular Automata 2 Distributed Petri Nets 4 Synthesis. INRIA-RR 6192 (2007).

[10] Desel, J., Reisig, W.: The Synthesis Problem of Petri Nets. *Acta Informatica* **33** (1996) 297-315.

[11] Duboc, C.: Mixed Product and Asynchronous Automata. *Theoretical Computer Science* **48**, 3 (1986) 183-199.

[12] Ehrenfeucht, A., Rozenberg, G.: Partial (Set) 2-Structures; *Part I:* Basic Notions and the Representation Problem. *Acta Informatica* **27** (1990) 315-342.

[13] Ehrenfeucht, A., Rozenberg, G.: Partial (Set) 2-Structures; *Part II:* State Spaces of Concurrent Systems. *Acta Informatica* **27** (1990) 343-368.

[14] Hoogers, P.W., Kleijn, H.C.M., Thiagarajan, P.S.: A Trace Semantics for Petri Nets. Proc. ICALP, LNCS **623** (1992) 595-604.

[15] Mukund, M.: Petri Nets and Step Transition Systems. *International Journal of Foundations of Computer Science* **3**,4 (1992) 443-478.

[16] Petrucci, L.: Modélisation, vérification et applications. Mémoire d'habilitation à diriger des recherches, Université d'Evry (2002).

[17] van Glabbeek, R.J., Weijland, W.P.: Branching time and abstraction in bisimulation semantics. Proc. IFIP Congress, North-Holland/IFIP (1989) 613-618.

On the Reducibility of Persistent Petri Nets

Eike Best[1] and Jörg Desel[2]

[1]*Parallel Systems, Department of Computing Science*
Carl von Ossietzky Universität Oldenburg
eike.best@informatik.uni-oldenburg.de

[2]*Lehrstuhl für Angewandte Informatik*
Katholische Universität Eichstätt-Ingolstadt
joerg.desel@ku-eichstaett.de

Abstract

This paper aims to find a transformation from persistent Petri nets, which are a general class of conflict-free Petri nets, into a more restricted class of nets called behaviourally conflict-free nets. In a persistent net, whenever two distinct transitions are simultaneously enabled, one cannot become disabled through the occurrence of the other. In a behaviourally conflict-free net, two distinct transitions which are simultaneously enabled do not share a common pre-place. Relying on a series of earlier results which characterise the cyclic structure of the reachability graphs of persistent nets, we present a partial solution for transforming persistent into behaviourally conflict-free nets.

Keywords: Petri nets, persistence, net transformations

1 Introduction

There exists a hierarchy of Petri net classes [3], all of which can intuitively be called 'conflict-free', and of which marked graphs [5, 6] are the smallest and persistent nets [8] the largest class. This paper is concerned with an intermediate class called *behaviourally conflict-free* Petri nets. We address the question whether a persistent net can be transformed into a behaviourally conflict-free net with isomorphic reachability graph. We prove that under some conditions, such a transformation can be found.

2 Definitions

A Petri net (S,T,F,M_0) consists of two finite and disjoint sets S (places) and T (transitions), a function $F:((S\times T)\cup(T\times S))\to\mathbb{N}$ (flow) and a marking M_0 (the initial marking). A marking is a mapping $M: S\to\mathbb{N}$. A Petri net is plain if the range of F is $\{0,1\}$, i.e., F is a relation. The pre-set ${}^\bullet x$ of a net element $x\in S\cup T$ is the set $\{y\in S\cup T \mid (y,x)\in F\}$. Similarly, the post-set of x is $x^\bullet=\{y\in S\cup T\mid (x,y)\in F\}$.

The incidence matrix C is an $S\times T$-matrix of integers where the entry corresponding to a place s and a transition t is, by definition, equal to the number $F(t,s)-F(s,t)$. A T-invariant J is a vector of integers with index set T satisfying $C\cdot J=\underline{0}$ where $\cdot$ is the inner (scalar) product, and $\underline{0}$ is the vector of zeros with index set S. J is called semipositive if $J(t)\geq 0$, for all $t\in T$. The support of a semipositive T-invariant J, written *supp*(J), is the set of transitions t for which $J(t)>0$. Two semipositive T-invariants J and J' are called transition-disjoint if $\forall t\in T: J(t)=0\vee J'(t)=0$, or, equivalently, if *supp*$(J)\cap$ *supp*$(J')=\emptyset$. For a sequence $\sigma\in T^*$ of transitions, the Parikh vector $\Psi(\sigma)$ is a vector of natural numbers with index set T, where $\Psi(\sigma)(t)$ equals $\#(t,\sigma)$, the number of occurrences of t in σ.

A transition t is enabled (or activated, or firable) in a marking M (denoted by $M[t\rangle$) if, for all places s, $M(s)\geq F(s,t)$. If t is enabled in M, then t can occur (or fire) in M, leading to the marking M' defined by $M'(s)=M(s)+F(t,s)-F(s,t)$ (notation: $M[t\rangle M'$). We apply definitions of enabledness and of the reachability relation to transition (or firing) sequences $\sigma\in T^*$, defined inductively: $M[\varepsilon\rangle$ and $M[\varepsilon\rangle M$ are always true; and $M[\sigma t\rangle$ (or $M[\sigma t\rangle M'$) iff there is some M'' with $M[\sigma\rangle M''$ and $M''[t\rangle$ (or $M''[t\rangle M'$, respectively).

A marking M is reachable (from M_0) if there exists a transition sequence σ such that $M_0[\sigma\rangle M$. The reachability graph of N, with initial marking M_0, is the graph whose vertices are the markings reachable from M_0 and where an edge (M,t,M') labelled with t leads from M to M', iff $M[t\rangle M'$. Figure 1 shows an example where on the right-hand side, M_0 denotes the marking shown in the Petri net on the left-hand side. The marking equation states that if $M[\sigma\rangle M'$, then $M'=M+C\cdot\Psi(\sigma)$. Thus, if $M[\sigma\rangle M$ then $\Psi(\sigma)$ is a T-invariant.

A Petri net with initial marking is k-bounded if in any reachable marking M, $M(s)\leq k$ holds for every place s, and bounded if there is some k such that it is k-bounded. A finite Petri net (and we consider only such nets in the sequel) is bounded if and only if the set of its reachable markings is finite. A net with initial marking is called reversible if its reachability graph is strongly connected.

A net $N = (S, T, F, M_0)$ is a marked graph if $\sum_{t \in T} F(s,t) \leq 1$ as well as $\sum_{t \in T} F(t,s) \leq 1$, for all places s; output-nonbranching (on) if $\sum_{t \in T} F(s,t) \leq 1$ for all places s; behaviourally conflict-free (bcf) if, whenever $M[t_1\rangle$ and $M[t_2\rangle$ for a reachable marking M and transitions $t_1 \neq t_2$, then ${}^\bullet t_1 \cap {}^\bullet t_2 = \emptyset$; and persistent, if whenever $M[t_1\rangle$ and $M[t_2\rangle$ for a reachable marking M and transitions $t_1 \neq t_2$, then $M[t_1 t_2\rangle$. Directly from this definition, we have

$$\text{marked graph} \Rightarrow \text{on} \Rightarrow \text{bcf} \Rightarrow \text{persistent}.$$

The aim of this paper is to show that under certain conditions, the last implication can 'essentially' be reversed. Consider two different transitions in the post-set of a place and a reachable marking. If the net is persistent, then

(a) either at most one of the transitions is enabled at the marking or

(b) both are enabled but the occurrence of any of the transitions does not disable the other transition (which is in particular the case if the transitions are concurrently enabled).

If the net is behaviourally conflict-free, then (b) is not possible, i.e., at most one of the transitions is enabled at the marking. So our aim is to investigate conditions, under which forward branching places with option (b) can be replaced in such a way that the resulting net has only forward branching places with option (a).

Throughout the paper, we assume all nets to be plain (no arc weights > 1), T-restricted (transitions have at least one input place and at least one output place), simply live (every transition can be fired at least once), and free of isolated places.

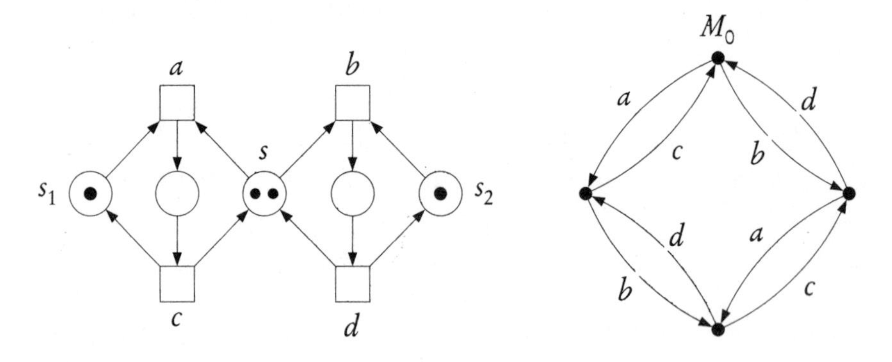

Figure 1: *A persistent Petri net and its reachability graph*

Figure 1 shows a Petri net which is persistent but not behaviourally conflict-free. Initially, a and b are enabled and share a common pre-place s. Observe that in the special case shown in this figure, place s can actually just be omitted without altering the net's behaviour. *Unaltered behaviour* here (and in the following) means reachability graph isomorphism. Alternatively, s could be split into two separate places, with one (or even two) tokens each, one of them connected only to the left-hand-side cycle, the other connected to the right-hand-side cycle. In both cases, the net is reduced to a (disconnected) marked graph. We investigate general circumstances under which such transformations are possible.

3 The Cyclic Behaviour of Persistent Nets

Two sequences $M[\sigma\rangle$ and $M[\sigma'\rangle$, firable from M and with $\sigma, \sigma' \in T^*$, are said to arise from each other by a transposition if they are the same, except for the order of an adjacent pair of transitions, thus:

$$\sigma \;=\; t_1 \dots t_k t\, t' \dots t_n \;\text{ and }\; \sigma' \;=\; t_1 \dots t_k t' t \dots t_n.$$

Two sequences $M[\sigma\rangle$ and $M[\sigma'\rangle$ are said to be permutations of each other (from M, written $\sigma \equiv_M \sigma'$) if they are both firable at M and arise out of each other through a (possibly empty) sequence of transpositions.

Theorem 1 KELLER [7]

Let N be a persistent net and let $\tau_1, \dots, \tau_m$ be m firing sequences starting from some reachable marking M. Then there is also a firing sequence $M[\tau\rangle$ such that $\forall t \in T : \Psi(\tau)(t) = \max_{1 \le j \le m} \Psi(\tau_j)(t)$.

A transition sequence $\tau \in T^*$ is called cyclic if its Parikh vector is a T-invariant (which is the case if and only if for all markings M, $M[\tau\rangle$ implies $M[\tau\rangle M$).

A cyclic transition sequence τ is called decomposable if $\tau = \tau_1 \tau_2$ such that τ_1 and τ_2 are cyclic and $\tau_1 \neq \varepsilon \neq \tau_2$. A firing sequence $M[\tau\rangle M'$ is called a cycle if τ is cyclic, i.e. if $M = M'$. A cycle $M[\tau\rangle M$ is called simple if there is no permutation $\tau' \equiv_M \tau$ such that τ' is decomposable. In other words, a non-simple sequence can be permuted such that the permuted sequence has a smaller cyclic subsequence leading from some marking back to the same marking. In Figure 1, for example, we have that:

$abcd$	is not decomposable
$acbd$	is decomposable, namely by $\tau_1 = ac$ and $\tau_2 = bd$
$M_0[ac\rangle M_0$	is simple
$M_0[abcd\rangle M_0$	is not simple, because of the permutation $M_0[acbd\rangle M_0$.

Figure 2 shows that a simple cycle can contain some transition more than once.

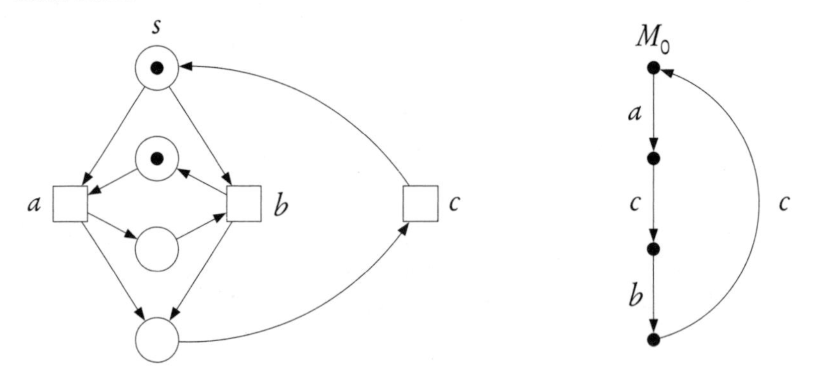

Figure 2: *Another persistent net and its reachability graph*

Theorem 2 A DECOMPOSITION THEOREM [1, 2]

Let N, with some initial marking, be bounded, reversible, and persistent. There is a finite set $\mathcal{B}$ of semipositive T-invariants such that any two of them are transition-disjoint and every cycle $M[\alpha\rangle M$ in the reachability graph decomposes up to permutations to some sequence of cycles $M[\alpha_1\rangle M[\alpha_2\rangle M \ldots [\alpha_n\rangle M$ with all Parikh vectors $\Psi(\alpha_i) \in \mathcal{B}$.

$\mathcal{B}$ can be constructed by picking simple cycles in the reachability graph and computing their T-invariants. For example, for Figure 1, $\mathcal{B} = \{J_1, J_2\}$ with

$$J_1(a)=J_1(c)=1,\ J_1(b)=J_1(d)=0;\ J_2(a)=J_2(c)=0,\ J_2(b)=J_2(d)=1.$$

For Figure 2, $\mathcal{B} = \{J\}$ with $J(a)=1$, $J(b)=1$, and $J(c)=2$.

Theorem 3 A CONVERSE OF THEOREM 2 [2]

Let N, with some initial marking, be bounded, reversible, and persistent. Let M be a reachable marking. Let $J_1, \ldots, J_m$ be (not necessarily mutually distinct) T-invariants from $\mathcal{B}$, as in Theorem 2. Then there is a cycle $M[\alpha\rangle M$ such that $\alpha = \alpha_1 \ldots \alpha_m$ and $\Psi(\alpha_j) = J_j$, for all $1 \le j \le m$.

Thus, the reachability graph is covered by simple cycles, any two of which are either transition-disjoint or have the same Parikh vector. Moreover, any such simple cycle can be fired from anywhere in the reachability graph, though the order of firing its transitions may vary from marking to marking. It is shown in [2] that reversibility is absolutely needed for these strong results, but that part of the decomposition properties can be recovered if the premise of reversibility is dropped.

4 Analysing Branching Places

From now on, let N be a bounded, reversible, persistent net with initial marking M_0.

Let us consider a place s and the set $s^\bullet$ of its output transitions. By the absence of isolated places and dead transitions and by reversibility, $s^\bullet \neq \emptyset \neq {}^\bullet s$ and every $a \in s^\bullet$ is in one of the simple cycles mentioned in the above theorem. Let $\mathcal{J}_s$ denote the set of T-invariants in $\mathcal{B}$ such that every one of them contains some transition in $s^\bullet$. We distinguish two cases.

Case 1:

$|\mathcal{J}_s| = 1$, that is, there is a simple cycle of the reachability graph containing *all* transitions in $s^\bullet$. This is certainly the case if $|s^\bullet| = 1$, but it may also be the case if $|s^\bullet| > 1$, as shown in Figure 2. As a consequence of the decomposability theorem, the simple cycle γ containing all transitions in $s^\bullet$ must also contain all transitions in ${}^\bullet s$ since if some of them were missing, there would be another simple cycle through them, which would have to contain at least one transition in $s^\bullet$ in order to restore the marking on s and would thus not be transition-disjoint with γ. This case will be investigated in section 5 below.

Case 2:

$|\mathcal{J}_s| > 1$, that is, all simple cycles contain proper subsets of $s^\bullet$, and therefore (with the same reasoning as above), also proper subsets of ${}^\bullet s$. An example is shown in Figure 1 where $|\mathcal{J}_s| = 2$.

Suppose that $\mathcal{J}_s = \{J_1, \ldots, J_m\}$ with $m > 1$. Our decomposition theorems imply that the set $\{supp(J_\ell) \cap s^\bullet \mid 1 \leq \ell \leq m\}$ partitions the set $s^\bullet$ and the set $\{supp(J_\ell) \cap {}^\bullet s \mid 1 \leq \ell \leq m\}$ partitions the set ${}^\bullet s$. Moreover, none of the sets in these sets is empty (which implies, in particular, that both $s^\bullet$ and ${}^\bullet s$ contain at least m transitions).

Now we will make a connection between finite *synchronic distances* and places. Informally, the (asymmetric) synchronic distance between two sets of transitions A and B indicates how far transitions from A can 'run ahead' of transitions from B. Formally:

$$asd_M(A,B) = \max\{(\sum_{t_1 \in A} \#(t_1, \tau)) - (\sum_{t_2 \in B} \#(t_2, \tau)) \mid \tau \in T^*, M[\tau\rangle\}.$$

If some sequence τ with $M[\tau\rangle$ actually satisfies

$$asd_M(A,B) = (\sum_{t_1 \in A} \#(t_1, \tau)) - (\sum_{t_2 \in B} \#(t_2, \tau)),$$

then we call it a *witness* for the 'gap' $asd_M(A,B)$.

The maximum can become infinite. For instance, $asd_{M_0}(\{a\},\{d\}) = \infty$ in Figure 1 and $asd_{M_0}(\{c\},\{a\}) = \infty$ in Figure 2. If the sets A and B are 'controlled' by place s and transition invariant J_ℓ, however, then their synchronic distance is always finite.

Lemma 4 CONTROLLED TRANSITION SETS BY INDUCE FINITE asd

Let N be a bounded, reversible, persistent net with initial marking M_0, let s be a place of N, let $J \in \mathcal{J}_s$ and let $A = supp(J) \cap (s^\bullet \setminus {}^\bullet s)$ and $B = supp(J) \cap ({}^\bullet s \setminus s^\bullet)$. Then both $asd_{M_0}(A,B)$ and $asd_{M_0}(B,A)$ are well-defined finite numbers.

Proof: By Theorem 3, there exists an infinite firing sequence

$$M_0[\tau\rangle M_0[\tau\rangle M_0[\tau\rangle M_0 \ldots$$

such that $\Psi(\tau) = J$. In this sequence, the only transitions putting tokens on s are those of B and the only transitions removing tokens from s are those of A. In case $asd_{M_0}(A,B) = \infty$, we get a contradiction to the fact that s contains finitely many tokens initially, and in case $asd_{M_0}(B,A) = \infty$ we get a contradiction to the boundedness of s.

□ 4

Using this lemma, and keeping $\mathcal{J}_s = \{J_1, \ldots, J_m\}$ in mind, we can define the numbers

$$L_\ell \quad = \quad asd_{M_0}(supp(J_\ell) \cap (s^\bullet \setminus {}^\bullet s), supp(J_\ell) \cap ({}^\bullet s \setminus s^\bullet))$$

for every $1 \leq \ell \leq m$. Let the net $N[s] = (T', S', F', M_0')$ be defined from $N = (S, T, F, M_0)$ as follows.

- The transitions of $N[s]$ are $T' = T$, the same as the transitions of N.

- The places of $N[s]$ are $S' = S \uplus \{s_1, \ldots, s_m\}$, i.e. the places of N plus m new places s_ℓ, one for each $\ell \in \{1, \ldots, m\}$.
- The initial marking M_0' is defined as follows: $M_0'(q) = M_0(q)$ for every $q \in S$, and $M_0'(s_\ell) = L_\ell$ for every $\ell \in \{1, \ldots, m\}$.
- The flow relation is extended as follows:

$$\begin{aligned} F' \;=\; F \;& \cup \{(s_\ell, t) \mid t \in supp(J_\ell) \cap s^\bullet, 1 \leq \ell \leq m\} \\ & \cup \{(t, s_\ell) \mid t \in supp(J_\ell) \cap {}^\bullet s, 1 \leq \ell \leq m\}. \end{aligned}$$

Theorem 5 PLACE COVERING

Let N be a bounded, reversible, persistent net with initial marking M_0, let s be a place with $|\mathcal{J}_s| > 1$ and let $N[s]$ be constructed as above. Then the reachability graphs of N and $N[s]$ are isomorphic. Moreover, with the numbers $L_1, \ldots, L_m$ defined above, we have $L_1 + \ldots + L_m \leq M_0(s)$.

Proof: For reachability graph isomorphism, we first note that by construction, every firing sequence of $N[s]$ is also a firing sequence of N (this is always the case when only places are added).

Conversely, assume that τ is a firing sequence of N which is not a firing sequence of $N[s]$ and assume that τ is a shortest such sequence. That is, $\tau = \tau' t$ such that τ' is a firing sequence both of N and of $N[s]$ leading to markings M in N (which enables t in N) and M' in $N[s]$ (which does not enable t in $N[s]$).

Because M' does not enable t in $N[s]$, there must be some place $q \in {}^\bullet t$ which is token-empty at M', and since by $M[t\rangle$ this is not true for any of the places of N, q must be one of the newly introduced places $q = s_\ell$. By the fact that $M_0'(q) = L_\ell$ and $M'(q) = 0$, transitions in $(q^\bullet \setminus {}^\bullet q)$ must have occurred L_ℓ times more often in τ' than transitions in $({}^\bullet q \setminus q^\bullet)$. But since t is also in $s^\bullet$ (but not in ${}^\bullet s$), τ is a sequence which is firable in N but contains transitions from $supp(J_\ell) \cap (s^\bullet \setminus {}^\bullet s)$ $L_\ell + 1$ times more often than transitions in $supp(J_\ell) \cap ({}^\bullet s \setminus s^\bullet)$, which contradicts the definition of L_ℓ as an asymmetric synchronic distance.

Thus, a firing sequence of N which is not also a firing sequence of $N[s]$ does not exist, and we have firing sequence equality, and hence also reachability graph isomorphism, between N and $N[s]$. The latter can be seen as follows. Let two nets $N, \widetilde{N}$ with the same transition set and the same set of firing sequences be given and construct a relation R between their respective sets of reachable markings by putting $(M, \widetilde{M}) \in R$ iff there is some τ with $M_0[\tau\rangle M$ and $\widetilde{M_0}[\tau\rangle \widetilde{M}$. Then R is surjective because any sequence is firable in $\widetilde{N}$ iff it is firable in N, and it is injective because the marking produced by a sequence is uniquely determined from the initial marking and the sequence itself.

Now we prove that the inequality $L_1 + \ldots + L_m \leq M_0(s)$ holds true. First, we will show that not only are there witnesses for the individual gaps $L_1,\ldots,L_m$, but there is even a witness realising all m gaps simultaneously.

Consider the first T-component, J_1, and consider any witness for L_1, that is, some sequence $M_0[\tau_1\rangle$ satisfying

$$\begin{array}{rcl} L_1 & = & \max\{\#(a,\tau_1) - \#(c,\tau_1) \\ & & \mid a \in supp(J_1) \cap (s^\bullet \setminus {}^\bullet s),\ c \in supp(J_1) \cap ({}^\bullet s \setminus s^\bullet)\}. \end{array} \quad (1)$$

Since τ_1 may contain transitions from other T-invariants $J_2,\ldots,J_m$, we will strive to 'remove' such transitions. Let $M_0[\tau\rangle M_0$ be a cycle whose Parikh vector is larger or equal to the Parikh vector of the sequence of non-J_1-transitions within τ_1. Such a cycle exists by Theorem 3. Since both τ_1 and τ are firable from M_0, Keller's theorem tells us that also $M_0[\tau\rangle M_0[(\tau_1 \dot{-} \tau)\rangle$. Let $\widetilde{\tau_1} = (\tau_1 \dot{-} \tau)$. By the fact that τ covers all non-J_1-transitions from τ_1 and because the T-invariants are transition-disjoint, $\widetilde{\tau_1}$ contains only transitions from J_1. Moreover, τ_1 is firable from M_0 and realises the gap L_1, since non-J_1-transitions do not contribute to formula (1) and $\widetilde{\tau_1}$ has exactly the same J_1-transitions as τ_1.

Repeating this procedure for all j from 2 to m, we find individual witnesses $\widetilde{\tau_1},\ldots,\widetilde{\tau_m}$ for $L_1,\ldots,L_m$, respectively, such that every $\widetilde{\tau_j}$ contains transitions from J_j only. By Theorem 1 again, there is some sequence τ with $\Psi(\tau)(t) = \max_{1\leq j\leq m} \Psi(\tau_j)(t)$ for all $t \in T$. Because all individual sequences are clean and because the J_ℓ are mutually transition-disjoint, the sequence τ realises all gaps $L_1,\ldots,L_m$ simultaneously.

Now we prove the desired inequality by contradiction. Assume otherwise, that is, assume that $L_1 + \ldots + L_m > M_0(s)$. By the above, we can find a witness τ realising all gaps $L_1,\ldots,L_m$ simultaneously. Since the sets $\{supp(J_\ell) \cap (s^\bullet \setminus {}^\bullet s) \mid 1 \leq \ell \leq m\}$ partition $s^\bullet \setminus {}^\bullet s$ and the sets $\{supp(J_\ell) \cap ({}^\bullet s \setminus s^\bullet) \mid 1 \leq \ell \leq m\}$ partition ${}^\bullet s \setminus s^\bullet$, this implies that transitions in $s^\bullet \setminus {}^\bullet s$ have occurred at least $M_0(s)+1$ times more often in τ than transitions in ${}^\bullet s \setminus s^\bullet$, creating a negative token count on s and leading to a contradiction.

Hence the assumption $L_1 + \ldots + L_m > M_0(s)$ is wrong and $L_1 + \ldots + L_m \leq M_0(s)$ is true instead. □ 5

According to the first part of Theorem 5, places $s_1,\ldots,s_m$ can be added to the net without altering its behaviour. According to the construction of the $s_1,\ldots,s_m$ and to the second part of Theorem 5, place s covers the sum of the places $s_1,\ldots,s_m$ in the sense that its F-connections are *exactly* the sum of the individual F'-connections of the s_ℓ and its initial marking is *equal to or larger* than the sum of the individual initial markings of the s_ℓ. Hence after $s_1, \ldots, s_m$ are added, s becomes a redundant place and can be omitted without altering the behaviour of the net. Altogether, we can replace place s by m places $s_1, \ldots, s_m$.

While the number of places properly increases by this transformation, the 'degree of conflict', that is, the number

$$\textit{conf-deg} = \sum_{t \in T} |({}^\bullet t)^\bullet|$$

decreases. If Case 1 never arises before or during this construction, *conf-deg* will eventually be down to $|T|$, that is, we will get a net which is output-nonbranching and hence behaviourally conflict-free.

Note that Theorem 5 is also true if $|\mathcal{J}_s| = 1$, but does not, in this case, lead to a reduction of *conf-deg*.

To sum this section up, if a place s is affected by m simple cycles, then it can be split into m places, each of which is 'responsible' for one of the cycles.

5 Branching Places Surrounded by a Single Simple Cycle

In Case 1, when there is a single simple cycle through all transitions bordering on a place s with two or more output transitions, a similar analysis and reduction may not necessarily be possible. For instance, consider the net shown in Figure 3. The transition inscriptions denote their values in the only minimal realisable T-invariant, defined by the transition counts on one of the cycles of the reachability graph. In state M', the output transitions of place s are concurrently enabled. We might consider the subset $A = \{t_6\}$ of $s^\bullet$ and the subset $B = \{t_5\}$ of ${}^\bullet s$ because t_5 and t_6 occur equally often in a cycle. Note that both $asd_{M_0}(A,B)$ and $asd_{M_0}(B,A)$ are finite. Moreover, for the complementary sets $A' = s^\bullet \setminus A = \{t_7\}$ and $B' = {}^\bullet s \setminus B = \{t_3, t_4\}$, the values of $asd_{M_0}(A',B')$ and $asd_{M_0}(B',A')$, as well as the other combinations, $asd_{M_0}(A,B')$, $asd_{M_0}(B',A)$, $asd_{M_0}(A',B)$ and $asd_{M_0}(B,A')$, are finite. We can thus add (up to eight) places reflecting these finite synchronic distances without changing the behaviour of the net. Nevertheless, even after addition of all these places, the place s cannot be removed without changing behaviour, because the sequence $t_1 t_5 t_6 t_8 t_{11} t_2 t_7$ is firable in the net so obtained. As can be seen from the reachability graph in Figure 3, this sequence is not firable in the original net.

The same net proves that, in a persistent Petri net, a forward branching place can have both options (a) and (b) (see section 2). There are reachable markings that enable only one of the output transitions of place s (namely, the markings reached after the occurrences of $t_1 t_5$ and after $t_2 t_5$, respectively), and there is a reachable marking (reached after the occurrence of t_4) that enables both output transitions concurrently.

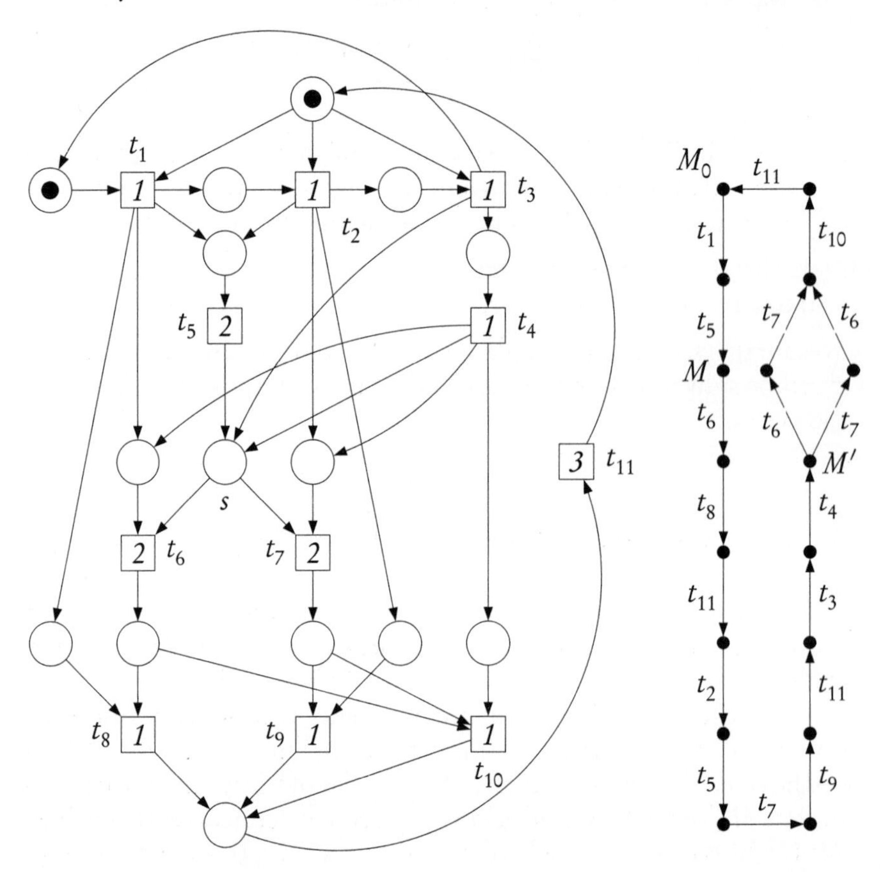

Figure 3: *A third persistent Petri net and its reachability graph*

The asd approach fails for another reason in the example shown in Figure 4. The only realisable (minimal and semipositive) T-invariant assigns *1* to transition c, *3* to transition d and *2* each to transitions a and b. More concretely, every simple cycle is a permutation of $adzbc'dxadybc$. Therefore, we do not find two proper nonempty subsets $A \subseteq s^\bullet$ and $B \subseteq {}^\bullet s$ such that transitions in A and transitions in B occur equally often in a cycle. For example, $asd_{M_0}(a, c)$ is infinite.

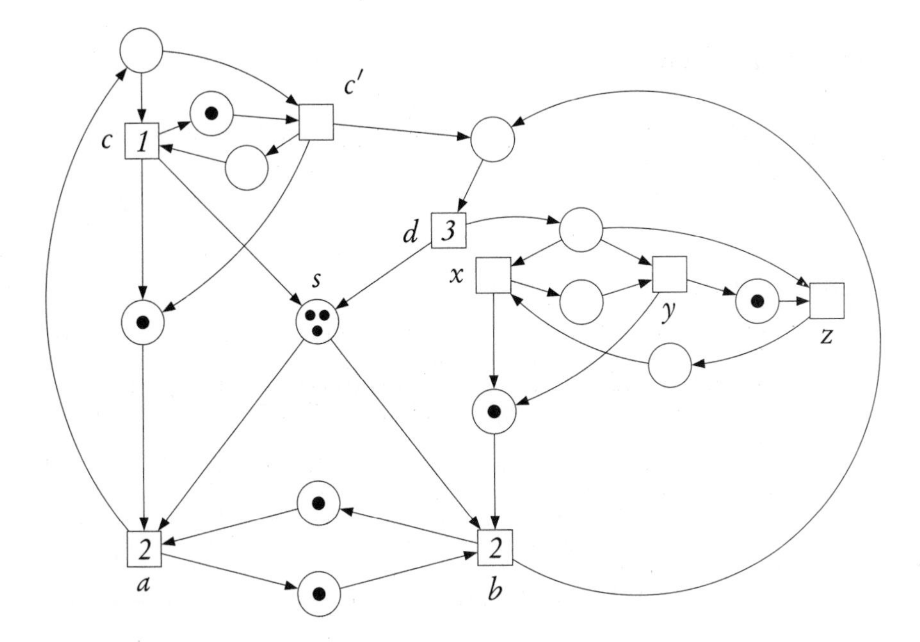

Figure 4: *A fourth persistent Petri net*

6 Concluding Remarks

This paper is part of a longer-term goal, namely to prove that bounded, reversible and persistent Petri nets are separable. Separability can be described informally as follows. Let k denote the greatest common divisor of the numbers $M_0(s)$, for places s and initial marking M_0 of some net. Then the net is (k-)separable if it behaves as k independent copies of the net arising when M_0 is replaced by M_0' with $M_0'(s) = M_0(s)/k$, for every place s.

We can presently imagine two possible ways of proving this conjecture. One possibility is a direct proof. Another possibility is using the results in [3, 4], where separability has been proved for live, bounded, and behaviourally conflict-free nets. In order to be able to make use of the second avenue, a transformation from persistent to behaviourally conflict-free nets such as explored in this paper is needed.

Independently of this connection, we are also looking at ways of adapting the construction by Ramamoorthy et al. [11] (mentioned also in Murata [10]) to yield transition-labelled bfc nets which are bisimilar [9] to a given persistent net.

Acknowledgment

The first author would like to thank Philippe Darondeau for objecting to a proof in one of the drafts of this paper.

References

[1] E. Best: A Note on Persistent Petri Nets. P. Degano et al. (eds.): Montanari Festschrift, LNCS 5065, pp. 427-438, 2008.
[2] E. Best, Ph. Darondeau: Decomposition Theorems for Bounded Persistent Petri Nets. K.M. van Hee and R. Valk (Eds.): PETRI NETS 2008, LNCS 5062, pp. 33-51, 2008.
[3] E. Best, Ph. Darondeau, H. Wimmel: Making Petri Nets Safe and Free of Internal Transitions. Fundamenta Informaticae 80, 75-90 (2007).
[4] E. Best, J. Esparza, H. Wimmel, K. Wolf: Separability in Conflict-free Petri Nets. In Proc. PSI'2006 (I. Virbitskaite, A. Voronkov, eds), LNCS Vol. 4378, Springer-Verlag, 1-18 (2006).
[5] F. Commoner, A.W. Holt, S. Even, A. Pnueli: Marked Directed Graphs. J. Comput. Syst. Sci. 5(5): 511-523 (1971).
[6] H.J. Genrich, K. Lautenbach: Synchronisationsgraphen. Acta Informatica 2(2), 143-161 (1973).
[7] R.M. Keller: A Fundamental Theorem of Asynchronous Parallel Computation. Parallel Processing, LNCS Vol. 24, Springer-Verlag, 102-112 (1975).
[8] L.H. Landweber, E.L. Robertson: Properties of Conflict-Free and Persistent Petri Nets. JACM 25(3), 352-364 (1978).
[9] R. Milner: Communication and Concurrency. Prentice Hall (1989).
[10] T. Murata: Petri Nets: Properties, Analysis and Applications. Proc. of the IEEE, Vol.77, No.4, 541-580 (1989).
[11] G.V. Ramamoorthy, C.S. Ho: Performance Evaluation of Asynchronous Concurrent Systems Using Petri Nets. IEEE Trans. Softw. Eng. SE-6 No.5, 440-449 (1980).

On Extensions of Timed Automata

Patricia Bouyer **Antoine Petit**
LSV, ENS Cachan, CNRS – France
{bouyer,petit}@lsv.ens-cachan.fr

Abstract

Since their definition in the early nineties, timed automata have been one of the most used and widely studied models for representing and analyzing real-time systems. In their seminal paper, Alur and Dill proved the probably most important property of timed automata: checking emptiness of the language accepted by a timed automaton, or equivalently checking a reachability property in a timed automaton, is decidable. This result relies on the construction of the so-called region automaton, which abstracts behaviours of a timed automaton into behaviours of a finite automaton. Since then, symbolic algorithms have been developed to solve that problem, several model-checkers have been implemented, and numerous case studies have been verified.
Lots of works have naturally aimed at proposing extensions of timed automata with new features, while preserving the above-mentioned fundamental decidability result. The motivation for these extensions is basically twofold. First it can increase the expressiveness of timed automata, allowing to model larger classes of systems. Then it can improve the conciseness (and hence the readability) of models by constructing more compact representations for a given system.
In this paper, we discuss and compare some of the most important extensions of timed automata that have been considered in the literature.

Keywords: Timed automata, real-time systems, automated verification

1 Introduction

Since their introduction by Alur and Dill [1, 2] in the early nineties, timed automata have been one of the most studied models for representing and analyzing real-time systems. This success is mostly based on the fundamental property that checking emptiness of the language accepted by a timed automaton, or equivalently checking a reachability property in a timed automaton, is decidable. Symbolic algorithms have been developed to solve that problem, several model-checkers have been implemented (for instance CMC [34], HyTech [30], Kronos [25], and Uppaal [15]) and numerous case studies have been treated (see the web pages of the tools for a list of them).

Lots of works have naturally aimed at proposing extensions of timed automata with new features, whilst preserving the fundamental decidability result. The motivation for these extensions is basically twofold. First it can increase the expressiveness of timed automata, allowing to model larger classes of systems. Then it can improve the conciseness (and hence the readability) of models by constructing more compact representations for a given system. Such conciseness can be very important to obtain more easily models of real systems in the same way that it is easier to write a program in an advanced programming language than in assembly language.

Recall that in a timed automaton *à la* Alur and Dill, a transition is made of:

- a guard that constrains the values of finitely many variables, which are called clocks,
- a letter corresponding to the action to be performed,
- a subset of clocks which are reset to zero when the transition is taken.

There are thus various ways to extend timed automata: either extend the types of guards that are allowed, or the way actions are taken into account, or extend the types of operations on clocks which are allowed.

It is well known that in the untimed framework, adding silent transitions (also known as ε-transitions or internal actions) to classical finite automata do not increase the expressive power of these automata and that they can be eliminated with no blowup in the number of states of the automaton. It turns out that the situation is very different in the timed framework and that silent transitions cannot be removed in general (see [19, 24] or [14] for a survey).

Concerning the guards (or constraints) on clocks, Alur and Dill allow to compare in their seminal papers (the value of) a clock, or the difference between two clocks, with a rational constant. A folklore result claims that to compare difference between clocks is even not necessary and that using diagonal-free constraints (where only the comparison between a clock and a constant is allowed) ensures the same expressive power. Nevertheless, it turns out that if difference between clocks can be removed (see for instance [14] for a construction), this cannot be done in general without an exponential blowup of the number of states, see [9]. The use of comparison of the sum of two clocks with a constant lead to an undecidable class of automata (see [2] but also [26, 10] for more precise results on the number of clocks that are used). Periodic clock constraints, as defined in [23], allow to express properties like "the value of a clock is even" or "the value of a clock lies in an interval of the form $(3n, 3n+1)$ where n is some integer". The corresponding class of automata is strictly more powerful than Alur and Dill's timed automata if silent transitions are not allowed but coincides otherwise.

The third natural possible extension concerns the operations on clocks, that we will call updates. "Deterministic" updates where a clock can be reset to a given constant, which has not anymore to be zero, or to the value of another clock or more generally to the sum of constant and of the value of another clock, are natural and are worth to be considered. Modelling telecommunication protocols (see e.g. the model of the ABR protocol proposed in [16, 18]) even requires the use of "non-deterministic" updates which allow for instance a clock to be reset to an arbitrary value lower than some fixed constant. It is easy to verify that such updates, even if we use only deterministic ones, lead to undecidable class of automata (*i.e.* automata for which checking emptiness of the accepted language is undecidable). Indeed, a two-counter machine (or Minsky machine) can easily be simulated. Nevertheless, it turns out that very interesting classes of updatable timed automata can be proved decidable (see [11] or [13] for a general presentation of so-called "updatable timed automata"). None of these classes increases the power of timed automata *à la* Alur and Dill if silent transitions are allowed (see [12, 13]). But, in most cases, these decidable classes are exponentially more concise, see [9].

The aim of this survey is not to propose original or unpublished contributions but rather to present the most important results about the three types of extensions presented just above, extensions which can be of course combined with sometimes surprising consequences. For instance, with more general updates than resets, it turns out that the decidability can depend on the clock constraints that are used — diagonal-free or not. For instance updates of the form $x := x + 1$ lead to an undecidable class of timed automata if arbitrary clock constraints are allowed but to a decidable class if only diagonal-free clock constraints are authorized. But automata with updates of the form $x := x - 1$ form undecidable classe whatever constraints, diagonal-free or linear, are used. Decidability is therefore often not far from undecidability and the frontier between the two worlds has to be carefully studied.

The paper is organized as follows. In section 2, we present basic definitions of timed words, clock constraints and updates. Timed automata, with a general definition, are introduced in section 3. Decidability results are presented in section 4, through the important notion of region automaton. In section 5, we present a comparison between the expressiveness and conciseness of the decidable classes described in section 4 and the class of timed automata *à la* Alur and Dill. Section 6 is devoted to further extensions, one where modulo constraints are allowed, and one where alternance is added to the model.

2 Definitions

2.1 Timed words and clocks

If Z is any set, let Z^* be the set of *finite* sequences of elements in Z. We consider as time domain $\mathbb{T}$ the set $\mathbb{Q}^+$ of non-negative rationals or the set $\mathbb{R}^+$ of non-negative reals and Σ as a finite set of *actions*. A *time sequence* over $\mathbb{T}$ is a finite nondecreasing sequence $\tau = (t_i)_{1\leq i\leq n} \in \mathbb{T}^*$. A *timed word* $\omega = (a_i, t_i)_{1\leq i\leq n}$ is an element of $(\Sigma \times \mathbb{T})^*$, also written as a pair $\omega = (\sigma, \tau)$, where $\sigma = (a_i)_{1\leq i\leq n}$ is a word in Σ^* and $\tau = (t_i)_{1\leq i\leq n}$ a time sequence in $\mathbb{T}^*$ of same length.

We consider a finite set X of variables, called *clocks*. A *clock valuation* over X is a mapping $v : X \to \mathbb{T}$ that assigns to each clock a time value. The set of all clock valuations over X is denoted $\mathbb{T}^X$. Let $t \in \mathbb{T}$, the valuation $v + t$ is defined by $(v+t)(x) = v(x) + t$, $\forall x \in X$.

2.2 Clock constraints

Given a set of clocks X, we introduce a set of linear clock constraints over X, that we denote $\mathscr{C}_{lin}(X)$, and which is formally defined by the following grammar:

$$\varphi \quad ::= \quad \alpha_1 x_1 + \ldots + \alpha_n x_n \sim c \quad | \quad \varphi \wedge \varphi$$

where $x_1, \ldots, x_n \in X$, $\alpha_1, \ldots, \alpha_n, c \in \mathbb{Z}$, and $\sim \in \{<, \leq, =, \geq, >\}$. A term of the form $\alpha_1 x_1 + \ldots + \alpha_n x_n \sim c$ is called a *simple (clock) constraint.*

We denote $\mathscr{C}(X)$ the restriction of $\mathscr{C}_{lin}(X)$ where simple constraints are only of the form $x \sim c$ or $x - y \sim c$. We further restrict the set of formulas that can be used, and define the set of *diagonal-free* clock constraints that is a subset of $\mathscr{C}(X)$ where diagonal constraints (*i.e.*, constraints of the form $x-y \sim c$) are forbidden. We write $\mathscr{C}_{df}(X)$ for this restricted set of clock constraints.

Clock constraints are interpreted over clock valuations. Let φ be a clock constraint in $\mathscr{C}_{lin}(X)$, and $v \in \mathbb{T}^X$ be a clock valuation. The satisfaction relation, denoted "$v \models \varphi$" if valuation v satisfies the clock constraint φ, is defined in a natural way:

$$\begin{cases} v \models \alpha_1 x_1 + \ldots \alpha_n x_n \sim c & \text{if } \alpha_1 v(x_1) + \ldots + \alpha_n v(x_n) \sim c \\ v \models \varphi_1 \wedge \varphi_2 & \text{if } v \models \varphi_1 \text{ and } v \models \varphi_2 \end{cases}$$

2.3 Updates

Clock constraints allow to test the values of the clocks. In order to change these values, we use the notion of *updates* which are functions from $\mathbb{T}^X$ to $\mathscr{P}(\mathbb{T}^X)$*. An update hence associates to each valuation a set of valuations.

In this paper, we focus on a rather small class of updates, the so-called *local updates*, constructed in the following way. We first define a *simple update* over a clock z as one of the two following functions:

$$up \quad ::= \quad z :\sim c \quad | \quad z :\sim y + d$$

where c, $d \in \mathbb{Z}$, $y \in X$, and $\sim \in \{<, \leq, =, \geq, >\}$.

Let v be a valuation and up be a simple update over z. A valuation v' is in $up(v)$ if $v'(y) = v(y)$ for any clock $y \neq z$ and if $v'(z)$ satisfies:

$$\begin{cases} v'(z) \sim c \wedge v'(z) \geq 0 & \text{if } up = z :\sim c \\ v'(z) \sim v(y) + d \wedge v'(z) \geq 0 & \text{if } up = z :\sim y + d \end{cases}$$

A *local update* over a set of clocks X is a collection $up = (up_i)_{1 \leq i \leq k}$ of simple updates, where each up_i is a simple update over some clock $x_i \in X$ (note that it may happen that $x_i = x_j$ for some $i \neq j$). Let $v, v' \in \mathbb{T}^X$ be two clock valuations. The valuation v' is in $up(v)$ if for every i, the set $up_i(v)$ contains the valuation v'' defined by

$$\begin{cases} v''(x_i) & = & v'(x_i) \\ v''(y) & = & v(y) \quad \text{for every } y \neq x_i \end{cases}$$

The terminology "*local*" comes from that $v'(x)$ does not depend on the other values $v'(y)$.

*$\mathscr{P}(\mathbb{T}^X)$ denotes the powerset of $\mathbb{T}^X$.

Example 1 Consider the local update $up = (x :> y, x :< 7)$, and two valuations v and v'. It holds that $v' \in up(v)$ if: $v'(x) > v(y) \wedge v'(x) \geq 0 \wedge v'(x) < 7 \wedge v'(y) = v(y)$.

Note that $up(v)$ may be empty. For instance, the local update $(x :< 1, x :> 1)$ leads to an empty set.

For any set of clocks X, we denote by $\mathcal{U}(X)$ the set of local updates over X. In this paper, we will simply call updates these local updates. The following subsets of $\mathcal{U}(X)$ need to be distinguished for the rest of the paper.

- $\mathcal{U}_0(X)$ is the set of reset updates. A *reset update* is a local update *up* such that each simple update defining *up* is of the form $x := 0$.
- $\mathcal{U}_{\text{cst}}(X)$ is the set of "constant" updates, that is the set of updates *up* such that each simple update defining *up* is of the form $x := c$ with $c \in \mathbb{N}$ (formally, $c \in \mathbb{Z}$, but if $c < 0$, the resulting set of valuations will be empty for any initial valuation v).
- $\mathcal{U}_{\text{det}}(X)$ is the set of deterministic updates. An update *up* is said *deterministic* if for any clock valuation v, there exists at most one valuation v' such that $v' \in up(v)$. It is immediate to check that a local update $up = (up_i)_{1 \leq i \leq k}$ is deterministic if all simple updates up_i are of one of the following form:

    1. $x := c$ with $x \in X$ and $c \in \mathbb{Z}$,
    2. $x := y$ with $x, y \in X$, or
    3. $x := y + c$ with $x, y \in X$ and $c \in \mathbb{Z} \setminus \{0\}$.

3 Timed Automata

We now define the central notion of timed automata. As we explain in details below, these automata extend the classical family of Alur and Dill's timed automata [1, 2], but we nevertheless stick to this terminology.

A *timed automaton* is a tuple $\mathcal{A} = (\Sigma, X, Q, T, I, F)$, where:

- Σ is a finite alphabet of actions,
- X is a finite set of clocks,
- Q is a finite set of states,
- $T \subseteq Q \times [\mathcal{C}(X) \times (\Sigma \cup \{\varepsilon\}) \times \mathcal{U}(X)] \times Q$ is a finite set of transitions,
- $I \subseteq Q$ is the subset of initial states,
- $F \subseteq Q$ is the subset of final states.

The special action ε is called *silent action* and a transition in $Q \times [\mathscr{C}(X) \times \{\varepsilon\} \times \mathscr{U}(X)] \times Q$ is called a *silent transition* or an ε*-transition*.

If $\mathscr{C} \subseteq \mathscr{C}(X)$ is a subset of clock constraints and $\mathscr{U} \subseteq \mathscr{U}(X)$ a subset of updates, the class ***T-Aut***$_\varepsilon(\mathscr{C}, \mathscr{U})$ denotes the set of all updatable timed automata in which transitions only use clock constraints in $\mathscr{C}$ and updates in $\mathscr{U}$. The subclass of automata which do not use silent transitions is simply written ***T-Aut***$(\mathscr{C}, \mathscr{U})$. Classical timed automata *à la* Alur and Dill [1, 2] correspond to the classes ***T-Aut***$_\varepsilon(\mathscr{C}_{df}(X), \mathscr{U}_0(X))$ and ***T-Aut***$(\mathscr{C}_{df}(X), \mathscr{U}_0(X))$ (where $\mathscr{C}_{df}(X)$ and $\mathscr{U}_0(X)$ are respectively the set of diagonal-free clock constraints and reset updates defined in section 2).

A behaviour in a timed automaton is obtained through the notion of paths and runs. Let us fix for the rest of this section a timed automaton $\mathscr{A}$. A *path* in $\mathscr{A}$ is a finite sequence of consecutive transitions:

$$P = q_0 \xrightarrow{\varphi_1, a_1, up_1} q_1 \xrightarrow{\varphi_2, a_2, up_2} q_2 \cdots q_{n-1} \xrightarrow{\varphi_n, a_n, up_n} q_n$$

where $(q_{i-1}, \varphi_i, a_i, up_i, q_i) \in T$ for every $0 < i \leq n$. The path is said to be *accepting* if it starts in an initial state ($q_0 \in I$) and ends in a final state ($q_n \in F$).

A *run* (over the time domain $\mathbb{T}$) through the path P from the clock valuation v_0, with $v_0(x) = 0$ for every clock x, is a sequence of the form:

$$\langle q_0, v_0 \rangle \xrightarrow[t_1]{a_1} \langle q_1, v_1 \rangle \xrightarrow[t_2]{a_2} \langle q_2, v_2 \rangle \ \ldots \ \langle q_{n-1}, v_{n-1} \rangle \xrightarrow[t_n]{a_n} \langle q_n, v_n \rangle$$

where $\tau = (t_i)_{1 \leq i \leq n}$ is a time sequence over $\mathbb{T}$, and $(v_i)_{0 \leq i \leq n}$ are clock valuations such that:

$$\begin{cases} v_{i-1} + (t_i - t_{i-1}) \models \varphi_i \\ v_i \in up_i\left(v_{i-1} + (t_i - t_{i-1})\right) \end{cases}$$

Note that any set $up_i(v_{i-1} + (t_i - t_{i-1}))$ of a run has to be non empty. In the following, to make the notations more compact , we will note such a run

$$\langle q_0, v_0 \rangle \xrightarrow[t_1]{\varphi_1, a_1, up_1} \langle q_1, v_1 \rangle \xrightarrow[t_2]{\varphi_2, a_2, up_2} \ \ldots \ \langle q_{n-1}, v_{n-1} \rangle \xrightarrow[t_n]{\varphi_n, a_n, up_n} \langle q_n, v_n \rangle$$

The label of such a run is the timed word $w = (a_1, t_1)(a_2, t_2) \ldots (a_n, t_n)$. If the path P is accepting, then this timed word is said to be accepted by $\mathscr{A}$. The set of all timed words accepted by $\mathscr{A}$ over the time domain $\mathbb{T}$ is denoted by $L(\mathscr{A}, \mathbb{T})$, or simply $L(\mathscr{A})$.

Example 2 Consider the following timed automaton.

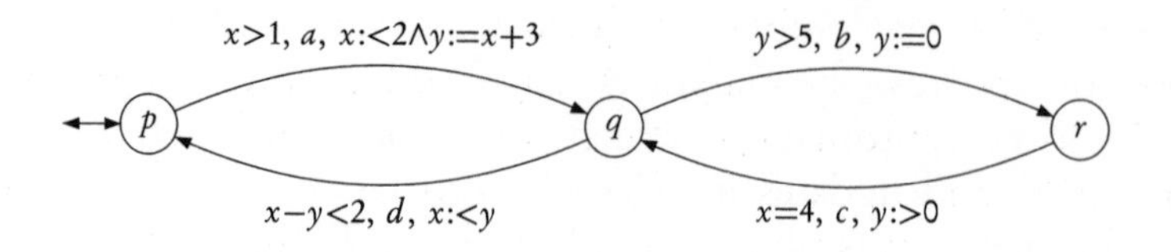

An accepting run in this automaton is:

$$\langle p,(0,0)\rangle \xrightarrow[1.3]{a} \langle q,(0.2,4.3)\rangle \xrightarrow[2.1]{b} \langle r,(1,0)\rangle$$

$$\xrightarrow[5.1]{c} \langle q,(4,3.1)\rangle \xrightarrow[9.6]{d} \langle p,(7.2,8.6)\rangle$$

Let us explain this run:

- the transition $\langle p,(0,0)\rangle \xrightarrow[1.3]{a} \langle q,(0.2,4.3)\rangle$ is possible because after having delayed 1.3 units of time in p, the value of both x and y is 1.3. The clock constraint $x > 1$ is thus satisfied and the valuation $(0.2,4.3)$ is in the image of the valuation $(1.3,1.3)$ by the update $x :< 2 \wedge y := x+3$;
- the transition $\langle q,(0.2,4.3)\rangle \xrightarrow[2.1]{b} \langle r,(1,0)\rangle$ is possible because after having delayed $2.1 - 1.3 = 0.8$ units of time in q, the value of x is 1 and the value of y is 5.1. The clock constraint $y > 5$ is thus satisfied and, after resetting y to 0, we get that the valuation $(1,0)$ can be reached;
- etc...

4 Decidability Results

In this section, we present a generic decidability result for timed automata, and then specialize the construction to subclasses of timed automata. The principle of this generic deep decidability result relies on the construction, for any timed automaton $\mathscr{A}$, of a finite untimed automaton which accepts precisely the (untimed) language $\textsc{Untime}(L(\mathscr{A}))$ where

$$\textsc{Untime}(L(\mathscr{A})) = \{\sigma \in \Sigma^* \mid \text{there exists a time sequence } \tau \text{ s.t. } (\sigma,\tau) \in L(\mathscr{A})\}$$

The emptiness of $L(\mathscr{A})$ is obviously equivalent to the emptiness of $\textsc{Untime}(L(\mathscr{A}))$, so the result follows from the decidability of the emptiness checking problem for untimed finite automata (see e.g. [33]).

We first define the notion of regions, region graphs, and region automata.

4.1 Regions, region graph, and region automaton

Let X be a finite set of clocks. We fix a finite partition $\mathscr{R}$ of $\mathbb{T}^X$. For each valuation $v \in \mathbb{T}^X$, the unique element of $\mathscr{R}$ that contains v is denoted by $[v]_{\mathscr{R}}$. We define the set of successors of R, $\mathsf{Succ}(R) \subseteq \mathscr{R}$, in the following natural way:

$$R' \in \mathsf{Succ}(R) \text{ if } \exists v \in R,\ \exists t \in \mathbb{T} \text{ s.t. } v+t \in R'$$

We say that such a finite partition is a *set of regions* whenever the following condition holds:

$$R' \in \mathsf{Succ}(R) \iff \forall v \in R,\ \exists t \in \mathbb{T} \text{ s.t. } v+t \in R' \tag{1}$$

This natural condition assesses that the equivalence relation defined by the partition $\mathscr{R}$ is stable with time elapsing.† Let us note that this condition is not satisfied by any finite partition of $\mathbb{T}^X$ as illustrated by the following counter-example.

Example 3 We consider the partition of $\mathbb{T}^2$ depicted on the figure below. Condition (1) is not satisfied by the gray region. Indeed, from valuation $(0.5, 1.8)$, when time elapses it is possible to reach the valuation $(0.7, 2)$ and thus the region defined by the constraints "$0 < x < 1 \wedge y = 2$". But this region cannot be reached from valuation $(0.5, 1.1)$ by letting time elapse.

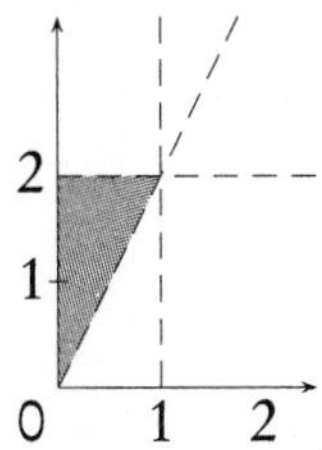

Let $\mathscr{U} \subseteq \mathscr{U}(X)$ be a finite set of updates. Each update $up \in \mathscr{U}$ naturally induces a function $up : \mathscr{R} \to \mathscr{P}(\mathscr{R})$ which maps any region R onto the set $\{R' \in \mathscr{R} \mid up(R) \cap R' \neq \varnothing\}$. The set of regions $\mathscr{R}$ is said *compatible* with $\mathscr{U}$ if the following holds: whenever a valuation $\overline{v}' \in R'$ is reachable from a valuation $\overline{v} \in R$ by some up, then R' is

†Note that time is however abstracted, in that for two different valuations v and v' in a region R, the time-elapsing witnesses $v+t$ and $v'+t'$ in R' may be such that t and t' do not coincide.

reachable from any $v \in R$ by the same *up*. Formally, the compatibility conditions for the updates can be written:

$$R' \in up(R) \implies \forall v \in R,\ \exists v' \in R' \text{ s.t. } v' \in up(v) \qquad (2)$$

Note that this condition has an interpretation similar to the one done for condition (1), except that it concerns discrete moves. Of course these conditions are related to some kind of bisimulation property, see the remark below.

Remarks: We write $\rho_{\mathcal{R}}$ for the relation defined as:

$$v\ \rho_{\mathcal{R}}\ v' \iff [v]_{\mathcal{R}} = [v']_{\mathcal{R}}$$

If the transition relation $\hookrightarrow$ on $\mathbb{T}^X$ is defined by

$$v \hookrightarrow v' \iff \exists t \in \mathbb{T} \text{ s.t. } v' = v + t$$

then condition (1) assesses that $\rho_{\mathcal{R}}$ is a bisimulation with respect to the relation $\hookrightarrow$.

Similarly, if the transition relations $(\hookrightarrow_{up})_{up}$ on $\mathbb{T}^X$ are defined by

$$v \hookrightarrow_{up} v' \iff v' \in up(v)$$

then condition (2) assesses that $\rho_{\mathcal{R}}$ is a bisimulation with respect to the relations $(\hookrightarrow_{up})_{up}$.

Whenever a set of regions $\mathcal{R}$ is compatible with a set of updates $\mathcal{U}$, we define the *region graph* associated with $\mathcal{R}$ and $\mathcal{U}$ as the graph whose set of nodes is $\mathcal{R}$ and whose edges are of two distinct types:

$$\begin{array}{ll} R \longrightarrow R' & \text{if } R' \in \mathsf{Succ}(R) \\ R \longrightarrow_{up} R' & \text{if } R' \in up(R) \end{array}$$

Example 4 Let us consider the set of four regions $\mathcal{R}$ defined by the following equations, and depicted below:

$$\overset{R_1}{\begin{pmatrix} 0 \le x < 1 \\ 0 \le y \le 1 \\ x < y \end{pmatrix}} \overset{R_2}{\begin{pmatrix} x \ge 0 \\ 0 \le y \le 1 \\ x \ge y \end{pmatrix}} \overset{R_3}{\begin{pmatrix} x > 1 \\ y > 1 \\ x \ge y \end{pmatrix}} \quad \overset{R_4}{\begin{pmatrix} x \ge 0 \\ y > 1 \\ x < y \end{pmatrix}}$$

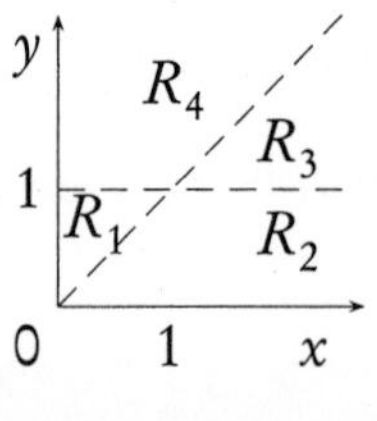

It is easy to verify that $\mathcal{R}$ is compatible with the set of updates $\mathcal{U} = \{x := 1, y := 0\}$. The region graph associated with $\mathcal{R}$ and $\mathcal{U}$ is represented below on Figure 1.

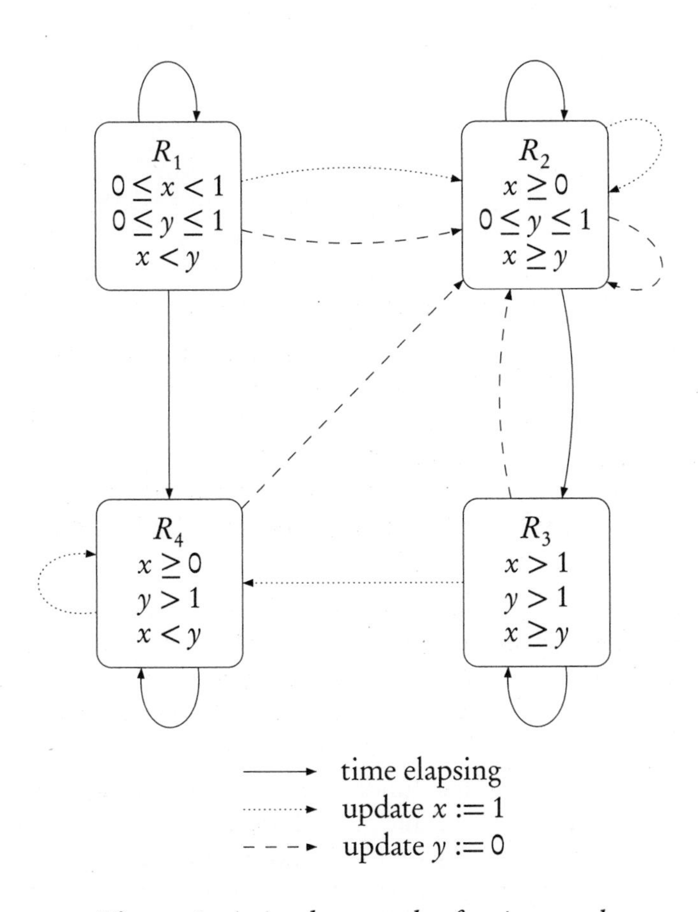

Figure 1: *A simple example of region graph*

Finally, let $\mathscr{C} \subseteq \mathscr{C}(X)$ be a finite set of clock constraints. A set of regions $\mathcal{R}$ is said to be *compatible* with $\mathscr{C}$ if for every clock constraint $\varphi \in \mathscr{C}$ and for every region R, either $R \subseteq \varphi$ or $R \subseteq \neg\varphi$.

We consider now a timed automaton $\mathscr{A} = (\Sigma, X, Q, T, I, F)$ in some class $\textbf{\textit{T-Aut}}_\varepsilon(\mathscr{C}, \mathcal{U})$, and we assume $\mathcal{R}$ is a set of regions compatible with $\mathscr{C}$ and $\mathcal{U}$. We define the *region automaton* $\Gamma_{\mathcal{R}}(\mathscr{A})$ associated with $\mathscr{A}$ and $\mathcal{R}$, as the following finite (untimed) automaton:

- Its set of states is $Q \times \mathcal{R}$:
  - the initial states are $(q_0, \mathbf{0})$ where $q_0 \in I$ is initial and $\mathbf{0}$

is the unique region containing the valuation where all clocks are set to zero;
- the final states are (f, R) where f is final in $\mathscr{A}$ and R is any region.

- Its transitions are defined by $(q, R) \xrightarrow{a} (q', R')$ if there exists a region $\widehat{R}$ and a transition $q \xrightarrow{\varphi, a, up} q'$ in $\mathscr{A}$ such that:
  - $R \rightarrow \widehat{R}$ is a transition of the region graph,
  - $\widehat{R} \subseteq \varphi$,
  - $\widehat{R} \rightarrow_{up} R'$ is a transition of the region graph.

Under conditions (1) and (2), the region automaton is an interesting abstraction of the original automaton in the following sense:

Proposition 1 *[[13]] Let $\mathscr{A}$ be a timed automaton in $\textbf{T-Aut}_\varepsilon(\mathscr{C}, \mathscr{U})$ where $\mathscr{C}$ (resp. $\mathscr{U}$) is a finite set of clock constraints (resp. of updates). Let $\mathscr{R}$ be a set of regions compatible with $\mathscr{C}$ and $\mathscr{U}$. Then the finite automaton $\Gamma_{\mathscr{R}}(\mathscr{A})$ accepts the language* UNTIME$(L(\mathscr{A}))$.
Since the emptiness checking problem for untimed automaton is decidable (see e.g. [33]), the previous proposition leads to the next theorem.

Theorem 2 *[[13]] Let $\mathscr{C}$ (resp. $\mathscr{U}$) be a finite set of clock constraints (resp. of updates). Assume there exists a set of regions $\mathscr{R}$, which is compatible with $\mathscr{C}$ and $\mathscr{U}$, and such that the region graph associated with $\mathscr{R}$ and $\mathscr{U}$ can be effectively constructed and the compatibility with $\mathscr{C}$ can be decided, then the class $\textbf{T-Aut}_\varepsilon(\mathscr{C}, \mathscr{U})$ is decidable.*

This theorem is of course fundamental, but it does not exhibit any real decidable class of timed automata. Indeed, we need to (effectively) construct sets of clock constraints $\mathscr{C}$ and sets of updates $\mathscr{U}$, together with sets of regions $\mathscr{R}$ such that $\mathscr{R}$ is compatible with both $\mathscr{C}$ and $\mathscr{U}$.

Remarks: The relation between $\mathscr{A}$ and $\Gamma_{\mathscr{R}}(\mathscr{A})$ is even stronger. Indeed, going further remark 4.1, we can prove that the equivalence relation $\equiv_{\mathscr{R}}$ over $Q \times \mathbb{T}^X$ defined by "$(q, v) \equiv_{\mathscr{R}} (q', v')$ iff $q = q'$ and $[v]_{\mathscr{R}} = [v']_{\mathscr{R}}$" is actually a *time-abstract bisimulation*.[‡] Then, $\Gamma_{\mathscr{R}}(\mathscr{A})$ is the quotient of $\mathscr{A}$ with respect to the equivalence relation $\equiv_{\mathscr{R}}$.

[‡]This means that whenever $(q_1, v_1) \equiv_{\mathscr{R}} (q_2, v_2)$ and there exists $t_1 \in \mathbb{T}$ such that $(q_1, v_1) \xrightarrow[t_1]{a} (q_1', v_1')$, then there exists $t_2 \in \mathbb{T}$ such that $(q_2, v_2) \xrightarrow[t_2]{a} (q_2', v_2')$ and $(q_1', v_1') \equiv_{\mathscr{R}} (q_2', v_2')$. This is very similar to classical bisimulation, except that the value of the delays before the action is taken needs not be the same, hence the terminology "time-abstract".

4.2 Back to the classical timed automata *à la* Alur and Dill

The original paper by Alur and Dill [1, 2] that proved the decidability of classical timed automata (*i.e.*, the class $\boldsymbol{T\text{-}Aut}(\mathscr{C}_{df}(X), \mathscr{U}_0(X))$) already relied on a (specialized) version of the generic construction we described earlier. We explain how this model actually fits in our generic framework.

We fix a timed automaton $\mathscr{A} = (\Sigma, X, Q, T, I, F)$, and we let M be the maximal constant appearing in one of the clock constraints of $\mathscr{A}$, *i.e.*

$$M = \max\{c \in \mathbb{N} \mid x \sim c \text{ constraint labelling a transition in } T\}^{\S}$$

Given two valuations v and v', we say that they are equivalent, and we write $v \cong_{X,M} v'$, whenever:

- for every clock $x \in X$, $v(x) > M$ iff $v'(x) > M$,
- for every clock $x \in X$, if $v(x) \leq M$, then $\lfloor v(x) \rfloor = \lfloor v'(x) \rfloor$,[¶] and $v(x) = \lfloor v(x) \rfloor$ iff $v'(x) = \lfloor v'(x) \rfloor$,
- for every pair of clocks $(x, y) \in X^2$, if $v(x) \leq M$ and $v(y) \leq M$ then $(\{v(x)\} \leq \{v(y)\}$ iff $\{v'(x)\} \leq \{v'(y)\})$.[‖]

We then define $\mathscr{R}_{X,M}$ as the quotient $\mathbb{T}^X/\cong_{X,M}$. It is rather tedious (but not really difficult) to prove that $\mathscr{R}_{X,M}$ is a set of regions compatible with the set of constraints used in $\mathscr{A}$ and with the updates $\mathscr{U}_0(X)$. The complete proof can be found in [13].

Example 5 As an example, we assume that the set of clocks is $X = \{x, y\}$, and that the maximal constant is 2. Then, the set of regions $\mathscr{R}_{X,2}$ is depicted below.

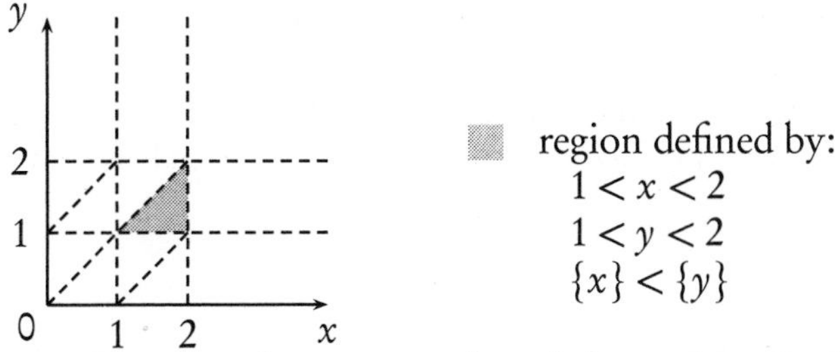

The set $\mathscr{R}_{X,M}$ can be computed, and the region graph $\Gamma_{\mathscr{R}_{X,M}}(\mathscr{A})$ as well. Hence all conditions of Theorem 2 are satisfied. Then note

§Without loss of generality we remove all constraints of the form $x \sim c$ with $c < 0$, because such a constraint is either trivially satisfied, or trivially not satisfied.

¶$\lfloor \alpha \rfloor$ denotes the integral part of α.

‖$\{\alpha\}$ denotes the fractional part of α.

that the number of equivalence classes of $\cong_{X,M}$ can be bounded by $2^{|X|}|X|!(2M+2)^{|X|}$. Hence the size of $\Gamma_{\mathscr{R}_{X,M}}(\mathscr{A})$ is at most exponential, and an on-the-fly non-deterministic algorithm allows to check for emptiness of timed automata in NPSPACE. Hence the problem is in PSPACE. Finally, the halting problem for a linearly bounded Turing machine is reducible to the emptiness of timed automata (see [3] for the detailed reduction), hence emptiness of timed automata is PSPACE-hard. We can thus summarize our discussion as follows.

Theorem 3 *[[1, 2]] Emptiness of classical timed automata is decidable, and PSPACE-complete.*

4.3 Summary of the decidability results

We have seen in the first subsection a generic construction that allows to prove decidability of classes of timed automata. We have also seen how classical timed automata (*à la* Alur and Dill) could fit that generic construction. It is first worth noticing that not all classes of timed automata satisfy the hypotheses of Theorem 2. This is for instance not the case of the class of classical timed automata in which we add the possibility to have clock constraints of the form $x+y=1$. This larger class of systems has indeed been proved undecidable in [10], as soon as the number of clocks becomes larger then four (note that the class is decidable when we restrict the number of clocks down to two). We will not detail all the decidability/undecidability results, but we better give a table (see Table 1) that summarizes most of the results which are known (unless specifically mentioned, the results are proven in [13]). All decidability results are obtained applying the generic construction (and hence Theorem 2), and the set of regions which is used is a refinement of the one given in Example 5 for classical timed automata. Undecidability results are obtained by reduction from the halting problem of a two-counter machine, which is known to be undecidable [38] (depending on the class which is considered, more or less clever encodings need to be used).

5 Expressiveness

For these definitions, we assume that $\mathscr{S}$ and $\mathscr{S}'$ are subclasses of timed systems, *i.e.* we can find a set X of clocks such that $\mathscr{S}, \mathscr{S}' \subseteq$ $\textit{\textbf{T-Aut}}_{\varepsilon}(\mathscr{C}(X), \mathscr{U}(X))$.

The class $\mathscr{S}$ is *at least as expressive as* $\mathscr{S}'$ whenever for every $\mathscr{A}'$ in $\mathscr{S}'$, there exists $\mathscr{A}$ in $\mathscr{S}$ which accepts the same language as $\mathscr{A}'$.

<table>
<tr><th></th><th>$\mathscr{U}_0(X) \cup \ldots$</th><th>Constraints $\mathscr{C}_{df}(X)$</th><th>Constraints $\mathscr{C}(X)$</th><th>Constraints $\mathscr{C}_{lin}(X)$</th></tr>
<tr><td>1</td><td>$x := c,\ x := y$</td><td rowspan="3">PSPACE-complete</td><td>PSPACE-complete [2]</td><td rowspan="4">Undecidable [10]</td></tr>
<tr><td>2</td><td>$x := x + 1$</td><td rowspan="3">Undecidable</td></tr>
<tr><td>3</td><td>$x := y + c$</td></tr>
<tr><td>4</td><td>$x := x - 1$</td><td>Undecidable</td></tr>
<tr><td>5</td><td>$x :< c$</td><td rowspan="4">PSPACE-complete</td><td>PSPACE-complete</td><td rowspan="5">Undecidable [10]</td></tr>
<tr><td>6</td><td>$x :> c$</td><td rowspan="4">Undecidable</td></tr>
<tr><td>7</td><td>$x :\sim y + c$</td></tr>
<tr><td>8</td><td>$y + c <: x :< y + d$</td></tr>
<tr><td>9</td><td>$y + c <: x :< z + d$</td><td>Undecidable</td></tr>
</table>

with $\sim \in \{\leq, <, >, \geq\}$ and $c, d \in \mathbb{Z}$.

Table 1: *Summary of the decidability results*

If moreover, $\mathscr{S}'$ is **not** at least as expressive as $\mathscr{S}$, then the class $\mathscr{S}$ is said *strictly more expressive* than $\mathscr{S}'$. Otherwise, the two class are said *equally expressive.*

Let $\mathscr{A}$ be a timed automaton. The size of $\mathscr{A}$, denoted $\mathsf{Size}(\mathscr{A})$, is the length of its encoding (states and transitions) on the tape of a Turing Machine (in particular we suppose a binary encoding for constants). The class $\mathscr{S}$ is said *exponentially more concise* than the class $\mathscr{S}'$ whenever there exists a sequence of timed automata $(\mathscr{A}_n)_{n \geq 0}$ in $\mathscr{S}$ of polynomial size in n such that for any sequence of timed automata $(\mathscr{B}_n)_{n \geq 0}$ such that $L(\mathscr{A}_n) = L(\mathscr{B}_n)$ in $\mathscr{S}'$, $\mathsf{Size}(\mathscr{B}_n)$ is at least exponential in n.

We have described in the previous section several subclasses of timed automata that are decidable. In this section, we will compare their relative expressiveness and conciseness.

5.1 Internal actions

For finite automata, it is well-known that *silent transitions* (also known as *ε-transitions* or *internal actions*) do not add expressive power to finite automata and that they can be eliminated with no blowup in the number of states of the automaton. In the timed framework, the situation is far from the one in the untimed framework.

Example 6 [[14]] Let us consider the two following languages of ***T-Aut***$_\varepsilon(\mathscr{C}(X), \mathscr{U}_0(X))$.

- $\{(a, t_1) \ldots (a, t_i) \cdots \mid \forall i,\ t_i \mod 2 = 0\}$ recognized by the timed automaton x=2,a,x:=0 x=2,ε,x:=0

- $\{(\alpha_1, t_1)\ldots(\alpha_i, t_i)\cdots \mid \alpha_i = a \text{ if } t_i = i \text{ and } \alpha_i = b \text{ if } i-1 < t_i < i\}$ recognized by the timed automaton

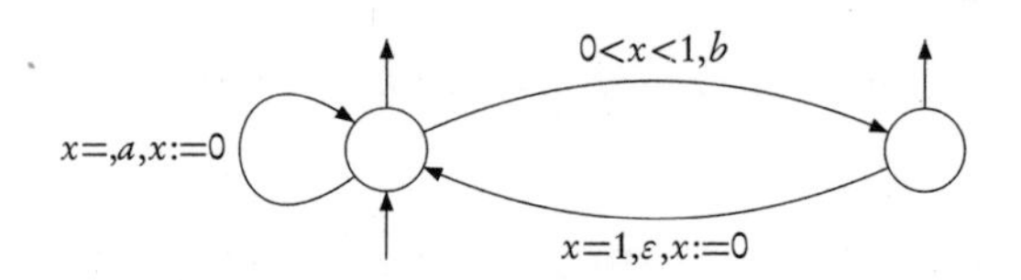

It can be proved that there do not exist timed automata without silent transitions (*i.e.* in ***T-Aut***$(\mathscr{C}(X), \mathscr{U}_0(X))$) which recognize the two above timed languages. If the proof of this claim is very easy for the first language, it is much more tricky for the second one. Indeed, there is no real criterion for a timed language to be recognized by a timed automaton without silent transitions (this problem is actually undecidable, see [20]). However, sufficient criteria are given in [14], for instance the one which follows.

Theorem 4 *[[14]] Let $\mathscr{A} \in$ **T-Aut**$_\varepsilon(\mathscr{C}(X), \mathscr{U}_0(X))$ be a timed automaton. Assume that in $\mathscr{A}$ there is no loop in which a clock is reset on an ε-transition. Then a timed automaton $\mathscr{A}'$ in **T-Aut**$(\mathscr{C}(X), \mathscr{U}_0(X))$ recognizing the same language can be effectively constructed.*

Note that this theorem can be in particular applied for timed automata where ε-transitions do not reset clocks. Also note that in general, we don't know anything concerning the conciseness of ***T-Aut***$_\varepsilon(\mathscr{C}(X), \mathscr{U}_0(X))$ compared with ***T-Aut***$(\mathscr{C}(X), \mathscr{U}_0(X))$.

5.2 Role of general updates *vs* resets to zero

In their seminal papers, Alur and Dill have considered timed automata where the only possible updates were resets, *i.e.*, the classes ***T-Aut***$(\mathscr{C}(X), \mathscr{U}_0(X))$ and ***T-Aut***$_\varepsilon(\mathscr{C}(X), \mathscr{U}_0(X))$, if we follow our notations. It turns out that considering more general updates does not increase the power of the model, at least if we want to keep the decidability of the model.

Theorem 5 *[[13]] Let $\mathcal{A}$ be a timed automaton of one of the decidable classes described in Table 1. Then a timed automaton $\mathcal{A}'$ in* $\textbf{T-Aut}_\varepsilon(\mathcal{C}(X), \mathcal{U}_0(X))$ *that recognizes the same language can be effectively constructed.*

Moreover if $\mathcal{A}$ is $\textbf{T-Aut}(\mathcal{C}(X), \mathcal{U}_{det}(X))$*, i.e. only uses deterministic updates, then the automaton $\mathcal{A}'$ can be constructed without ε-transitions, i.e. belongs to the class* $\textbf{T-Aut}(\mathcal{C}(X), \mathcal{U}_0(X))$*.*

In the proof of this theorem, the way $\mathcal{A}'$ is constructed from $\mathcal{A}$ depends naturally of the type of updates used in $\mathcal{A}$. We propose below some hints to explain some of these constructions, and illustrate some of their difficulties.

We first consider the case where any update used in $\mathcal{A}$ is of the form $\{x := d \mid x \in X \text{ and } d \in \mathbb{Z}\}$. For every tuple $\alpha = (\alpha_x)_{x \in X} \in \mathbb{Z}^X$ such that $x := \alpha_x$ is an update for clock x that appears in $\mathcal{A}$, we construct a copy of the automaton $\mathcal{A}$, that we denote by $\mathcal{A}_\alpha$. Intuitively, in the automaton $\mathcal{A}_\alpha$, the value of the clock x is what the value should be in $\mathcal{A}$ decremented by α_x (α corresponds to a shift of the clocks, comparing with what their values should be in the initial automaton).

If $q \xrightarrow{\varphi, a, up} q'$ is a transition of $\mathcal{A}$, for every α, there will be a transition $q_\alpha \xrightarrow{\varphi_\alpha, a, up_\alpha} q'_{\alpha'}$ where:

- $\varphi_\alpha = \varphi[x \leftarrow x + \alpha_x]$ (we replace x by $x + \alpha_x$),
- $up_\alpha = up[x := 0$ instead of $x := c]$,
- $\alpha'_x = c$ if $x := c$ is part of the update up, $\alpha'_x = \alpha_x$ otherwise.

There are finitely many such tuples $\alpha = (\alpha_x)_{x \in X}$, we thus only construct finitely many copies of the initial automaton. We denote by $\mathcal{A}'$ the union of all these automata $\mathcal{A}_\alpha$. The automaton $\mathcal{A}'$ is obviously in $\textbf{\textit{T-Aut}}_\varepsilon(\mathcal{C}(X), \mathcal{U}_0(X))$ and it can be shown that $\mathcal{A}'$ recognizes the same language than $\mathcal{A}$. Furthermore, if $\mathcal{A}$ had initially no ε-transitions, then so has $\mathcal{A}'$.

Example 7 We consider the automaton $\mathcal{A}$ depicted on the left. The construction gives the automaton on the right: we only need two copies of the automaton because only clock x is set to a value different from 0, hence there are two tuples: $(0, 0)$ and $(1, 0)$.

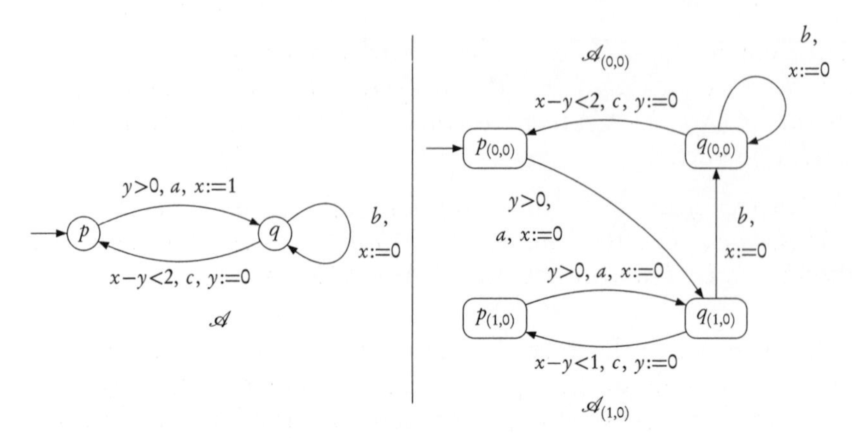

If non-deterministic updates are allowed, the constructions become much more tricky. Let us try to illustrate the difficulties and the techniques used on two following toy examples.

Example 8 The following automaton, which uses a non-deterministic update $x :< 1$, recognizes the timed language $\{(a,t)(b,t') \mid 0 \leq t < 2 \text{ and } 0 < t' - t < 1\}$.

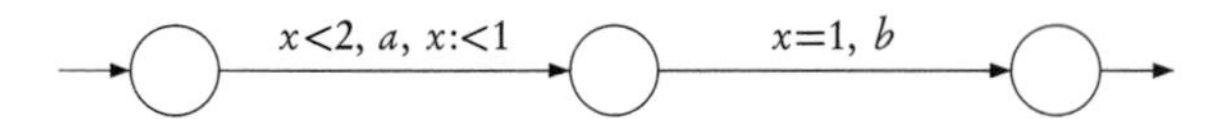

This language is also recognized by the following automaton, which only uses deterministic updates:

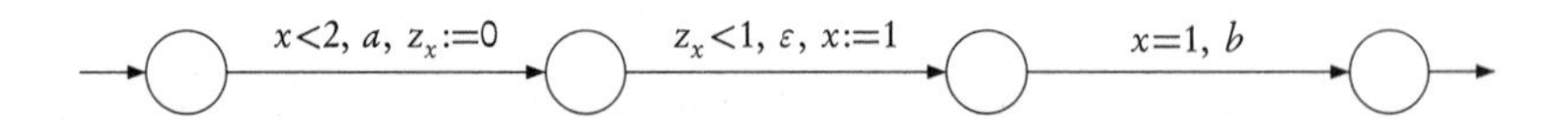

The non-deterministic update of the first automaton has been replaced by a silent action. The clock z_x which has been added represents the fractional part of x and thus checks whether it does not become larger than 1.

Example 9 The next automaton, which uses the non-deterministic update $x :< y$, recognizes the language $\{(a,t)(b,t') \mid t < 1 \text{ and } t' > 2\}$.

y<1, a
x:<y∧y:=0
x=2, b

A first idea, which is wrong, is to do the same transformation as in the previous example, *i.e.* to consider the following timed automaton:

$y<1,\ a$ — $z_x:=0 \wedge y:=0$ — $z_x<1,\ \varepsilon$ — $x:=1 \wedge z_x:=0$ — $x=2,\ b$

However, this automaton accepts the timed word $(a, 0.5)(b, 1.8)$, which was not recognized by the initial automaton.

To avoid this problem, we can add a new clock, $w_{x,y}$ which aims at keeping in mind that, when x has been reset, the value of x was less than the value of y. We thus contruct the following timed automaton:

$y<1,\ a$ — $z_x:=0 \wedge w_{x,y}:=z_y \wedge y:=0$ — $w_{x,y}>1 \wedge z_x<1,\ \varepsilon$ — $x:=1 \wedge z_x:=0$ — $x=2,\ b$

It is easy to verify that this automaton recognizes the same timed language as the initial automaton.

The proof of Theorem 5 proposed in [13] relies on a generalization of the constructions proposed in the two previous examples.

In Example 9, we have started with a timed automaton which does not use ε-transitions *i.e.* in the class ***T-Aut***$(\mathscr{C}(X), \mathscr{U}(X))$, and we have constructed a timed automaton with ε-transitions but with reset updates only, that recognizes the same language. In general, the use of ε-transitions canno be avoided, as illustrated by the next automaton, which recognizes the same language as the second automaton of Example 6 (and we said that no timed automaton in ***T-Aut***$(\mathscr{C}(X), \mathscr{U}_0(X))$ recognizes this timed language).

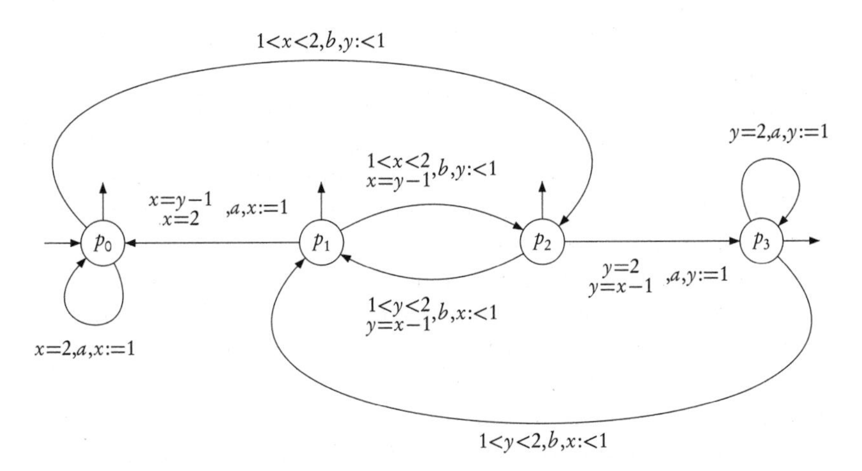

The constructions described briefly above lead to an exponential blow-up of the size automata. It turns out that is blow-up is unavoidable. We do not detail why this is actually the case, we better refer to the next subsection and Example 10. Constructions similar to the one mentioned in Example 10 can be done for all decidable subclasses mentioned in Table 1 that use more complex updates than resets to zero. Hence, we can finally state the following result.

Theorem 6 *[[9]] Every decidable subclass* $\textbf{\textit{T-Aut}}(\mathscr{C}_{df}(X),\mathscr{U})$ *mentioned in Table 1 (where* $\mathscr{U}$ *is some set of updates labelling a line of the table) is exponentially more concise than* $\textbf{\textit{T-Aut}}(\mathscr{C}_{df}(X),\mathscr{U}_0(X))$.

Note that we however don't know if the conciseness due to updates in the class $\textbf{\textit{T-Aut}}(\mathscr{C}_{df}(X),\mathscr{U})$ could be lost by considering diagonal constraints, *i.e.*, what is the relative conciseness of the classes $\textbf{\textit{T-Aut}}(\mathscr{C}_{df}(X),\mathscr{U})$ and $\textbf{\textit{T-Aut}}(\mathscr{C}_{df}(X),\mathscr{U}_0(X))$.

5.3 Role of diagonal constraints

Recall that diagonal constraints are constraints of the form $x-y\sim c$, where $x,y\in X$, and $c\in\mathbb{Z}$. We have seen in section 4 that with more general updates than resets to zero, the use of diagonal constraints can change the decidability of the emptiness problem. Nevertheless, if only resets to zero are allowed, it was known as a folklore result that diagonal constraints can be eliminated, *i.e.* that they do not add expressiveness to timed automata *à la* Alur and Dill. A formal proof of this result has been given in [14].

Proposition 7 *[[2, 14]] For every* $\mathscr{A}\in\textbf{\textit{T-Aut}}(\mathscr{C}(X),\mathscr{U}_0(X))$, *there exists* $\mathscr{B}\in\textbf{\textit{T-Aut}}(\mathscr{C}_{df}(X),\mathscr{U}_0(X))$ *such that* $L(\mathscr{A})=L(\mathscr{B})$. *Note that* $\mathscr{B}$ *is even* ***strongly bisimilar***[**] *to* $\mathscr{A}$.

The construction of this equivalent automaton is illustrated on Figure 2. Each diagonal is eliminated one by one. For example, for eliminating a diagonal $x-y\leq c$, two copies of the automaton are constructed, one copy in which the constraints $x-y\leq c$ (implicitely) holds and the other one in which the constraint $x-y>c$ (implicitely) holds. Note that the truth of a constraint $x-y\sim c$ is invariant by time elapsing. It is thus sufficient to check the truth of such a constraint when one of the clock involved in the diagonal constraint is reset, which can be done with simple (non-diagonal) constraints: the constraint $x-y\sim c$ is equivalent to $x\sim c$ when y is reset to 0.

This construction leads to an exponential (in the number of diagonal constraints) blowup of the number of states of the automaton (to

[**] Which means they are bisimilar (in a classical way) for actions taken in $\Sigma\cup\mathbb{T}$: if a system can do action, then so can also the other system, and if a system can wait d units of time, then so can also the other system.

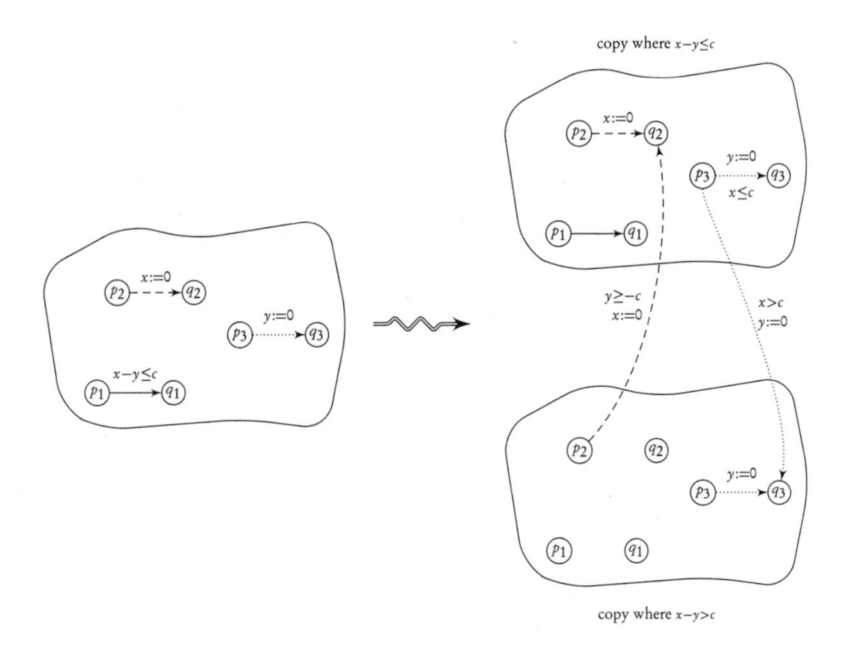

Figure 2: *Erasing diagonal constraint $x - y \leq c$, assuming $c \geq 0$*

remove one diagonal constraint, we duplicate the automaton). This blowup is in general unavoidable as shows the following example [9].

Example 10 [[9]] For every $n \in \mathbb{N}$, we consider the timed language

$$L_n = \{(a^{2^n}, \tau) \mid 0 < \tau_1 < \tau_2 < \cdots < \tau_{2^n} < 1\}$$

This timed language cannot be recognized by any timed automaton in ***T-Aut***$(\mathscr{C}_{df}(X), \mathscr{U}_0(X))$ that has not exponential size (in n). On the contrary, L_n is recognized by a small (polynomial size) timed automaton $\mathscr{A}_n$ in ***T-Aut***$(\mathscr{C}(X), \mathscr{U}_0(X))$. The shape of automaton $\mathscr{A}_n$

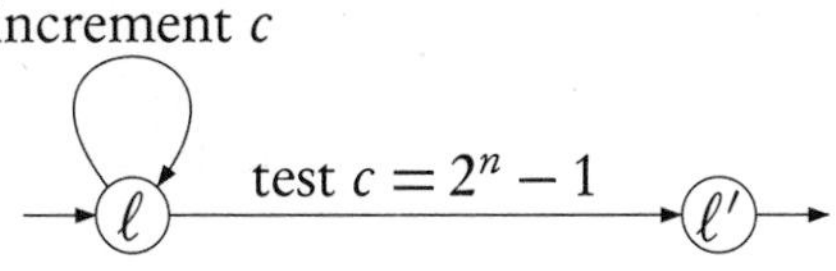

increment c

test $c = 2^n - 1$

ℓ ℓ'

Figure 3: *Shape of automaton $\mathscr{A}_n$ recognizing L_n*

is depicted on Figure 3. It basically manipulates a counter that will increase up to $2^n - 1$. We will use $2n + 2$ clocks $X = \{x_i, y_i \mid 1 \leq i \leq n\} \cup \{y, z\}$. The clock y is tested ($y > 0$) and reset on each transition and ensures thus that the time sequence is (strictly) increasing. The clock z is never reset but tested ($z < 1$) on the last transition to ensure that the global time be bounded by 1.

The clocks $\{x_i, y_i \mid 1 \leq i \leq n\}$ are used to encode the counter c. The binary encoding $b_1 b_2 \ldots b_n$ of c (least significant bit first) is given by: $b_i = 1$ if $x_i - y_i > 0$ while $b_i = 0$ if $x_i - y_i = 0$ (the two conditions are invariant when time elapses). Hence, the constraint on the transition between ℓ and ℓ' is $_{1 \leq i \leq n} x_i - y_i > 0$. To increment the counter, we will somehow implement the incrementation by adding loops on ℓ of the following kind: fixing some index $0 \leq j < n$, there is a loop whose constraint is $(y > 0) \wedge \; _{1 \leq i \leq j}(x_i - y_i > 0) \wedge (x_{j+1} - y_{j+1} = 0)$, and whose reset is $(y := 0) \wedge \; _{1 \leq i \leq j}(x_i := 0 \wedge y_i := 0) \wedge (y_{j+1} := 0)$. This encodes that if the j first bits are 1, and the $j+1$-th bit is 0, then we set the j first bits to 0 and the $j+1$-th bit to 1.

6 Some Other Extensions

6.1 Adding modulo constraints

An extension of timed automata with *periodic clock constraints* has been considered in [23]. In this model, clock constraints are generated by the following grammar:

$$\varphi \quad ::= \quad x \in I \mod k \quad | \quad x - y \in I \mod k \quad | \quad \varphi \wedge \varphi$$

where $x, y \in X$ are clocks, I an interval with integral bounds, and $k \in \mathbb{N}$ is an integer. We write $\mathscr{C}_{mod}(X)$ for this set of clock constraints. The classical satisfaction relation is extended as follows, if $v \in \mathbb{T}^X$ is a valuation, and $\varphi \in \mathscr{C}_{mod}(X)$:

$$\begin{array}{ll} v \models (x \in I \mod k) & \text{if there exists } h \in \mathbb{Z} \text{ s.t. } v(x) - kh \in I \\ v \models (x - y \in I \mod k) & \text{if there exists } h \in \mathbb{Z} \text{ s.t.} \\ & \qquad (v(x) - v(y)) - kh \in I \\ v \models \varphi_1 \wedge \varphi_2 & \text{if } v \models \varphi_1 \text{ and } v \models \varphi_2 \end{array}$$

Note that $k = 0$ implies that the constraint ($x \in I \mod 0$) is the classical constraint ($x \in I$). In that sense, $\mathscr{C}_{mod}(X)$ extends the set $\mathscr{C}(X)$. A constraint ($x \in I \mod k$) or ($x - y \in I \mod k$) with $k > 0$ is called a *modulo constraint*.

These extended clock constraints allow to express e.g. the first timed language mentioned in Example 6, which cannot be recognized by a timed automaton in ***T-Aut***$(\mathscr{C}(X), \mathscr{U}_0(X))$:

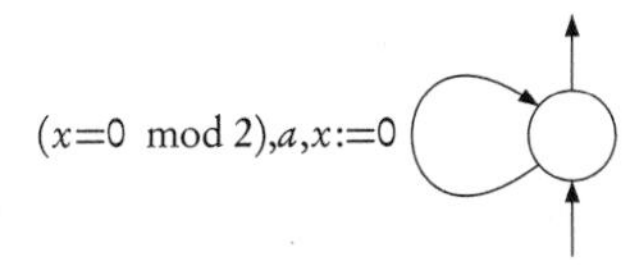

It is first not difficult to be convinced that diagonal constraints (even with modulos) can be removed, following a procedure similar to the one described in subsection 5.3. Furthermore, modulo constraints can be removed from the transformed timed automaton (with no diagonal constraints) by adding ε-transitions: the first timed automaton of Example 6 is the result of this transformation for the above automaton. This can easily be generalized by adding copies of clocks that are regularly reset to 0 (for checking modulo k constraints, we reset the copy of the clock every k time units), and the modulo constraint is then checked on the copy of the clock. We can summarize our discussion as follows.

Proposition 8 *[[23]] The two classes* $\textbf{\textit{T-Aut}}_\varepsilon(\mathcal{C}_{mod}(X), \mathcal{U}_0(X))$ *and* $\textbf{\textit{T-Aut}}_\varepsilon(\mathcal{C}_{df}(X), \mathcal{U}_0(X))$ *are equally expressive.*

The class $\textbf{\textit{T-Aut}}(\mathcal{C}_{mod}(X), \mathcal{U}_0(X))$ *is strictly more expressive than the class* $\textbf{\textit{T-Aut}}(\mathcal{C}(X), \mathcal{U}_0(X))$*.*

It is worth noticing that the second transformation (removing modulos from diagonal-free constraints) is not expensive, the size of the transformed automaton is polynomial in the size of the original timed automaton. Hence there is no concision that is due to modulo constraints.

From the above transformations, we get that the emptiness problem is decidable for the class $\textbf{\textit{T-Aut}}_\varepsilon(\mathcal{C}_{mod}(X), \mathcal{U}_0(X))$. However we can even prove that the general framework developed in section 4 also applies for that class. We briefly describe the way to fit the class $\textbf{\textit{T-Aut}}_\varepsilon(\mathcal{C}_{mod}(X), \mathcal{U}_0(X))$ to the general framework. Let $\mathcal{A} \in \textbf{\textit{T-Aut}}_\varepsilon(\mathcal{C}_{mod}(X), \mathcal{U}_0(X))$. Given two valuations v and v', we say that they are equivalent, and we write $v \cong_X v'$, whenever:

- for every clock $x \in X$, $\lfloor v(x) \rfloor = \lfloor v'(x) \rfloor$, and $v(x) = \lfloor v(x) \rfloor$ iff $v'(x) = \lfloor v'(x) \rfloor$,
- for every pair of clocks $(x, y) \in X^2$, $\{v(x)\} \leq \{v(y)\}$ iff $\{v'(x)\} \leq \{v'(y)\}$.

This equivalence relation generalizes the equivalence $\cong_{X,M}$ that had been defined for classical timed automata, by removing the maximal constant to which clocks are compared to. Obviously, $\cong_X$'s index is not finite. We then write M for the maximal constant to which clocks are compared in $\mathcal{A}$, and K for the lcm of all k's that are used in modulo constraints of $\mathscr{A}$. We then define the equivalence relation $\equiv_{X,M,K}$ by $v \cong_{X,M,K} v'$ whenever

- for every clock $x \in X$, $v(x) \leq M$ implies $v \cong_X v'$,
- there exists $h_x \in \mathbb{Z}$ for every $x \in X$ (with $h_x = 0$ whenever $v(x) \leq M$) such that $v_h \cong_X v'$ where $v_h(x) = v(x) - Kh_x$.

It is not hard to be convinced that this equivalence relation defines a set of regions for $\mathscr{A}$. Furthermore, the size of the set of regions is not much larger than the size of the original set of regions, and is still exponential. Hence we get the following result.

Theorem 9 *Emptiness of* $\textbf{T-Aut}_\varepsilon(\mathscr{C}_{mod}(X), \mathscr{U}_0(X))$ *is decidable, and PSPACE-complete.*

6.2 Adding alternation

As for finite automata, we can add alternation to the model. However this is only recently that this model has been considered [36, 39]. We do not define it formally, but give an example below.

In this example, we have the obvious interpretation that any time an a is done (within the two first time units), we fork a new thread which will check that a b appears one time unit later. A behaviour of this alternating timed automaton is an unbounded tree, and it is not obvious that it is possible to check for emptiness of such a system. Indeed, checking emptiness of alternating timed automata is decidable only for one clock over finite timed words, any slight extension (infinite timed words, two clocks, ε-transitions) leading to undecidability [37, 40].

We explain how the decidability of this model can be understood [36, 39]. Consider the timed word $(c, 0.6)(a, 0.7)(a, 1.5)(b, 1.7)$. The execution of the above alternating timed automaton on that timed word can be depicted as the following tree, which is not accepting as one of the branches (the second one on the picture) is not accepting (accepting states are underlined).

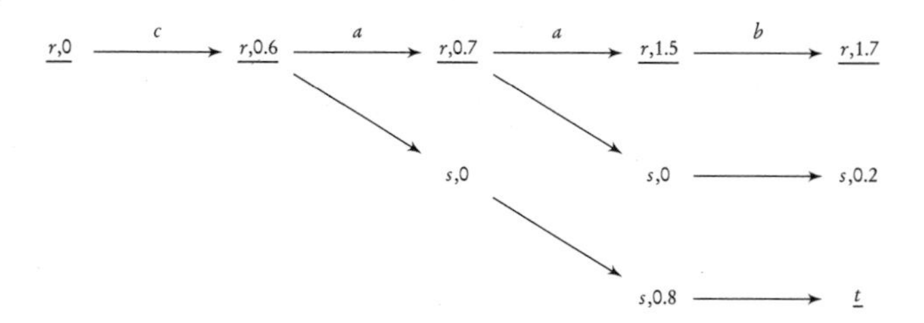

A configuration of the alternating timed automaton is a slice of the tree. For instance, the set $\{(r,1.5),(s,0),(s,0.8)\}$ is a configuration. Because we consider finite words, there is no need to consider the tree structure of the execution, but we can reason globally on configurations of the automaton. There are infinitely many such configurations, but as for the region automaton construction for timed automata [2], the precise values of the clocks is not really relevant, and the things which are important in a configuration are the integral parts of the clocks and the relative order of the fractional parts of the clocks. For instance, for the above-mentioned configuration, we only need to know that there is a state $(s,0)$ with fractional part 0, and two other states $(s,0)$ and $(r,1)$ such that the fractional part for $(s,0)$ is greater than the fractional part for $(r,1)$. For all configurations with the same abstraction, the possible future behaviours are the same, in a time-abstract bisimulation sense [40, 37]. Unfortunately, the set of abstractions of possible configurations of the alternating timed automaton is also infinite. The most important property is then that there is a *well-quasi-order*[††] on the set of abstractions of configurations, and that we can use it to provide an algorithm to decide emptiness [28]! This briefly sketches an algorithm (that is actually not primitive recursive)[‡‡] for deciding the emptiness of single-clock alternating timed automata over finite timed words.

Theorem 10 *[[36, 39]] Emptiness of single-clock alternating timed automata is decidable, but non-primitive recursive.*

What about expressiveness?

Alternating timed automata (with no restriction over the number of clocks) are obvious more expressive than timed automata (straightforward inclusion). However, it is interesting to see that the class of single-clock alternating timed automata and the class of classical timed automata are actually incomparable.

[††]A preorder $\preceq$ over set S is a well-quasi-order if for every infinite sequence $(s_i)_{i\geq 0}$ (with $s_i \in S$ for every i), there exist $i < j$ such that $s_i \preceq s_j$ (every sequence in S is saturating).

[‡‡]It is however optimal...

Example 11 The following alternating timed automaton recognizes the timed language

$$\{(a,\tau_1)\dots(a,\tau_n) \mid \forall i,\, \forall j,\, j > i \Rightarrow \tau_j - \tau_i \neq 1\}$$

which is known not to be recognizable by a classical timed automaton [2].

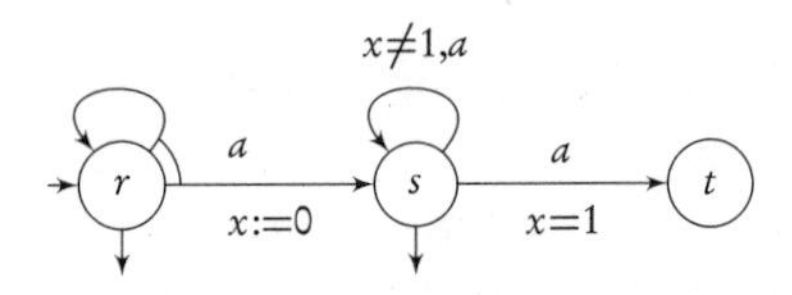

On the other hand, the timed language

$$\{(a,\tau_1)(a,\tau_2)(a,\tau_3) \mid 0 < \tau_1 < \tau_2 < 1 \text{ and } \tau_1 + 1 < \tau_3 < \tau_2 + 1\}$$

can be recognized by a classical timed automaton, but it can be proved that it can not be recognized by a single-clock alternating timed automaton [36].

7 Conclusion

In this paper, we have presented several extensions of timed automata *à la* Alur and Dill, that have been considered in the literature. We have mostly focused on the decidability of the emptiness problem, which is fundamental for the use of the model for verification purposes, and also on the expressiveness and conciseness of the different classes of systems.

Several other exotic extensions have been proposed among which we mention the work [27] where a subset of clocks can be "freezed". Due to restrictions on those subsets of clocks, the decidability of the emptiness problem is preserved, and is based on the construction of a set of regions.

We should also mention the class of (linear) hybrid automata [31, 29, 32], that can also be seen as extensions of timed automata, even though they have been developed simultaneously and independently. Linear hybrid automata are undecidable in general, and even the simplest linear hybrid automata which are timed automata where clocks can be stopped (that is thus a generalization of [27]), are undecidable. Decidable subclasses have been investigated, like initialized rectangular automata [32], o-minimal hybrid automata [35, 22], weighted timed automata [4, 17], lazy rectangular automata [7], or strongly-reset hybrid automata [8], *etc.*

Timed automata are a very active field of research as attested by the 2008 CAV award given to Rajeev Alur and David L. Dill for their seminal article [1]. There are therefore lots of works on timed automata, and not all of them have been adressed in this rather short survey. For instance, concerning other decision problems like inclusion checking, we point to [5]. Also, we do not discuss the important question of the various characterizations of the languages accepted by timed automata using other formalisms, like the logical characterization of [41], or the algebraic characterization of [21]. For such questions, we refer to [6].

References

[1] RAJEEV ALUR AND DAVID DILL. Automata for modeling real-time systems. In *Proc. 17th International Colloquium on Automata, Languages and Programming (ICALP'90)*, volume 443 of *Lecture Notes in Computer Science*, pages 322–335. Springer, 1990.

[2] RAJEEV ALUR AND DAVID DILL. A theory of timed automata. *Theoretical Computer Science*, 126(2):183–235, 1994.

[3] LUCA ACETO AND FRANÇOIS LAROUSSINIE. Is your model-checker on time? On the complexity of model-checking for timed modal logics. *Journal of Logic and Algebraic Programming*, 52–53:7–51, 2002.

[4] RAJEEV ALUR, SALVATORE LA TORRE, AND GEORGE J. PAPPAS. Optimal paths in weighted timed automata. In *Proc. 4th International Workshop on Hybrid Systems: Computation and Control (HSCC'01)*, volume 2034 of *Lecture Notes in Computer Science*, pages 49–62. Springer, 2001.

[5] RAJEEV ALUR AND P. MADHUSUDAN. Decision problems for timed automata: A survey. In *Proc. 4th International School on Formal Methods for the Design of Computer, Communication and Software Systems: Real Time (SFM-04:RT)*, volume 3185 of *Lecture Notes in Computer Science*, pages 122–133. Springer, 2004.

[6] EUGENE ASARIN. Challenges in timed languages: From applied theory to basic theory. *The Bulletin of the European Association for Theoretical Computer Science*, (83), 2004.

[7] MANINDRA AGRAWAL AND P.S. THIAGARAJAN. Lazy rectangular hybrid automata. In *Proc. 7th International Workshop on Hybrid Systems: Computation and Control (HSCC'04)*, volume 2993 of *Lecture Notes in Computer Science*, pages 1–15. Springer, 2004.

[8] PATRICIA BOUYER, THOMAS BRIHAYE, MARCIN JURDZIŃSKI, RANKO LAZIĆ, AND MICHAŁ RUTKOWSKI. Average-price and reachability-price games on hybrid automata with strong resets. In *Proc. 6th International Conference on Formal Modeling and Analysis of Timed Systems (FORMATS'08)*, Lecture Notes in Computer Science. Springer, 2008. To appear.

[9] PATRICIA BOUYER AND FABRICE CHEVALIER. On conciseness of extensions of timed automata. *Journal of Automata, Languages and Combinatorics*, 10(4):393–405, 2005.

[10] BÉATRICE BÉRARD AND CATHERINE DUFOURD. Timed automata and additive clock constraints. *Information Processing Letters*, 75(1–2):1–7, 2000.

[11] PATRICIA BOUYER, CATHERINE DUFOURD, EMMANUEL FLEURY, AND ANTOINE PETIT. Are timed automata updatable? In *Proc. 12th International Conference on Computer Aided Verification (CAV'00)*, volume 1855 of *Lecture Notes in Computer Science*, pages 464–479. Springer, 2000.

[12] PATRICIA BOUYER, CATHERINE DUFOURD, EMMANUEL FLEURY, AND ANTOINE PETIT. Expressiveness of updatable timed automata. In *Proc. 25th International Symposium on Mathematical Foundations of Computer Science (MFCS'00)*, volume 1893 of *Lecture Notes in Computer Science*, pages 232–242. Springer, 2000.

[13] PATRICIA BOUYER, CATHERINE DUFOURD, EMMANUEL FLEURY, AND ANTOINE PETIT. Updatable timed automata. *Theoretical Computer Science*, 321(2–3):291–345, 2004.

[14] BÉATRICE BÉRARD, VOLKER DIEKERT, PAUL GASTIN, AND ANTOINE PETIT. Characterization of the expressive power of silent transitions in timed automata. *Fundamenta Informaticae*, 36(2–3):145–182, 1998.

[15] GERD BEHRMANN, ALEXANDRE DAVID, KIM G. LARSEN, JOHN HÅKANSSON, PAUL PETTERSSON, WANG YI, AND MARTIJN HENDRIKS. Uppaal 4.0. In *Proc. 3rd International Conference on the Quantitative Evaluation of Systems (QEST'06)*, pages 125–126. IEEE Computer Society Press, 2006.

[16] BÉATRICE BÉRARD AND LAURENT FRIBOURG. Automated verification of a parametric real-time program: the ABR conformance protocol. In *Proc. 11th International Conference on Computer Aided Verification (CAV'99)*, volume 1633 of *Lecture Notes in Computer Science*, pages 96–107. Springer, 1999.

[17] GERD BEHRMANN, ANSGAR FEHNKER, THOMAS HUNE, KIM G. LARSEN, PAUL PETTERSSON, JUDI ROMIJN, AND FRITS VAANDRAGER. Minimum-cost reachability for priced timed automata. In *Proc. 4th International Workshop on Hybrid Systems: Computation and Control (HSCC'01)*, volume 2034 of *Lecture Notes in Computer Science*, pages 147–161. Springer, 2001.

[18] BÉATRICE BÉRARD, LAURENT FRIBOURG, FRANCIS KLAY, AND JEAN-FRANÇOIS MONIN. A compared study of two correctness proofs for the standardized algorithm of abr conformance. *Formal Methods in System Design*, 22(1):59–86, 2003.

[19] BÉATRICE BÉRARD, PAUL GASTIN, AND ANTOINE PETIT. On the power of non-observable actions in timed automata. In *Proc. 13th Annual Symposium on Theoretical Aspects of Computer Science (STACS'96)*, volume 1046 of *Lecture Notes in Computer Science*, pages 257–268. Springer, 1996.

[20] PATRICIA BOUYER, SERGE HADDAD, AND PIERRE-ALAIN REYNIER. Undecidability results for timed automata with silent transitions. Technical Report LSV-07-12, Laboratoire Spécification et Vérification, ENS Cachan, France, 2007.

[21] PATRICIA BOUYER, ANTOINE PETIT, AND DENIS THÉRIEN. An algebraic approach to data languages and timed languages. *Information and Computation*, 2002.

[22] THOMAS BRIHAYE. Words and bisimulation of dynamical systems. *Discrete Mathematics and Theoretical Computer Science*, 9(2):11–31, 2007.

[23] CHRISTIAN CHOFFRUT AND MASSIMILIANO GOLDWURM. Timed automata with periodic clock constraints. *Journal of Automata, Languages and Combinatorics*, 5(4):371–404, 2000.

[24] VOLKER DIEKERT, PAUL GASTIN, AND ANTOINE PETIT. Removing ε-transitions in timed automata. In *Proc. 14th Annual Symposium on Theoretical Aspects of Computer Science (STACS'97)*, volume 1200 of *Lecture Notes in Computer Science*, pages 583–594. Springer, 1997.

[25] CONRADO DAWS, ALFREDO OLIVERO, STAVROS TRIPAKIS, AND SERGIO YOVINE. The tool Kronos. In *Proc. Hybrid Systems III: Verification and Control (1995)*, volume 1066 of *Lecture Notes in Computer Science*, pages 208–219. Springer, 1996.

[26] CATHERINE DUFOURD. Une extension d'un résultat d'indécidabilité pour les automates temporisés. In *Proc. 9th Rencontres Francophones du Parallélisme (RenPar'97)*, 1997.

[27] FRANÇOIS DEMICHELIS AND WIESLAW ZIELONKA. Controlled timed automata. In *Proc. 9th International Conference on Concurrency Theory (CONCUR'98)*, volume 1466 of *Lecture Notes in Computer Science*, pages 455–469. Springer, 1998.

[28] ALAIN FINKEL AND PHILIPPE SCHNOEBELEN. Well structured transition systems everywhere! *Theoretical Computer Science*, 256(1–2):63–92, 2001.

[29] THOMAS A. HENZINGER. The theory of hybrid automata. In *Proc. 11th Annual Symposim on Logic in Computer Science (LICS'96)*, pages 278–292. IEEE Computer Society Press, 1996.

[30] THOMAS A. HENZINGER, PEI-HSIN HO, AND HOWARD WONG-TOI. HyTech: A model-checker for hybrid systems. *Journal on Software Tools for Technology Transfer*, 1(1–2):110–122, 1997.

[31] THOMAS A. HENZINGER, PETER W. KOPKE, ANUJ PURI, AND PRAVIN VARAIYA. What's decidable about hybrid automata? In *Proc. 27th Annual ACM Symposium on the Theory of Computing (STOC'95)*, pages 373–382. ACM, 1995.

[32] THOMAS A. HENZINGER, PETER W. KOPKE, ANUJ PURI, AND PRAVIN VARAIYA. What's decidable about hybrid automata? *Journal of Computer and System Sciences*, 57(1):94–124, 1998.

[33] JOHN E. HOPCROFT AND JEFFREY D. ULLMAN. *Introduction to Automata Theory, Languages and Computation*. Addison-Wesley, 1979.

[34] FRANÇOIS LAROUSSINIE AND KIM G. LARSEN. CMC: A tool for compositional model-checking of real-time systems. In *Proc. IFIP Joint International Conference on Formal Description Techniques & Protocol Specification, Testing, and Verification (FORTE-PSTV'98)*, pages 439–456. Kluwer Academic, 1998.

[35] GERARDO LAFFERRIERE, GEORGE J. PAPPAS, AND SHANKAR SASTRY. O-minimal hybrid systems. *Mathematics of Control, Signals, and Systems*, 13(1):1–21, 2000.

[36] SLAWOMIR LASOTA AND IGOR WALUKIEWICZ. Alternating timed automata. In *Proc. 8th International Conference on Foundations of Software Science and Computation Structures (FoSSaCS'05)*, volume 3441 of *Lecture Notes in Computer Science*, pages 250–265. Springer, 2005.

[37] SLAWOMIR LASOTA AND IGOR WALUKIEWICZ. Alternating timed automata. *ACM Transactions on Computational Logic*, 9(2), 2008. Paper 10.

[38] MARVIN MINSKY. *Computation: Finite and Infinite Machines*. Prentice Hall International, 1967.

[39] JOËL OUAKNINE AND JAMES WORRELL. On the decidability of Metric Temporal Logic. In *Proc. 20th Annual Symposium on Logic in Computer Science (LICS'05)*, pages 188–197. IEEE Computer Society Press, 2005.

[40] JOËL OUAKNINE AND JAMES WORRELL. On the decidability and complexity of metric temporal logic over finite words. *Logical Methods in Computer Science*, 3(1), 2007. Paper 8.

[41] THOMAS WILKE. Specifying timed state sequences in powerful decidable logics and timed automata. In *Proc. 3rd International Symposium on Formal Techniques in Real-Time and Fault-Tolerant Systems (FTRTFT'94)*, volume 863 of *Lecture Notes in Computer Science*, pages 694–715. Springer, 1994.

Scheduling Stochastic Branching Processes *

Tomáš Brázdil, Javier Esparza,
Stefan Kiefer, Michael Luttenberger
Institut für Informatik, Technische Universität München,
85748 Garching, Germany
{brazdil,esparza,kiefer,luttenbe}@model.in.tum.de

Abstract

In probability theory, branching processes are processes that can die and reproduce, but for which only stochastic information on the number of produced offspring is available. Traditionally, the theory considers the synchronous case, in which all processes reproduce at the same time. We study the asynchronous case in the following setting: a computing agent (a microprocessor or an operating system) executes threads, which can generate new threads, and a scheduler determines which thread to execute next. We obtain results on the maximal size of the queue of non-terminated threads, or the probability that it exceeds a given bound.

Keywords: Branching processes, stochastic processes, process scheduling

1 Introduction

Branching stochastic processes[†] are a well established topic of probability theory. Their origin can be traced back to the following problem, formulated by Francis Galton back in the 19th century [8], and quoted by Thomas Harris in his classical text [6]:

> Let $p_0, p_1, p_2 \ldots$ be the respective probabilities that a man has 0, 1, 2, ... sons, let each son have the same probability for sons of his own, and so on. What is the probability that the male line is extinct after r generations, and more

*This work was partially supported by the DFG project *Algorithms for Software Model Checking*.

†Not to be confused with branching processes as defined by Engelfriet in [1], although, as we shall see, there is a connection.

> generally what is the probability for any given number of descendants in the male line in any given generation?

Since Galton's problem, the study of branching processes has expanded into a general theory of stochastic processes that can "die" and "reproduce", and has been applied to the study of nuclear reactions in physics (where a particle can decay into others), or to the study of populations in biology.

In this note we look at branching processes from a computer science point of view. We assume that a process corresponds to a *computational* process like a task in an operating system, or a thread in a program. A task can generate new tasks (for instance, the user can decide to open a new window from an existing one) and a thread can generate new threads. While the theory of branching processes assumes that a process generates offspring only when it dies, which is not the case of tasks or threads, this is a minor modelling issue: we can conceptually assume that the parent thread dies and generates children, one of which corresponds to the continuation of the parent's process.

The original problem by Galton asks for the number of descendants after some number n of generations. This corresponds to a *synchronous model*, in which all processes of a generation simultaneously make a move (dying and generating offspring). In this note we consider the *asynchronous* paradigm, where we assume that only one process makes a move at a time. This is the natural paradigm for our application: we assume that tasks or threads are executed by one microprocessor, and the next task to be executed is chosen according to some scheduling strategy.

We are interested in the *maximal* number of threads or tasks that can be simultaneously active. The motivation for this is that keeping a queue of threads or tasks awaiting execution consumes resources, since their states must be stored.

We study the simplest possible setting, in which all processes have the same probability to generate a given number of children, and the scheduler does not have information on the future of the computation (i.e., it does not know which processes will generate how many children). We call such a scheduler a *black-box* scheduler. We show that all black-box schedulers are equivalent, i.e., for every black-box scheduler the maximal number of active threads has the same probability distribution. To determine the quality of the black-box schedulers, we compare them with an *optimal scheduler*, which has complete information about the future of the computation (i.e., the scheduler has access to an oracle that knows how the probabilistic choices will be resolved). We also study the *pessimal*scheduler, and obtain an upper bound on the resources needed when the scheduler is under the control of an adversary.

2 Preliminaries

We formalize the behaviour of processes and schedulers. Our study is independent of the nature of processes: as mentioned in the introduction, they could be tasks of an operating system, threads executed in a microprocessor, individuals of a population in biology, etc. However, in order to fix ideas in what follows we choose to look at processes as threads.

2.1 Assumptions on Thread Behaviour

We assume that every thread eventually terminates, and generates upon termination either 0 or 2 threads. In the first case the thread *dies*, in the second case it *branches*. In the branching case, we assume that one of the children is in fact the continuation of the parent thread[‡].

Threads are executed by a microprocessor. We assume that only one thread can be executed at a time, i.e., that there is no true parallelism[§]. Threads awaiting execution are stored in a queue. A scheduler determines the next thread to be executed.

We model the behaviour of the system from the point of view of an observer that does not have access to the code of the threads, and so cannot determine with certainty whether it will die or branch. However, we assume that the observer has the following stochastic information:

(a) Threads branch and die with probabilities p and $q := (1 - p)$, respectively;
(p can be thought to stand for procreate, q for quit)
(b) The probability that a thread dies or branches is independent of its history and its context (the other threads awaiting execution).

We are aware that the second assumption may not always hold. In particular, there are scenarios in which a thread has a smaller probability of dying than its parent (e.g., a thread modelling a call of a divide-and-conquer procedure that splits the input into two parts, and branches

[‡]As an example consider forking in Unix.

[§]This assumption could be relaxed, and allow for at most k threads to be concurrently executed but this extension is beyond the scope of this paper.

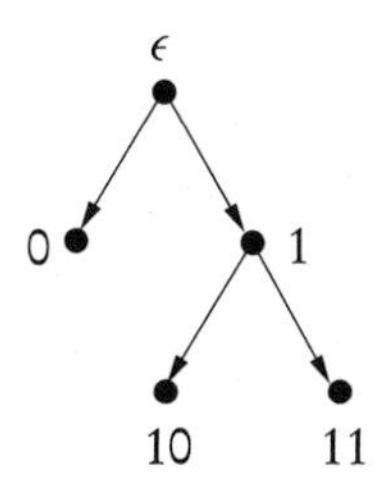

Figure 1: *A family tree*

into two different threads). In other scenarios (e.g., threads modelling biological populations) the assumption is more plausible.

2.2 Formal Model

With the assumptions above we can give our system a partial order semantics in terms of a discrete probability space $(\mathcal{T}, \Pr)$, whose set of elementary events is the set $\mathcal{T}$ of family trees[¶]. A *family tree* is a finite tree whose nodes, called *thread identifiers* or just *threads*, are elements of $\{0,1\}^*$. The root is ϵ, and every node has 0 or 2 children. If a node w has two children, then its left child is $w0$ and its right child is $w1$. Notice that a family tree is completely determined by its nodes. The family tree of Figure 1 corresponds to the following history: the initial thread ϵ branches into threads 0 and 1. The thread 0 dies, the thread 1 branches into two threads $10, 11$, which both die. We say that the thread *goes extinct.*

The probability function Pr assigns to each family tree its probability. Because of assumption (b) above, the probability of a family tree T with k leaves (dying threads) and $k-1$ inner nodes (branching threads) is $\Pr[T] = p^{(k-1)}q^k$.

The attentive reader will notice that $(\mathcal{T}, \Pr)$ is not a probability space for every value of p, because the probabilities of the elementary events do not add up to 1. For instance, for $p = 1$ the probability of every family tree is 0. It is not difficult to see that $(\mathcal{T}, \Pr)$ is a probability space only for $p \leq 1/2$:

Proposition 1 $\Pr[\mathcal{T}] = \sum_{T \in \mathcal{T}} \Pr[T]$ *is the least non-negative fixed point of the polynomial*

$$f(x) := px^2 + q. \tag{1}$$

[¶]For readers familiar with partial order semantics, family trees are very similar to finite nonsequential processes.

Proof Let $\mathcal{T}_h$ be set of all trees having height at most h so that $\mathcal{T}_0$ only includes of the binary tree consisting of a single node. Obviously, $(\mathcal{T}_h)_{h\in\mathbb{N}}$ is a strictly increasing sequence of sets converging to $\mathcal{T}$. We show inductively that $\Pr\left[\mathcal{T}_h\right] = f^{h+1}(0)$ where $f^{k+1}(0) := f(f^k(0))$ and $f^1(0) := q$. The base case $h = 0$ is trivial as there is only one tree of height 0 having weight exactly q. For the inductive step consider the following equation:

$$f^{h+2}(0) = f(f^{h+1}(0)) = f(\Pr\left[\mathcal{T}_h\right]) = q + p \cdot \Pr\left[\mathcal{T}_h\right]^2$$
$$= \Pr\left[\mathcal{T}_0\right] + \sum_{T,T' \in \mathcal{T}_h} p \cdot \Pr\left[T\right] \cdot \Pr\left[T'\right].$$

But the right most sum simply states that we can obtain any tree of $\mathcal{T}_h \setminus \mathcal{T}_0$ by taking any two trees from $\mathcal{T}_h$ and connecting them by a common root. So, $\Pr\left[\mathcal{T}_h\right] = f^{h+1}(0)$ follows.

Now, as f is monotone, i.e. for all $0 \leq x \leq y$ we have $f(x) \leq f(y)$, the sequence $(f^h(0))_{h\in\mathbb{N}}$ is monotonically increasing. Further, we have $f^h(0) \leq 1$ for all h as $f(1) = 1$. Thus, $\mu f := \lim_{h\to\infty} f^h(0)$ exists and is equal to $\lim_{h\to\infty} \Pr\left[\mathcal{T}_h\right] = \Pr\left[\mathcal{T}\right]$. As f is continuous we also have $f(\mu f) = \mu f$, i.e. μf is the least non-negative fixed point of f. ⊣

Corollary 2 $\Pr\left[\mathcal{T}\right] = 1$ *if and only if* $p \in [0, \frac{1}{2}]$.

Proof If $p = 0$, we are done. Thus, assume $p \in (0, \frac{1}{2}]$. The least non-negative fixed point of f is the least non-negative zero of $f(x) - x = 0$. We have

$$x \in \left\{ \frac{1 \pm \sqrt{1 - 4pq}}{2p} \right\} = \left\{ 1, \frac{q}{p} \right\}.$$

So, only for $p \leq q$ the least fixed point of f is 1. ⊣

Intuitively, for $p \leq 1/2$ a thread almost surely goes extinct. In this note we restrict our attention to this case. The case $p > 1/2$, which requires to explicitly deal with infinite family trees, is left for future research.

2.3 Derivations

A family tree does not contain information on the order in which threads are executed, apart from the causal information that a thread ends before its children start. For this purpose we introduce derivations.

Recall that $\{0,1\}^*$ is the set of thread identifiers. A *state* is a subset of $\{0,1\}^*$. We define the relation $\Rightarrow$ between states as follows: For every state s and every $w \in s$,

$$s \Rightarrow s \setminus \{w\} \quad \text{and} \quad s \Rightarrow (s \setminus \{w\}) \cup \{w0, w1\}\,.$$

A *partial derivation* is a sequence $s_1 \Rightarrow \ldots \Rightarrow s_n$ of states. A *derivation* is a partial derivation such that $s_n = \emptyset$. A derivation induces a family tree, namely the tree that contains the node w if and only if w occurs in the derivation. We define the *derivations of a tree* T as the set of all derivations that induce T. Observe that a tree may have multiple derivations. The *states of* T are the states that occur in the derivations of T. For example, the family tree of Figure 1 has 8 derivations, including the following:

$$\begin{array}{l}\{\epsilon\} \Rightarrow \{0,1\} \Rightarrow \{1\} \Rightarrow \{10,11\} \Rightarrow \{10\} \Rightarrow \emptyset \\ \{\epsilon\} \Rightarrow \{0,1\} \Rightarrow \{0,10,11\} \Rightarrow \{10,11\} \Rightarrow \{10\} \Rightarrow \emptyset \\ \{\epsilon\} \Rightarrow \{0,1\} \Rightarrow \{0,10,11\} \Rightarrow \{0,11\} \Rightarrow \{0\} \Rightarrow \emptyset\end{array}$$

A *partial derivation of a tree* T is a partial derivation that can be prolonged to a derivation of T.

2.4 Schedulers

A *scheduler* is a mapping σ that assigns to a family tree T and a partial derivation $d = (s_1 \Rightarrow \ldots \Rightarrow s_n)$ of T, $s_n \neq \emptyset$, a thread $w \in s_n$. Intuitively, $\sigma(T,d) = w$ means that the scheduler selects w as the next thread to be executed. A scheduler determines for each T and d a *successor state* s such that $s_n \Rightarrow s$, defined as follows: if w is a leaf of T, then $s = s_n \setminus \{w\}$, otherwise $s = (s_n \setminus \{w\}) \cup \{w0, w1\}$. It follows that σ also determines a derivation of T, namely the one in which each state is the successor state of the previous one. Abusing notation, we denote this derivation by $\sigma(T)$.

Recall that, loosely speaking, a black-box scheduler is a scheduler that does not have access to the code of the threads, and so it does not know whether threads will branch or die. Not every scheduler is a black-box scheduler. For instance, consider the scheduler σ defined as follows: given a tree T and a partial derivation d ending at a state $s = \{w_1, \ldots, w_k\}$, let $\sigma(T,d)$ be the thread w_i of s such that the subtree of T rooted at w_i has a minimal number of nodes. Loosely speaking, σ always selects the thread that *in the future of* T will generate the least number of threads. So σ has access to an oracle that determines whether a thread will branch or die (the oracle could be an algorithm that is given the code of the thread and returns the family tree it will produce).

Formally, a scheduler $\sigma(T,d)$ is a *black-box scheduler* if $\sigma(T,d)$ is independent of T, i.e., if $\sigma(T,d) = \sigma(T',d)$ for every two trees T, T'.

Quality of a Scheduler

Recall that a state contains the set of threads awaiting to be scheduled. Since storing this set consumes resources, we are interested in the maximal size it reaches during the execution of a derivation. So we define the *width* of a derivation $d = (s_1 \Rightarrow \ldots \Rightarrow s_n)$ as $wd(d) = \max\{|s_1|, \ldots, |s_k|\}$, where $|s_i|$ denotes as usual the number of elements of the set s_i.

Since a scheduler assigns to each family tree one of its derivations, we can compare the quality of different schedulers on a particular family tree. Consider for instance any two black-box schedulers σ_0 and σ_1 that assign to the partial derivation $\epsilon \Rightarrow 0, 1$ the threads 0 and 1, respectively. On the family tree of Figure 1, these schedulers induce derivations of width 2 and 3, and so σ_0 is better than σ_1 in this case. For each family tree there is also an optimal and a pessimal width, given by the widths of derivations with minimal and maximal width, respectively. For the tree of Figure 1 the optimal and pessimal widths are 2 and 3.

Taking into account that family trees have a probability distribution, we define a random variable W^σ, the *random width*, on the probability space of family trees. The variable assigns to a tree T the width of the derivation $\sigma(T)$, i.e., $W^\sigma(T) = wd(\sigma(T))$.

In Section 3 we study black-box schedulers. We first show that their random widths have all the same probability mass function. Intuitively, since all threads have the same probability of branching or dying, and this probability is independent of the past or the present context of a thread, no black-box scheduler has an advantage over any other. Since two schedulers with the same mass function are for the purposes of this paper equivalent, we denote by W_{bb} the random width of an arbitrary scheduler, and study its properties.

A scheduler that has information on whether a given thread will die or branch (for instance by inspecting the code) can apply it to reduce the width. Since the inspection also consumes resources, it is important to evaluate whether the reduction in width will be worth the cost. For this we investigate the optimal random width W_{opt}, defined as the random width of an optimal scheduler that always assigns to a family tree a derivation of minimal width. In Section 4 we study the random variable W_{opt} for the optimal scheduler (again, there may be more than one optimal scheduler, but for our purposes they are equivalent).

Finally, in Section 5 we study W_{pes}, the pessimal random width, i.e., the random width of a pessimal scheduler. This corresponds to the case in which the scheduler is controlled by an adversary having perfect information on the threads.

3 Black-Box Schedulers

In this section we show that all black-box schedulers σ yield the same probability mass function of the variable W^σ. As a consequence, the expectation of the variable W^σ is the same for all black-box schedulers σ. This allows us to denote by W_{bb} the random variable W^σ for an arbitrary (but fixed) scheduler σ. We show that there is a closed arithmetic expression for the probability mass function of W_{bb}, and that its expectation can be approximated. We also provide an error estimate for this approximation.

We start by proving that the probability mass function of W^σ does not depend on σ as long as σ is black-box (Theorem 3 below). To prove this we make use of a correspondence between black-box schedulers and a (one-dimensional) random walk with an absorbing barrier. The random walk can be defined as follows. Consider a graph whose nodes are the natural numbers $\mathbb{N}_0$ and for every $n \geq 1$ there is an edge to $n+1$ with probability p and an edge to $n-1$ with probability q. The intuition is that the node n corresponds to states with n threads. A step corresponds to the branching of a thread (step from n to $n+1$) or its death (n to $n-1$). The initial node is 1. Given a finite path w of the graph starting at 1, we denote by $\Pr[w]$ the product of probabilities of all edges of w. Intuitively, $\Pr[w]$ is the probability that the walk follows precisely the path w.

Given $n \neq m$, we denote by A^n_m the set of all paths from n to m that enter m only in the last step. Given a set of paths $A \subseteq A^n_m$, we define its probability $\Pr[A] = \sum_{w \in A} \Pr[w]$. It is a well-known fact that $\Pr\left[A^k_0\right] = 1$ for every $k \geq 0$ iff $p \leq q$ (see e.g. [7]).

Remember that W_{bb} denotes the variable W^σ where σ is an arbitrary (but fixed) black-box scheduler.

Theorem 3 *For every black-box scheduler σ and every $k \geq 0$ we have that*

$$\Pr[W^\sigma \geq k] = \Pr\left[A^1_k\right] = \Pr\left[W_{bb} \geq k\right]$$

Consequently, for $p < q$ we have

$$\Pr\left[W_{bb} \geq k\right] = \Pr\left[A_k^1\right] = \frac{(1-\frac{p}{q})(\frac{p}{q})^{k-1}}{1-(\frac{p}{q})^k}$$

and for $p = q$ we have $\Pr\left[W_{bb} \geq k\right] = \Pr\left[A_k^1\right] = \frac{1}{k}$.

Proof We define a bijective mapping Θ from the space of family trees to the set of paths A_0^1 such that Θ preserves probabilities, and moreover, such that for every tree T we have that $W^\sigma(T)$ is equal to the maximal number visited by the path $\Theta(T)$.

Let T be a family tree and let $\sigma(T) = (s_1 \Rightarrow \cdots \Rightarrow s_n)$ be its derivation under σ. We define $\Theta(T) = (k_1 \to k_2 \to \cdots \to k_n)$, where $k_i = |s_i|$ for every $1 \leq i \leq n$. Clearly, $\Theta(T)$ is a path in the random walk graph and the probability assigned to this path is equal to the probability of the tree T. Also, by definition, $W^\sigma(T)$ is equal to the maximal number visited by $\Theta(T)$. Now it suffices to show that for every path $w \in A_0^1$ there is a tree T such that $\Theta(T) = w$ (the injectivity then follows from the fact that $\Pr\left[A_0^1\right] = 1$ and that $\Pr[T] = \Pr[\Theta(T)] > 0$ for all trees T). Let $i_1 \to i_2 \to \ldots \to i_n$ be a path in the random walk such that $i_1 = 1$ and $i_n = 0$. Let us define a derivation $s_1 \Rightarrow \cdots \Rightarrow s_n$ as follows. First, $s_1 = \epsilon$. Second, assume that $s_\ell = \{w_1, \ldots, w_k\}$ and that $\sigma(T, s_1 \Rightarrow \cdots \Rightarrow s_\ell) = w_j$ for all trees T. If $i_{\ell+1} - i_\ell = 1$, then we define $s_{\ell+1} = (s_\ell \setminus \{w_j\}) \cup \{w_j 0, w_j 1\}$, otherwise we define $s_{\ell+1} = s_\ell \setminus \{w_j\}$. Clearly, there is a family tree T such that $\sigma(T) = (s_1 \Rightarrow \cdots \Rightarrow s_n)$ and thus $\Theta(T) = (i_1 \to \cdots \to i_n)$.

It follows that

$$\begin{aligned}\Pr[W^\sigma \geq k] = \Pr\left[\Theta(T) \in A_k^1 A_0^k\right] &= \Pr\left[A_k^1 A_0^k\right] \\ &= \Pr\left[A_k^1\right] \Pr\left[A_0^k\right] = \Pr\left[A_k^1\right].\end{aligned}$$

Here $A_k^1 A_0^k$ is the set of all paths of A_0^1 that reach the state k before reaching 0. It remains to derive the corresponding arithmetic expressions for $\Pr\left[A_k^1\right]$. Denoting $p_{i,k} = \Pr\left[A_k^i\right]$, we obtain the following

$$p_{0,k} = 0 \qquad p_{k,k} = 1$$

and for $0 < i < k$ we have

$$p_{i,k} = p \cdot p_{i+1,k} + q \cdot p_{i-1,k}\,.$$

Moreover, the probabilities $\Pr\left[A^i_k\right]$ constitute the unique solution of this recurrence relation. Assuming that $p < q$, the solution is given by

$$p_{i,k} = \frac{(1-\frac{p}{q})(\frac{p}{q})^{k-i}}{1-(\frac{p}{q})^k}$$

and in particular

$$\Pr\left[A^1_k\right] = p_{1,k} = \frac{(1-\frac{p}{q})(\frac{p}{q})^{k-1}}{1-(\frac{p}{q})^k}$$

If $p = q$, then the solution to the above recurrence relation is $p_{i,k} = \frac{i}{k}$ and in particular $\Pr\left[A^1_k\right] = p_{1,k} = \frac{1}{k}$. ⊣

Corollary 4 $\mathbb{E}\left[W_{bb}\right] = \sum_{k=1}^{\infty} \Pr\left[A^1_k\right]$. *Hence, for* $p < q$, *we have that*

$$\mathbb{E}\left[W_{bb}\right] = \sum_{k=1}^{\infty} \frac{(1-\frac{p}{q})(\frac{p}{q})^{k-1}}{1-(\frac{p}{q})^k}$$

and for $p = q$ *we have* $\mathbb{E}\left[W_{bb}\right] = \sum_{k=1}^{\infty} \frac{1}{k} = \infty$.

Thus, if $p < q$, then the expectation $\mathbb{E}\left[W_{bb}\right]$ can be approximated by the partial sums $\sum_{k=1}^{i} \Pr\left[A^1_k\right]$. The error of i'th approximation can be estimated as follows.

Lemma 5 *If* $p < q$, *then for all* $i \geq 1$ *we have that*

$$\mathbb{E}\left[W_{bb}\right] - \sum_{k=1}^{i} \Pr\left[A^1_k\right] \leq \frac{(\frac{p}{q})^i}{1-(\frac{p}{q})^{i+1}}$$

Proof By Corollary 4,

$$\mathbb{E}\left[W_{bb}\right] - \sum_{k=1}^{i} \Pr\left[A^1_k\right] = \sum_{\ell=i+1}^{\infty} \frac{(1-\frac{p}{q})(\frac{p}{q})^{\ell-1}}{1-(\frac{p}{q})^{\ell}} \leq \frac{(1-\frac{p}{q})(\frac{p}{q})^{i}}{1-(\frac{p}{q})^{i+1}} \sum_{\ell=1}^{\infty} (\frac{p}{q})^{\ell-1}$$

$$= \frac{(\frac{p}{q})^i}{1-(\frac{p}{q})^{i+1}}$$

4 Optimal Schedulers

Recall that W_{opt} is the random variable that assigns a tree the minimal width of its derivations. We call $W_{opt}(T)$ the *optimal width* of T. The following proposition characterizes the optimal width of a tree in terms of the optimal width of its children.

Proposition 6 *Let T be a family tree with root ϵ. If ϵ has children, let T_0, T_1 be the family trees rooted at them*[||]. *Then*

$$W_{opt}(T) = \begin{cases} \min \left\{ \begin{array}{l} \max\{W_{opt}(T_0)+1, W_{opt}(T_1)\}, \\ \max\{W_{opt}(T_0), W_{opt}(T_1)+1\} \end{array} \right\} & \text{if } \epsilon \text{ has children} \\ 1 & \text{if } \epsilon \text{ has no children} \end{cases}$$

Proof (Sketch) We define a scheduler σ for T, show that σ is optimal and that the width of the produced derivations satisfies the relation given in the proposition. We proceed by induction on the tree structure. If T has no child, then the only possible derivation has width 1. So assume in the following that T has two children. Hence, any derivation starts with $\{\epsilon\} \Rightarrow \{0, 1\}$. An optimal scheduler has two options, at least one of which is optimal.

- It derives T_0 completely and in an optimal way before it touches the thread 1. Then it derives T_1 in an optimal way. While it derives T_0, the thread 1 sticks around, so the resulting derivation has a width of $\max\{W_{opt}(T_0)+1, W_{opt}(T_1)\}$.
- The same with the threads 0 and 1 exchanged.

The optimal scheduler picks the one with smaller width, so $W_{opt}(T) = \min\{\max\{W_{opt}(T_0) + 1, W_{opt}(T_1)\}, \max\{W_{opt}(T_0), W_{opt}(T_1)+1\}\}$. ⊣

We are interested in the probabilities $\Pr[W_{opt} \leq i]$ for $i \geq 1$. Using Proposition 6 we could obtain a recurrence relation for them. However, using results of [3, 2] we can improve on this by exhibiting a surprising connection between these probabilities and the function $f(x) = px^2 + (1-p)$ of Proposition 1. The proposition states that $\Pr[\mathcal{T}]$ is equal to the least fixed point of $f(x)$. By analytically solving

[||]Strictly speaking, T_0 and T_1 are not family trees, because, for instance, their roots are labelled by 0 and 1, not by ϵ; however, they become family trees after removing the initial 0 or the initial 1 from all labels, and we silently assume this has been done.

the equation $x = f(x)$, we obtain $\Pr[\mathcal{T}] = 1$ for the case $p \leq 1/2$. But, since $\mathcal{T}$ is the set of finite family trees, in this case we have $\Pr[\mathcal{T}] = \Pr\left[W_{opt} < \infty\right]$, and so $\Pr\left[W_{opt} < \infty\right]$ is also equal to the least fixed point of $f(x)$. Imagine now that $x = f(x)$ could *not* be solved analytically. It would still be possible to *numerically* approximate its least solution. This can be achieved by applying Newton's method, which approximates a zero of a differentiable function g by the *Newton sequence* $(\nu_i)_{i \in \mathbb{N}}$, defined as follows:

$$\nu_0 = 0 \quad \text{and} \quad \nu_{i+1} = \nu_i + \Delta_{i+1} \tag{2}$$

where

$$\Delta_{i+1} = -\frac{g(\nu_i)}{g'(\nu_i)} .$$

Under certain conditions, the Newton sequence converges to the zero of g. For $g(x) = f(x) - x = px^2 + (1-p) - x$ (a g for which these conditions hold), we obtain:

$$\Delta_{i+1} = \frac{p\nu_i^2 - \nu_i + 1 - p}{1 - 2p\nu_i} , \tag{3}$$

and so

$$\nu_0 = 0 \quad \text{and} \quad \nu_{i+1} = \nu_i + \frac{p\nu_i^2 - \nu_i + 1 - p}{1 - 2p\nu_i} = \frac{1 - p - p\nu_i^2}{1 - 2p\nu_i} .$$

Usually one is only interested in the limit of the Newton sequence, not in the sequence itself. However, in our case the sequence turns out to contain precisely the information we are looking for:

Theorem 7 $\Pr\left[W_{opt} \leq i\right] = \nu_i$ *for every* $i \geq 0$.

Proof By induction on i. The base case $i = 0$ is easy. Let $i \geq 0$ and let T be a family tree with $W_{opt}(T) = i + 1$. So the root of T has children. Let T_0 and T_1 be the family trees rooted at them. It follows from Proposition 6 that either $W_{opt}(T_0) = W_{opt}(T_1) = i$ or $W_{opt}(T_0) < W_{opt}(T_1) = i + 1$ or $W_{opt}(T_1) < W_{opt}(T_0) = i + 1$. Let us inductively define the function L with

$$L(T) = \begin{cases} 0 & \text{if } W_{opt}(T_0) = W_{opt}(T_1) = i \\ L(T_1) + 1 & \text{if } W_{opt}(T_0) < W_{opt}(T_1) = i + 1 \\ L(T_0) + 1 & \text{if } W_{opt}(T_1) < W_{opt}(T_0) = i + 1 . \end{cases}$$

We have to show $\Pr\left[W_{opt} = i+1\right] = \Delta_{i+1}$ where

$$\Delta_{i+1} = -\frac{g(\nu_i)}{g'(\nu_i)} = \frac{f(\nu_i) - \nu_i}{1 - f'(\nu_i)} = \sum_{k=0}^{\infty} f'(\nu_i)^k \cdot (f(\nu_i) - \nu_i)\,.$$

We show the following stronger claim:

$$\Pr\left[W_{opt}(T) = i+1,\ L(T) = k\right] = f'(\nu_i)^k \cdot (f(\nu_i) - \nu_i)$$

We proceed by an (inner) induction on k. For the case $k = 0$ we have

$$\begin{aligned}
&\Pr\left[W_{opt}(T) = i+1,\ L(T) = 0\right] \\
&\quad = \Pr[T \text{ has children with} \\
&\qquad\quad W_{opt}(T_0) = W_{opt}(T_1) = i] && \text{(def. of } L) \\
&\quad = \Pr[T \text{ has children with} \\
&\qquad\quad W_{opt}(T_0) \le i,\ W_{opt}(T_1) \le i] \\
&\qquad + \Pr\left[T \text{ has no children}\right] && \text{(Prop. 6)} \\
&\qquad - \Pr\left[W_{opt}(T) \le i\right] \\
&\quad = p\nu_i\nu_i + q - \nu_i && \text{(ind. hyp. on } i) \\
&\quad = f(\nu_i) - \nu_i\,.
\end{aligned}$$

Let $k \ge 0$. Then

$$\begin{aligned}
&\Pr\left[W_{opt}(T) = i+1,\ L(T) = k+1\right] \\
&\quad = \Pr[W_{opt}(T) = i+1,\ L(T) = k+1, \\
&\qquad\quad T \text{ has children } T_0, T_1] && \text{(def. of } L) \\
&\quad = p \cdot \big(\Pr[W_{opt}(T_0) < W_{opt}(T_1) = i+1, \\
&\qquad\qquad L(T_1) = k] \quad + \\
&\qquad\quad \Pr[W_{opt}(T_1) < W_{opt}(T_0) = i+1, \\
&\qquad\qquad L(T_0) = k]\big) && \text{(def. of } L) \\
&\quad = 2p \cdot \Pr\left[W_{opt}(T_0) \le i\right] \cdot \\
&\qquad \Pr\left[W_{opt}(T_1) = i+1,\ L(T_1) = k\right] && \text{(independence)} \\
&\quad = 2p\nu_i \cdot f'(\nu_i)^k \cdot (f(\nu_i) - \nu_i) && \text{(ind. hyp. on } i, k)
\end{aligned}$$

$$= f'(\nu_i)^{k+1} \cdot (f(\nu_i) - \nu_i).$$

So we have $\Pr[W_{opt} = i] = \Delta_i$ for every $i \geq 1$. This allows to numerically approximate the probability mass function of W_{opt} and its expectation. Before studying the convergence order of this technique, let us consider an example.

Example 8 *The special case $p = 1/2$*
From (3) we obtain $\Delta_{i+1} = (1 - \nu_i)/2$, so by induction we get $\nu_i = 1 - \left(\frac{1}{2}\right)^i$ and $\Delta_i = \left(\frac{1}{2}\right)^i$. The expectation is

$$\mathbb{E}\left[W_{opt}\right] = \sum_{i=0}^{\infty} i \cdot \Pr\left[W_{opt} = i\right] = \sum_{i=0}^{\infty} i \cdot \Delta_i = \sum_{i=1}^{\infty} i \cdot \left(\frac{1}{2}\right)^i$$
$$= \frac{1/2}{(1-1/2)^2} = 2.$$

For general p we do not get a closed form for $\mathbb{E}\left[W_{opt}\right]$. However, we can approximate its value by $\sum_{i=0}^{k} i \cdot \Delta_i$ for finite k. The following proposition provides an upper bound for the error of this approximation as a function of k:

Proposition 9 *Let $0 \leq p < 1/2$ and define $Q := 2p/(1-2p)$ and $e_k := 1 - \nu_k$.*

(a) $1 \geq e_k > e_{k+1} \geq 0$ for all $k \in \mathbb{N}$, and the sequence $(e_k)_{k \in \mathbb{N}}$ converges to 0. Moreover, for every k such that $Q \cdot e_k < 1$,

$$\left| \mathbb{E}\left[W_{opt}\right] - \sum_{i=1}^{k} i \cdot \Delta_i \right| \leq \frac{(k+1) \cdot e_k}{(1 - Q \cdot e_k)^2}. \tag{4}$$

(b) Define $k_0 := \left\lceil \log_2 \frac{8p}{1-2p} \right\rceil$. For every $k \geq 0$,

$$\left| \mathbb{E}\left[W_{opt}\right] - \sum_{i=1}^{k_0+k} i \cdot \Delta_i \right| \leq 4 \cdot (k_0 + k + 1) \cdot 2^{-2^k}. \tag{5}$$

Proof The statement about the convergence of the e_k follows from [2]. The left hand side of (4) equals $\sum_{i=k+1}^{\infty} i \cdot \Delta_i$. Set $e_i := 1 - \nu_i$ for all i. We have $\Delta_i \leq e_{i-1}$, because $\nu_{i-1} + \Delta_i = \nu_i \leq 1$. Furthermore, using standard arguments about the quadratic convergence behaviour of Newton's method we have

$$e_{k+i} \leq (Q \cdot e_k)^{2^i} / Q\,. \tag{6}$$

Combining those facts we get:

$$\begin{aligned}
\sum_{i=k+1}^{\infty} i \cdot \Delta_i &\leq \sum_{i=k+1}^{\infty} i \cdot e_{i-1} && (\Delta_i \leq e_{i-1}) \\
&\leq \sum_{i=0}^{\infty} (k+i+1) \cdot (Q \cdot e_k)^{2^i} / Q && (\text{by (6)}) \\
&\leq \frac{1}{Q} \sum_{i=0}^{\infty} (k+i+1) \cdot (Q \cdot e_k)^{i+1} && (Q \cdot e_k < 1, 2^i \geq i+1) \\
&= \frac{k+1}{Q} \cdot \frac{Q \cdot e_k}{(1 - Q \cdot e_k)^2} && (\sum_{i=0}^{\infty} i \cdot q^i = \frac{q}{(1-q)^2}) \\
&= \frac{(k+1) \cdot e_k}{(1 - Q \cdot e_k)^2}
\end{aligned}$$

This gives us statement (a) of the proposition.

For the proof of statement (b), notice that $e_i \leq 1/2^i$ for all $i \geq 0$. This follows from the fact that $e_i = 1/2^i$ holds for $p = 1/2$ (see Example 8) and that this is an upper bound on e_i for $p \leq 1/2$. The latter can be concluded from the fact that the derivative of (3) w.r.t. p is negative for all $0 \leq p < 1/2$ and $0 \leq v_i < 1$.

So we have $Q \cdot e_{k_0 - 1} \leq Q \cdot (1/2)^{k_0 - 1} = Q \cdot 2^{1-k_0} \leq Q \cdot \frac{1-2p}{4p} \leq 1/2$. For the error bound we get:

$$\begin{aligned}
\sum_{i=k_0+k+1}^{\infty} i \cdot \Delta_i &\leq \sum_{i=k_0+k+1}^{\infty} i \cdot e_{i-1} && (\Delta_i \leq e_{i-1}) \\
&\leq \sum_{i=0}^{\infty} (k_0+k+i+1) \cdot \frac{(Q \cdot e_{k_0-1})^{2^{k+1+i}}}{Q} && (\text{by (6)}) \\
&\leq \sum_{i=0}^{\infty} (k_0+k+i+1) \cdot 2^{1-2^{k+1+i}} && (Q \cdot e_{k_0-1} \leq \frac{1}{2}) \\
&\leq (k_0+k+1) \cdot \frac{2^{1-2^{k+1}}}{(1-2^{1-2^{k+1}})^2}
\end{aligned}$$

$$\leq 4 \cdot (k_0 + k + 1) \cdot 2^{-2^k} \qquad (2^{-2^k} \leq 1/2)$$

⊣

Example 10 We illustrate Proposition 9 for the special case $p = 1/3$. Then $Q = 2$. We compute $\sum_{i=1}^{4} i \cdot \Delta_i = 1.40386$, and wish to bound the error of this approximation of the expectation. We have $e_4 = 1 - \nu_4 = 0.0000152589$. Proposition 9 gives us an error bound of $\frac{5 \cdot e_4}{(1-2 \cdot e_i)^2} < 0.00008$. Hence, $1.40386 \leq \mathbb{E}\left[W_{opt}\right] \leq 1.40394$.

5 Pessimal Schedulers

Obviously, the worst case that can happen when scheduling a given instance of a process is that the first leaf terminates only after all child processes have become leaves themselves. The pessimal random width W_{pes} is, thus, given by the number of leaves of a family tree.

Proposition 11 *Let T be a family tree with root ϵ. If ϵ has children, let T_0, T_1 be the trees rooted at the children. Then*

$$W_{pes}(T) = \begin{cases} W_{pes}(T_0) + W_{pes}(T_1) & \text{if } \epsilon \text{ has children} \\ 1 & \text{if } \epsilon \text{ has no children} \end{cases}$$

As for $p = 0$ we only have to consider the tree consisting of a single node, we assume in the following that $p \in (0, \frac{1}{2}]$.

Counting the leaves of (randomly generated) ordered binary trees is a well-known problem, dating back to 1844 when Eugene Catalan showed that the number of ordered binary trees having exactly $n + 1$ leaves is given by

$$C_n := \frac{1}{n+1}\binom{2n}{n}.$$

Using these Catalan numbers we can immediately state the probability mass function of W_{pes}:

Theorem 12 *The probability mass function of W_{pes} is given by*

$$\Pr\left[W_{pes} = n\right] = C_{n-1} \cdot p^{n-1} q^n \text{ for } n = 1, 2, 3, \ldots$$

In contrast to W_{bb} and W_{opt}, we can determine closed forms for the expectation and the variance of W_{pes}. For this, we write $\mathbb{E}_p\left[W_{pes}\right]$, resp. $\mathrm{Var}_p\left[W_{pes}\right]$ for the expectation, resp. variance of W_{pes} if we want to emphasize the dependency on the parameter p.

Theorem 13 *For $p \in [0, \frac{1}{2})$ we have*

$$\mathbb{E}_p\left[W_{pes}\right] = \frac{q}{q-p} \quad \textit{resp.} \quad \mathrm{Var}_p\left[W_{pes}\right] = \frac{pq}{(q-p)^3}$$

while $\mathbb{E}_{\frac{1}{2}}\left[W_{pes}\right]$ is unbounded.

In the remainder of this section, we sketch how these formulas for expectation, resp. variance of W_{pes} can be obtained using *probability-generating functions* (cf. [5]). Let us therefore recall the definition of *probability-generating function*:

> Let $X : \Omega \to \mathbb{N}$ be a *discrete random variable*. Its *probability-generating function* G_X is defined by
>
> $$G_X(x) := \sum_{n \in \mathbb{N}} \Pr[X = n] \cdot x^n$$
>
> where $\Pr[X = n]$ denotes the probability that X takes the value n.

It is not hard to prove that G_X converges at least for $x \in [0, 1]$ with $G_X(1^-) := \lim_{x \nearrow 1} G_X(x) = 1$, and that $G'_X(1^-)$ is the expectation of X. Similarly, higher order moments of X can be extracted from G_X by calculating higher derivatives of G_X.

As $\Pr[W_{pes} = 0] = 0$, the generating function $G_{pes}(x)$ of W_{pes} is

$$G_{pes}(x) := \sum_{n \geq 1} \Pr\left[W_{pes} = n\right] \cdot x^n = \sum_{n \geq 1} C_{n-1} \cdot p^{n-1} \cdot q^n \cdot x^n$$

where we assume $C_{-1} := 0$. Working with this explicit representation of G_{pes} is quite cumbersome. We therefore show that G_{pes} has to satisfy the following equation

$$0 = p \cdot G_{pes}(x)^2 + q \cdot x - G_{pes}(x). \tag{7}$$

where we prove equality by equating coefficients. We only give an informal proof neglecting concerns regarding convergence. Using the Cauchy-product we may write $G_{pes}(x)^2$ as

$$G_{pes}(x)^2 = \sum_{n \geq 1} x^n \cdot \sum_{k=1}^{n} \Pr\left[W_{pes} = k\right] \cdot \Pr\left[W_{pes} = n - k\right]. \tag{8}$$

Note that the sum yielding the coefficient of the monomial x^n in $G_{pes}(x)^2$ simply enumerates all possible cases for generating an ordered binary tree having exactly $n \geq 2$ leaves where k can be imagined to be the number of leaves of the left subtree. So, this sum is almost the n-th coefficient $\Pr\left[W_{pes} = n\right]$ of $G_{pes}(x)$, all that is missing is the factor p stating that we split the root into two subtrees. Finally, we have to add $q \cdot x = \Pr\left[W_{pes} = 1\right] \cdot x$ as the term for $W_{pes} = 1$ is missing in $p \cdot G_{pes}(x)^2$.

We now can solve Eq. (7) for G_{pes}. This equation has two possible solutions but only the following satisfies the necessary side condition $G_{pes}(1^-) = 1$ for $p \leq q$:

Theorem 14 *The probability generating function G_{pes} of W_{pes} has the closed form*

$$G_{pes}(x) = \frac{1 - \sqrt{1 - 4pq \cdot x}}{2p}. \tag{9}$$

As mentioned, the expectation $\mathbb{E}_p\left[W_{pes}\right]$ of W_{pes} is is given by $G'_{pes}(1^-)$ yielding

$$\mathbb{E}_p\left[W_{pes}\right] = \frac{q}{\sqrt{1 - 4pq}} = \frac{q}{\sqrt{(p+q)^2 - 4pq}} = \frac{q}{q - p}. \tag{10}$$

Obviously, this equality also holds for $p = 0$.

For $p = \frac{1}{2}$ we note that $p^{n-1}q^n \leq 2^{-2n+1}$ holds for all $p \in [0, \frac{1}{2}]$ and $n \geq 1$. So we have for all $k \in \mathbb{N}$:

$$\sum_{n=0}^{k} n \cdot C_{n-1} \cdot p^{n-1} \cdot q^n \leq \sum_{n=0}^{k} n \cdot C_{n-1} \cdot 2^{-2n+1}.$$

Letting k go to infinity, the right hand side becomes $\mathbb{E}_p\left[W_{pes}\right] = \frac{q}{q-p}$ for $p \in [0, \frac{1}{2})$ while the right hand side converges to $\mathbb{E}_{p=\frac{1}{2}}\left[W_{pes}\right]$. Thus, $\frac{q}{q-p} \leq \mathbb{E}_{p=\frac{1}{2}}\left[W_{pes}\right]$ for all $p \in [0, \frac{1}{2})$, i.e. $\mathbb{E}_{\frac{1}{2}}\left[W_{pes}\right]$ is unbounded.

In the case of the variance, one can show that the variance of a random variable X with generating function G_X is given by

$$\mathrm{Var}\left[X\right] = G''_X(1^-) + G'_X(1^-) - (G'_X(1^-))^2. \tag{11}$$

In the case of $p \in (0, \frac{1}{2})$ this gives us after simplification

$$\mathrm{Var}_p\left[W_{pes}\right] = \frac{pq}{(q-p)^3}. \tag{12}$$

Again, it is easily checked that this also holds for $p = 0$.

6 Comparison

We have obtained results on the expected value $\mathbb{E}[W]$ and on the probabilities $\Pr[W = k]$ for $W \in \{W_{opt}, W_{bb}, W_{pes}\}$. The results on the expected value are of different nature. For W_{opt} we obtain a characterization of $\Pr[W \leq k]$ as the k-th Newton approximant of a simple polynomial. While this does not provide a closed form, it allows to numerically estimate the probabilities and the expected value very efficiently, due to the very fast convergence of Newton's method. For W_{bb} we obtain a closed form for the probabilities, and using them we can express the expected value as a series. Again, while the series seems to have no closed form, its value can be easily approximated by its partial sums. Finally, for W_{pes} we can compute the generating function, and so obtain a very simple closed form for the expected value and variance, and more complicated closed forms for the probabilities.

Figure 2 compares the expectations of W_{opt}, W_{bb}, W_{pes} in the range $0 \leq p \leq 1/2$. All three values coincide for $p = 0$, and stay very close until approximately $p = 0.3$. From that moment on, W_{bb} and W_{pes} grow very fast, and diverge for $p = 1/2$.

Figure 3 shows the 95% quantile of the variables. Loosely speaking, this is the least amount of thread states we must be able to store in order to have 95% confidence that we will be able to carry out the execution of a given thread. Again, in the range $0 \leq p \leq 0.3$ the quantiles do not differ very much. So in this range it probably does not pay off to invest in designing a scheduler that examines the thread and improves on the black-box scheduler.

7 Conclusions

We have presented a computer science application of the theory of stochastic branching processes. Processes are threads, all of them executed by a computing agent, and a scheduler determines the next thread to be executed. We have studied the performance of different

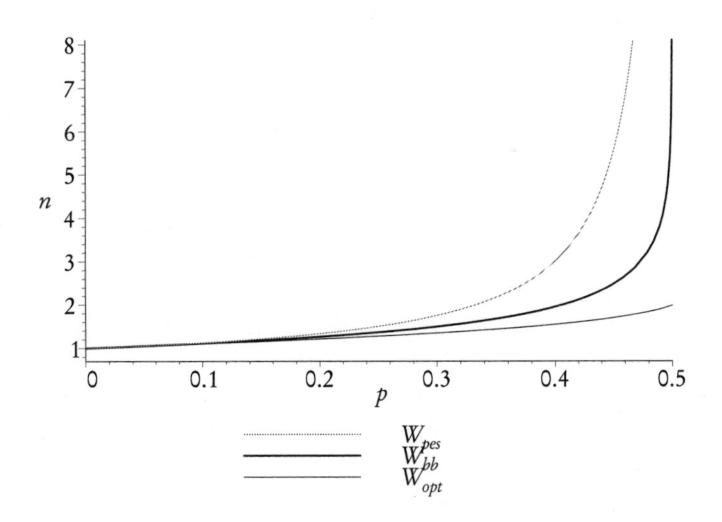

Figure 2: *Expectations of* W_{opt}, W_{bb}, W_{pes} *as a function of the branching probability* p.

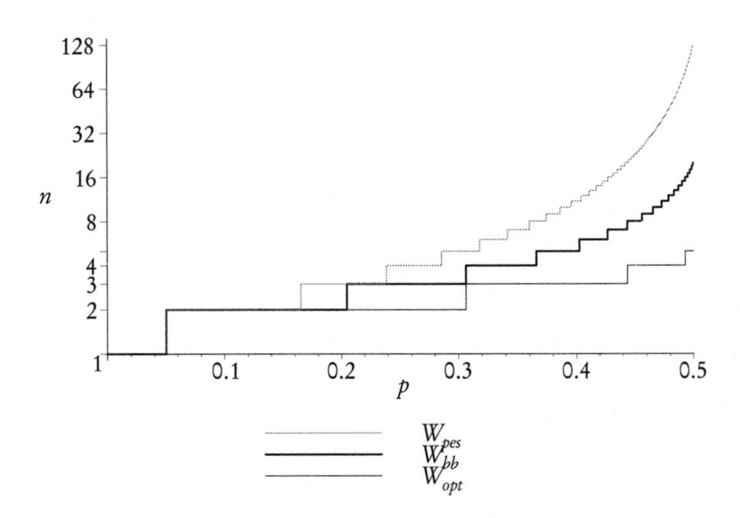

Figure 3: *The least number* n *such that* $\Pr[X \leq n] \geq 0.95$ *for* $X \in \{W_{pes}, W_{bb}, W_{opt}\}$. *Notice the logarithmic scale.*

schedulers under some stochastic assumptions on the probability that threads die and reproduce.

Our work can be extended in many ways, which we leave for future research:

- *True parallelism* We have assumed that there is only one computing agent. An interesting extension is to consider a bounded number of agents, as is for instance the case in multicore architectures.
- *Processes with extinction probability different from 1* If the branching probability p exceeds $1/2$, then the probability of extinction becomes smaller than 1. From a modelling point of view, $p \geq 1/2$ can be meaningful for systems that are not supposed to terminate like an operating system. Obtaining stochastic information on the width requires in this case to extend the probability space in order to allow infinite derivations (this is standard), and study their width. We have some preliminary results on this question.
- *Multitype threads* A more realistic model of thread behaviour should allow for a finite number of types of threads having different probabilities of branching and dying. It is easy to see that in this case not all black-box schedulers are equivalent, and the question of finding the optimal scheduler is certainly interesting.
- *Competitive analysis* The stochastic competitive ratio of the black-box scheduler is the random variable W_{bb}/W_{opt}. We also plan to study this variable.

We finish with a remark on the relation of our work to the theory of concurrency. The conventional way of assigning a semantics to our model is not in terms of family trees, but in terms of Markov decision processes (MDPs). The states of the MDP associated to our model are of two types: states of the first type are sets of threads, those of the second type are sets of threads with a distinguished element; intuitively, the distinguished element is the one selected by the scheduler. If the system is at a state of the first type, the successor state is chosen nondeterministically, otherwise stochastically. For instance, at the state $\{0, 1\}$ a nondeterministic choice takes place leading to tho states $\{\overline{0}, 1\}$, $\{0, \overline{1}\}$ of the second type, where the overline indicates the choice of the scheduler. At the state $\{\overline{0}, 1\}$ a stochastic choice takes place, leading to the states $\{00, 01, 1\}$ and $\{1\}$ with probabilities p and q. A scheduler is defined in this model as a mapping that assigns to a partial derivation ending at a nondeterministic state a successor state.

Unfortunately, the MDP semantics does not account for schedulers having information not only about the past of the derivation, but also about its future. It cannot naturally deal with the optimal scheduler, which can predict whether the threads will branch or die. For this reason we have resorted to our partial order semantics in terms of family trees, which allow to easily define the optimal and pessimal schedulers. From a concurrency theory point of view, family trees are just the finite nonsequential processes of the system. One of us already used them (although not in the stochastic case) in [4].

The connection between the MDP and our semantics is interesting. In the MDP semantics, one first resolves the nondeterministic choices by means of a scheduler, which transforms the MDP into a Markov chain, and the derivations are obtained by resolving the stochastic choices. In the family tree semantics, we first resolve the stochastic choices, which yields the set of family trees, and the derivations are obtained by fixing the scheduler.

References

[1] J. Engelfriet. Branching processes of petri nets. *Acta Informatica*, 28(6), 1991.

[2] J. Esparza, S. Kiefer, and M. Luttenberger. An extension of Newton's method to ω-continuous semirings. In *DLT'07*, LNCS 4588, pages 157–168. Springer, 2007.

[3] J. Esparza, S. Kiefer, and M. Luttenberger. On fixed point equations over commutative semirings. In *STACS'07*, LNCS 4397, pages 296–307. Springer, 2007.

[4] J. Esparza and A. Kiehn. On the model checking problem for branching time logics and basic parallel processes. In *CAV'95*, LNCS 939. Springer, 1995.

[5] W. Feller. *An introduction to probability theory and its applications*, volume I. John Wiley & Sons, 1968.

[6] T.E. Harris. *The Theory of Branching Processes*. Springer, 1963.

[7] J. R. Norris. *Markov Chains*. Oxford University Press, first edition, 1998.

[8] H.W. Watson and F. Galton. On the probability of the extinction of families. *J. Anthropol. Inst. Great Britain and Ireland*, 4:138–144, 1874.

Eliminating Past Operators in Metric Temporal Logic

Deepak D'Souza[1], **Raj Mohan M**[1], **and Pavithra Prabhakar**[2]
[1]*Dept. of Computer Science & Automation*
Indian Institute of Science, Bangalore 560012, India.
{deepakd,raj}@csa.iisc.ernet.in
[2]*Dept. of Computer Science*
University of Illinois at Urbana-Champaign, USA.
pprabha2@uiuc.edu

Abstract

We consider variants of Metric Temporal Logic (MTL) with the past operators S and S_I. We show that these operators can be "eliminated" in the sense that for any formula in these logics containing the S and S_I modalities, we can give a formula over an extended set of propositions which does not contain these past operators, and which is equivalent to the original formula modulo a projection onto the original set of propositions. These results have implications with regard to decidability and closure under projection for some well-known real-time logics in the literature.

Keywords: Real-time systems, metric temporal logic, expressiveness, decidability

1 Introduction

Metric Temporal Logic (MTL) introduced by Koymans [11] is a popular logic for specifying quantitative timing properties of real-time systems. It is interpreted over timed behaviours and extends the "until" modality U of LTL [13] with an interval index, allowing formulas of the form $\varphi U_I \psi$ which say that ψ is satisfied in the future at a time point whose distance lies in the interval I, and φ is held on to at all time points in between. The modalities S and S_I express the symmetric properties in the past. What is considered a possible timepoint to assert a formula depends on whether one considers the "pointwise" or "continuous" interpretation of the logic. In the pointwise semantics we are allowed to make assertions only at time points corresponding to event occurrences, while in the continuous semantics we are allowed to make assertions at arbitrary timepoints. In general, the continuous semantics is strictly more expressive than the pointwise semantics [2, 14].

In this paper we show some results about the equivalence of various fragments of MTL in terms of satisfiability-preserving translations. As a basic stepping stone we first show that the formulas of MTL can be "flattened" in the sense that for any formula in the logic (possibly with the past modality S_I) we can construct a satisfiability-equivalent formula which has no occurrences of nested U_I, S_I, or even S formulas. In fact, the only subformulas involving the above modalities are of the form pU_Iq, pS_Iq, or pSq, where p and q are propositions. We call this the "flat" or "non-recursive" version of MTL. The idea we use is quite simple: to flatten U_I formulas for example, we introduce new propositions p_0 and p_1 for each subformula of the form $\varphi U_I \psi$, replace each occurrence of $\varphi U_I \psi$ by $p_0 U_I p_1$, and add formulas which ensure that p_0 and p_1 correctly capture the truth of φ and ψ along the model. As a simple illustrative example, the flattened form of the formula $(pUq)S_{(0,1)}(p \wedge (qUr))$ is $(p_0 S_{(0,1)} p_1) \wedge \Box(p_0 \Leftrightarrow (pUq)) \wedge \Box(p_1 \Leftrightarrow (p \wedge (qUr)))$ (here $\Box\varphi$ stands for "always φ" or $\neg(\top U \neg\varphi)$). This result is shown for both the pointwise and continuous versions of the logic. To point out a simple consequence of this result, we recall that the pointwise version of MTL over infinite models was shown to be undecidable [12] via a reduction from channel systems to the general (recursive) version of MTL. From our result above it now follows that the corresponding non-recursive fragment of the logic is also undecidable.

Many real-time logics have classical temporal logic (in both the pointwise and continuous semantics) as the base logic to which distance operators are added. We show that for any formula in this base logic extended with the S modality, we can "eliminate" S subformulas from this formula in the sense that we can transform it to a formula over an extended set of propositions, which does not contain any S subformulas, and is equivalent to the original formula modulo a projection to the original set of propositions. This result holds for both the pointwise and continuous semantics. The technique used is to first flatten the S subformulas, then replace each pSq subformula by a new proposition r, and finally add formulas which force r to reflect correctly the truth of pSq along the model. For the pointwise case this last part of the formula is easy to construct. The continuous case is a little less obvious, and one has to consider points of discontinuity for p and q in the model, and ensure that r is updated correctly in the intervals between these points.

Among the implications of the above result is that adding the past modality S to a decidable real-time temporal logic, cannot lead to undecidability. Thus the logics MITL^c (continuous MTL in which only non-singular intervals are allowed) over both finite and infinite words, and MTL^{pw} (pointwise MTL) over finite words, which were shown to be decidable in [1] and [12] respectively, remain decidable even when we add the S modality.

Next, using similar techniques, we show that in continuous time we can eliminate S_I subformulas using U and the distance operator $\Diamond_I$ (which is the same as $\top U_I -$). This gives us the result that adding the S_I modality to a decidable variant of MTL in the continuous semantics, cannot lead to undecidability. A similar result cannot be obtained for the pointwise semantics, since it is known that introducing S_I in pointwise MTL over finite words makes the logic undecidable [6].

Finally, one of our goals was to show that we can eliminate S_I subformulas from the logic MITL$^c_{S_I}$ in a similar manner. The transformation for eliminating S_I subformulas above does not work here, since it may introduce singular intervals even when there were none in the original formula to begin with. However we show that it is still possible to give a satisfiability-preserving transformation which eliminates S_I subformulas when I is a non-singular interval, without introducing any singular intervals. More precisely, we show that for a given MITL$^c_{S_I}$ formula we can construct an MITL^c formula over an extended set of propositions, whose set of models is the same as the set of models of the original formula, modulo a projection to the original set of propositions followed by a truncation of a fixed length prefix from the models. In particular the transformation is satisfiability-preserving. As a result, we can conclude that the logic MITL$^c_{S_I}$ remains decidable. This decidability result is not new as it follows from the work of Henzinger et. al. in [8] where they show that Recursive Event-clock logic is equal in expressiveness to MITL$^c_{S_I}$ in the continuous semantics. Nonetheless, our construction gives a different and a more direct proof of this fact.

We should point out here that our translations only preserve the equivalence of models up to projection onto the original set of propositions (and hence satisfiability), and not expressiveness in general. In fact, each of the logics MTL^{pw}, MTL^c, and MITL^c are known to be be strictly less expressive than their counterparts with the S operator [2, 14]. However, this fact together with our elimination results, tells us something about the class of languages definable in these logics: namely, that none of the logics MTL^{pw}, MTL^c, and MITL^c are closed under the operation of projection.

It also follows that the class of languages definable by $\mathrm{MITL}^c_{S_I}$ are contained in the class of languages definable by (continuous time) Alur-Dill timed automata. This follows since MITL^c was shown to be translatable to the class of continuous timed automata [1], which in turn are closed under projection. In a similar way, it also follows that MTL^{pw}_S over finite words is contained the class of languages definable by 1-clock alternating timed automata.

In related research, the well-known work of [7] shows how to eliminate S from pointwise classical LTL, without expanding the set of propositions, thus preserving expressiveness in addition to satisfiability. However to the best of our knowledge, no similar result is known for continuous time. The elimination of S can also be seen to follow from the connection between finite-state automata and monadic second order (MSO) logics, in both pointwise and continuous time ([3, 4]). This is because one can go from LTL with S to MSO (in fact its first-order fragment), then to automata, and then back to an existentially quantified LTL formula (without S).

In another related piece of work Hirshfeld and Rabinovich [9] show how to eliminate the future distance operator $\Diamond_I$ (with I non-singular) using existentially quantified "timer" formulas which can express the past distance operator $\ominus_I$.

In the rest of this paper we concentrate on the continuous semantics. The pointwise case is similar and easier to handle. Further details can be found in the technical report [5].

2 MTL in the Continuous Semantics

We denote the set of non-negative real numbers by $\mathbb{R}_{\geq 0}$. We use the standard notation to represent intervals, which are convex subsets of $\mathbb{R}_{\geq 0}$. For example $[2,\infty)$ denotes the set $\{t \in \mathbb{R}_{\geq 0} \mid 2 \leq t\}$. We use $\mathcal{I}_{\mathbb{Q}}$ to denote the set of intervals whose bounds are either rational or ∞. For any interval $I \in \mathcal{I}_{\mathbb{Q}}$, let $l(I)$ be the left limit and $r(I)$ be the right limit of I respectively. Then we denote the *length of* I, i.e. $r(I) - l(I)$ by *len*(I). We also denote by $t + I$ the interval I' such that $t' \in I'$ iff $t' - t \in I$.

In the continuous semantics MTL is typically interpreted over "timed state sequences". Before we define these, let us first introduce the notion of a finitely varying function. Let B be a finite non-empty set, and let $f : \mathbb{R}_{\geq 0} \to B$. Then $t \in \mathbb{R}_{\geq 0}$ is a point of *left discontinuity* for f if there does not exist an $\epsilon > 0$ such that f is constant in the interval $(t - \epsilon, t]$. Similarly, $t \in \mathbb{R}_{\geq 0}$ is a point of *right discontinuity* for f if there does not exist an $\epsilon > 0$ such that f is constant in the interval $[t, t+\epsilon)$. The point t is a point of *discontinuity* for f if it is a point of left or right discontinuity. The function f is called *finitely varying* if it has only a finite number of points of discontinuity in any bounded interval in its domain.

Let P be a finite set of propositions. A *timed state sequence* τ over P is a finitely varying map $\tau : \mathbb{R}_{\geq 0} \to 2^P$. An equivalent definition (as given in [1]) is that there exists a sequence of subsets of propositions $s_0, s_1, \ldots$ and a sequence of intervals $I_0, I_1, \ldots$ satisfying:

1. I_0 is of the form $[0, r\rangle$ for some $r \in \mathbb{R}_{\geq 0}$ where we use '$\rangle$' to stand for the bracket ')' or ']'.
2. Every pair of intervals I_j and I_{j+1} are *adjacent* in the sense that I_j and I_{j+1} are disjoint, and $I_j \cup I_{j+1}$ forms an interval.
3. The sequence of intervals is "progressive" in that for every $t \in \mathbb{R}_{\geq 0}$, there exists $j \in \mathbb{N}$ such that $t \in I_j$.
4. The function τ is constant and equal to s_j in each I_j.

We call the sequence $(s_0, I_0)(s_1, I_1) \cdots$ above an *interval representation* of the function τ. It is easy to see that a timed state sequence τ has a "canonical" interval representation of the form $(s_0, I_0)(s_1, I_1) \cdots$ where the I_i's are an alternating sequence of singular and open intervals (i.e. for each i, I_{2i} is of the form $[l, l]$ and I_{2i+1} is of the form (l, r)), where the singular intervals are precisely the points of discontinuity of τ. We denote the set of timed state sequences over P by *TSS*(P).

The continuous version of MTL will be denoted by MTL^c. The syntax of MTL^c formulas over a set of propositions P is given by:

$$\varphi ::= p \mid \varphi U_I \varphi \mid \neg \varphi \mid \varphi \vee \varphi,$$

where $p \in P$ and I is an interval in $\mathcal{I}_{\mathbb{Q}}$.

The formulas of MTL^c above are interpreted over timed state sequences over P. Let τ be a timed state sequence over P, and let $t \in \mathbb{R}_{\geq 0}$. Then the satisfaction relation $\tau, t \models \varphi$ is given by:

$$
\begin{array}{lll}
\tau, t \models p & \text{iff} & p \in \tau(t) \\
\tau, t \models \psi U_I \eta & \text{iff} & \exists t' \geq t : \tau, t' \models \eta,\ t' - t \in I, \text{ and} \\
 & & \forall t'' : t < t'' < t',\ \tau, t'' \models \psi \\
\tau, t \models \neg \psi & \text{iff} & \tau, t \not\models \psi \\
\tau, t \models \psi \vee \eta & \text{iff} & \tau, t \models \psi \text{ or } \tau, t \models \eta.
\end{array}
$$

We say that a timed word τ satisfies a MTL^c formula φ, written $\tau \models \varphi$, if and only if $\tau, 0 \models \varphi$, and set $L(\varphi) = \{\tau \in TSS(P) \mid \tau \models \varphi\}$.

We can also consider a version of MTL with the past modality S_I whose semantics is given by:

$$\tau, t \models \psi S_I \eta \quad \text{iff} \quad \begin{array}{l} \exists t' \leq t : \tau, t' \models \eta,\ t - t' \in I, \text{ and} \\ \forall t'' : t' < t'' < t,\ \tau, t'' \models \psi. \end{array}$$

We denote this logic by MTL$^c_{S_I}$.

We define the standard temporal abbreviations as follows: $\psi U \eta \equiv \psi U_{[0,\infty)} \eta$, $\psi S \eta \equiv \psi S_{[0,\infty)} \eta$, $\Diamond \psi \equiv \top U \psi$, $\Box \psi \equiv \neg \Diamond \neg \psi$, $\Diamond_I \psi \equiv \top U_I \psi$, $\Box_I \psi \equiv \neg \Diamond_I \neg \psi$.

It will be convenient to work with a slightly different presentation of MTL. Let us define a base logic which is similar to classical continuous time LTL [10], and which we denote by LTL^c. The syntax of the logic (over the set of propositions P) is given by $\varphi ::= p \mid \varphi U^s \varphi \mid \neg \varphi \mid \varphi \vee \varphi$, and is interpreted over timed words in the same way as MTL^c above, with the modality U^s interpreted as $U_{(0,\infty)}$. Thus U^s is a "strict" until modality, which is strict in both its arguments. This is the natural choice for the until modality in continuous time and in the absence of an interval constraint. We note that the non-strict modality U is expressible using U^s, as $\psi U \eta \equiv \eta \vee (\psi U^s \eta)$, but not vice-versa. We define the derived modalities $\Diamond^s \psi$ and $\Box^s \psi$ to be: $\Diamond^s \psi \equiv \top U^s \psi$ and $\Box^s \psi \equiv \neg(\Diamond^s \neg \psi)$.

To this base logic we can add the past-time modalities S^s (for "strict since") and the distance operators $\Diamond$ and $\ominus$ to get the logic LTL$^c(S^s, \Diamond, \ominus)$, whose syntax is given by

$$\varphi ::= p \mid \varphi U^s \varphi \mid \neg \varphi \mid \varphi \vee \varphi \mid \varphi S^s \varphi \mid \Diamond_I \varphi \mid \ominus_I \varphi.$$

The modality S^s is interpreted as $S_{(0,\infty)}$. We also denote the "non-recursive" versions of these logics by *nr*-LTL^c (with the appropriate arguments), in which we restrict the use of the distance operators to the propositions in P, i.e. we allow only distance subformulas of the form $\Diamond_I p$ and $\ominus_I p$, with $p \in P$.

The logic MTL^c can be seen to be expressively equivalent to the logic LTL$^c(\Diamond)$ as the U_I modality of MTL^c can be expressed in terms of U^s and $\Diamond$. For example, if $I = [l, l]$, then $\psi U_I \eta = (\Box_{(0,l)} \psi) \wedge (\Diamond_{[l,l]} \eta)$; and if $I = (l, r)$ then $\psi U_I \eta = (\Box_{(0,l]} \psi) \wedge (\Diamond_{[l,l]} (\psi U \eta)) \wedge (\Diamond_{(l,r)} \eta)$. Similarly, the logic MTL$^c_{S_I}$ can be seen to be equivalent to the logic LTL$^c(S^s, \Diamond, \ominus)$.

Below we give some definitions which we will use in later sections. For any formula φ in LTL$^c(S^s, \Diamond, \ominus)$ and a timed state sequence τ

we define the characteristic function for φ in τ, $f_{\varphi,\tau} : \mathbb{R}_{\geq 0} \rightarrow \{\top, \bot\}$, given by $f_{\varphi,\tau}(t) = \top$ if $\tau, t \models \varphi$ and $\bot$ otherwise. We note that the function $f_{\varphi,\tau}$ is a finitely varying function. This follows from the argument in [1] which says that for every timed state sequence τ and every $\mathrm{LTL}^c(\Diamond)$ formula φ, there is an equivalent interval representation of τ (i.e. denoting the same function as τ) which is "φ-fine" – i.e. φ is constant throughout each interval in the interval sequence. We say that a point t is a point of (right) discontinuity in τ w.r.t. the formula φ, if it is a point of (right) discontinuity of the function $f_{\varphi,\tau}$.

As an example of what we can say in the base logic LTL^c, we define the "macro" formula $rd(\varphi)$ that will be of use later in the paper, which characterises points in a timed state sequence at which φ is true and which are points of right discontinuities w.r.t. φ. We define $rd(\varphi) = \varphi \wedge ((\neg\varphi) U^s (\neg\varphi))$.

3 Flattening MTL^c

We now show that each of the sublogics of $\mathrm{LTL}^c(S^s, \Diamond, \ominus)$ can be flattened to its non-recursive version. We show that every $\mathrm{LTL}^c(\Diamond)$ formula (equivalently an MTL^c formula) over a set of propositions P, can be flattened to an nr-$\mathrm{LTL}^c(\Diamond)$ formula over a set of propositions P' which is an extension of P. We assume the standard notion of subformulas: thus the subformulas of the formula $\varphi = pU(\Diamond_{[1,2]}(q \vee \Diamond_{(0,\infty)} r))$ are φ, p, $\Diamond_{[1,2]}(q \vee \Diamond_{(0,\infty)} r)$, $q \vee \Diamond_{(0,\infty)} r$, q, $\Diamond_{(0,\infty)} r$ and r. The *distance subformulas* of a formula are all its subformulas of the form $\Diamond_I \psi$.

We define the *level* of an $\mathrm{LTL}^c(\Diamond)$ formula φ as a measure of the nesting depth of distance subformulas in φ. Inductively, the level of a formula without any distance subformulas is 0; the level of a formula φ is $i+1$ if it has a distance subformula of the form $\Diamond_I \psi$ with ψ a level i formula, and no distance subformula of the form $\Diamond_J \eta$ with the level of η more than i. A *top-level* distance subformula of φ is a distance subformula which has at least one occurrence outside the scope of any other distance subformula. More formally, the set of top-level distance subformulas of φ, denoted $top\text{-}dsf(\varphi)$ is defined inductively as:

$$\begin{array}{lcl} top\text{-}dsf(p) & = & \{\} \\ top\text{-}dsf(\psi U^s \eta) & = & top\text{-}dsf(\psi) \cup top\text{-}dsf(\eta) \\ top\text{-}dsf(\psi \vee \eta) & = & top\text{-}dsf(\psi) \cup top\text{-}dsf(\eta) \\ top\text{-}dsf(\Diamond_I \psi) & = & \{\Diamond_I \psi\}. \end{array}$$

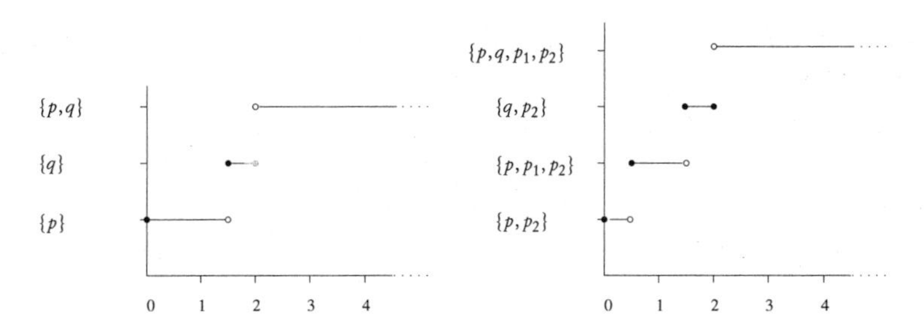

Figure 1: *TSS τ and its canonical extension w.r.t. X.*

Let us call a set of formulas X *closed* if for every distance subformula $\Diamond_I \psi$ of a formula in X, the formula ψ also belongs to X.

We fix a set of propositions P for the rest of this section.

Let $X = \{\psi_0, \ldots, \psi_n\}$ be a closed set of formulas, in increasing order of level. Let $\tau = (s_0, I_0)(s_1, I_1)\cdots$ be a timed state sequence over P. Let P' be the extended set of propositions $P \cup \{p_0, \ldots, p_n\}$, where each proposition p_j is meant to capture the truth of ψ_j. Then we define the *canonical extension* of τ w.r.t. X, denoted by $\textit{can-ext}_X(\tau)$, to be the timed state sequence over P', given by $\textit{can-ext}_X(\tau)(t) = \tau(t) \cup \{p_j \mid \tau, t \models \psi_j\}$.

We need to argue that $\textit{can-ext}_X(\tau)$ is indeed a timed state sequence, in that it is finitely varying. But this is true since as observed earlier, the characteristic function $f_{\tau,\psi}$, for each ψ in τ is finitely varying.

For a timed state sequence τ' over P', let us denote the timed state sequence over P obtained by projecting each $\tau'(t)$ to P by $\tau' \lceil P$. Thus $(\tau' \lceil P)(t) = \tau'(t) \cap P$. We extend the notion of projection to a set of timed state sequences over P' in the natural way.

As an example, consider the formula $\varphi = \Diamond_{(0,1)} \psi_2$ where $\psi_2 = \Diamond_{[0,1]} \psi_1$ and $\psi_1 = p \wedge \Diamond_{[1,1]} q$. Let us take X to be the set $\{q, \psi_1, \psi_2\}$. Then the canonical extension of the timed state sequence

$$\tau = (\{p\}, [0, 1.5))(\{q\}, [1.5, 2])(\{p, q\}, (2, \infty))$$

with respect to X is shown in Fig. 1.

Let us now define the "canonical translation" of a formula φ. Let X be any closed finite set of formulas containing the set of formulas ψ such that φ has a distance subformula of the form $\Diamond_I \psi$. Then the *canonical translation* of φ w.r.t. X, denoted $\textit{can-tr}_X(\varphi)$, is obtained from φ by replacing every top-level distance subformula of the form $\Diamond_I \psi_k$ by $\Diamond_I p_k$. We note that for any φ, $\textit{can-tr}_X(\varphi)$ is always a formula in nr-LTL$^c(\Diamond)$ over P'.

Lemma 1 *Let φ and X be as above, and let τ be a timed state sequence over P. Then $\tau, t \models \varphi$ iff $can\text{-}ext_X(\tau), t \models can\text{-}tr_X(\varphi)$.*
Proof The proof follows easily by induction on the structure of φ. For the base case, clearly $\tau, t \models p$ iff $can\text{-}ext_X(\tau), t \models p$. For the inductive step, the only interesting case is $\Diamond_I \psi_j$. But $\tau, t \models \Diamond_I \psi_j$ iff $can\text{-}ext_X(\tau), t \models \Diamond_I p_j$, since ψ_j is true in τ precisely when the proposition p_j is true in $can\text{-}ext_X(\tau)$. ⊣

We show that the property of an extended timed state sequence being a canonical extension of a timed state sequence w.r.t. a closed set of formulas X is definable in *nr*-LTL$^c(\Diamond)$. Let $X = \{\psi_0, \ldots, \psi_n\}$ be a closed set of formulas, with $\psi_0, \ldots, \psi_n$ being in increasing order of level. The formula ν^X_{can} basically says that the value of each p_i at a point in the extended word is correct. We define

$$\nu^X_{can} = \Box(\bigwedge_{j=0}^{n}(p_j \Leftrightarrow can\text{-}tr_X(\psi_j))).$$

Lemma 2 *Let X be as above. Let $P' = P \cup \{p_0, \ldots, p_n\}$ and let τ' be a timed state sequence over P'. Then τ' is the canonical extension of $\tau' \lceil P$ w.r.t. X iff $\tau' \models \nu^X_{can}$.*
Proof Let τ' be a canonical extension of $\tau' \lceil P$. Then we can argue by induction on j going from 0 to n that $\tau' \lceil P, t \models \psi_j$ iff $\tau', t \models can\text{-}tr_X(\psi_j)$. This gives us that $\tau' \models \nu^X_{can}$.

Conversely, suppose $\tau' \models \nu^X_{can}$. Then once again we argue by induction on j going from 0 to n that $\tau' \lceil P, t \models \psi_j$ iff $\tau', t \models can\text{-}tr_X(\psi_j)$. From the definition of ν^X_{can} we then have that $\tau', t \models can\text{-}tr_X(\psi_j)$ iff $\tau', t \models p_j$. Hence $\tau' \lceil P, t \models \psi_j$ iff $\tau', t \models p_j$, and therefore τ' is a canonical extension w.r.t. X. ⊣

We can now construct the required flattening of a formula φ in LTL$^c(\Diamond)$. Let X be the set of all ψ such that there is a distance subformula of φ of the form $\Diamond_I \psi$. Note that X is closed. We now define φ' to be $can\text{-}tr_X(\varphi) \wedge \nu^X_{can}$. By Lemmas 1 and 2, it follows that $L(\varphi) = L(\varphi') \lceil P$. To summarize:

Theorem 3 *Let $\varphi \in$ LTL$^c(\Diamond)$ over the set of propositions P. Then we can construct a formula $\varphi' \in$ nr-LTL$^c(\Diamond)$ over an extended set of propositions P', such that $L(\varphi) = L(\varphi') \lceil P$. In particular φ is satisfiable iff φ' is.* ⊣

We also observe that the flattening carried out here can be extended to the past operator $\diamondminus$, as well as the S modality, meaning that given a formula φ we give a projection equivalent formula φ' over P' in which only S formulas are of the form pSq, where $p, q \in P'$.

4 Eliminating S^s in LTL^c

In this section we show how to go from a formula in $\mathrm{LTL}^c(S^s)$ to a satisfiability-equivalent formula in LTL^c over an extended alphabet.

Let φ be an $\mathrm{LTL}^c(S^s)$ formula over the set of propositions P. Without loss of generality we assume that φ has been flattened, and thus the only occurrences of S^s formulas are of the form pS^sq with $p, q \in P$. The aim is to eliminate the formula pS^sq by introducing a new proposition r and adding formulas which make sure that r holds precisely at the points where pS^sq holds. The idea is to consider intervals in the model in which the truth of p and q is constant and to ensure that r is updated correctly in these intervals based on the values of p and q in the intervals.

Let $\tau \in TSS(P)$ and let t be a point in τ. Let $t' > t$ be the next point of discontinuity in τ w.r.t. p and q (i.e. a point of discontinuity of either $f_{p,\tau}$ or $f_{q,\tau}$), if such a point exists. Then Table 1 summarizes the truth value of the formula pS^sq in the interval (t, t') and at t', depending on the values of pS^sq, p and q at t, and the values of p and q in the interval (t, t'). The table for the case when there does not exist such a point t' is similar.

No.	t	p, q in (t, t')	pS^sq in (t, t')	pS^sq at t'
1	δ	$p \wedge q$	$true$	$true$
2	$(((pS^sq) \wedge p) \vee q)$	$p \wedge \neg q$	$true$	$true$
3	$\neg(((pS^sq) \wedge p) \vee q)$	$p \wedge \neg q$	$false$	$false$
4	δ	$\neg p \wedge \neg q$	$false$	$false$
5	δ	$\neg p \wedge q$	$false$	$false$

Table 1: *The value of pS^sq in (t, t') and at t'.*

The entries in Table 1 can be explained as follows. The δ in the table represents "don't care". If $p \wedge q$ is true throughout the interval (t, t') then clearly pS^sq is true throughout the interval (t, t') and also at t'. If $p \wedge \neg q$ is true throughout the interval (t, t') then the truth value of pS^sq in the interval depends on the truth values of pS^sq, p and q at t. It is easy to see that if q is true at t then pS^sq is true everywhere in the interval (t, t') and at t'. Similarly if $(pS^sq) \wedge p$ is true at t then $pS^s q$ is true throughout the interval (t, t') and at t'. The remaining entries can be similarly explained.

Now let $X = \{p_0 S^s q_0, \ldots, p_n S^s q_n\}$ be all the S^s subformulas in φ. We introduce a proposition r_j for each $p_j S^s q_j$ which is meant to capture the truth of $p_j S^s q_j$, and call this extended set of propositions P'. The canonical extension of τ in $TSS(P)$ is τ' in $TSS(P')$, given by $\tau'(t) = \tau(t) \cup \{r_j \mid \tau, t \models p_j S^s q_j\}$. The translation of φ obtained by replacing every occurrence of $p_j S^s q_j$ by r_j, clearly preserves the truth of φ in the canonical extension. It is now sufficient if we can define a formula ν_{can} in LTL^c over P' which describes precisely the timed state sequences over P' which are canonical extensions w.r.t. X.

It will be convenient to use the macro $\alpha(\psi, \mu)$ defined below, which is true at a point t in a model τ iff either there is a subsequent point of discontinuity t' in τ w.r.t. ψ such that ψ is true throughout the interval (t, t'), and μ is true throughout the interval $(t, t']$; or ψ is true throughout (t, ∞) and so is μ. We define $\alpha(\psi, \mu)$ to be:

$$((\psi U^s((\neg\psi) \vee rd(\psi))) \Rightarrow \\ ((\psi \wedge \mu) U^s(((\neg\psi) \vee rd(\psi)) \wedge \mu))) \wedge ((\Box^s \psi) \Rightarrow (\Box^s \mu)).$$

We now define ν_{can} as follows:

$$\nu_{can} = \bigwedge_{j=0}^{n} ((\neg r_j) \wedge \Box(\varphi_{j1} \wedge \varphi_{j2} \wedge \varphi_{j3} \wedge \varphi_{j4} \wedge \varphi_{j5}))$$

where

1. $\varphi_{j1} : \alpha(p_j \wedge q_j, r_j)$
2. $\varphi_{j2} : ((r_j \wedge p_j) \vee q_j) \Rightarrow \alpha(p_j \wedge \neg q_j, r_j)$
3. $\varphi_{j3} : \neg((r_j \wedge p_j) \vee q_j) \Rightarrow \alpha(p_j \wedge \neg q_j, \neg r_j)$
4. $\varphi_{j4} : \alpha(\neg p_j \wedge \neg q_j, \neg r_j)$
5. $\varphi_{j5} : \alpha(\neg p_j \wedge q_j, \neg r_j)$

Lemma 4 *Let $\tau' \in TSS(P')$. Then $\tau' \models \nu_{can}$ iff τ' is a canonical extension w.r.t. X.*

Proof Let τ' be a canonical extension w.r.t. X. Then we argue that τ' satisfies ν_{can}. Consider any $p_j S^s q_j$ in X. Since $p_j S^s q_j$ is not satisfied at time 0 in τ', we have $\neg r_j$ true at time 0 in τ'. To show that the other conjuncts in ν_{can} are satisfied, let t be a point in τ'. Now, there are two cases: either there is a next point of discontinuity t' of p_j and q_j in τ' after t, or there is none. Let us consider the first case. Then exactly one of $p_j \wedge q_j$, $p_j \wedge \neg q_j$, $\neg p_j \wedge \neg q_j$, $\neg p_j \wedge q_j$ is true throughout the interval (t, t'). Say for example that $p_j \wedge q_j$ was true. Then, from the table 1, we must have $\alpha(p_j \wedge q_j, r_j)$ true at t, and hence φ_{j1} is satisfied at time t in τ'. The formulas φ_{j2} to φ_{j5} are vacuously satisfied at t. Similarly, the other cases can be handled. The second case when there is no point of discontinuity after t is also handled similarly.

For the converse direction let $\tau' \models \nu_{can}$. We argue that τ' is a canonical extension w.r.t. X, i.e for every $p_j S^s q_j \in X$ and for every $t \in \mathbb{R}_{\geq 0}, \tau', t \models r_j$ iff $\tau', t \models p_j S^s q_j$. Let $t_0, t_1, \ldots$ be the (finite or infinite) sequence of discontinuities in τ' w.r.t. p_j and q_j. Let $I_0 = [t_0, t_0], I_1 = (t_0, t_1], I_2 = (t_1, t_2]$ and so on. We use induction on i to prove that the value of r_j is correctly updated in the interval I_i.

Base case: Since $\tau' \models \nu_{can}$, at time 0 we have r_j is not true. Since $p_j S^s r_j$ is also not true at time 0, the value of r_j is correctly updated in I_0.

Induction step: Let us assume that r_j is correctly updated in all the intervals up to I_i. We now argue that for all $t \in I_{i+1}$ we have $\tau', t \models p_j S^s q_j$ iff $\tau', t \models r_j$. There are two cases: either there exists a point of discontinuity t_{i+1}, or $I_{i+1} = (t_i, \infty)$.

For the case there exists t_{i+1}: Then exactly one these $p_j \wedge q_j$, $p_j \wedge \neg q_j$, $\neg p_j \wedge \neg q_j$, $\neg p_j \wedge q_j$ holds in (t_i, t_{i+1}). According to the table 1 the value of $p_j S^s q_j$ is fixed in the interval I_{i+1}. Further, by the induction hypothesis the value of r_j is correctly updated at the point t_i. Hence, the formulas φ_{j1} to φ_{j5} ensure that the value of r_j is correctly updated in I_{i+1}.

For the case there does not exist t_{i+1}, the argument is again similar. $\dashv$

5 Eliminating $\ominus$ in LTL^c

We now show how we can remove the past distance modality $\ominus$ from the continuous version of MTL, while preserving satisfiability of the original formula.

Let φ be a formula in any of the versions of our continuous logic LTL^c with the $\ominus$ operator. By the results of Section 3 we can assume the $\ominus$ subformulas in φ are of the form $\ominus_I p$ with p a proposition in

P. We essentially show how to express formulas of the form $\ominus_I p$ in terms of new propositions, the U modality, and $\Diamond$ operator.

Let $\ominus_{I_0} p_0, \ldots, \ominus_{I_n} p_n$ be all the $\ominus_I$ subformulas in φ. Let τ be a timed state sequence over P. Let P' be the extended set of propositions $P \cup \{q_0, \ldots, q_n\} \cup \{r_0, \ldots, r_n\}$, where q_j and r_j are new propositions associated with each $\ominus_{I_j} p_j$. The proposition q_j is meant to capture the truth of $\ominus_{I_j} p_j$, and r_j is meant to capture the fact that we have seen a p_j sometime strictly in the past. We define the canonical extension of τ (w.r.t. $\{\ominus_{I_0} p_0, \ldots, \ominus_{I_n} p_n\}$) to be the timed state sequence τ' over P', given by

$$\tau'(t) = \tau(t) \cup \{q_j \mid \tau, t \models \ominus_{I_j} p_j\} \cup \{r_j \mid \exists t' < t : \tau, t' \models p_j\}.$$

The canonical translation of φ is obtained by simply replacing each $\ominus_{I_j} p_j$ by q_j. It is clear that $\tau, t \models \varphi$ iff the canonical translation of φ is satisfied at t in the canonical extension of τ.

We now define the formula ν_{can} which characterises canonical extensions. The formula ν_{can} is the formula below

$$\bigwedge_{j=0}^{n} ((\neg r_j) \wedge \Box(p_j \Rightarrow \Box_{(0,\infty)} r_j) \wedge \neg((\neg p_j) \wedge ((\neg p_j) U r_j)))$$

in conjunction with a formula ψ_j for each j as below:

- If I_j is of the form $[l, l]$ take ψ_j to be $\Box(p_j \Leftrightarrow \Diamond_{[l,l]} q_j)$;
- If I_j is of the form (l, r) take ψ_j to be $\Box((\Diamond_{[r,r]} q_j) \Leftrightarrow \Diamond_{(0,r-l)} p_j)$;
- If I_j is of the form (l, ∞) take ψ_j to be $\Box(r_j \Leftrightarrow \Diamond_{[l,l]} q_j)$;
- and if I_j is of the form $[l, \infty)$ take ψ_j to be $\Box((r_j \vee p_j) \Leftrightarrow \Diamond_{[l,l]} q_j)$.

We note that we have introduced the proposition r_j to avoid using a formula of the form $\top S p_j$ in the translation. The construction here is simpler than if we had used S and then eliminated it using the results of Section 4.

6 Eliminating $\ominus$ from $\mathrm{MITL}^c_{S_I}$

In this section we show that an $\mathrm{MITL}^c_{S_I}$ formula can be reduced to satisfiability equivalent $\mathrm{LTL}^c(\Diamond)$ formula. In section 2 it has been shown that an $MITL^c_{S_I}$ formula can be reduced to an $LTL^c(S^s, \Diamond, \ominus)$ formula. In section 3 we show that for given a recursive $LTL^c(S^s, \Diamond, \ominus)$ formula we can construct a satisfiability equivalent non-recursive $LTL^c(S^s, \Diamond, \ominus)$ formula over an extended alphabet. In section 4 we show that for given a non-recursive $LTL^c(S^s, \Diamond, \ominus)$ formula we can construct a satisfiability equivalent non-recursive $LTL^c(\Diamond, \ominus)$ formula, once again over an extended alphabet. We also note that the $LTL^c(\Diamond, \ominus)$ formula so obtained also does not have any singular intervals since none of the above translations introduce singular intervals if the original formula is singular interval free.

We now show how to go from a formula φ in non-recursive $LTL^c(\Diamond, \ominus)$ with non singular intervals to a satisfiability equivalent $LTL^c(\Diamond)$ formula over an extended set of propositions. Let $X = \{\ominus_{I_0} p_0, \ldots, \ominus_{I_n} p_n\}$ be the set of all past distance subformulas used by φ. The idea is introduce a proposition q_j for each $\ominus_{I_j} p_j \in X$ which is meant to be true precisely at the points were $\ominus_{I_j} p_j$ is true. Once again we define the canonical extension τ' of timed sequence $\tau \in TSS(P)$ w.r.t. X. We give a formula ν which characterises the canonical extensions of τ modulo a prefix. We also note that the formula ν does not introduce any singular intervals if the original formula is over non singular intervals.

We define ν as follows:

$$\nu = \bigwedge_{j=0}^{j=n} \nu_j$$

where ν_j makes sure that q_j is true precisely at the points were $\ominus_{I_j} p_j$ is true. The formula ν_j depends on whether I_j is left closed, right closed or unbounded. Below we give the formula ν_j for each case.

Case 1: If I_j is of the form (l, r) then we define ν_j as follows:

$$\nu_j = \Box(\varphi_1 \wedge \varphi_2 \wedge \varphi_3 \wedge \varphi_4 \wedge \varphi_5)$$

Then the formulas $\varphi_1, \ldots, \varphi_5$ are as given below:

1. $\varphi_1 : p_j \Rightarrow \Box_I q_j$
2. $\varphi_2 : \Box \neg q_j \vee (\neg q_j \Rightarrow \neg q_j U \Box_{(0,r-l)} q_j)$
3. $\varphi_3 : q_j \Rightarrow (q_j U \neg q_j) \vee \Box q_j$
4. $\varphi_4 : (\Box_{(l,r)} q_j \wedge \Diamond_{[l,r)} \neg q_j) \Rightarrow p_j U p_j$
5. $\varphi_5 : \Box_{[r,2r-l)} q_j \Rightarrow (\Diamond_{(0,r-l)} p_j \wedge \neg \Diamond_{(0,r-l)} \Box_{(0,r-l)} \neg p_j)$.

Case 2: If I_j is of the form $[l, r)$ then we define ν_j as follows:

$$\nu_j = \Box(\varphi_1 \wedge \varphi_2 \wedge \varphi_3 \wedge \varphi_4 \wedge \varphi_5)$$

where the formulas $\varphi_1, \ldots, \varphi_5$ are as given below:

1. $\varphi_1 : p_j \Rightarrow \Box_I q_j$
2. $\varphi_2 : \Box\neg q_j \vee (\neg q_j \Rightarrow \neg q_j U \Box_{(0,r-l)} q_j)$
3. $\varphi_3 : q_j \Rightarrow (q_j U \neg q_j) \vee \Box q_j$
4. $\varphi_4 : (\Box_{(l,r)} q_j \wedge \Diamond_{[l,r)} \neg q_j) \Rightarrow p_j U p_j$
5. $\varphi_5 : \Box_{[r,2r-l)} q_j \Rightarrow (\Diamond_{(0,r-l]} p_j \wedge \neg\Diamond_{(0,r-l)} \Box_{(0,r-l]} \neg p_j)$.

Case 3: If I_j is of the form $(l, r]$ then we define ν_j as follows:

$$\nu_j = \Box(\varphi_1 \wedge \varphi_2 \wedge \varphi_3 \wedge \varphi_4)$$

where the formulas $\varphi_1, \ldots, \varphi_4$ are as given below:

1. $\varphi_1 : p_j \Rightarrow \Box_I q_j$
2. $\varphi_2 : \Box\neg q_j \vee (\neg q_j \Rightarrow \neg q_j U \Box_{(0,r-l]} q_j)$
3. $\varphi_3 : (\Box_{(l,r]} q_j \wedge \Diamond_{[l,r]} \neg q_j) \Rightarrow p_j U p_j$
4. $\varphi_4 : \Box_{[r,2r-l]} q_j \Rightarrow (\Diamond_{[0,r-l)} p_j \wedge \neg\Diamond_{(0,r-l]} \Box_{[0,r-l)} \neg p_j)$.

Case 4: If I_j is of the form $[l, r]$ then we define ν_j as follows:

$$\nu = \Box(\varphi_1 \wedge \varphi_2 \wedge \varphi_3 \wedge \varphi_4)$$

where the formulas $\varphi_1, \ldots, \varphi_4$ are as given below:

1. $\varphi_1 : p_j \Rightarrow \Box_I q_j$
2. $\varphi_2 : \Box\neg q_j \vee (\neg q_j \Rightarrow \neg q_j U \Box_{(0,r-l]} q_j)$
3. $\varphi_3 : (\Box_{(l,r]} q_j \wedge \Diamond_{[l,r]} \neg q_j) \Rightarrow p_j U p_j$
4. $\varphi_4 : \Box_{[r,2r-l]} q_j \Rightarrow (\Diamond_{[0,r-l]} p_j \wedge \neg\Diamond_{(0,r-l]} \Box_{[0,r-l]} \neg p_j)$.

In the next two cases we handle the unbounded cases.
Case 5: If I_j is of the form (l, ∞) then we define ν_j as follows:

$$\nu_j = \Box(\varphi_1 \wedge \varphi_2 \wedge \varphi_3 \wedge \varphi_4)$$

where the formulas $\varphi_1, \ldots, \varphi_4$ are as given below:

1. $\varphi_1 : p_j \Rightarrow \Box_I q_j$
2. $\varphi_2 : \Box\neg q_j \vee (\neg q_j \Rightarrow \neg q_j U \Box_{(0,\infty)} q_j)$

3. $\varphi_3 : (\Box_{(l,\infty)} q_j \wedge \Diamond_{[l,\infty)} \neg q_j) \Rightarrow p_j U p_j$
4. $\varphi_4 : \neg(\neg q_j U \Box_{[l,\infty)} q_j)$

Case 6: If I_j is of the form $[l,\infty)$ then we define ν_j as follows:

$$\nu_j = \Box(\varphi_1 \wedge \varphi_2 \wedge \varphi_3 \wedge \varphi_4)$$

where the formulas $\varphi_1, \ldots, \varphi_4$ are as given below:

1. $\varphi_1 : p_j \Rightarrow \Box_I q_j$
2. $\varphi_2 : \Box \neg q_j \vee (\neg q_j \Rightarrow \neg q_j U \Box_{(0,\infty)} q_j)$
3. $\varphi_3 : (\Box_{(l,\infty)} q_j \wedge \Diamond_{[l,\infty)} \neg q_j) \Rightarrow p_j U p_j$
4. $\varphi_4 : (\neg q_j U \Box_{[l,\infty)} q_j) \Rightarrow p_j$

For each $1 \leq j \leq n$, we define d_j as follows:

$$d_j = \begin{cases} 2r(I_j) - l(I_j) & \text{if } r(I_j) < \infty \\ l(I_j) & \text{otherwise.} \end{cases}$$

Let d_{max} be that the maximum value among all the d_j and let $P' = P \cup \{q_j | 1 \leq j \leq n\}$.

Lemma 5 *Let $\tau' \in TSS(P')$. If τ' is a canonical extension of τ w.r.t. $\ominus_{I_j} p_j$, then $\tau' \models \nu_j$. Conversely if $\tau' \models \nu_j$, then for all $t \geq d_{max}$, $\tau', t \models \ominus_{I_j} p_j$ iff $\tau', t \models q_j$.*

Proof For convenience we assume that I_j is of the form (l, r). We can prove the lemma for other cases along the similar lines.

Let τ' be a canonical extension w.r.t. $\ominus_{I_j} p_j$ and let t be a point in τ'. Then we argue that τ at t satisfies the formulas $\varphi_1, \ldots, \varphi_5$. Hence it follows that $\tau' \models \nu_j$. We now explain the formulas $\varphi_1, \ldots, \varphi_5$ in the proof given below. These formulas say that ν_j characterises the canonical extensions w.r.t. $\ominus_{I_j} p_j$ except possibly for a prefix.

Let p_j be true at t in τ'. Then $\ominus_{I_j} p_j$ is true in the interval $t + I_j$. Since τ' is a canonical extension w.r.t. $\ominus_{I_j} p_j$ we have that q_j is true throughout the interval $t + I_j$. Hence $\tau', t \models \varphi_1$. Pictorially:

p $\longleftarrow q \longrightarrow$

Let t be a point such that $\tau', t \models \neg q_j$. Then we have two cases:

Case 1: $\neg q_j$ true always in the interval $[t,\infty)$ in which case $\tau', t \models \Box\neg q_j$.

Case 2: There exists a $t' > t$ such that $\tau', t' \models q_j$. Since τ' is a canonical extension it should be the case that there exists a point t'' in τ' such that $\tau', t'' \models p_j$ and $t' \in t'' + I_j$. Since $\tau', t \models \neg q_j$ and $\tau', t \models \varphi_1$, we can easily argue that $t < t'' + l(I_j)$. Since $\ominus_{I_j} p_j$ is true throughout the interval $t'' + I_j$ we have $\tau', t \models \varphi_2$.

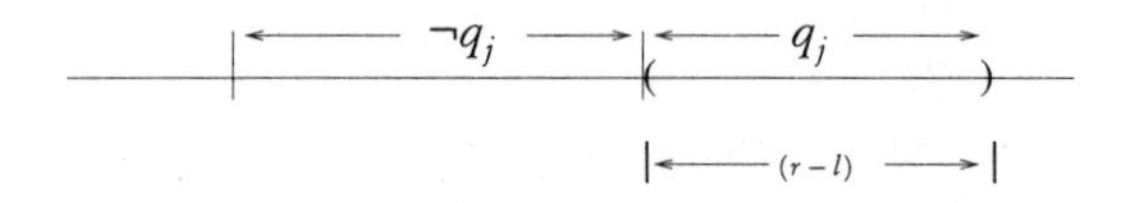

For any formula ψ in $\mathrm{LTL}^c(\Diamond, \ominus)$ and a timed state sequence τ, we say that an interval J is *ψ-interval* in τ iff for each $t \in J$, $f_{\tau,\psi}(t) = \top$ and the interval J is said to be a *maximal ψ-interval* in τ iff there does not exist an interval J' such that $J \subset J'$ and for each $t' \in J'$, $f_{\tau,\psi}(t') = \top$.

Let $\tau', t \models q_j$. We now argue that $\tau', t \models \varphi_3$. So let J be the maximal q_j-interval such that $t \in J$. Since the interval I_j is both left and right open, any maximal $\ominus_{I_j} p_j$-interval should be both left and right open. Since τ' is a canonical extension w.r.t. $\ominus_{I_j} p_j$ we have that any maximal q_j-interval is also both left and right open and therefore $\tau', t \models \varphi_3$. Pictorially:

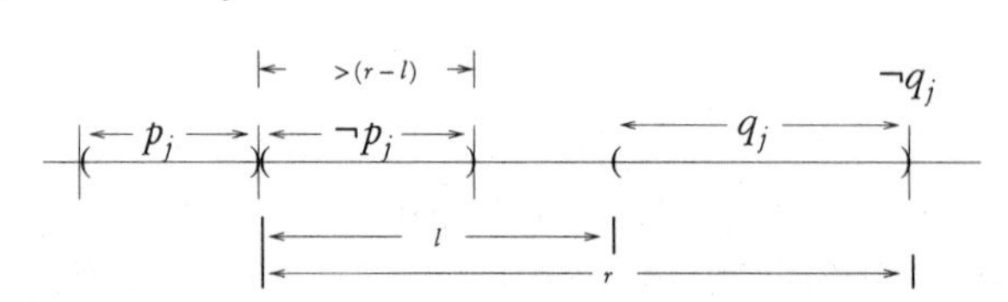

We argue that $\tau', t \models \varphi_4 \wedge \varphi_5$. Let $\tau', t \models q_j$. Since $\tau', t \models \varphi_1 \wedge \varphi_2 \wedge \varphi_3$ we can easily argue that there exists an q_j-interval J, which is both left and right open, such that $t \in J$ and $len(J) = len(I_j) = r - l$.

Now there are two cases:

Case 1: Let $\tau', l(J) \models \neg q_j$ and let $t' = l(J) - l$. Then the formula φ_4 is applicable only if $t' \geq 0$, so we assume this is so. Then we have that $\tau', t' \models \Box_{(l,r)} q_j \wedge \Diamond_{[l,r)} q_j$. We now argue that $\tau', t' \models p_j U p_j$. Suppose $\tau', t' \not\models p_j U p_j$. This implies that there exists an $\epsilon > 0$ such that for every $t'' \in [t', t' + \epsilon)$, $\tau', t'' \models \neg p_j$. Since $\neg q_j$ is true at $l(J)$ it follows that p_j is not true in the $(t' - (r - l), t')$ and therefore everywhere in the interval $(t' - (r - l), t' + \epsilon)$ we have $\neg p_j$ true. But this implies q_j is not true throughout in the interval J, a contradiction. The two cases where the formula $p_j U p_j$ is true at t' are shown side by side in the figure given below:

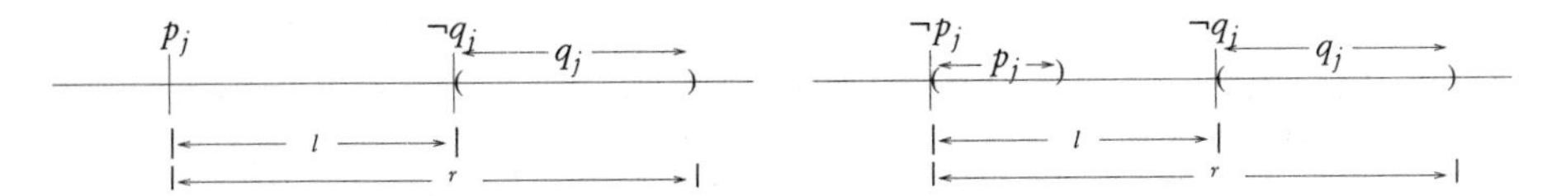

Case 2: Let $\tau', l(J) \models q_j$ and let $t' = l(J) - r$. Then the formula φ_5 is applicable only if $t' \geq 0$, so we assume this is so. Then we have that $\tau', t' \models \Box_{[r,2r-l)} q_j$. We now argue $\tau', t' \models \Diamond_{(0,r-l)} p_j \wedge \neg\Diamond_{(0,r-l)} \Box_{(0,r-l)} \neg p_j$. Since q_j is true at $l(J)$ it should be that case that there exists a p_j in the interval $(l(J) - r, l(J) - l)$, i.e. in the interval $(t', t' + (r - l))$. Hence $\tau', t' \models \Diamond_{(0,r-l)} p_j$. Again if we have a point $t'' \in (t', t' + (r - l))$ such that $\tau', t'' \models \Box_{(0,r-l)} \neg p_j$ then we have that $\neg q_j$ is true at the point $t'' + r$. Since $l(J) - r < t'' < l(J) - l$, we have $l(J) < t'' + r < r(J)$, a contradiction. Pictorially:

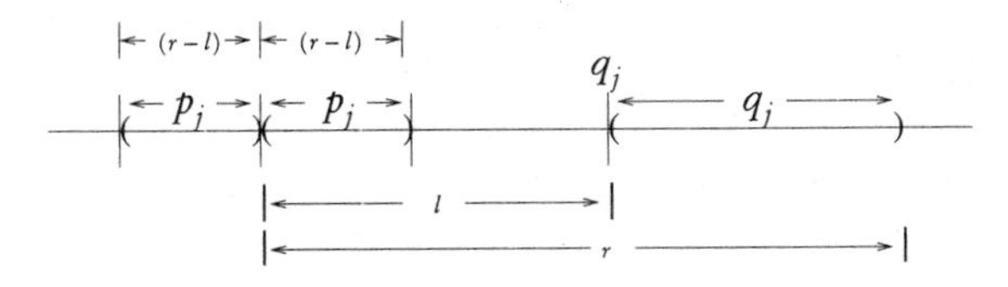

If $\tau', t \models \neg q_j$, then we can argue similarly that $\tau', t \models \varphi_4 \wedge \varphi_5$.

For the converse direction let $\tau' \models \nu$. We argue that for all $t \geq 2r - l$, $\tau', t \models \ominus_{I_j} p$ iff $\tau', t \models q$. By the first formula if $\tau', t \models \ominus_{I_j} p_j$ then $\tau', t \models q_j$. We now argue that if $\tau', t \models q_j$, then $\tau', t \models \ominus_{I_j} p_j$.

So let $\tau', t \models q_j$. Since $\tau', t \models \varphi_1 \wedge \varphi_2 \wedge \varphi_3$ we can easily argue that there exists a q_j-interval J, which is both left and right open, such that $t \in J$ and $len(J) = len(I_j) = r - l$. There are two cases:

1. $\tau', l(J) \not\models q_j$: This implies that $\tau', l(J) - l \models \Box_{(l,r)} q_j \wedge \Diamond_{[l,r)} \neg q_j$ (note that $t \geq 2r - l$, so $l(J) - l > 0$). Since $\tau' \models \varphi_4$ it follows that $\tau', l(J) - l \models p_j U p_j$ and therefore $\tau', t \models \ominus_{I_j} p_j$.
2. $\tau', l(J) \models q_j$: The proof is similar to the one given above which uses formula φ_5 to ensure that $\tau', t \models \ominus_{I_j} p_j$. ⊣

Corollary 6 *If $\tau \in TSS(P)$ is a model such that p_j is not true in the interval $[0,d)$, for any $d \geq 2r-l$, i.e. $p_j \notin \tau(t)$ for all $t \in [0,d)$ then $\tau' \in TSS(P')$ is a canonical extension of τ w.r.t. $\ominus_{I_j} p_j$ iff τ' satisfies $\nu \wedge \Box_{[0,d)} \neg q_j$.* ⊣

In order to take the advantage of Corollary 6 we first shift the models to the right. Let $\tau \in TSS(P)$ and let $d \geq 0$. Let c be a proposition such that $c \notin P$ and let $P' = P \cup \{c\}$. Then τ shifted by d time units to the right, called $\tau_d \in TSS(P')$, is defined as follows:

$$\tau_d(t) = \begin{cases} \tau(t-d) \cup \{c\} & \text{if } t \geq d \\ \phi & \text{otherwise.} \end{cases}$$

We define the *translated* version of a formula μ, called $tr(\mu)$, which essentially forces μ to be true at that points where c is true. $tr(\mu)$ is defined inductively as follows:

$$\begin{array}{lcl} tr(p) & = & p, \text{where } p \in P \\ tr(\neg\psi) & = & \neg tr(\psi) \\ tr(\psi \vee \eta) & = & tr(\psi) \vee tr(\eta) \\ tr(\psi U \eta) & = & tr(\psi) U tr(\eta) \\ tr(\Diamond_I \psi) & = & \Diamond_I (tr(\psi)) \\ tr(\ominus_I \psi) & = & \ominus_I (c \wedge tr(\psi)) \end{array}$$

Now we define the formula μ_d which is satisfied precisely by the models of μ shifted by d time units to the right:

$$\mu_d = (\Box_{[0,d)} \bigwedge_{p \in P} \neg p) \wedge (\Box_{[0,d)} \neg c) \wedge (\Box_{[d,\infty)} c) \wedge (\neg c\, U (c \wedge tr(\mu))).$$

For any timed state sequence τ and $d \geq 0$, let $tail_d(\tau)$ be the timed state sequence τ' such that $\forall t \in \mathbb{R}_{\geq 0}, \tau'(t) = \tau(d+t)$. We extend the function $tail_d$ in the natural way over the sets of timed state sequences.

Lemma 7 *Let μ be a formula. Then $L(\mu) = tail_d(L(\mu_d) \upharpoonright P)$.*
Proof Let $\tau \models \mu$. It is easy to see that $\tau_d \models (\Box_{[0,d)} \bigwedge_{p \in P} \neg p) \wedge (\Box_{[0,d)} \neg c) \wedge (\Box_{[d,\infty)} c)$. We can now argue using structural induction on the formula $tr(\mu)$ to show that if $\tau, t \models \mu$ then $\tau_d, t+d \models tr(\mu)$ and therefore it follows that $\tau_d \models \mu_d$. Since $\tau = tail_d(\tau_d) \upharpoonright P$ we have $L(\mu) \subseteq tail_d(L(\mu_d) \upharpoonright P)$.

Base case: It is easy to see that if μ is an atomic formula and $\tau, t \models \mu$ then $\tau_d, t+d \models tr(\mu)$.

Induction step: Let $\tau \models \ominus_I \psi$. Then there exists a $t' \geq 0$ such that $t - t' \in I$ and $\tau, t' \models \psi$. Then by induction hypothesis it follows that $\tau_d, t' + d \models tr(\psi)$ and since $(t+d) - (t'+d) \in I$ we have $\tau_d, t + d \models \ominus_I(tr(\psi))$. Since $t' + d \geq d$ we have that $\tau_d, t' + d \models c$. So we have $\tau_d, t + d \models \ominus_I(c \wedge tr(\psi))$ and therefore $\tau_d, t + d \models tr(\ominus_I \psi)$. Similarly we can argue for other cases that if $\tau, t \models \mu$ then $\tau_d, t + d \models tr(\mu)$.

For the converse direction let $t \geq 0$ and $\tau', t + d \models \mu_d$. We now argue that $tail_d(\tau') \restriction P, t \models \mu$. Then it follows that $\tau' \models \mu_d$ implies $tail_d(\tau'), t \models \mu$. The proof is similar to the one given above which uses structural induction on the formula μ to show that if $\tau', t + d \models tr(\mu)$ then $\tau, t \models \mu$. Thus we have $tail_d(L(\mu_d) \restriction P) \subseteq L(\mu)$. ⊣

Returning now to our original formula $\varphi \in nr\text{-LTL}^c(\Diamond, \ominus)$ which uses the past formulas $\ominus_{I_0} p_0, \ldots, \ominus_{I_n} p_n$, we first go over to the formula $\varphi_{d_{max}}$ which is satisfied precisely by the models of φ shifted by d_{max} time units. Let $\varphi' = \varphi_{d_{max}}[p'_j/(c \wedge p_j)] \wedge (p'_j \Leftrightarrow (c \wedge p_j))$ be the flattened version of $\varphi_{d_{max}}$. Now define

$$\widehat{\varphi} = \varphi'[q_j / \ominus_{I_j} p'_j] \wedge (\nu[p'_j/p_j] \wedge (\bigwedge_{j=0}^{j=n} \Box_{[0,d_{max})} \neg q_j)).$$

Then by the Corollary 6 and Lemma 7 we have:

Theorem 8 *Let $\varphi \in \text{LTL}^c(\Diamond, \ominus)$ over the set of propositions P. Then we can construct a formula $\widehat{\varphi} \in \text{LTL}^c(\Diamond)$ over an extended set of propositions P' such that $L(\varphi) = tail_{d_{max}}(L(\widehat{\varphi}) \restriction P)$.* ⊣

References

[1] R. Alur, T. Feder, and T. A. Henzinger. The benefits of relaxing punctuality. *J. ACM*, 43(1):116–146, 1996.

[2] P. Bouyer, F. Chevalier, and N. Markey. On the expressiveness of TPTL and MTL. In Ramanujam and Sen [15], pages 432–443.

[3] J.R. Büchi. On a decision method in restricted second order arithmetic. In *Z. Math. Logik Grundlag. Math.*, pages 66–92, 1960.

[4] F. Chevalier, D. D'Souza, and P. Prabhakar. On continuous timed automata with input-determined guards. In S. Arun-Kumar and N. Garg, editors, *FSTTCS*, volume 4337 of *Lecture Notes in Computer Science*, pages 369–380. Springer, 2006.

[5] D. D'Souza, Raj Mohan M., and P. Prabhakar. Flattening Metric Temporal Logic. Technical Report IISc-CSA-TR-2006-11, Indian Institute of Science, Bangalore 560012, India, September 2006. URL: `http://archive.csa.iisc.ernet.in/TR/2006/11/`.

[6] D. D'Souza and M. R. Mohan. Eventual Timed Automata. In Ramanujam and Sen [15], pages 322–334.

[7] D. M. Gabbay, A. Pnueli, S. Shelah, and J. Stavi. On the Temporal Basis of Fairness. In *POPL*, pages 163–173, 1980.

[8] T. A. Henzinger, J. -F. Raskin, and P. -Y. Schobbens. The Regular Real-Time Languages. In Kim Guldstrand Larsen, Sven Skyum, and Glynn Winskel, editors, *ICALP*, volume 1443 of *Lecture Notes in Computer Science*, pages 580–591. Springer, 1998.

[9] Y. Hirshfeld and A. M. Rabinovich. Logics for Real Time: Decidability and Complexity. *Fundam. Inform.*, 62(1):1–28, 2004.

[10] J. A. W. Kamp. *Tense Logic and the Theory of Linear Order*. PhD thesis, University of California, Los Angeles, California, 1968.

[11] R. Koymans. Specifying Real-Time Properties with Metric Temporal Logic. *Real-Time Systems*, 2(4):255–299, 1990.

[12] J. Ouaknine and J. Worrell. On Metric Temporal Logic and Faulty Turing Machines. In Luca Aceto and Anna Ingólfsdóttir, editors, *FoSSaCS*, volume 3921 of *Lecture Notes in Computer Science*, pages 217–230. Springer, 2006.

[13] A. Pnueli. The Temporal Logic of Programs. In *FOCS*, pages 46–57. IEEE, 1977.

[14] P. Prabhakar and D. D'Souza. On the Expressiveness of MTL with Past Operators. In Eugene Asarin and Patricia Bouyer, editors, *FORMATS*, volume 4202 of *Lecture Notes in Computer Science*, pages 322-336. Springer, 2006.

[15] R. Ramanujam and S. Sen, editors. *FSTTCS 2005: Foundations of Software Technology and Theoretical Computer Science, 25th International Conference, Hyderabad, India, December 15-18, 2005, Proceedings*, volume 3821 of *Lecture Notes in Computer Science*. Springer, 2005.

Local Safety and Local Liveness for Distributed Systems*

Volker Diekert[1] **Paul Gastin**[2]
[1] *FMI, Universität Stuttgart,*
Universitätsstraße 38, D-70569 Stuttgart
Volker.Diekert@fmi.uni-stuttgart.de
[2] *LSV, ENS Cachan, CNRS*
61, av. du Prés. Wilson, F-94230 Cachan, France
Paul.Gastin@lsv.ens-cachan.fr

Abstract

We introduce local safety and local liveness for distributed systems whose executions are modeled by Mazurkiewicz traces. We characterize local safety by local closure and local liveness by local density. Restricting to first-order definable properties, we prove a decomposition theorem in the spirit of the separation theorem for linear temporal logic. We then characterize local safety and local liveness by means of canonical local temporal logic formulae.

Keywords: Local temporal logics, safety, liveness, Mazurkiewicz traces, concurrency.

1 Introduction

Distributed systems are widely used nowadays in almost all application fields of computer science such as telecomunication systems or embedded systems. Since most of these are safety critical systems it is important to develop theory and tools to formally specify and verify them.

Abstract models of distributed systems such as Petri nets have been introduced. Concurrent executions of such systems are naturally described with partial orders such as Mazurkiewicz traces [20, 21] or event structures [30] and Thiagarajan with other authors [17, 25–27]

*This work has been partially supported by projects ANR-06-SETIN-003 DOTS, and P2R MODISTE-COVER/Timed-DISCOVERI.

described the relationships between Petri nets and Mazurkiewicz traces or event structures.

When it comes to specification languages, temporal logics are amongst the best formalisms both because they are intuitive and enjoy good algorithmic properties [19]. It turns out that any property can be written as a conjunction of some *safety* and some *liveness* properties [4]. For sequential systems, we can characterize safety and liveness by topological closure and density as well as by canonical temporal logic formulae [1, 2, 4].

When dealing with distributed systems, temporal logics should be extended in order to specify properties of partial orders instead of linear orders. Several temporal logics were introduced [18, 23] but with a global semantics and an existential until these logics are undecidable. One of the first decidable temporal logic for traces was introduced by Ebinger [11] in his PhD-Thesis and another one (TrPTL) was introduced by Mukund and Thiagarajan [22]. Characterizing the expressive power of temporal logics for traces was a major open problem for several years until Thiagarajan with Walukiewicz showed that the global temporal logic LTrL with universal semantics and some past constants has the same expressive power as first-order logic [28]. Unfortunately, the satisfiability problem for this logic is non-elementary [29]. The expressivity result was later improved with algebraic techniques by showing that the pure future global temporal logic is also expressively complete for first-order logic [7]. This opened the way to the characterization of safety and liveness properties for trace languages with global temporal logic [8].

The global semantics means that we are interested in properties a system may satisfy at global configurations corresponding to possible snapshots or global views. Another interesting approach is to look at local configurations, which describe what a local process may know about the current state of the system. A global configuration correspond to some finite partial order possibly having several maximal events whereas local configurations only have a single maximal event, and can thus be identified with events of the partial order describing an execution. Several local temporal logics for traces were introduced and studied [3, 6]. A major achievement was to establish that the simplest pure future local temporal logic is expressively complete for first-order logic [9]. Contrary to global temporal logics for which the satisfiability problem is either undecidable or non-elementary, local temporal logics enjoy much better algorithmic properties since both satisfiability and model checking are decidable in PSPACE [14, 15].

In the present paper, we introduce and study local safety and local liveness for distributed systems. Intuitively, a system is *locally safe* if all local configurations it can reach are "good" configurations. We characterize local safety by *local closure*. A local configuration is *locally live* with respect to some property if it can be extended to a distributed execusion which meets the property. Local liveness properties (those for which all local configurations are locally live) are characterized by *local density*. In order to obtain local temporal logic characterizations, we restrict to first-order definable properties. Building on the expressive completeness of local temporal logic for traces [9], one of our main result is a *decomposition* theorem for all first-order properties in the spirit of the separation theorem of [13] for linear temporal logic. We also generalize this separation theorem to local temporal logic over traces. Using our decomposition theorem, we are able to characterize local safety and local liveness by canonical local temporal logic formulae. We also introduce a stronger notion of local liveness and give a characterization with special local decomposition formulae.

The paper is organized as follows. In the first section we give some general remarks, next we recall basic definitions on Mazurkiewicz traces. In Section 4 we define local temporal logic and recall the main result of [9] on which this work is based. Our local decomposition theorem and the generalization of the separation theorem to local temporal logic are established in Section 5. The last two sections are devoted to the definitions and characterizations of local safety and local liveness respectively.

2 General remarks

Before we dive into our specific setting, let us try to explain some basic ideas from a general viewpoint. Some background is useful to understand the following lines. But the reader is free to skip this section since its aim is only to put our work in perspective and it is not needed to understand the rest or the paper.

If X is a topological space, then the topology can be defined in terms of the closure operator $L \mapsto \overline{L}$. Clearly, for $L \subseteq X$ we have $L = \overline{L} \cap (L \cup (X \setminus \overline{L}))$. Hence, in a topological space, every set is the intersection of a closed set and a dense set, because $\overline{L}$ is closed by definition and $L \cup (X \setminus \overline{L})$ is dense, because its closure contains at least $\overline{L} \cup (X \setminus \overline{L})$, hence its closure is X.

Now, if $\mathscr{C}$ is a family of subsets of X which forms a Boolean algebra and which contains $\overline{L}$ for every $L \in \mathscr{C}$, then, trivially, every $L \in \mathscr{C}$

can be written as an intersection $L = A \cap B$ where $A, B \in \mathscr{C}$ and A is closed and B is dense.

In the setting of infinite words and temporal logic, X is the (compact) Cantor space Σ^ω and $\mathscr{C}$ is the set of LTL-definable languages. Closed LTL-definable languages are called *safety properties* and dense LTL-definable languages are called *liveness properties*. Safety properties play an important role in applications. In some sense they are simpler to handle, e.g., they can be recognized by deterministic Büchi automata. We note that actually we could replace $\mathscr{C}$ by other varieties like FO^2-definable languages and we would have a similar statement.

Now, everything transfers smoothly to Mazurkiewicz traces, provided we use the Scott topology which is defined by saying that infinite traces are (very) close, if they agree on (very) long finite prefixes. This means however that we work with a global semantics where we have control over prefixes. This is not natural in the setting of Mazurkiewicz traces, because events are ordered by the information flow and the model should reflect this. As a consequence in a purely distributed setting we do not know the global state and therefore we cannot control global prefexes. This is one of the reasons to favor a local temporal logic. Another one is that these logics allow much better complexities for model checking. From an abstract viewpoint this means that we have to switch from a topology setting to a domain theoretical setting. The *local closure* of a language L now includes all traces where every finite prefix which has a single maximal vertex is a prefix of some element in L. By the very definition every locally closed language is closed in the Scott topology, but the converse does not hold. Consider $\{a^\infty\}$ and $\{b^\infty\}$. Both sets are locally closed, but the union is not, as soon as a and b are independent. Thus, a *local safety property* is a stronger condition than a *global safety property* and cannot be investigated from a purely topological viewpoint. Things become more complicate and we need some careful analysis. On the other hand, knowing that a language satisfies a *local safety property* makes a distributed model checking possible. Our results show that, indeed, local safety is a robust and natural concept.

Having defined local safety properties it turns out that there is also a natural notion of *local liveness* which can also be characterized in terms of *local density* and such that every locally specified temporal property is the intersection of a local safety and a local liveness property. This notion of local liveness is inherently bound to an "optimistic" viewpoint, where from the local view on the past, we cannot exclude that one continuation is as desired (and so may hope for it). The more strict viewpoint is that, from the local view on the past, we can enforce locally that at least one continuation is as desired. We call this *strong local liveness*. It is however not true that every locally specified temporal property is the intersection of a local safety and a strong local liveness property.

3 Mazurkiewicz Traces

We recall some standard notations from trace theory which will be used in the paper. The reader is refered to [10] for more details.

A *dependence alphabet* is a pair (Σ, D) where the alphabet Σ is a finite set and the *dependence relation* $D \subseteq \Sigma \times \Sigma$ is reflexive and symmetric. The *independence relation* I is the complement of D. For $a \in \Sigma$, we let $I(a) = \{b \in \Sigma \mid (a,b) \in I\}$ be the set of letters independent from a.

A *Mazurkiewicz trace* is an equivalence class of a labelled partial order $t = [V, \leq, \lambda]$ where V is a set of vertices labelled by $\lambda : V \to \Sigma$ and $\leq$ is a partial order over V satisfying the following conditions: For all $x \in V$, the downward set $\downarrow x = \{y \in V \mid y \leq x\}$ is finite, and for all $x, y \in V$ we have that $(\lambda(x), \lambda(y)) \in D$ implies $x \leq y$ or $y \leq x$, and that $x \lessdot y$ implies $(\lambda(x), \lambda(y)) \in D$, where $\lessdot \, = \, < \setminus <^2$ is the immediate successor relation in t. For $x \in V$, we also define $\uparrow x = \{y \in V \mid x \leq y\}$.

The trace t is finite if V is finite and we denote the set of finite traces by $\mathbb{M}(\Sigma, D)$ (or simply $\mathbb{M}$). By $\mathbb{R}(\Sigma, D)$ (or simply $\mathbb{R}$), we denote the set of finite or infinite traces (also called *real traces*). We write $\mathrm{alph}(t) = \lambda(V)$ for the alphabet of t and we let $\mathrm{alphinf}(t) = \{a \in \Sigma \mid \lambda^{-1}(a) \text{ is infinite}\}$ be the set of letters occurring infinitely often in t.

We define the concatenation for traces $t_1 = [V_1, \leq_1, \lambda_1]$ and $t_2 = [V_2, \leq_2, \lambda_2]$, provided $\mathrm{alphinf}(t_1) \times \mathrm{alph}(t_2) \subseteq I$. It is given by $t_1 \cdot t_2 = [V, \leq, \lambda]$ where V is the disjoint union of V_1 and V_2, $\lambda = \lambda_1 \cup \lambda_2$, and $\leq$ is the transitive closure of the relation $\leq_1 \cup \leq_2 \cup (V_1 \times V_2 \cap \lambda^{-1}(D))$. The set $\mathbb{M}$ of finite traces is then a monoid with the empty trace $1 = (\emptyset, \emptyset, \emptyset)$ as unit.

The concatenation of two trace languages $K, L \subseteq \mathbb{R}$ is $K \cdot L = \{r \cdot s \mid r \in K, s \in L \text{ and } \mathrm{alphinf}(r) \times \mathrm{alph}(s) \subseteq I\}$.

Let $r, t \in \mathbb{R}$ be traces. We say that r is a *prefix* of t and we write $r \leq t$ if $t = r \cdot s$ for some $s \in \mathbb{R}$. Prefixes of $t = [V, \leq, \lambda]$ can be identified with downward closed subsets $U = \downarrow U$ of V. We denote by $\mathrm{Pref}(t)$ the set of *finite prefixes* of t. This notation is extended to languages in the obvious way. The set $\mathbb{R}$ endowed with the prefix partial order relation is a coherently complete domain. In particular, any real trace t is the *least upper bound* of its finite prefixes: $t = \sqcup \mathrm{Pref}(t)$.

A trace p is called *prime*, if it is finite and has a unique maximal element. The maximal event of a prime trace t is denoted $\max(t)$. The set of all prime traces in $\mathbb{R}$ is denoted by $\mathbb{P}$. The set of *prime prefixes* of a trace t is denoted $\mathbb{P}\mathrm{ref}(t) = \mathrm{Pref}(t) \cap \mathbb{P}$. Prime prefixes of $t = [V, \leq, \lambda]$ can be identified with downward closures $\downarrow x$ of events $x \in V$. Any real trace t is the *least upper bound* of its prime prefixes: $t = \sqcup \mathbb{P}\mathrm{ref}(t)$.

4 Local Temporal Logic

Our characterization of local safety and local liveness will be in terms of temporal logic formulae under a *local* semantics. Contrary to the *global* semantics which defines when a formula holds at some *global configuration* in some trace, the *local* semantics characterizes the *local events* in a trace satisfying some temporal formulae. In this section, we recall the definition and semantics of local temporal logic over traces, together with the main expressivity result from [9] which is needed for our characterizations of local safety and local liveness.

The syntax of local temporal logic $\mathrm{LocTL}_\Sigma[\mathsf{EX}, \mathsf{U}, \mathsf{EY}, \mathsf{S}]$ is given by

$$\varphi ::= \top \mid a \mid \neg\varphi \mid \varphi \vee \varphi \mid \mathsf{EX}\,\varphi \mid \varphi\,\mathsf{U}\,\varphi \mid \mathsf{EY}\,\varphi \mid \varphi\,\mathsf{S}\,\varphi$$

where a ranges over Σ and $\top$ denotes *true*. Here EX denotes the usual (existential) *next*-operator and U means *until*. Their past versions are EY and S meaning *Yesterday* and *Since*.

Formally, the locally defined semantics of $\mathrm{LocTL}_\Sigma[\mathsf{EX}, \mathsf{U}, \mathsf{EY}, \mathsf{S}]$ is given as follows. Let $t = [V, \leq, \lambda] \in \mathbb{R} \setminus \{1\}$ be a real trace and $x \in V$ be a *local event* (we also write $x \in t$ instead of $x \in V$). Then we define:

$$\begin{array}{lll}
t, x \models \top & & \\
t, x \models a & \text{if} & \lambda(x) = a \\
t, x \models \neg\varphi & \text{if} & t, x \not\models \varphi \\
t, x \models \varphi \vee \psi & \text{if} & t, x \models \varphi \text{ or } t, x \models \psi \\
t, x \models \mathsf{EX}\,\varphi & \text{if} & \exists y \in t\,(x \lessdot y \text{ and } t, y \models \varphi) \\
t, x \models \varphi\,\mathsf{U}\,\psi & \text{if} & \exists z \in t\,(x \leq z \text{ and } t, z \models \psi \text{ and} \\
 & & \forall y \in t\,(x \leq y < z \Rightarrow t, y \models \varphi)) \\
t, x \models \mathsf{EY}\,\varphi & \text{if} & \exists y \in t\,(y \lessdot x \text{ and } t, y \models \varphi) \\
t, x \models \varphi\,\mathsf{S}\,\psi & \text{if} & \exists z \in t\,(z \leq x \text{ and } t, z \models \psi \text{ and} \\
 & & \forall y \in t\,(z < y \leq x \Rightarrow t, y \models \varphi)).
\end{array}$$

As ususal, we use $\mathsf{F}\,\varphi$ as an abbreviation for $\top\,\mathsf{U}\,\varphi$ and $\mathsf{G}\,\varphi = \neg\mathsf{F}\neg\varphi$. The meaning of $\mathsf{F}\,\varphi$ is that somewhere in the future φ holds, whereas $\mathsf{G}\,\varphi$ means that always in the future φ holds. By $\perp$ we mean $\neg\top$, which denotes *false*.

Henceforth formulae in $\mathrm{LocTL}_\Sigma[\mathsf{EY},\mathsf{S}]$ using only the past modalities are called *past formulae* whereas we refer to formulae in $\mathrm{LocTL}_\Sigma[\mathsf{EX},\mathsf{U}]$ using only future modalities as *future formulae*. It is easy to see by structural induction that if φ is a past formula then for all $t \in \mathbb{R}$ and $x \in t$ we have $t, x \models \varphi$ if and only if $\downarrow x, x \models \varphi$. Similarly, if φ is a future formula then for all $t \in \mathbb{R}$ and $x \in t$ we have $t, x \models \varphi$ if and only if $\uparrow x, x \models \varphi$.

Formulae of the form $\mathsf{F}\,\varphi$ and $\mathsf{G}\,\varphi$ with $\varphi \in \mathrm{LocTL}_\Sigma[\mathsf{EX},\mathsf{U},\mathsf{EY},\mathsf{S}]$ are called F and G formulae respectively. By definition, a formula in $\mathrm{LocTL}_\Sigma[\mathsf{EX},\mathsf{U},\mathsf{EY},\mathsf{S}]$ can be viewed as a first-order formula in one free variable. However, for F and G formulae, it is also natural to give a direct interpretation on traces. Let $t \in \mathbb{R}$, we define:

$$\begin{array}{lll} t \models_\ell \mathsf{F}\,\varphi & \text{if} & \exists x \in t,\ t, x \models \varphi \\ t \models_\ell \mathsf{G}\,\psi & \text{if} & \forall x \in t,\ t, x \models \psi. \end{array}$$

More generally, if γ is any Boolean combination of F and G formulae, then we extend the meaning of $t \models_\ell \gamma$ in the obvious way; and this defines a language $\mathcal{L}(\gamma) = \{t \in \mathbb{R} \mid t \models_\ell \gamma\}$. Note that the empty trace $1 \models \mathsf{G}\,\varphi$ but $1 \not\models \mathsf{F}\,\varphi$ for all $\varphi \in \mathrm{LocTL}_\Sigma$.

We let $\mathbb{R}^1$ be the set of real traces t with exactly one minimal event, denoted $\min(t)$. Our results are consequences of the following theorem.

Theorem 1 *[9] Let $L \subseteq \mathbb{R}$ be a first-order definable real trace language. Then there is a future formula $\varphi \in \mathrm{LocTL}_\Sigma[\mathsf{EX},\mathsf{U}]$ such that*

$$L \cap \mathbb{R}^1 = \{t \in \mathbb{R}^1 \mid t, \min(t) \models \varphi\}.$$

By duality, Theorem 1 implies the following corollary where, as introduced above, $\mathbb{P}$ is the set of finite traces t with exactly one maximal event, denoted $\max(t)$.

Corollary 2 *Let $L \subseteq \mathbb{R}$ be a first-order definable real trace language. Then there is a past formula $\psi \in \mathrm{LocTL}_\Sigma[\mathsf{EY},\mathsf{S}]$ such that*

$$L \cap \mathbb{P} = \{t \in \mathbb{P} \mid t, \max(t) \models \psi\}.$$

Proof It is clear that $\mathbb{P}$ is first-order definable. Hence, if L is first-order definable, then so is $L \cap \mathbb{P}$. Now, we *reverse* all traces, i.e., we read traces from right-to-left (formally, we replace $\leq$ by $\geq$). For the reverse language of $L \cap \mathbb{P}$ (which is a first-order definable subset in $\mathbb{R}^1 \cap \mathbb{M}$) we obtain a future formula $\varphi \in \mathrm{LocTL}_\Sigma[\mathsf{EX},\mathsf{U}]$ by Theorem 1. We obtain ψ by replacing in φ all occurrences of EX by EY and all occurrences of U by S. $\dashv$

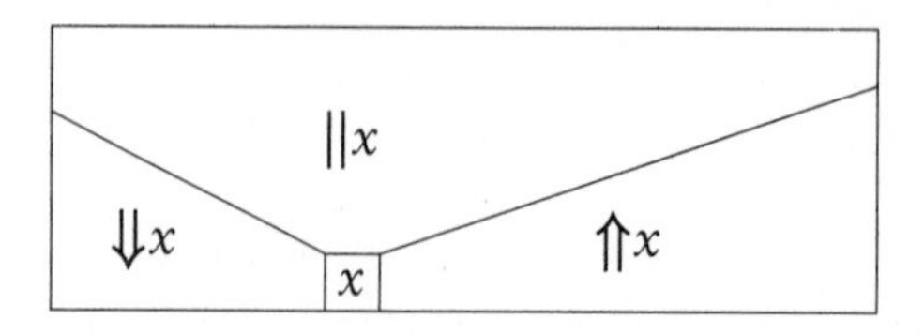

Figure 1: *Canonical decomposition of a trace*

5 Local Decomposition of First-Order Languages

In this section, we establish a *decomposition* theorem based on the local semantics of temporal logic. This decomposition theorem is in the spirit of the *separation* theorem for temporal logic over words [13]. Since the proof techniques are quite similar, we also extend this *separation* theorem to the local semantics of temporal logic over traces. Very little is known for the *complexity of separation* over traces and it seems difficult to derive any reasonable complexity bound. However, such bounds are certainly of interest and research on this topic might be a useful future project.

Given a trace $t = [V, \leq, \lambda] \in \mathbb{R}$ and a vertex $x \in V$, we are interested in the sets of vertices that are strictly below x or strictly above x or concurrent to x:

$$\Downarrow x = \{y \in V \mid y < x\}$$
$$\Uparrow x = \{y \in V \mid x < y\}$$
$$\| x = \{y \in V \mid x \not\leq y \text{ and } y \not\leq x\}.$$

By slight abuse of notation, we also denote by $\Downarrow x$, $\Uparrow x$ and $\| x$ the corresponding factors of t. We have a canonical decomposition of t, depicted in Figure 1:

$$\begin{aligned} V &= \Downarrow x \cup \{x\} \cup \| x \cup \Uparrow x \\ t &= (\Downarrow x) \cdot \lambda(x) \cdot (\| x) \cdot (\Uparrow x) \end{aligned}$$

Let us introduce a new local modality CO which talks about the part $\| x$ of the trace which is concurrent to the current vertex x. By a *concurrent formula* we mean a formula of type $\mathsf{CO}\,\gamma$, where γ is any Boolean combination of F and G formulae. The semantics of a concurrent formula is given by

$$t, x \models \mathsf{CO}\,\gamma \quad \text{if} \quad \| x \models_\ell \gamma.$$

The main result of this section associates with any first-order definable language some *decomposition formula*. A decomposition formula is a disjunction

$$\delta = \bigvee_{j \in J} a_j \wedge \psi_j \wedge \varphi_j \wedge \mathsf{CO}\,\gamma_j$$

where J is some finite index set, and for each $j \in J$, $a_j \in \Sigma$ is a letter, $\psi_j \in \mathrm{LocTL}_\Sigma(\mathsf{EY}, \mathsf{S})$ is a past formula, $\varphi_j \in \mathrm{LocTL}_\Sigma(\mathsf{EX}, \mathsf{U})$ is a future formula, and γ_j is an F or G formula. Note that, if $J = \emptyset$ then we get $\delta = \bot$ by convention. We can now state the *decompotion theorem.*

Theorem 3 *Let $L \subseteq \mathbb{R}$ be a first-order definable real trace language. There exists a* decomposition formula $\delta = \bigvee_{j \in J} a_j \wedge \psi_j \wedge \varphi_j \wedge \mathsf{CO}\,\gamma_j$ *such that*

(i) $L \cup \{1\} = \mathscr{L}(\mathsf{G}\,\delta)$,
(ii) $L \setminus \{1\} = \mathscr{L}(\mathsf{F}\,\delta)$,
(iii) $\mathbb{P}\mathrm{ref}(L) = \{r \in \mathbb{P} \mid r, \max(r) \models \bigvee_{j \in J} a_j \wedge \psi_j\}$,
(iv) for each $j \in J$, the formula $a_j \wedge \psi_j \wedge \varphi_j \wedge \mathsf{CO}\,\gamma_j$ is satisfiable.

Proof The proof is by induction on the size of the alphabet. If $\Sigma = \emptyset$ then we let $J = \emptyset$ so that $\delta = \bot$. We have $\mathscr{L}(\mathsf{F}\,\delta) = \emptyset$ and $\mathscr{L}(\mathsf{G}\,\delta) = \{1\}$. Statements (i-iv) are satisfied since either $L = \emptyset$ or $L = \{1\}$ and in any cases $\mathbb{P}\mathrm{ref}(L) = \emptyset$.

Assume now that $\Sigma \neq \emptyset$. Since L is first-order, it is also aperiodic [12]. Let $h : \mathbb{M}(\Sigma, D) \to S$ be a morphism to some finite aperiodic monoid S which recognizes† L. Without loss of generality, we assume that h is *alphabetic*, i.e., $h(r) = h(s)$ implies $\mathrm{alph}(r) = \mathrm{alph}(s)$ for all $r, s \in \mathbb{M}(\Sigma, D)$.

Let $t \in L$ be a nonempty trace and let $x \in t$. The h-equivalence classes $[\Downarrow x]$, $[\Uparrow x]$ and $[\| x]$ are recognized by h, hence are first-order definable trace languages. Note also that all vertices in $\| x$ are labeled with letters independent from $\lambda(x)$. Since h is alphabetic, we deduce that the trace language $[\| x]$ is over a *strict* sub-alphabet of Σ.

We define the index set

$$J = \{(\lambda(x), [\Downarrow x], [\| x], [\Uparrow x]) \mid t \in L \setminus \{1\} \text{ and } x \in t\}$$

which is finite since there are finitely many h-equivalence classes. Fix now some index $j = (a_j, L_j^{\Downarrow}, L_j^{\|}, L_j^{\Uparrow}) \in J$. By Theorem 1 and Corollary 2

†The reader is refered to [24] for recognizability by morphisms. Here we will only use the fact that the morphism h induces an equivalence relation on $\mathbb{R}$ with finitely many classes and which saturates L ($t \in L$ implies $[t] \subseteq L$) and satisfies $[r] \cdot [s] \subseteq [r \cdot s]$ whenever the product $r \cdot s$ is defined. Moreover, each h-equivalence class is first-order definable.

we find a future formula φ_j and a past formula ψ_j such that

$$\begin{aligned} a_j \cdot L_j^{\Uparrow} \cap \mathbb{R}^1 &= \{s \in \mathbb{R}^1 \mid s, \min(s) \models \varphi_j\} \\ L_j^{\Downarrow} \cdot a_j \cap \mathbb{P} &= \{r \in \mathbb{P} \mid r, \max(r) \models \psi_j\}. \end{aligned}$$

Now, $L_j^{||}$ is over a strict sub-alphabet of Σ and by induction we find a decomposition formula δ_j for this language. Depending on whether $1 \in L_j^{||}$ or not we let $\gamma_j = \mathsf{G}\,\delta_j$ or $\gamma_j = \mathsf{F}\,\delta_j$ so that $L_j^{||} = \mathscr{L}(\gamma_j)$. We claim that the decomposition formula $\delta = \bigvee_{j \in J} a_j \wedge \psi_j \wedge \varphi_j \wedge \mathsf{CO}\,\gamma_j$ satisfies statements (i-iv).

First, let $t \in L \setminus \{1\}$ and let $x \in t$. Fix the index

$$j = (a_j, [\Downarrow x], [||x], [\Uparrow x]) \in J$$

with $a_j = \lambda(x)$. We have $\downarrow x = \Downarrow x \cdot a_j$ and $\uparrow x = a_j \cdot \Uparrow x$, hence by definition of φ_j and ψ_j we get $\downarrow x, x \models \psi_j$ and $\uparrow x, x \models \varphi_j$ and $||x \models \gamma_j$. Since ψ_j is a past formula and φ_j is a future formula, we deduce that $t, x \models a_j \wedge \psi_j \wedge \varphi_j \wedge \mathsf{CO}\,\gamma_j$, which is therefore a satisfiable formula. We have also shown $t \models \mathsf{G}\,\delta$ and $t \models \mathsf{F}\,\delta$ for all $t \in L \setminus \{1\}$ and that $r, \max(r) \models \bigvee_{j \in J} a_j \wedge \psi_j$ for all $r \in \mathbb{P}\mathrm{ref}(L)$. Hence, we have proved statement (iv) and that the left hand side is contained in the right hand side for statements (i-iii).

Conversely, assume that $t \models \mathsf{F}\,\delta$ or that $t \neq 1$ and $t \models \mathsf{G}\,\delta$. Let $x \in t$ and $j \in J$ be such that $t, x \models a_j \wedge \psi_j \wedge \varphi_j \wedge \mathsf{CO}\,\gamma_j$. For clarity, we write below r, s and u for the factors $\Downarrow x$, $\Uparrow x$ and $||x$ of t. Since ψ_j is a past formula and φ_j is a future formula, we deduce that $ra_j, x \models \psi_j$ and $a_j s, x \models \varphi_j$ and $u \models \gamma_j$ and $\lambda(x) = a_j$. Let $t' \in L \setminus \{1\}$ and $x' \in t'$ be such that $j = (\lambda(x'), [\Downarrow x'], [||x'], [\Uparrow x'])$. As above, we write r', s' and u' for the factors $\Downarrow x'$, $\Uparrow x'$ and $||x'$ of t'. By definition of the formulae ψ_j, φ_j and γ_j we have $r \in [r']$, $s \in [s']$ and $u \in [u']$. We deduce that $t = r \cdot a_j \cdot u \cdot s \in [t']$. Since $t' \in L$ and h recognizes L we obtain $t \in L$.

For the converse inclusion of (iii), let $j \in J$ and $ra_j \in \mathbb{P}$ be such that $ra_j, \max(ra_j) \models a_j \wedge \psi_j$. Let $t' \in L \setminus \{1\}$ and $x' \in t'$ be such that $j = (\lambda(x'), [\Downarrow x'], [||x'], [\Uparrow x'])$. As above, we write r', s' and u' for the factors $\Downarrow x'$, $\Uparrow x'$ and $||x'$ of t'. We obtain $r \in [r']$. Let $t = r \cdot a_j \cdot u' \cdot s'$. We have $t \in [t']$ and we get $t \in L$. Therefore, $ra_j \in \mathbb{P}\mathrm{ref}(L)$. $\dashv$

A similar proof allows to generalize the separation theorem from words to traces with the local semantics of temporal logic.

Theorem 4 *Let $\varphi \in \mathrm{FO}_\Sigma(<)$ be a first-order formula with one free variable. Then there exists a* decomposition formula

$$\delta = \bigvee_{j \in J} a_j \wedge \psi_j \wedge \varphi_j \wedge \mathsf{CO}\,\gamma_j$$

such that for all $t \in \mathbb{R} \setminus \{1\}$ and all $x \in t$ we have

$$t, x \models \varphi(x) \quad \textit{if and only if} \quad t, x \models \delta.$$

6 Local Safety Properties

In this section, we investigate local safety for distributed systems. Intuitively, a *safety property* is defined by a set P of *safe partial executions*. A finite or infinite execution is *safe* if all its partial executions are in the given set P. The difference between global safety and local safety stems from the semantics of *partial execution*. For local safety, the partial executions of a trace $t \in \mathbb{R}$ are its *prime prefixes* $\mathbb{P}\mathrm{ref}(t)$. The advantage of local safety is that it can be locally enforced by processes of the system: before executing an action, a local process makes sure that the partial execution having this action as maximal event is in the set P of safe partial executions.

Below, we first define local safety. Then we characterize local safety by means of local closure. Next, for local safety properties which are first-order definable, we give a characterization by local temporal logic formulae of the form $\mathsf{G}\,\psi$ where ψ is a past formula.

We say that a trace $t \in \mathbb{R}$ is *locally safe* with respect to some set $P \subseteq \mathbb{P}$ if $\mathbb{P}\mathrm{ref}(t) \subseteq P$. A trace language $L \subseteq \mathbb{R}$ is said to be a *local safety property*, if we can find some set $P \subseteq \mathbb{P}$ such that

$$L = \{t \in \mathbb{R} \mid \mathbb{P}\mathrm{ref}(t) \subseteq P\}.$$

Example 5 For instance, if $\Sigma = \{a, b, c\}$ and $I = \{(a, b), (b, a)\}$ then the language L of traces $t \in \mathbb{R}$ such that $t = ucrcscv$ with $|r|_c = |s|_c = 0$ implies $|r|_a + |r|_b$ is different from $|s|_a + |s|_b$ modulo 2 is a local safety property. In order to ensure locally this property, the process executing a counts the number of a's modulo 2 since the last c and similarly for the process executing b. Now, whenever a c is about to occur, these numbers are checked by the process in charge of c and depending on whether the safety property is satisfied or not, the action c is enabled or not.

A subset $K \subseteq \mathbb{R}$ is called *coherent* if for all $r, s \in K$ there is some $t \in \mathbb{R}$ such that r and s are prefixes of t. Since $\mathbb{R}$ is coherently complete [16], the least upper bound of any coherent set exists. Hence, $K \subseteq \mathbb{R}$ is coherent if and only if it is bounded from above, i.e., if all elements of K are prefixes of some $t \in \mathbb{R}$. In particular, $\mathbb{P}\mathrm{ref}(t)$ is coherent for all $t \in \mathbb{R}$.

We say that a trace language L is *locally closed* if it is closed under prime prefixes and under least upper bounds of coherent subsets: $\mathbb{P}\mathrm{ref}(L) \subseteq L$ and $\sqcup K \in L$ for any coherent set $K \subseteq L$. Note that, if L is locally closed, then it is also closed under prefixes: it $s \leq t \in L$ then $\mathbb{P}\mathrm{ref}(s) \subseteq \mathbb{P}\mathrm{ref}(t) \subseteq L$ and we get $s = \sqcup \mathbb{P}\mathrm{ref}(s) \in L$.

The *local closure* $\overline{L}^{\ell}$ of a language $L \subseteq \mathbb{R}$ is the smallest set which is locally closed and contains L. Note that $1 = \sqcup \emptyset \in \overline{L}^{\ell}$ for all $L \subseteq \mathbb{R}$. We have

$$\overline{L}^{\ell} = \{t \in \mathbb{R} \mid \mathbb{P}\mathrm{ref}(t) \subseteq \mathbb{P}\mathrm{ref}(L)\}.$$

Indeed, we have $L \subseteq L' = \{t \in \mathbb{R} \mid \mathbb{P}\mathrm{ref}(t) \subseteq \mathbb{P}\mathrm{ref}(L)\}$. Next, L' is locally closed since $s \leq t$ implies $\mathbb{P}\mathrm{ref}(s) \subseteq \mathbb{P}\mathrm{ref}(t)$ and $t = \sqcup X$ for some $X \subseteq L'$ implies $\mathbb{P}\mathrm{ref}(t) = \mathbb{P}\mathrm{ref}(X) \subseteq \mathbb{P}\mathrm{ref}(L)$. Finally, assume that $L \subseteq K$ for some locally closed set K. Let $t \in L'$. We get $\mathbb{P}\mathrm{ref}(t) \subseteq \mathbb{P}\mathrm{ref}(L) \subseteq \mathbb{P}\mathrm{ref}(K) \subseteq K$ and $t = \sqcup \mathbb{P}\mathrm{ref}(t) \in K$ since $\mathbb{P}\mathrm{ref}(t)$ is coherent. Therefore, $L' \subseteq K$.

Proposition 6 *A trace language $L \subseteq \mathbb{R}$ is a local safety property if and only if it is locally closed.*

Proof Assume that $L = \{t \in \mathbb{R} \mid \mathbb{P}\mathrm{ref}(t) \subseteq P\}$ for some $P \subseteq \mathbb{P}$. Let $s \leq t \in L$. We have $\mathbb{P}\mathrm{ref}(s) \subseteq \mathbb{P}\mathrm{ref}(t) \subseteq P$ hence $s \in L$. Let now $t = \sqcup K$ for some $K \subseteq L$ coherent. We have $\mathbb{P}\mathrm{ref}(t) = \mathbb{P}\mathrm{ref}(K) \subseteq \mathbb{P}\mathrm{ref}(L) \subseteq P$. Hence $t \in L$.

Conversely, assume that L is locally closed. Then, $L = \overline{L}^{\ell}$ is a local safety property defined by $\mathbb{P}\mathrm{ref}(L)$. ⊣

We turn now to the characterization of local safety by means of temporal logic formulae. A formula in $\mathrm{LocTL}_{\Sigma}[\mathsf{EX}, \mathsf{U}, \mathsf{EY}, \mathsf{S}]$ is called a *canonical local safety formula* if it can be written as $\mathsf{G}\,\psi$ where $\psi \in \mathrm{LocTL}_{\Sigma}[\mathsf{EY}, \mathsf{S}]$ is a past formula. We show that local safety properties which are first order definable can be characterized by canonical local safety formulae.

Theorem 7 *A first-order definable real trace language is a local safety property if and only if it can be expressed by a canonical local safety formula. More precisely:*

(i) Each language defined by a canonical local safety formula is locally closed.
(ii) The local closure of a first-order definable language can be expressed by a canonical local safety formula.

Proof (i) Let $L = \mathcal{L}(\mathsf{G}\,\psi)$ where ψ is a past formula. Let $t \in \mathbb{R}$ with $\mathbb{P}\mathrm{ref}(t) \subseteq \mathbb{P}\mathrm{ref}(L)$. For all $x \in t$ we have $r = \downarrow x \in \mathbb{P}\mathrm{ref}(t)$ hence we find $s \in L$ such that $r \in \mathbb{P}\mathrm{ref}(s)$. Since ψ is a past formula, we have $t, x \models \psi$ if and only if $r, x \models \psi$ if and only if $s, x \models \psi$, which holds since $s \in L$. Therefore, $t \models \mathsf{G}\,\psi$ and $t \in L$. Therefore, $L = \overline{L}^{\ell}$ is locally closed.

(ii) Let $L \subseteq \mathbb{R}$ be a first-order definable language. Consider a decomposition formula $\delta = \bigvee_{j \in J} a_j \wedge \psi_j \wedge \varphi_j \wedge \mathsf{CO}\,\gamma_j$ for L as given by Theorem 3 and let $\psi = \bigvee_{j \in J} a_j \wedge \psi_j$. For all $r \in \mathbb{P}$, we have $r \in \mathbb{P}\mathrm{ref}(L)$ if and only if $r, \max(r) \models \psi$. We show that $\overline{L}^{\ell} = \mathcal{L}(\mathsf{G}\,\psi)$.

First, let $t \in \mathcal{L}(\mathsf{G}\,\psi) \setminus \{1\}$. Let $r = \downarrow x$ with $x \in t$ be a prime prefix of t. We have $t, x \models \psi$ and since ψ is a past formula we deduce $r, \max(r) \models \psi$ and then $r \in \mathbb{P}\mathrm{ref}(L)$. Therefore, $\mathbb{P}\mathrm{ref}(t) \subseteq \mathbb{P}\mathrm{ref}(L)$ and we obtain $t \in \overline{L}^{\ell}$.

Conversely, let $t \in \mathbb{R}$ with $\mathbb{P}\mathrm{ref}(t) \subseteq \mathbb{P}\mathrm{ref}(L)$. For all $r \in \mathbb{P}\mathrm{ref}(t)$, we get $r, \max(r) \models \psi$ by Theorem 3(iii). Since ψ is a past formula, we deduce that $t, x \models \psi$ for all $x \in t$. Therefore, $t \in \mathcal{L}(\mathsf{G}\,\psi)$. $\dashv$

Example 8 The language L defined in Example 5 is a local safety property but is not first-order definable. On the other hand, with the same dependence alphabet, the language L' of traces $t \in \mathbb{R}$ such that $t = ucrcv$ with $|r|_c = 0$ implies $|r|_a \leq 2$ and $|r|_b \leq 2$ is a local safety property which is first-order definable. It is defined by the canonical local safety formula

$$\mathsf{G}\big(c \wedge \mathsf{EY}(\top \mathsf{S} c) \longrightarrow \neg\,\mathsf{EY}(a \wedge \mathsf{EY}(a \wedge \mathsf{EY}\,a)) \wedge \neg\,\mathsf{EY}(b \wedge \mathsf{EY}(b \wedge \mathsf{EY}\,b))\big).$$

We conclude this section by a comparison between global safety and local safety. With the global semantics, a partial execution is an arbitrary prefix, not necessarily a prime. Hence, a trace $t \in \mathbb{R}$ is *globally safe* with respect to some set $M \subseteq \mathbb{M}$ if $\mathrm{Pref}(t) \subseteq M$. Moreover, a language $L \subseteq \mathbb{R}$ is a *global safety property* if $L = \{t \in \mathbb{R} \mid \mathrm{Pref}(t) \subseteq M\}$ for some set $M \subseteq \mathbb{M}$. It was shown in [8] that a language $L \subseteq \mathbb{R}$ is a global safety property if and only if it is *Scott closed*, i.e., closed under prefixes and under least upper bounds of directed sets. The Scott closure of a set $L \subseteq \mathbb{R}$ is $\overline{L}^{\sigma} = \{t \in \mathbb{R} \mid \mathrm{Pref}(t) \subseteq \mathrm{Pref}(L)\}$. It follows immediately that $\overline{L}^{\sigma} \subseteq \overline{L}^{\ell}$ and if L is locally closed then it is also Scott closed. Therefore, any local safety property is also a global safety property.

The complement of Scott closed sets are called Scott open sets and they form a topology. In particular, the union of two Scott closed sets is also Scott closed. But a union of two locally closed sets needs not be locally closed. Indeed, let $\Sigma = \{a, b\}$ with $(a, b) \in I$ and consider the set $L = a^{\infty} \cup b^{\infty}$. This set is Scott-closed, but its local closure is $\mathbb{R}$. Therefore the global safety property $a^{\infty} \cup b^{\infty}$ is not necessarily a local safety property: locally safety is a stronger requirement than global safety.

As another example, let $\Sigma = \{a, b, c\}$ with $I = \{(a, b), (b, a)\}$ and consider the set L of traces $t \in \mathbb{R}$ such that $t = ucrv$ with $|r|_c = 0$ implies $|r|_a + |r|_b \leq 3$. Then L is a first-order definable global safety property but not a local safety property. In order to enforce the global safety property L we need a synchronization between the processes executing a and b, which is not possible in a distributed system since a and b are independent. In other words, any asynchronous (cellular) automaton [5, 31] for L has *deadlocks*: it is not possible to have a *safe (without deadlocks)* distributed implementation for L. On the other hand, any first-order definable local safety property can be implemented by a deterministic asynchronous cellular automaton without deadlocks.

7 Local Liveness Property

We turn now our attention to local liveness for distributed systems. A partial execution is *live* with respect to a set L of desired behaviors if it can be extended to some element in L. Again, the difference between local liveness and global liveness stems from the semantics of *partial execution*. In our local paradigm, a partial execution is a prime trace r and it is live with respect to L if $r \in \mathbb{P}\text{ref}(L)$.

Below, we formally define local liveness and we show that it is characterized by local density. We establish a natural characterization by local temporal logic formulae using the decomposition formulae introduced in Section 5. Then we define and characterize *strong local liveness* where the possibility to *extend* a partial execution to a desired one is strongly limited.

A language $L \subseteq \mathbb{R}$ is a *local liveness property* if each partial execution can be extended to some trace in L, i.e., if $\mathbb{P}\text{ref}(L) = \mathbb{P}$.

With the global semantics, a partial execution is simply a finite trace and not necessarily a prime. Hence a language $L \subseteq \mathbb{R}$ is a *global liveness property* if $\mathbb{M} = \mathrm{Pref}(L)$. Global liveness implies local liveness since $\mathbb{P}\mathrm{ref}(L) = \mathrm{Pref}(L) \cap \mathbb{P}$.

Example 9 Let $\Sigma = \{a, b\}$ with $(a, b) \in I$. Then the language $L = \{a^\omega, b^\omega\}$ is a local liveness property since we have $\mathbb{P} = a^+ \cup b^+ = \mathbb{P}\mathrm{ref}(L)$. But L is not a global liveness property since $\mathrm{Pref}(L) = \mathbb{P}\mathrm{ref}(L) \neq \mathbb{M}$. On the other hand, the language $L = \{(ab)^\omega\}$ is a global liveness property, hence also a local liveness property.

Let us call a language $L \subseteq \mathbb{R}$ *locally dense* if all traces are in the local closure of L, i.e., if $\overline{L}^\ell = \mathbb{R}$.

Proposition 10 *A trace language $L \subseteq \mathbb{R}$ is a local liveness property if and only if it is locally dense.*
Proof Assume first that L is a local liveness property and let $t \in \mathbb{R}$. We have $\mathbb{P}\mathrm{ref}(t) \subseteq \mathbb{P} = \mathbb{P}\mathrm{ref}(L)$, hence $t \in \overline{L}^\ell$. Conversely, assume that L is locally dense. Let $r \in \mathbb{P} \subseteq \overline{L}^\ell$. We have $r \in \mathbb{P}\mathrm{ref}(r) \subseteq \mathbb{P}\mathrm{ref}(L)$. Hence $\mathbb{P} = \mathbb{P}\mathrm{ref}(L)$. ⊣

It follows by some purely formal argument that every language $L \subseteq \mathbb{R}$ is the intersection of a locally closed language and a locally dense one. Indeed, $\overline{L}^\ell$ is locally closed, $L \cup \mathbb{R} \setminus \overline{L}^\ell$ is locally dense and we have

$$L = \overline{L}^\ell \cap (L \cup \mathbb{R} \setminus \overline{L}^\ell).$$

We deduce that every trace language is the intersection of a *local safety property* with a *local liveness property*, which extends a classical result on words.

We turn now to the characterization of first-order definable local liveness properties by means of local temporal logic formulae. This will be based on the decomposition formulae introduced in Section 5.

Consider a first-order language $L \subseteq \mathbb{R}$ and a decomposition formula

$$\delta = \bigvee_{j \in J} a_j \wedge \psi_j \wedge \varphi_j \wedge \mathsf{CO}\,\gamma_j$$

for L as given by Theorem 3. In particular, with $\psi = \bigvee_{j \in J} a_j \wedge \psi_j$ we have

$$\mathbb{P}\mathrm{ref}(L) = \{r \in \mathbb{P} \mid r, \max(r) \models \psi\}$$

and we deduce that a partial execution $r \in \mathbb{P}$ is *live* with respect to L if $r, \max(r) \models \psi$. Note that this can be checked *locally* by a deterministic distributed automaton. More precisely, there is a deterministic asynchronous cellular automaton [5, 31] such that whenever executing an event e labeled a, the local process in charge of a knows whether the partial execution $r = \downarrow e$ with e as maximal event is live, i.e., satisfies the past formula ψ.

The second consequence is that the language L is a local liveness property if and only if the formula ψ is *valid*, i.e., if for all non empty trace $t \in \mathbb{R}$ and all $x \in t$ we have $t, x \models \psi$. Indeed, if ψ is valid then $r, \max(r) \models \psi$ for all $r \in \mathbb{P}$ and we deduce that $\mathbb{P}\mathrm{ref}(L) = \mathbb{P}$. Conversely, if L is a local liveness property then for all $t \in \mathbb{R} \setminus \{1\}$ and all $x \in t$ we must have $\downarrow x, x \models \psi$ since $\downarrow x$ is prime. We deduce that $t, x \models \psi$ since ψ is a past formula.

This motivates the following definition. A *canonical local liveness formula* is an F formula $\mathsf{F}\,\delta$ where

$$\delta = \bigvee_{j \in J} a_j \wedge \psi_j \wedge \varphi_j \wedge \mathsf{CO}\,\gamma_j$$

is a decompotion formula such that $\psi = \bigvee_{j \in J} a_j \wedge \psi_j$ is *valid* and $a_j \wedge \varphi_j \wedge \mathsf{CO}\,\gamma_j$ is satisfiable for all $j \in J$.

Proposition 11 *Let* $\mathsf{F}\,\delta$ *be a canonical local liveness formula. Then the language* $L = \mathcal{L}(\mathsf{F}\,\delta)$ *is a local liveness property.*

Proof We use the notations above for δ and ψ. Let $r \in \mathbb{P}$. Since ψ is valid we have $r, \max(r) \models a_j \wedge \psi_j$ for some $j \in J$. Now, $a_j \wedge \varphi_j \wedge \mathsf{CO}\,\gamma_j$ is satisfiable and we find $t \in \mathbb{R} \setminus \{1\}$ and $x \in t$ such that $t, x \models a_j \wedge \varphi_j \wedge \mathsf{CO}\,\gamma_j$. For clarity, we write s and u for the factors of t corresponding to $\|x$ and $\Uparrow x$. Let $t' = r \cdot s \cdot u$ and $y = \max(r) \in t'$. We know that $s \models \mathsf{CO}\,\gamma_j$ and $s = \|y$ in t'. Since φ_j is a future formula we deduce from $t, x \models a_j \wedge \varphi_j$ that $a_j u, y \models a_j \wedge \varphi_j$ and also $t', y \models a_j \wedge \varphi_j$. Finally, ψ_j being a past formula we get $t', y \models \psi_j$. Therefore, $t', y \models a_j \wedge \psi_j \wedge \varphi_j \wedge \mathsf{CO}\,\gamma_j$ and we obtain $t' \in L = \mathcal{L}(\mathsf{F}\,\delta)$. Hence, $r \in \mathbb{P}\mathrm{ref}(L)$ and we have shown $\mathbb{P} = \mathbb{P}\mathrm{ref}(L)$ as desired. ⊣

We have already explained above that, conversely, any first-order definable local liveness property $L \subseteq \mathbb{R} \setminus \{1\}$ can be described by a canonical local liveness formula. The following theorem is more precise.

Theorem 12 *Let* $L \subseteq \mathbb{R}$ *be a first-order definable real trace language and let*

$$\delta = \bigvee_{j \in J} a_j \wedge \psi_j \wedge \varphi_j \wedge \mathsf{CO}\,\gamma_j$$

be a decomposition formula for L given by Theorem 3. Let also $\psi = \bigvee_{j \in J} a_j \wedge \psi_j$. *Then we have*

(i) $\overline{L}^{\ell} = \mathcal{L}(\mathsf{G}\,\psi)$.
(ii) *If L is a local liveness property, then ψ is a valid formula and $L \setminus \{1\} = \mathcal{L}(\mathsf{F}\,\delta)$ is defined by a canonical local liveness formula.*
(iii) $\mathsf{F}(\neg\psi \vee \delta)$ *is a canonical local liveness formula.*
Hence $\widetilde{L} = \mathcal{L}(\mathsf{F}(\neg\psi \vee \delta))$ is a local liveness property.
Moreover, $\widetilde{L} = (L \setminus \{1\}) \cup (\mathbb{R} \setminus \overline{L}^{\ell})$ and $\widetilde{L}$ is the largest set K such that $L \setminus \{1\} = \overline{L}^{\ell} \cap K$.

Proof (i) This was already shown in the proof of Theorem 7(ii).

(ii) We have seen above that if L is a local liveness property then ψ is valid. We know that $L \setminus \{1\} = \mathcal{L}(\mathsf{F}\,\delta)$ by Theorem 3(ii).

(iii) Note that $\neg\psi = \bigvee_{a\in\Sigma} a \wedge \neg\psi \wedge \top \wedge \mathsf{CO}\,\mathsf{G}\,\top$, therefore, $\delta' = \neg\psi \vee \delta$ is a decomposition formula. Also, $\neg\psi \vee \psi$ is valid, hence $\mathsf{F}\,\delta'$ is a canonical local liveness formula. We deduce that $\widetilde{L}$ is a local liveness property by Proposition 11.

By Theorem 3 we have $\mathcal{L}(\mathsf{F}\,\delta) = L \setminus \{1\}$. Next, $\mathcal{L}(\mathsf{F}\,\neg\psi) = \mathbb{R} \setminus \overline{L}^{\ell}$ by (i). Therefore, $\widetilde{L} = \mathcal{L}(\mathsf{F}(\delta \vee \neg\psi)) = (L \setminus \{1\}) \cup (\mathbb{R} \setminus \overline{L}^{\ell})$. It follows that $\widetilde{L} \cap \overline{L}^{\ell} = L \setminus \{1\}$. Conversely, if $\overline{L}^{\ell} \cap K = L \setminus \{1\}$ then $K \subseteq (L \setminus \{1\}) \cup (\mathbb{R} \setminus \overline{L}^{\ell}) = \widetilde{L}$. ⊣

We introduce now a stronger notion of local liveness. We first motivate why a stronger notion may be interesting. If L is a local liveness property then each partial execution which is prime, i.e., corresponding to the local view of some process, may be extended to a behavior in L. But based on its local view a process does not know whether the current global execution can be extended to some behavior in L. For instance, if $\Sigma = \{a, b\}$ with $(a, b) \in I$, the language $L = \{a^{\omega}, b^{\omega}\}$ is a local liveness property. Assume that we have two processes, one executing a and the other executing b. After the execution of ab the local views of the two processes are a and b respectively. Based on its local view, each process may think that the current execution is live although it is not possible to extend it to some trace in L.

Alternatively, we may think that a computation is locally live with respect to some language L if each process has the possibility to locally initiate a computation reaching the language L whatever the current local states of concurrent processes are. This is a much stronger notion of local liveness that can be formalized as follows.

A language $L \subseteq \mathbb{R}$ is a *strong local liveness property* (SLLP) if it is a local liveness property (LLP) such that for all $t = raus \in \mathbb{R} \setminus \{1\}$ with $ra \in \mathbb{P}$, $a \in \Sigma$, $as \in \mathbb{R}^1$ and $\mathrm{alph}(u) \subseteq I(a)$ we have $raus \in L$ if and only if $ras \in L$. Intuitively, ra is the local view of some process, as is the computation that this process may initiate in order to reach L and $raus$ is a possible resulting behavior including u which is the part executed independently by the other processes. The additional condition makes sure that reaching L does not depend on what the concurrent processes may have already performed.

For instance, with $\Sigma = \{a, b\}$ and $(a, b) \in I$, the language $L = a^\omega b^\infty \cup a^\infty b^\omega$ is a strong local liveness property.

Note that if L is a strong local liveness property then it is also a global liveness property (GLP). Indeed, any nonempty finite trace may be written rau with $ra \in \mathbb{P}$, $a \in \Sigma$ and $\mathrm{alph}(u) \subseteq I(a)$. Since L is a local liveness property, we find $vs \in \mathbb{R}$ such that $ravs \in L$, $\mathrm{alph}(v) \subseteq I(a)$ and $as \in \mathbb{R}^1$. Since L is a SLLP we deduce that $ras \in L$ and then $raus \in L$. Therefore, any finite trace may be extended to some trace in L which is the definition of a GLP. But not all GLP are SLLP. Consider again $\Sigma = \{a, b\}$ with $(a, b) \in I$. If L is a SLLP over this dependence alphabet then $L \cap a^\infty \neq \emptyset$. Indeed, $a \in \mathbb{P}$ and L is a LLP hence we find $us \in \mathbb{R}$ with $aus \in L$, $as \in \mathbb{R}^1$ and $\mathrm{alph}(u) \subseteq I(a)$. Since L is a SLLP we deduce that $as \in L$. But since $as \in \mathbb{R}^1$ we get $\mathrm{alph}(as) = \{a\}$. Therefore, the singleton $\{(ab)^\omega\}$ is not a SLLP although it is a GLP. Recall also that any GLP is also a LLP but not all LLP are GLP:

$$\mathrm{SLLP} \subsetneq \mathrm{GLP} \subsetneq \mathrm{LLP}.$$

We conclude by giving a temporal logic characterization of first-order definable SLLP.

Theorem 13 *A language $L \subseteq \mathbb{R}$ is a first-order definable strong local liveness property if and only if there is a decomposition formula*

$$\delta = \bigvee_{j \in J} a_j \wedge \psi_j \wedge \varphi_j$$

with J finite, $\psi = \bigvee_{j \in J} a_j \wedge \psi_j$ valid, and for all $j \in J$, $a_j \in \Sigma$ is a letter, $\psi_j \in \mathrm{LocTL}_\Sigma(\mathsf{EY}, \mathsf{S})$ is a past formula, $\varphi_j \in \mathrm{LocTL}_\Sigma(\mathsf{EX}, \mathsf{U})$ is a future formula, $a_j \wedge \psi_j \wedge \varphi_j$ is satisfiable and such that

$$L \setminus \{1\} = \mathcal{L}(\mathsf{F}\,\delta) \qquad \textit{and} \qquad L \cup \{1\} = \mathcal{L}(\mathsf{G}\,\delta).$$

Proof Assume first that L is a first-order definable SLLP. The proof is similar to that of Theorem 3. Let $h : \mathbb{M}(\Sigma, D) \to S$ be a morphism to some finite aperiodic monoid S which recognizes L. We define the index set

$$J = \{(a, [r], [s]) \mid ras \in L \text{ with } ra \in \mathbb{P}, a \in \Sigma \text{ and } as \in \mathbb{R}^1\}$$

which is finite since there are finitely many h-equivalence classes. Fix now some index $j = (a_j, L_j^{\Downarrow}, L_j^{\Uparrow}) \in J$. By Theorem 1 and Corollary 2 we find a future formula φ_j and a past formula ψ_j such that

$$\begin{aligned} a_j \cdot L_j^{\Uparrow} \cap \mathbb{R}^1 &= \{s \in \mathbb{R}^1 \mid s, \min(s) \models \varphi_j\} \\ L_j^{\Downarrow} \cdot a_j \cap \mathbb{P} &= \{r \in \mathbb{P} \mid r, \max(r) \models \psi_j\}. \end{aligned}$$

We claim that the decomposition formula $\delta = \bigvee_{j \in J} a_j \wedge \psi_j \wedge \varphi_j$ satisfies the requirements.

By definition of J each formula $a_j \wedge \psi_j \wedge \varphi_j$ is satisfiable. Let now $t \in \mathbb{R} \setminus \{1\}$ and $x \in t$. We write r for the factor $\Downarrow x$ in t and we let a be the label of x. Then $ra \in \mathbb{P}$. Since L is a LLP we find $us \in \mathbb{R}$ such that $raus \in L$ with $\mathrm{alph}(u) \subseteq I(a)$ and $as \in \mathbb{R}^1$. We deduce that $ras \in L$ since this language is a SLLP. With $j = (a, [r], [s]) \in J$ we obtain $ra, \max(ra) \models a_j \wedge \psi_j$ and since ψ_j is a past formula it follows $t, x \models a_j \wedge \psi_j$. Therefore, ψ is valid.

Next, let $t \in L \setminus \{1\}$ and let $x \in t$. We write r, u and s the factors of t corresponding to $\Downarrow x$, $\|x$ and $\Uparrow x$ and let a be the label of x. We have $ra \in \mathbb{P}$, $\mathrm{alph}(u) \subseteq I(a)$ and $as \in \mathbb{R}^1$. Since L is a SLLP and $t = raus \in L$ we obtain $ras \in L$. Hence we can consider the index $j = (a, [r], [s]) \in J$. It is easy to check that $raus, x \models a_j \wedge \psi_j \wedge \varphi_j$. We deduce that $t \models \mathsf{G}\,\delta$ and $t \models \mathsf{F}\,\delta$. Therefore, $L \setminus \{1\} \subseteq \mathcal{L}(\mathsf{F}\,\delta)$ and $L \cup \{1\} \subseteq \mathcal{L}(\mathsf{G}\,\delta)$.

Conversely, assume that $t \models \mathsf{F}\,\delta$ or that $t \neq 1$ and $t \models \mathsf{G}\,\delta$. Let $x \in t$ and $j \in J$ be such that $t, x \models a_j \wedge \psi_j \wedge \varphi_j$. We write r, u and s for the factors $\Downarrow x$, $\|x$ and $\Uparrow x$ of t. Since ψ_j is a past formula and φ_j is a future formula, we deduce that $ra_j, x \models \psi_j$ and $a_j s, x \models \varphi_j$ and $\lambda(x) = a_j$. Let $r'as' \in L$ with $r'a \in \mathbb{P}$, $a = a_j$ and $as' \in \mathbb{R}^1$ be such that $j = (a, [r'], [s'])$. By definition of the formulae ψ_j and φ_j we have $r \in [r']$ and $s \in [s']$. We deduce that $ras \in [r'as']$ and since $r'as' \in L$ and h recognizes L we obtain $ras \in L$. Finally, since L is a SLLP we get $t = raus \in L$. Therefore, $L \setminus \{1\} \supseteq \mathcal{L}(\mathsf{F}\,\delta)$ and $L \cup \{1\} \supseteq \mathcal{L}(\mathsf{G}\,\delta)$.

We turn now to the proof of the "if" part of Theorem 13. So let δ be a decomposition formula with the notations and properties stated in the theorem. Since $\top = \mathsf{CO}\,\mathsf{G}\,\top$, the formula $\mathsf{F}\,\delta$ is a canonical local liveness formula and we deduce from Proposition 11 that $L \setminus \{1\} = \mathcal{L}(\mathsf{F}\,\delta)$ is a LLP. Hence, L is also a LLP.

Let now $t = raus \in \mathbb{R} \setminus \{1\}$ with $ra \in \mathbb{P}$, $a \in \Sigma$, $as \in \mathbb{R}^1$ and $\mathrm{alph}(u) \subseteq I(a)$. For all $j \in J$ we have $raus, \max(ra) \models a_j \wedge \psi_j \wedge \varphi_j$ if and only if $ras, \max(ra) \models a_j \wedge \psi_j \wedge \varphi_j$ since ψ_j is a past formula and φ_j is a future formula. Therefore, $raus \in L \subseteq \mathcal{L}(\mathsf{G}\,\delta)$ implies $ras \in \mathcal{L}(\mathsf{F}\,\delta) \subseteq L$ and conversely, $ras \in L \subseteq \mathcal{L}(\mathsf{G}\,\delta)$ implies $raus \in \mathcal{L}(\mathsf{F}\,\delta) \subseteq L$. ⊣

Remarks: We have seen two notions of local liveness: LLP and SLLP. We have chosen LLP as the standard notion since it is equivalent with local density. Note that, if we wish that every language is the intersection of a local safety property and a local liveness property then each locally dense language must be called locally live. Indeed, assume that $L = K_1 \cap K_2$ where L is locally dense and K_1 is a local safety property. Then $\mathbb{R} = \overline{L}^{\ell} \subseteq \overline{K}_1^{\ell} = K_1$ and we deduce $K_2 = L$.

References

[1] B. Alpern and F.B. Schneider. Defining liveness. *Information Processing Letters*, 21:181–185, 1985.

[2] B. Alpern and F.B. Schneider. Recognizing safety and liveness. *Distributed Computing*, 2:117–126, 1987.

[3] R. Alur, D. Peled, and W. Penczek. Model-checking of causality properties. In *Proc. of LICS'95*, pages 90–100. IEEE Computer Society Press, 1995.

[4] E. Chang, Z. Manna, and A. Pnueli. Characterization of temporal property classes. In W. Kuich, editor, *Proc. of ICALP'92*, number 623 in LNCS, pages 474–486. Springer Verlag, 1992.

[5] R. Cori, Y. Métivier, and W. Zielonka. Asynchronous mappings and asynchronous cellular automata. *Information and Computation*, 106:159–202, 1993.

[6] V. Diekert and P. Gastin. Local temporal logic is expressively complete for cograph dependence alphabets. In *Proc. of LPAR'01*, number 2250 in LNAI, pages 55–69. Springer Verlag, 2001.

[7] V. Diekert and P. Gastin. LTL is expressively complete for Mazurkiewicz traces. *Journal of Computer and System Sciences*, 64:396–418, 2002. A preliminary version appeared at ICALP'00, LNCS 1853, pages 211-222, Springer Verlag.

[8] V. Diekert and P. Gastin. Safety and liveness properties for real traces and a direct translation from LTL to monoids. In *Formal and Natural Computing — Essays Dedicated to Grzegorz Rozenberg*, number 2300 in LNCS, pages 26–38. Springer Verlag, 2002.

[9] V. Diekert and P. Gastin. Pure future local temporal logics are expressively complete for Mazurkiewicz traces. *Information and Computation*, 204:1597–1619, 2006. A preliminary version appeared at LATIN'04, LNCS 2976, pages 232–241, Springer Verlag.

[10] V. Diekert and G. Rozenberg, editors. *The Book of Traces*. World Scientific, Singapore, 1995.

[11] W. Ebinger. *Charakterisierung von Sprachklassen unendlicher Spuren durch Logiken*. Dissertation, Institut für Informatik, Universität Stuttgart, 1994.

[12] W. Ebinger and A. Muscholl. Logical definability on infinite traces. *Theoretical Computer Science*, 154:67–84, 1996.

[13] D. Gabbay, A. Pnueli, S. Shelah, and J. Stavi. On the temporal analysis of fairness. In *Proc. of PoPL'80*, pages 163–173, 1980.

[14] P. Gastin and D. Kuske. Satisfiability and model checking for MSO-definable temporal logics are in PSPACE. In *Proc. of CONCUR'03*, number 2761 in LNCS, pages 222–236. Springer Verlag, 2003.

[15] P. Gastin and D. Kuske. Uniform satisfiability in PSPACE for local temporal logics over Mazurkiewicz traces. *Fundamenta Informaticae*, 80(1-3):169–197, November 2007. A preliminary version appeared at CONCUR'05, LNCS 3653, pages 533–547, Springer Verlag.

[16] P. Gastin and A. Petit. Infinite traces. In V. Diekert and G. Rozenberg, editors, *The Book of Traces*, chapter 11, pages 393–486. World Scientific, Singapore, 1995.

[17] P.W. Hoogers, H.C.M. Kleijn, and P.S. Thiagarajan. A trace semantics for petri nets. In W. Kuich, editor, *Proc. of ICALP'92*, number 623 in LNCS, pages 595–604. Springer Verlag, 1992.

[18] S. Katz and D. Peled. Interleaving set temporal logic. *Theoretical Computer Science*, 75:21–43, 1991.

[19] Z. Manna and A. Pnueli. *The temporal logic of reactive and concurent systems: Specification*. Springer Verlag, 1992.

[20] A. Mazurkiewicz. Concurrent program schemes and their interpretations. DAIMI Rep. PB 78, Aarhus University, Aarhus, 1977.

[21] A. Mazurkiewicz. Traces, histories, graphs: Instances of a process monoid. In M.P. Chytil et al., editors, *Proc. of MFCS'84*, number 176 in LNCS, pages 115–133. Springer Verlag, 1984.

[22] M. Mukund and P.S. Thiagarajan. Linear time temporal logics over Mazurkiewicz traces. In *Proc. of MFCS'96*, number 1113 in LNCS, pages 62–92. Springer Verlag, 1996.

[23] W. Penczek. On undecidability of temporal logics on trace systems. *Information Processing Letters*, 43:147–153, 1992.

[24] D. Perrin and J.-E. Pin. *Infinite words*, volume 141 of *Pure and Applied Mathematics*. Elsevier, 2004.

[25] G. Rozenberg and P.S. Thiagarajan. Petri nets: Basic notions, structure and behaviour. In *Current Trends in Concurrency*, number 224 in LNCS, pages 585–668. Springer Verlag, 1986.

[26] B. Rozoy and P.S. Thiagarajan. Event structures and trace monoids. *Theoretical Computer Science*, 91(2):285–313, 1991.

[27] P.S. Thiagarajan. Elementary net systems. In W. Brauer, editor, *Petri nets: central models and their properties; advances in Petri nets Vol. 1*, number 254 in LNCS, pages 26–59. Springer Verlag, 1986.

[28] P.S. Thiagarajan and I. Walukiewicz. An expressively complete linear time temporal logic for Mazurkiewicz traces. In *Proc. of LICS'97*, pages 183–194, 1997.

[29] I. Walukiewicz. Difficult configurations - on the complexity of LTrL. In *Proc. of ICALP'98*, number 1443 in LNCS, pages 140–151. Springer Verlag, 1998.

[30] G. Winskel. Event structures. In W. Brauer, editor, *Petri nets: central models and their properties; advances in Petri nets Vol. 2*, number 255 in LNCS, pages 325–392. Springer Verlag, 1986.

[31] W. Zielonka. Notes on finite asynchronous automata. *R.A.I.R.O. — Informatique Théorique et Applications*, 21:99–135, 1987.

Static Deadlock Prevention in Dynamically Configured Communication Networks

Manuel Fähndrich[1], Sriram K. Rajamani[2] and Jakob Rehof[3]

[1]*Microsoft Research Redmond*
maf@microsoft.com

[2]*Microsoft Research India*
sriram@microsoft.com

[3]*Technische Universität Dortmund and Fraunhofer-ISST*
rehof@do.isst.fhg.de

Abstract

We propose a technique to avoid deadlocks in a system of communicating processes. Our network model is very general. It supports dynamic process and channel creation and the ability to send channel endpoints over channels, thereby allowing arbitrary dynamically configured networks.

Deadlocks happen in such networks if there is a cycle created by a set of channels, and processes along the cycle circularly wait for messages from each other. Our approach allows cycles of channels to be created, but avoids circular waiting by ensuring that for every cycle C, some process P breaks circular waits by selecting to communicate on both endpoints involved in the cycle C at P. We formalize this strategy as a calculus with a type system. Our type system keeps track of markers called *obstructions* where wait cycles are intended to be broken. Programmers annotate message types with design decisions on how obstructions are managed. Using these annotations, our type checker works modularly and independently on each process, without suffering any state space explosion.

We prove the soundness of the analysis (namely deadlock freedom) on a simple but realistic language that captures the essence of such communication networks. We also describe how the technique can be applied to a substantial example.

Keywords: Deadlock prevention, communicating systems, reconfigurable systems

1 Introduction

In this paper we study a problem in modular specification and analysis of concurrent, communicating processes with mobile channels, i.e., communication channels that can be passed in messages as formalized in, e.g., the pi-calculus [11]. Such systems are difficult to specify and analyze, because channel mobility allows programs to communicate in a dynamically evolving network topology.

Communicating systems based on message passing have been gaining practical importance over the past few years due to the emergence of web services, peer-to-peer algorithms, and in general by the desire to construct more loosely coupled and isolated systems. Our motivation for looking at message passing systems comes from working on the Singularity project [8, 7]. The Singularity project investigates ways to build reliable systems based on modern programming languages and sophisticated program analyses. The operating system is built almost entirely in an extension of C#. Processes in this system are strongly isolated (no shared memory) and communicate solely via message passing.

We wish to avoid deadlocks in such systems. One approach is to use model checking [3] to systematically explore the state space of the entire system and ensure absence of deadlocks, but the state spaces of such programs are infinite due to dynamic channel and process creation capabilities, preventing direct use of model checking. Even if we restrict attention to finite number of channels and processes, the state space is exponential, limiting scalability. Thus, we need a *modular* design and analysis technique to avoid deadlocks for such programs.

If we restrict our attention to systems with static communication topologies, and assume that we do not have cycles created by channels, then we can specify communication contracts at module boundaries, and check modularly if processes conform to their contracts (see for example, [5]).

However both the above assumptions (static topologies, and absence of cycles) are too restrictive. As we illustrate below, it is impossible to write operating system components such as name servers without allowing dynamic topologies, and without allowing cycles of channels. We present an approach that allows cycles of channels to be created in dynamic topologies, but avoids circular waiting by ensuring that for every cycle C, some process P breaks circular waits by selecting to communicate on on both endpoints involved in the cycle C at P. We formalize this strategy as a calculus with a type system. Our type system keeps track of markers called *obstructions* where wait cycles are intended to be broken. Programmers annotate message types with design decisions on how obstructions are managed. Using these

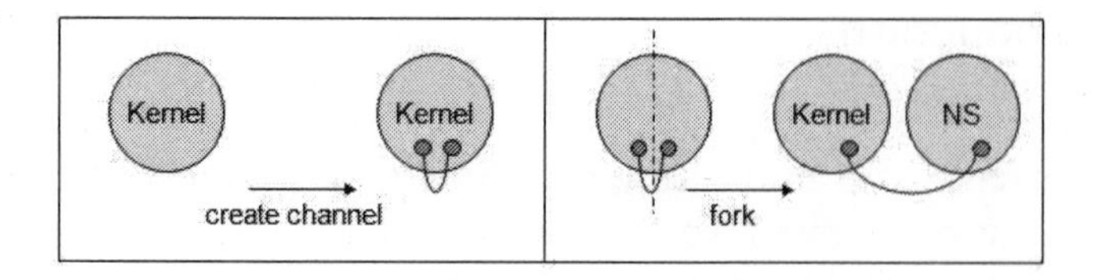

Figure 1: *Graphical illustration of channel creation and fork operations*

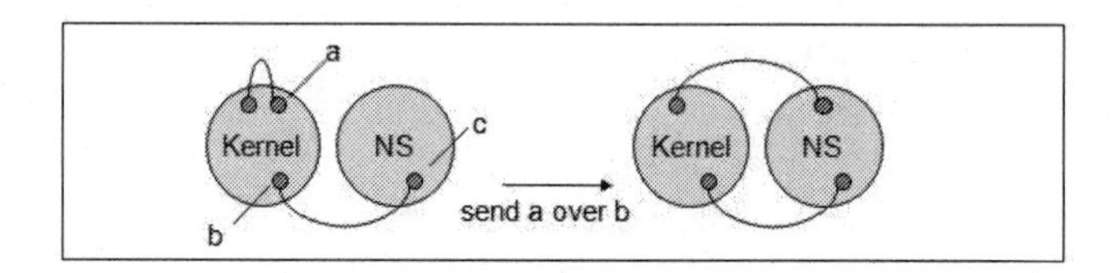

Figure 2: *Graphical illustration of endpoint send*

annotations, our type checker works modularly and independently on each process, without suffering any state space explosion.

We first provide an informal explanation of our communication model and how to prevent deadlocks using obstructions. In sections 2 and 3 we formalize a small programming language and its instrumented operational semantics. Section 4 presents our type system for static deadlock prevention, followed by a soundness theorem in Section 5. Section 6 describes an example to illustrate how obstructions and our type system can guide the design of a name service. Section 7 contains further discussion of the technical motivation behind our system and of future work. Sections 8 and 9 contain a discussion of related work and conclusions. Proofs of the main result are in the appendix.

1.1 Communication model

The communication model we are investigating in this paper is based on channels, where each channel consists of exactly two endpoints, and each endpoint is owned and operated on by at most one process at any time. This is in contrast to traditional π-calculus [11] style systems, where a shared name is all that is needed to communicate. Our endpoint-based approach is motivated by the fact that it is simpler for modular static verification and also corresponds more closely to actual implementations of message passing systems.

The communication model can be easiest understood using a graphical notation. A configuration consists of processes denoted by large circles and endpoints denoted by small circles. Channels are simply edges between endpoints. Endpoints appear nested within the boundary of the unique process that owns it.

There are three primary operations acting on a configuration: 1) channel creation, 2) process forking, and 3) sending of an endpoint over an existing channel. Other operations do not affect network topology and thus cycles and deadlocks, so we ignore them here. In Figure 1 on the left, we see a configuration consisting of a single process named Kernel. The Kernel process then creates a new channel, resulting in the second configuration, where the process owns both endpoints of the newly created channel. Observe that channel creation is a purely local operation.

On the right of Figure 1 we see how a process can fork (along the dotted line) resulting in a configuration with two distinct processes (here named Kernel and NS for name service). Note that a fork operation partitions the endpoints of the forking process so that each endpoint ends up in exactly one of the two resulting processes. Finally, Figure 2 shows how in the previous configuration, an endpoint a moves over the channel formed by endpoints (b,c). We consider moves to be atomic, corresponding to a synchronous communication model, where a send and corresponding receive operation synchronize. In Section 7, we briefly discuss how our approach extends to asynchronous communication models.

Through repeated applications of channel creation, forking, and sending of endpoints over channels, we can achieve arbitrary configurations of processes and endpoints. Figures 3 and 4 continue the progression of configurations from Figure 2, showing how the kernel can create processes with channels to the name server and how process P1 can obtain a direct channel to P2 by forwarding an endpoint of a locally created channel through the name service via two send operations.

1.2 Deadlock

The example configurations of the previous section make it clear that cycles in the process-channel graph arise both temporarily and sometimes as permanent parts of the configuration (e.g., the cycle P1, NS, P2). Thus, our strategy for avoiding deadlock is not to restrict networks to trees, or to restrict communication to a subset of tree edges of the graph. Instead, we allow arbitrary graphs, but track

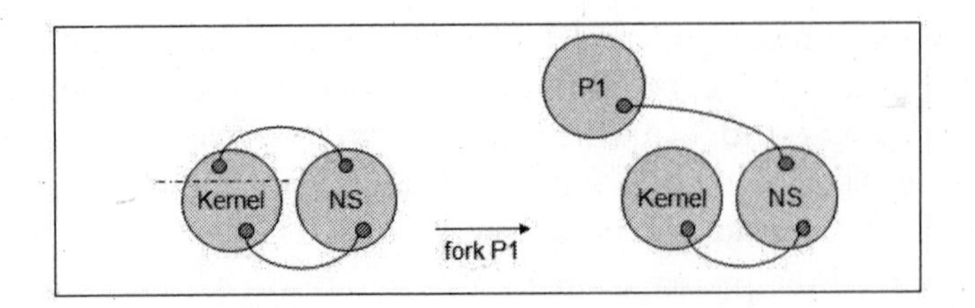

Figure 3: *Process creation with name service channel*

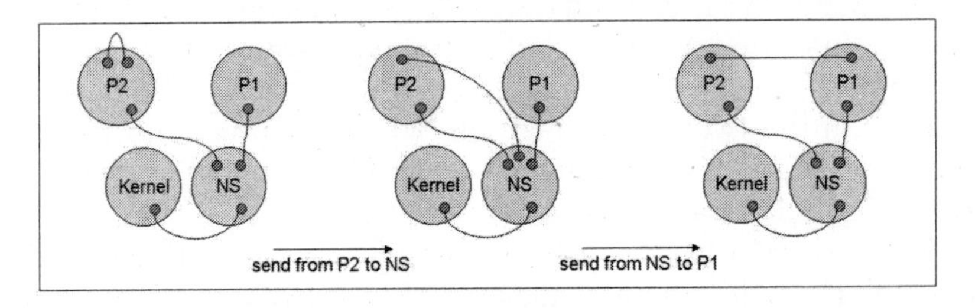

Figure 4: *Forwarding a channel through name server*

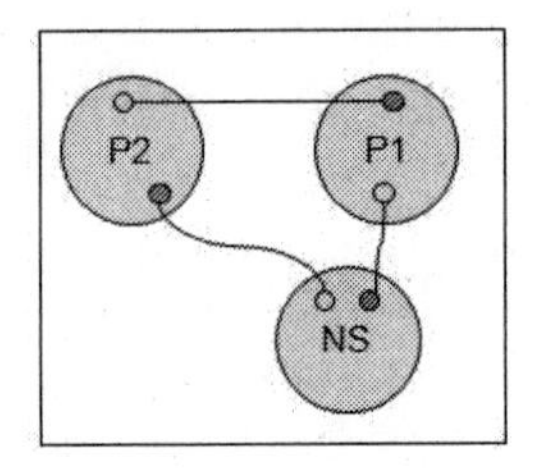

Figure 5: *A deadlock*

potential cycles and force processes to communicate in a way that prevents deadlock.

Definition 1 *[Path] A path is a sequence of segments $s_1..s_n$, where each segment s_i is of the form (a_i, p_i, b_i), and a_i, b_i are endpoints owned by process p_i, and each (b_i, a_{i+1}) forms a channel (for $i = 1..n-1$).*

We restrict our attention to primitive paths where no process (or endpoint) appears more than once.

Definition 2 *[Cycle] A cycle γ is a path $(a_1, p_1, b_1)...(a_n, p_n, b_n)$ where additionally, (b_n, a_1) forms a channel.*

A channel is enabled for communication if both its endpoints are selected for communication by their respective owning processes. A set of processes is deadlocked if each process p_i wants to communicate on a nonempty set S_i of endpoints it owns, but no channel is enabled for communication. In other words, a deadlock arises if there exists no channel (a, b), such that both a and b are selected by their respective owning processes, even though each process selects a nonempty set of endpoints to communicate.

Figure 5 shows a configuration with processes P1, P2, and the name service, where filled circles denote selected endpoints, and unfilled circles denote non-selected endpoints. No channel is enabled, since there is no channel with both endpoints being filled, even though each process has selected one endpoint. Therefore this configuration is deadlocked.

1.3 Obstructions

The main idea and contribution of the paper is to statically track potential cycles in the configuration and to prevent deadlock by forcing some process on each cycle to select enough endpoints for communication so as to guarantee the existence of an enabled channel.

We track potential cycles via *obstructions*. An obstruction is a marker in the form of a pair of endpoints (a, b) owned by the same process p. An obstruction summarizes the potential existence of a cycle containing segment (a, p, b). We say that an endpoint is obstructed if it participates in an obstruction.

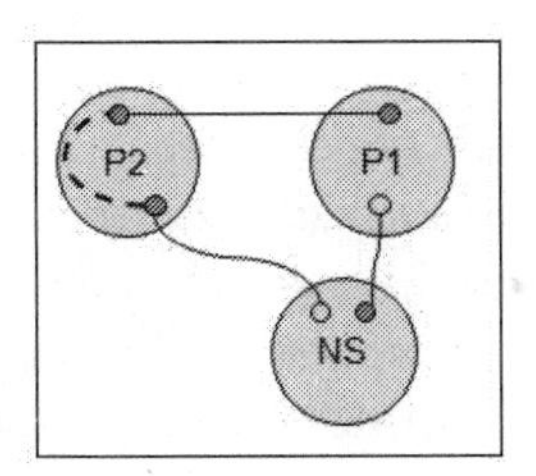

Figure 6: *P2 with obstruction and valid selection*

Figure 6 shows the same configuration as Figure 5, but where the cycle is covered by an obstruction maintained by process P2 between its two endpoints (depicted via a dashed edge). We can now prevent deadlocks by the following strategy for selecting endpoints:

Definition 3 *[Valid selection] A process p wanting to communicate has to select at least one unobstructed endpoint or at least one obstructed endpoint a and all obstruction peers $\{b \mid (a, b)$ is an obstruction in $p\}$.*

Process P2 in Figure 6 is thus forced by the obstruction on its endpoints to select both endpoints for communication, even if it wants to communicate only with NS. As a result, the channel between P1 and P2 is now enabled and the configuration can evolve a step.

We prove two main parts in this paper: 1) the operational rules for maintaining local obstructions (and the corresponding type rules) are sound, meaning that in every reachable configuration, all cycles are covered by an obstruction. 2) given that all potential cycles are covered by some obstruction, the obstruction observing selection strategy is sufficient to guarantee that processes don't deadlock.

1.4 Limitations

Though our type system is modular, it makes heavy use of programmer supplied annotations. We require programmers to annotate message protocols, we also require programmers to provide obstruction annotations (see Section 3) whenever endpoints are sent and received over other endpoints. We are aware that supplying these annotations requires a deep understanding of our strategy for propagating obstructions in the operational semantics. We leave it to future work to come up with automatically inferring these annotations, and decrease the annotation burden on the programmer.

1.5 Outline

The remainder of the paper makes these ideas more precise and provides the following contributions: 1) an operational semantics instrumented with obstructions, 2) a static type system for tracking obstructions and restricting processes to obstruction observing communications, and 3) proofs of soundness of the obstruction tracking and selection strategy.

2 Language

Figure 7 defines a small language. A program consists of a set of protocol state declarations, message sequence declarations, process declarations, and a starting process term P.

A protocol state specifies a set of possible message sequences. A state with an empty set is a terminating state signaling the end of the protocol. A message declaration $m : \tau^t \rightarrow \sigma$ specifies that message m carries an argument of type τ with obstruction tag t. Furthermore,

Program	::=	*StateDecl** *SeqenceDecl** *ProcDecl** P	*program*
StateDecl	::=	$State = \{m_1, m_2, \ldots, m_n\}$	*protocol state*
SeqenceDecl	::=	$m : \tau^t \rightarrow \sigma$	*sequence declaration*
τ	::=	**int** $\mid \sigma$	*parameter type*
σ	::=	$!State \mid ?State$	*protocol*
t	::=	$r \mid s$	*obstruction tag*
State	::=	*Name*	*state name*
m	::=	*Name*	*message name*
ProcDecl	::=	$PN(x_1 : \tau_1, \ldots, x_n : \tau_n) : O = P$	*process declaration*
P	::=	$(\texttt{new}\ x : \sigma, y); P \mid \texttt{fork}\, P_1, P_2$ $\mid PN(E_1, E_2, \ldots, E_n)$ $\mid \texttt{select} \sum_{i \in I} G_i \mid (\texttt{free}\, E); P$ $\mid \texttt{halt}$	*process*
O	::=	$\{(x_1, x_2), (x_3, x_4), \ldots\}$	*obstructions*
G	::=	$E?m[x].P$	*input guard*
	$\mid$	$E_1!m[E_2].P$	*output guard*
E	::=	$x \mid i \mid v$	*expressions*
PN	::=	*Name*	*process name*
x	::=	*Name*	*variable*
v	::=	*Name*	*value*
i	::=	integer	*integer*

Figure 7: *Process terms and types*

it specifies the continuation σ of the protocol for the sender of m. Thus, a message and a message sequence are isomorphic in our formalization. We use the notation next(m) to refer to the continuation protocol of m.

The parameter type τ specifies the message argument to be either a value of type **int** or a channel endpoint with the given protocol σ. The protocol σ of an endpoint describes the remaining interactions on the endpoint. A protocol σ is either !*State*, specifying a send of any message in set *State*, or ?*State*, specifying a receive of any message in set *State*.

By convention, the protocol continuation for any message is written from the perspective of the sender. The message sequences for the matching receiver is obtained by dualizing a protocol $\sim\sigma$ (turning sends into receives and vice-versa):

$$\begin{aligned} \sim !M &= ?M \\ \sim ?M &= !M \end{aligned}$$

Processes P can create new channels via `new` or `fork` into two separate processes. Processes recurse by invoking process definitions. Select offers a set of communication alternatives, each with its own continuation. Endpoints are explicitly freed via `free`, and `halt` allows a process to terminate. We sometimes write $m![b]$ as a shorthand for $\texttt{select}\, m![b]$, and similarly $m?[x]$ as a shorthand for $\texttt{select}\, m?[x]$.

Obstructions O are unordered pairs of endpoints. The obstruction set O on process definitions is used solely by the static type system and specifies under what obstruction assumptions to type P and applications of P.

The obstruction tag t on message parameter types specifies whether some obstructions known to the sender on the endpoint argument are maintained by the sender $t = s$, or passed to the receiver $t = r$ after the exchange. The need for these tags is explained in the operational semantics.

Our language has two binders: (1) `new` and (2) message receive. We say that names x and y are bound in the process $(\texttt{new}\ x : \sigma, y); P$. We say that name x is bound in the process $E?m[x].P$. A name that occurs in a process P, but is not bound in P is said to be *free* in P. For a process P, we use $\mathrm{fn}(P)$ to denote the set of all free names in P.

Given a process P, we use the notation $P[a/x]$ to denote the process obtained by substituting all free occurrences of x in P with a.

3 Operational Semantics

This section describes an operational semantics for processes that is instrumented to track the cycle obstructions our static type system reasons about. For each possible reduction, we show how the obstructions in the pre-state are related to the obstructions in the post-state. The semantics is presented as small step rewrite rules.

We use $x, y, z \in \mathscr{X}$ to range over bound variables, $u, v, w \in \mathscr{V}$ to range over values (or endpoints of channels), and $a, b, c \in \mathscr{X} \cup \mathscr{V}$ to range over either variables or values. The domain of values is needed to generate fresh names during application of the **new** operator while describing the operational semantics.

$\langle \text{peer}, O, (\texttt{new}\ a:\sigma, b); P \mid \Pi\rangle$ $\longrightarrow \langle \text{peer}', O', P \mid P_{\sim\sigma}(c,d) \mid \Pi\rangle$	$\text{peer}' = \text{peer} \cup \{(a,c),(c,a),(b,d),(d,b)\}$ $a,b,c,d \notin \text{dom}(\text{peer})$ $O' = O \cup \{(a,b)\}$
$\langle \text{peer}, O, \texttt{fork}\, P_1, P_2 \mid \Pi\rangle$ $\longrightarrow \langle \text{peer}, O', P_1 \mid P_2 \mid \Pi\rangle$ **requires** $\text{fn}(P_1) \cap \text{fn}(P_2) = \{\}$	$O' = (O \setminus O_{cut}) \cup O_{cl}$ $O_{cut} = \{(a,b) \mid a \in \text{fn}(P_1) \wedge b \in \text{fn}(P_2)$ $\wedge\, (a,b) \in O\}$ $O_{cl} = \{(b,d) \mid (a,b) \in O_{cut} \wedge (c,d) \in O_{cut}$ $\wedge\, b \neq d\}$
$\langle \text{peer}, O, \texttt{select}\ b!m[a].P_1 + S_1$ $\mid \texttt{select}\ c?m[x].P_2 + S_2 \mid \Pi\rangle$ $\longrightarrow \langle \text{peer}, O', P_1 \mid P_2[a/x] \mid \Pi\rangle$ **requires** $(a,b) \notin O$	$\text{peer}(b) = c \qquad m: \tau^t \to \sigma$ $O' = O \setminus a \cup O_{\text{sender}} \cup O_{\text{recvr}} \cup O_m$ $O_{\text{sender}} = \{(y,z) \mid (a,y) \in O \wedge (b,z) \in O\}$ $O_{\text{recvr}} = \{(a,z) \mid (c,z) \in O\}$ $O_m = \begin{cases} \{(c,a)\} & t = r \\ \{(b,y) \mid (a,y) \in O\} & t = s \end{cases}$
$\langle \text{peer}, O, p(a_1 \ldots a_n) \mid \Pi\rangle$ $\longrightarrow \langle \text{peer}, O, P[a_1/x_1 \ldots a_n/x_n] \mid \Pi\rangle$ **requires** a_i disjoint and $a_i \notin \text{fn}(P)$	$p(x_1 \ldots x_n) \triangleq P$
$\langle \text{peer}, O, (\texttt{free}\, a); P \mid \Pi\rangle$ $\longrightarrow \langle \text{peer}', O', P \mid \Pi\rangle$ **requires** $a \in \text{dom}(\text{peer})$	$\text{peer}' = \text{peer} \setminus a \qquad O' = O \setminus a$
$\langle \text{peer}, O, \texttt{halt} \mid \Pi\rangle \longrightarrow \langle \text{peer}, O, \Pi\rangle$	

Figure 8: *Small-step operational semantics*

3.1 Configurations

Machine configurations are tuples of the form $\langle \text{peer}, O, \Pi\rangle$. The function peer: $\mathcal{V} \to \mathcal{V}$ is a bijection that associates endpoints with their channel peer, i.e., if $\text{peer}(a) = b$, then (a,b) is a channel and $\text{peer}(b) = a$. The set $O \subseteq \mathcal{V} \times \mathcal{V}$ is the set of obstructions of the configuration. Finally, Π is a parallel composition of processes, $\Pi = P_1 \mid \ldots \mid P_n$, considered modulo reordering of its components.

We only consider configurations that satisfy the invariant that any free name occurring in $\Pi = P_1 \mid \ldots \mid P_n$ occurs in exactly one process P_i. We write $\text{proc}(a)$ to denote the process in which the free name a occurs. Furthermore, we are only interested in configurations where obstructions O are local to a process, i.e., an obstruction is between endpoints a, and b belonging to the same process.

$$\forall a,b.(a,b) \in O \implies \text{proc}(a) = \text{proc}(b) \qquad (1)$$

The motivation for this restriction is due to the fact that the type system performs a per process modular analysis and thus cannot keep track of obstructions if they are not local to a process. Naturally, the type system in the next section enforces these invariants.

3.2 Reduction rules

Figure 8 contains small-step reduction rules for our language. We write dom(R) to denote the domain of a binary relation R. If $O \subseteq \mathcal{V} \times \mathcal{V}$ is a set of pairs and $a \in \mathcal{V}$, then we write $O \setminus a$ for $O \setminus \{(a,b) \mid (a,b) \in O\}$.

If we ignore the treatment of obstructions, the reduction rules are essentially standard process reduction rules. The only complication arises in the creation of new channels. Instead of creating a direct channel between the two endpoints a and b of the `new` term, the operational semantics introduces a size one buffer process $P_{\sim\sigma}(c,d)$ (definition below) connected via channels (a,c) and (b,d). The buffer process executes message forwarding on c according to protocol $\sim\sigma$ and on d according to protocol σ. We chose to introduce a buffer process to prevent a process from trying to synchronize with itself. With buffer processes, we have the invariant that each process of the original program only communicates with buffer processes and vice versa. Thus, processes can never try to synchronize with themselves. In Figures, we continue to omit the buffer processes.

$$P_{?\{\}}(c,d) \triangleq \texttt{free}\, c; \texttt{free}\, d; \texttt{halt}$$
$$P_{!\{\}}(c,d) \triangleq \texttt{free}\, c; \texttt{free}\, d; \texttt{halt}$$
$$P_{?M}(c,d) \triangleq \texttt{select} \textstyle\sum_{m \in M} c?m[x].d!m[x].P_{\sim \text{next}(m)}(c,d)$$
$$P_{!M}(c,d) \triangleq \texttt{select} \textstyle\sum_{m \in M} d?m[x].c!m[x].P_{\text{next}(m)}(c,d)$$

The next section provides insight into how the treatment of obstructions by the reduction rules provides strong enough invariants to prevent deadlock.

3.3 Properties

We first relate configurations C to the graphs shown in the introduction.

Definition 4 *[Induced graph] The induced graph $G(C)$ of a configuration*

$$C = \langle \text{peer}; O; P_1 \mid \ldots \mid P_n \rangle$$

is the graph (N,E), where the set of vertices N consists of a vertex per process and a vertex per endpoint, and the set of edges consists of an edge per channel $(a, \text{peer}(a))$, and an edge linking each endpoint a to its owning process proc(a).

Recall that the purpose of obstructions is to characterize all cycles in a configuration. The next definition makes this precise.

Definition 5 *[Cycle coverage] Given an induced graph G of configuration* $\langle \text{peer}; O; P_1 \mid \ldots \mid P_n \rangle$, *we say that the* configuration is cycle covered, *if for each cycle* γ *in G, there exists a segment* $(a, i, b) \in \gamma$*, such that* $(a, b) \in O$.

We now inspect some of the reduction rules of Figure 8 in more detail to see how they transform a configuration that is cycle covered into a new configuration that is also cycle covered, while maintaining only local obstructions.

Note that every initial configuration is of the form $\langle \emptyset; \emptyset; P \rangle$, where P contains no free names. Initial configurations are thus cycle free and consequently cycle ccvered.

Rule `new` is the base case for creating a cycle. The new channels and the buffer process form a cycle containing two segments $(a, P, b)(d, P_{\sim\sigma}, c)$. In order to cover this cycle in the resulting configuration, obstruction (a, b) is added. Every other cycle in the resulting configuration exists in the prior configuration and is thus covered.

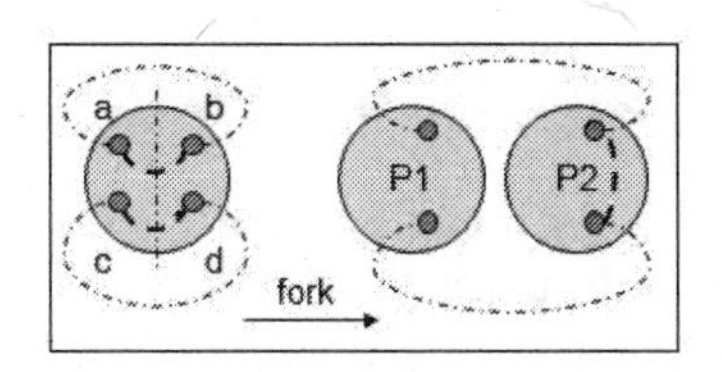

Figure 9: *Obstructions prior and after fork*

Rule `fork` has to consider what happens to the obstructions that are no longer local to a process after the split, namely those obstructions (a, b), where P_1 retains a and P_2 retains b. Figure 9 shows the situation for two such pairs of endpoints, (a, b) and (c, d). The dashed lines inside processes represent obstructions, the dot-dashed lines outside processes represent the potential path completing the cycle the obstruction covers. We see that after the fork, the two potential cycles have been turned into a larger single cycle. To cover it, one of the two processes needs to have an obstruction between its endpoints. The reduction rule for `fork` chooses (b, d) in P_2 as shown in the picture. In general, cutting n obstructions via `fork`, results in $\frac{n \cdot (n-1)}{2}$ new cycles and thus obstructions, one between every pair of paths connecting the two processes.

The most complicated case is the reduction for `select`. As can be seen from the rule in Figure 8, the obstructions O' after the move of endpoint a can be characterized by four groups. First, $O \setminus a$ are the obstructions not changed by the move. All obstructions in which a is involved are removed because they are no longer local to the sending process when a changes owner. The groups O_{sender} and O_{recvr} are closure rules that add obstructions in the sender and receiver necessary in the worst case to cover new potential cycles. To gain insight into these, we invite the reader to inspect the proof in the appendix. Here, we will focus on the last component, O_m.

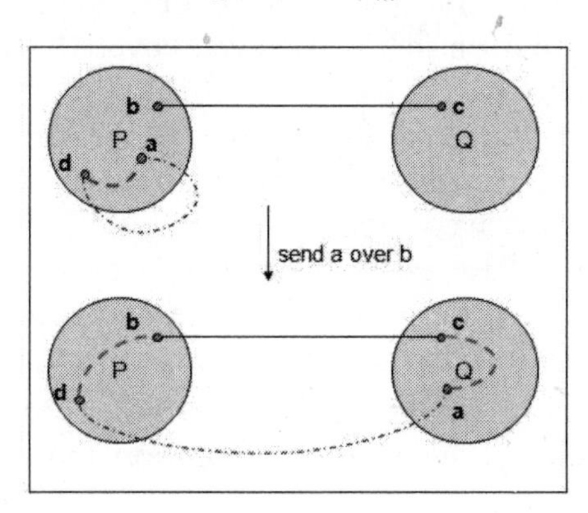

Figure 10: *Obstructions O_m for send*

Consider the top configuration in Figure 10. We are about to send a over b. It so happens that (a,d) is an obstruction, and that there might thus be a path from a to d in the graph. Consider that same path after the move (bottom of figure). We observe that we have added the channel edge (b,c) to the cycle. Since (a,d) is no longer a valid obstruction, we must find a new obstruction that covers this cycle. We now have two options to cover this cycle: either by obstruction (d,b) maintained by the sender, or (c,a) by the receiver. Both are highlighted in the figure, but only one is needed. The choice between these to possibilities is under programmer control. Message signatures contain an obstruction tag $t \in r,s$ on the message argument type. This tag provides the contract between the sender and receiver as to who witnesses the new cycle via an obstruction. If the tag is s, only obstruction (b,d) is added; if the tag is r, only obstruction (c,a) is added. In both cases, the cycle in question is covered.

3.4 Stuck configurations

A configuration C is stuck, if there does not exist C', such that $C \longrightarrow C'$ by one of the reduction rules. We group stuck configurations into two groups: *locally stuck* processes and *globally stuck* configurations. Locally stuck processes are stuck independently of other processes, whereas globally stuck configurations are stuck due to the interaction of processes. By inspection of the rules, we find that processes can get locally stuck due to the side-conditions in the reduction rules that check non-sharing of endpoints between processes and rule out sending an endpoint over a channel with which it is obstructed.

A configuration is globally stuck, if no process is locally stuck, and the configuration isn't terminal (no more processes). In that case, all processes are of the form `select`. Configurations are globally stuck on `select` for the following reasons:

1. Deadlock: no two processes want to communicate on the same channel
2. Wrong message: there exists a channel with both sides ready to communicate, but the sides do not agree on a common message.

The deadlock case naturally includes the situation where a channel is orphaned, namely one side deletes its endpoint or terminates, but the other still tries to communicate. Note that due to our channel buffer processes, it is not possible for a process to get stuck with itself.

While the type system presented in the next section guarantees stuck-freedom from all of the above, the formal treatment in this paper shows that our novel use of obstructions avoids deadlock. The remaining cases are treated using established techniques, such as linear typing of endpoints.

4 Obstruction Type System

The type system validates judgments of the form

$$E;O \vdash P$$

meaning that, in endpoint environment E, and obstructions O, the process term P is well typed.

We use M,N to denote state names and use them interchangeably with the message sets they denote. We use δ as a meta-variable for either ! or ?. Environments

$$E ::= x{:}\tau \mid E,E \mid \cdot$$

map names to types τ, which can be either base type (here just **int**), or an endpoint in a particular protocol state σ. In the latter case, σ

describes the remaining conversation on the endpoint. In order to track endpoint ownership by processes in the type system, we treat environments using a linear typing discipline. Every endpoint in the environment is owned. In principle, non-endpoint names do not need to be treated linearly. To avoid making the formalisms more complex however, we focus mostly on endpoints and thus all values are treated linearly. In practice, the environment would be split into a linear part for endpoints, and a non-linear part for other values. Obstructions

$$O ::= (x,y) \mid O,O \mid \cdot$$

correspond to the obstructions in the operational semantics, with a subtle difference that in the case of judgments, obstructions are kept in terms of variables in the process instead of endpoint values as in the operational semantics.

Figure 11 shows the type rule for each process term. It is interesting to note that the type system models the operational semantics precisely, except that it abstracts away the channel edges and actual values generated at runtime. In other words, the type system does not contain any information about which endpoints are peers, and the type system does not keep track of fresh values generated by the **new** operator.

The notation $ep(G)$ represents the endpoint and message on which G communicates, i.e., $ep(a?m[x].P) = (a,m)$ and $ep(a!m[b].P) = (a,m)$. We use $dom(R)$ for the domain of a binary relation R and $R(a)$ the range of R w.r.t. a. This notation is used in the T-SELECT rule.

We briefly explain each rule. The `halt` process is well-typed only in the empty environment. This guarantees that processes do not halt while holding some endpoint on which the peer expects more communication. Freeing an endpoint requires that the endpoint is owned and has no conversation left (type $\delta\{\}$). Note that by rules T-HALT and T-FREE the type system thus enforces that all endpoints are freed explicitly.

The modularity of the analysis is most explicit in the rule T-FORK. After the fork, the analysis of the two sub-processes is completely independent and governed only by the concise endpoint types and message declarations that programmers can write. Rule T-FORK splits the environment into E_1 and E_2, thereby guaranteeing a partitioning of the endpoints owned by the two resulting processes. Similarly, we partition the obstructions into O_1, O_2, and O_{cut}, where O_i are all obstructions on endpoints in E_i $(i = 1,2)$. After the fork, obstructions O_{cut} would no longer correspond to local obstructions, since they relate endpoints of distinct processes. They are thus removed from the

$$\cdot;\cdot \vdash \texttt{halt} \qquad \text{[T-HALT]}$$

$$\frac{E, x{:}\sigma, y{:}{\sim}\sigma; O, (x,y) \vdash P}{E; O \vdash (\texttt{new}\, x{:}\sigma, y).P} \qquad \text{[T-NEW]}$$

$$\frac{\begin{array}{c} m \in M \quad m : \tau^t \to \sigma \\ O_{\text{recvr}} = \{(b,x) \mid (b,a) \in O\} \\ O_m = \begin{cases} \emptyset & \text{if } t = s \\ \{(a,x)\} & \text{if } t = r \end{cases} \\ E, x{:}\tau, a{:}{\sim}\sigma; O \cup O_{\text{recvr}} \cup O_m \vdash P \end{array}}{E, a{:}?M; O \vdash a?m[x].P} \qquad \text{[T-INP]}$$

$$\frac{\begin{array}{c} m \in M \quad m : \tau^t \to \sigma \quad (a,b) \notin O \\ O_{\text{sender}} = \{(y,z) \mid (a,y) \in O \wedge (b,z) \in O\} \\ O_m = \begin{cases} \{(b,y) \mid (a,y) \in O\} & \text{if } t = s \\ \emptyset & \text{if } t = r \end{cases} \\ E, b{:}\sigma; O \setminus a \cup O_{\text{sender}} \cup O_m \vdash P \end{array}}{E, a{:}\tau, b{:}!M; O \vdash b!m[a].P} \qquad \text{[T-OUT]}$$

$$\frac{\begin{array}{c} \delta \in \{!, ?\} \\ E; O \setminus a \vdash P \end{array}}{E, a{:}\delta\{\}; O \vdash \texttt{free}\, a.P} \qquad \text{[T-FREE]}$$

$$\frac{\begin{array}{c} O_{\text{cut}} \subseteq \text{fn}(E1) \times \text{fn}(E2) \\ O_{\text{cl}} = \{(b,d) \mid (a,b) \in O_{\text{cut}} \wedge (c,d) \in O_{\text{cut}} \wedge b \neq d\} \\ E_1; O_1 \vdash P_1 \quad \text{fn}(O_1) \subseteq \text{fn}(E_1) \\ E_2; O_2, O_{\text{cl}} \vdash P_2 \quad \text{fn}(O_2) \subseteq \text{fn}(E_2) \end{array}}{E_1, E_2; O_1, O_2, O_{\text{cut}} \vdash \texttt{fork}\, P_1, P_2} \qquad \text{[T-FORK]}$$

$$\frac{\begin{array}{c} R = \bigcup_i ep(G_i) \quad S \subseteq dom(R) \\ ValidSelection(S, O) \\ \forall a \in S.\ Exhaustive(R(a), E(a)) \\ E; O \vdash G_i \quad (i = 1 \ldots n) \end{array}}{E; O \vdash \texttt{select} \sum_i G_i} \qquad \text{[T-SELECT]}$$

$$\begin{array}{l} ValidSelection(S, O) = \exists a \in S.\ a \notin O \vee \forall b.(a,b) \in O \implies b \in S \\ Exhaustive(N, !M) = true \ \text{ if } N \subseteq M \\ Exhaustive(N, ?M) = true \ \text{ if } \mathrm{N} = \mathrm{M} \end{array}$$

$$\frac{\begin{array}{c} p(x_1{:}\tau_1..x_n{:}\tau_n) : O \triangleq P \\ x_1{:}\tau_1, .., x_n{:}\tau_n; O \vdash P \\ O' \subseteq O[a_i/x_i] \end{array}}{a_1{:}\tau_1, .., a_n{:}\tau_n; O' \vdash p(a_1..a_n)} \qquad \text{[T-INVOKE]}$$

Figure 11: *Obstruction type rules*

obstruction set. Instead, we compute a new set of obstructions O_{cl} for process P_2 covering all resulting cycles (see also Figure 9). The choice of obstructions O_{cl} in the type rule is just one of many possible designs. Alternative type rules could equally well compute a set of obstructions for P_1, or distribute obstructions among P_1 and P_2. We have not yet investigated the need for such alternatives.

Rule T-SELECT checks that the process communicates on a valid selection of endpoints S that is obstruction observing. The *ValidSelection* predicate formalizes Definition 3. For each endpoint a in S, the rule then enforces that receives are exhaustive, namely that the process is ready to receive all possible messages that the type of the endpoint (that is, the corresponding message state) specifies. For sends, it suffices to choose any subset of possible messages that the type of the endpoint specifies. Sends in message sequence types correspond thus to internal choice, receives to external choice.

Rule T-INP checks that m is a valid message to receive given the type $?M$ of the receiving endpoint a. This is checked by ensuring that there is a message $m : \tau^t \to \sigma$ in the message state M. Since the message sequence for state M is written (by convention) from the perspective of the sender, and endpoint a is a receiver with type $?M$, we have to complement the polarity of σ to obtain $\sim\sigma$ and type the remaining process P where endpoint a has type $\sim\sigma$. We also check that the actual argument x has type τ as specified by the message signature. Obstructions consist of prior obstructions O plus obstructions O_{recvr} required on the receiver side between the received endpoint x and obstruction peers of the receiving endpoint a. In addition, if the message obstruction tag $t = r$, then the receiver keeps track of the obstruction (a, x) between the received and receiving endpoints.

Rule T-OUT checks that m is part of M, the messages that can be sent on b. The remaining process P is typed in the environment where b has type σ corresponding to the remaining conversation after sending m. Endpoint a is no longer in the environment, since ownership of a has passed to the owner of the peer of b. Obstructions consist of prior obstructions O, albeit without any obstructions mentioning a, followed by necessary obstructions O_{sender} between obstruction peers of a and b. If the obstruction tag $t = s$, then the sender additionally turns obstructions on a into obstructions on b (see also Figure 10).

The condition $(a, b) \notin O$ on T-OUT enforces that we can never send a over b, if a is obstructed with b. Were we to permit such an operation, it would have the effect of shortening a potential cycle in the graph by the one segment that contains the obstruction, but without being able to add the obstruction to any remaining segments. The receiving process would have to assume that the received endpoint is obstructed with all other endpoints owned by the process. We chose to simply disallow such sends rather than adding this conservative assumption.

5 Soundness

We present a soundness theorem establishing that if a configuration is typable by the obstruction typing rules then it cannot deadlock.

Recall that all cycles considered in this paper are primitive. We do not need to consider non-primitive cycles, since every dead-lock configuration exhibits a primitive wait cycle.

Definition 6 *[Valid configuration] A configuration* $\langle \text{peer}; O_1..O_n; P_1 \mid \ldots \mid P_n \rangle$ *is valid if*

1. *there exist environments* $E_1..E_n$*, such that* $E_i; O_i \vdash P_i$
2. *for any channel* (a, b)*, such that* $\text{peer}(a) = b$*,* $a \in E_i$ *and* $b \in E_j$*, the endpoint states agree, i.e., if* $E_i(a) = !M$*, then* $E_j(b) = ?M$*, and if* $E_i(a) = ?M$*, then* $E_j(b) = !M$*.*
3. *the configuration is cycle covered (Definition 5)*

Lemma 7 *[Preservation] Given a valid configuration* C_1 *and a reduction step* $C_1 \to C_2$*, configuration* C_2 *is valid.*
Proof By case analysis over the possible redexes of a configuration. Preservation of points 1, and 2 of definition 6 is straight-forward. Appendix A.1 contains the graph theoretic argument for preservation of cycle coverage.

The following lemma states that given a valid configuration where all processes want to communicate, there exist two processes connected by a channel and both processes are ready to exchange a message over that channel.

Lemma 8 *[Progress] Given a valid configuration* C_1*, then either* C_1 *is final and of the form* $\langle \emptyset; \emptyset; \emptyset \rangle$*, or there exists a configuration* C_2*, such that* $C_1 \to C_2$*.*
Proof By case analysis over stuck configurations. Appendix A.2 contains the proof.

Theorem 9 *[Soundness] Every closed program* P *that is well typed* $\cdot; \cdot \vdash P$ *either reduces to the final configuration, or it reduces ad-infinitum.*
Proof By induction over the reduction steps and Lemmas 7 and 8.

6 Name Server Example

A driving motivator for the present work is the design of a realistic name service for our research operating system. The name service process is a natural convergence point of many channels. It is also common to have cycles involving the name server as shown in the final configuration of Figure 4.

```
CLIENT = {NewClient, Bind}
ACKCLIENT = {AckClient}
NewClient : (?CLIENT)^r →?ACKCLIENT
Bind      : (?SERVICE)^r →?ACKCLIENT
AckClient : void → ?CLIENT

REGISTRAR = {NewReg, Register}
NewReg    : (?REGISTRAR)^s →?{AckReg}
Register  : (?SERVICEPROVIDER)^s →?{AckReg}
AckReg    : void → ?REGISTRAR

SERVICEPROVIDER = {Connect}
Connect   : (?SERVICE)^r →?{AckConnect}
AckConnect: void → ?SERVICEPROVIDER
```

Figure 12: *Contracts for the name server endpoints*

In designing the name service, one could attempt to have the name service handle all obstructions to cover such cycles and be responsive enough so that other clients need not deal with obstructions. It turns out, though, that such a strategy is only partially feasible. Keeping all obstructions on the name server leads to a situation where all its client endpoints are obstructed with all service provider endpoints. This is not a desirable situation, since it prevents the name server from forwarding connection endpoints it receives over a client channel to a service, due to the fact that the connection endpoint will be obstructed with all services upon receipt.

Instead, we use the design shown graphically in Figure 14. Figure 12 shows the contract definitions for the end points of the name server and Figure 13 contains the server code itself.

The name server maintains four sets of end points, called **clients, ackclients, registrants,** and **serviceproviders**. Sets are not part of our core language but are straight-forward to implement. The sets collect end points that share the same type (message state). Endpoints can be added to sets and selections can involve sets of endpoints. For example, the guard `case e.NewClient?( newclient ) in clients` is triggered by a **NewClient** message from any member of the set **clients**. The particular endpoint `e` on which the message is received is bound in the `case` block and removed from the set. We use the type **void** for message arguments when the message does not carry any value although this is not part of the formalism.

The main function of the name server is to bind service endpoints to service providers. The name server supports this on endpoints in the `client` set. These endpoints have type `?CLIENT`, specifying the

```
1  set<?CLIENT> clients ;
2  set<!ACKCLIENT> ackclients;
3  set<?REGISTRAR> registrants ;
4  set<?SERVICEPROVIDER> serviceproviders;
5  ...
6  /*  initialize  clients , ackclients , registrants and service providers */
7  ...
8  while(true) {
9    select {
10     case e.NewClient?( newclient ) in clients :
11       // e, newclient , clients , and ackclients are mutually obstructed
12        clients .Add( newclient );
13        ackclients .Add(e );
14       break;
15
16     case e.AckClient !() in ackclients :
17       // e, clients , and ackclients are mutually obstructed
18        clients .Add(e );
19       break;
20
21     case e.Bind ?( service ) in clients :
22       // e, service , clients , and ackclients are mutually obstructed
23        ackclients .Add(e);
24       // service , clients , and ackclients are mutually obstructed
25        serviceprovider = GetServiceProvider ( serviceproviders , ...);
26       // serviceprovider is not obstructed with any endpoints
27        serviceprovider .Connect!( service );
28       // clients , and ackclients are mutually obstructed
29        serviceprovider .AckConnect ?();
30        serviceproviders .Add( serviceprovider );
31       break;
32
33     case e.NewReg?(newreg) in registrants :
34       // neither e nor newreg is obstructed with any other endpoints
35       e.AckReg !(); // can send AckReg on e alone
36        registrants .Add(newreg);
37        registrants .Add(e );
38       break;
39
40     case e. Register ?( serviceprovider ) in registrants :
41       // neither e nor serviceprovider is obstructed
42       e.AckReg !(); // can send AckReg on e alone
43        serviceproviders .Add( serviceprovider );
44        registrants .Add(e );
45       break;
46   }
47 }
```

Figure 13: *Name server code*

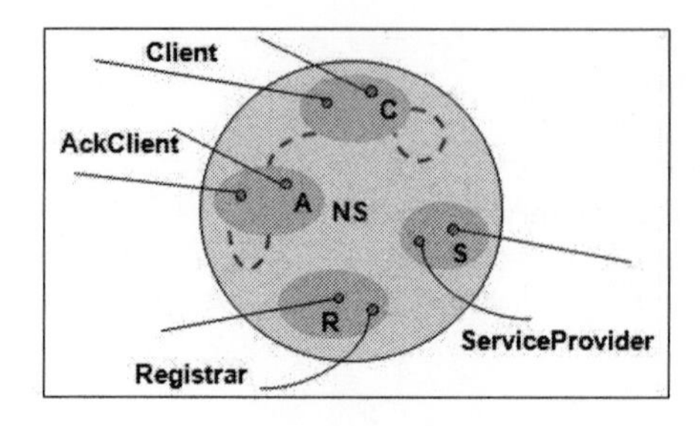

Figure 14: *Stable configuration of nameserver*

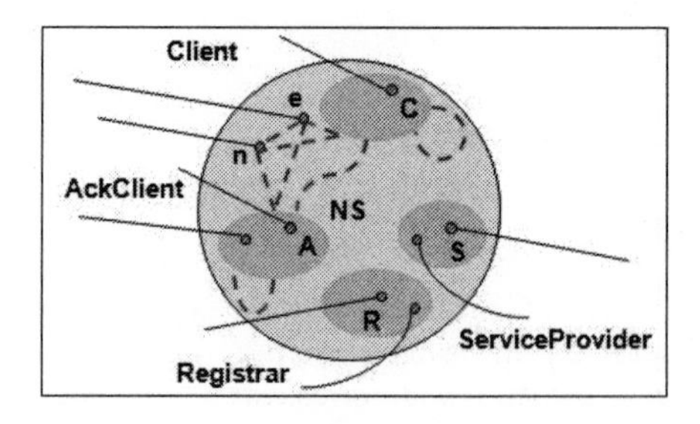

Figure 15: *Nameserver after receiving client request*

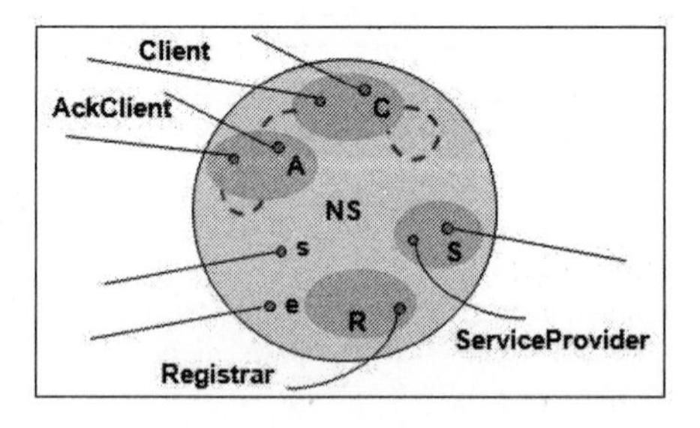

Figure 16: *Nameserver after receiving registrar request*

arrival of either message NewClient to add a new client, or message Bind to bind a service endpoint to a service provider. Both messages specify that obstructions on the message argument are handled by the receiver (the r superscript). Thus, processes can obtain duplicate name service channels or bind to service providers without being encumbered by obstructions on their side.

Registrations of service providers happen on a different set of endpoints stored in the registrars set. These endpoints have type

?REGISTRAR and the name server receives either message NewReg to add a new registrar client, or message Register to register a new service provider. In contrast to the messages on the clients set, the messages received from registrants are tagged with superscript s, making it part of the channel contract that obstructions on these message arguments are kept on the sender.

Service providers are kept in set serviceproviders and have type ?SERVICEPROVIDER. They expect a single message Connect which passes a service endpoint to the service provider. The final set ackclients will be explained below.

Figure 14 shows the four sets as shaded areas labeled C, R, S, and A. The figure shows that our design allows endpoints in clients and ackclients to be mutually obstructed, but registrar or service provider endpoints are not obstructed with any other endpoint.

Consider the first case of the **select** statement on line 10 in the name server implementation. It handles NewClient messages carrying a new endpoint newClient arriving on some endpoint e in set clients. Because the message declaration is tagged with r, endpoints e and newClient are obstructed after the receive. Additionally, due to the receiver closure rule, newClient is also obstructed with all obstruction peers of e, namely all endpoints in sets clients and ackclients. This is graphically shown in Figure 15 where newClient is abbreviated with n. Next, the newClient endpoint is added to the clients set (it has the same type and obstructions as all other endpoints in that set). Now the nameserver needs to send an acknowledgment message on e as prescribed by the message declaration. However, offering to send only on e at this point is not allowed by the obstruction rules (not a valid selection), since e is obstructed with all other endpoints in sets clients and ackclients. In order to offer the acknowledgment on e, we also have to simultaneously accept messages on all client endpoints. We achieve this by storing e in set ackclients and re-executing the select statement. The set ackclients thus contains all client endpoints whose peers may be expecting the AckClient message. This is handled by the case on line 16.

After receipt of Bind on line 21, the obstruction situation is analogous to the previous case shown in Figure 15 but where n represents the service endpoint. The service endpoint is forwarded to some service provider from the service provider set. Since the service provider is not obstructed, we can send the Connect message and receive the subsequent AckConnect without having to consider offering to handle messages on other endpoints. No obstructions from the service endpoint are retained by the name server, since the Connect message specifies that the receiver (the service provider) handles obstructions.

Thus, after adding e to the ackclient set and the serviceprovider back into the serviceproviders set, we have again reached the stable situation in Figure 14 and can re-execute the loop.

Figure 16 shows the obstruction state that holds after receiving a Register message on endpoint e with service endpoint argument s (line 39). In this case, no obstructions are present among these endpoints. Thus, we can send the acknowledgment on e directly without the need for a set like ackclients . The stable state is reached again by adding endpoint e back to the set registrars , and adding s to set serviceproviders .

6.1 Implementation

We have implemented a name service according to the design described in Section 6, as part of the Singularity operating system [8, 7]. The name service provides a uniform name space for all services on a system, including device drivers, network connections and file system. Singularity is written in an extension to C# which includes message-passing primitives and contract specifications as discussed in this paper. Our notion of obstructions was motivated by our experiences in programming Singularity and will be implemented as an extension to the current type system.

7 Discussion and Future Work

The type system proposed in this paper has two aspects, message signatures and obstructions. Message signatures can be used to encode simple contracts (stateful communication protocols, somewhat in the style of session types [6]) specifying temporal ordering of types of messages sent and received along a pair of endpoints. In this paper, we wish to focus on the use of obstructions and have deliberately chosen a very simple contract format, in which a specification τ has a natural inverse, $\sim\tau$. In principle, contracts over message types could be much more expressive. For example, channel contracts used in [9, 2] are CCS processes.

Local and modular reasoning about deadlock in the presence of channel passing is a fundamental challenge. The present work arose partly out of an attempt to apply the theory of conformance developed in [5] to programs with dynamic communication topology. Conformance theory [5] allows us to reason compositionally about deadlock and unhandled message types arising from misuse of contracts, but the theory is restricted to CCS (static topology). Type systems in the style of [9, 2] can be used to extract CCS models from π-calculus expressions, which could in turn be processed by a model checker. Hence, by composing the theory of conformance with such type systems we could hope to achieve a modular system for typing deadlock-free components with very rich contracts and dynamic topology. It remains an important challenge for future work to devise a system in which a theory of contract conformance such as [5] is integrated with the obstruction discipline.

In this paper we have chosen to factor the deadlock problem into two orthogonal subproblems, namely, deadlock problems arising due to cyclic structures created dynamically by channel passing, and deadlock problems arising from misuse of contracts. Our basic rationale has been that we want contracts to be local to a channel (pair of endpoints). In contrast, the systems [9, 2] admit cross-channel contracts that can express constraints on messaging actions that take place on multiple channels. The drawback with such contracts from a pragmatic perspective is that (i) they may need to span arbitrarily large parts of a system in practice, conflicting with the desire for modularity, and (ii) it is difficult, in general, to find a natural place to state them in programs. It is possible that cross-channel contracts will be needed in practice to accommodate more flexible programming and a wider span of communication topologies, in which actions on multiple channels are correlated in ways that are not allowed by our type system. However, even so, we feel that it is valuable to have a basic, simpler system that works well for channel-local contract specifications, and it has been the goal of this paper to provide such a system.

7.1 Asynchronous communication

If we extend the buffer processes we introduce on our channels to buffer arbitrary numbers of messages instead of just one, we obtain an encoding for an asynchronous communication model of the programmer specified processes. Since the asynchrony is simulated by a synchronous system, the deadlock prevention result of this paper extends to asynchronous systems as well.

8 Related Work

The most novel aspect of our system is the use of obstructions to prevent global deadlock caused by dynamically created communication cycles by enforcing only local conditions (observation of the obstruction rule), while still allowing cycles in the communication topology to occur during computation.

In the context of the π-calculus [11], type systems have been proposed for analyzing and controlling the communication structure of mobile processes. Some of these systems capture partial orderings between communication actions and can be used to reason about deadlock. However, we have found no system that reduces the problem of global deadlock detection to locally checkable conditions on processes, and we have not been able to find a concept similar to that of obstructions in the literature. Moreover, our system controls the use of individual channel endpoints rather than π-calculus channels.

Our notion of contracts bears resemblance to session types [6]. Yoshida's graph types [12] are abstractions of the communication structure of a process in which nodes represent communication actions and edges represent synchronization orderings between actions. Kobayashi et.al. [9, 10] further develop process types to reason about deadlock via partial ordering constraints extracted from communication actions. Further comparisons between our system and the π-calculus type systems [9, 2] have been given in Section 7.

Flanagan and Abadi's type system [4] prevents deadlock due to locking in programs written in Java-like languages. The system is based on the specification of lock orderings that rule out cyclic wait conditions on locks.

Interaction categories [1] have been used to define type systems that enforce sufficient local control of communication to ensure deadlock-freedom. However, the type discipline does not in general handle global communication cycles.

9 Conclusion

We have developed a novel way to modularly characterize potential cycles in communication networks through endpoint obstruction pairs along with a strategy for communicating that is informed by such obstructions. Together, these techniques provably guarantee that a system of processes will not deadlock. We have used the notion of obstructions as a design guide for implementing a name service in the Singularity operating system, and we will extend our contract type checker to include obstructions.

Acknowledgments

We are indebted to Bill McCloskey for his initial ideas on tainted endpoints.

References

[1] S. Abramsky, S. Gay, and R. Nagarajan. A type-theoretic approach to deadlock-freedom of asynchronous systems. In *Theoretical Aspects of Computer Software, LNCS Vol. 1281*, pages 295–320, 1997.

[2] S. Chaki, S. K. Rajamani, and J. Rehof. Types as models: Model checking message-passing programs. In *POPL 02: ACM Principles of Programming Languages*. ACM, 2002.

[3] E.M. Clarke, O. Grumberg, and D. Peled. *Model Checking*. MIT Press, 1999.

[4] C. Flanagan and M. Abadi. Types for safe locking. In *ESOP 99: European Symposium on Programming*, LNCS 1576, pages 91–108. Springer-Verlag, 1999.

[5] C. Fournet, C.A.R. Hoare, S.K. Rajamani, and J. Rehof. Stuck-free conformance. In *CAV 04: Computer-Aided Verification*, LNCS. Springer-Verlag, July 2004.

[6] S. Gay and V. T. Vasconcelos. Session types for inter-process communication. Technical Report 2003–133, Department of Computing, University of Glasgow, 2003.

[7] G. Hunt, J. Larus, M. Abadi, M. Aiken, P. Barham, M. Fähndrich, C. Hawblitzel, O. Hodson, S. Levi, N. Murphy, B. Steensgaard, D. Tarditi, T. Wobber, and B. Zill. An overview of the Singularity project. Technical Report MSR-TR-2005-135, Microsoft Research, 2005.

[8] G. C. Hunt, J. R. Larus, D. Tarditi, and T. Wobber. Broad new OS research: challenges and opportunities. In *Proceedings of Tenth Workshop on Hot Topics in Operating Systems*. USENIX, June 2005.

[9] A. Igarashi and N. Kobayashi. A generic type system for the Pi-calculus. In *POPL 01: Principles of Programming Languages*, pages 128–141. ACM, 2001.

[10] N. Kobayashi. A type system for lock-free processes. *Information and Computation*, 177:122–159, 2002.

[11] R. Milner. *Communicating and Mobile Systems: the* π*-Calculus*. Cambridge University Press, 1999.

[12] N. Yoshida. Graph types for monadic mobile processes. In *FSTTCS: Software Technology and Theoretical Computer Science*, LNCS 1180, pages 371–387. Springer-Verlag, 1996.

Appendix

A Proofs

A.1 Lemma 7 (Preservation)

We prove that `new`, `fork`, and `select` reductions preserve cycle coverage. The other reductions are trivial. Assume $C_1 = \langle \text{peer}_1; O_1; \Pi_1 \rangle$

and $C_2 = \langle \text{peer}_2; O_2; \Pi_2 \rangle$

Case
$\Pi_1 = (\texttt{new}\, a{:}\sigma, b); P \mid \Pi$

Recall that channel creation creates a buffer process $P_{\sim\sigma}(c,d)$ as defined in Section 3 in order to avoid having any process communicate with itself. Thus $\Pi_2 = P \mid P_{\sim\sigma}(c,d) \mid \Pi$. Let γ be a cycle in C_2. If γ does not contain any of a,b,c,d, then γ exists also in C_1 and is therefore covered. Otherwise, $\gamma = (a,p,b)(d,q,c)$, where q is the buffer process created for the channel. Since $(a,b) \in O_2$, the cycle is covered.

Case
$\Pi_1 = \texttt{fork}\, P_1, P_2 \mid \Pi$

Let r be the process doing the fork, and let $p = P_1$ and $q = P_2$ be the resulting processes. Let $A = \text{fn}(P_1)$ and $B = \text{fn}(P_2)$. Let γ be a cycle in C_2. We proceed with three cases: 1) $\neg\exists a \in A$ and $a \in \gamma$, 2) $\neg\exists b \in B$ and $b \in \gamma$, and 3) γ contains an endpoint of A and one of B.
Case 1: We have cycle $\gamma' = \gamma[r/q]$ in C_1 and an obstruction $(c,d) \in O_1$, where $c,d \notin A$. Thus $(c,d) \in O_2$, since the only obstructions removed by the step contain an endpoint from A.
Case 2: as case 1.
Case 3: $a,b \in \gamma$ **where** $a \in A$ **and** $b \in B$. Assume γ is not covered in C_2, otherwise we are done. Without loss of generality, $\gamma = (a,p,x)\gamma_1(b,q,y)\gamma_2$. We have cycle $\gamma_1(b,r,x)$ in C_1 and by assumption it must be covered by $(b,x) \in O_{\text{cut}} \subseteq O_1$. Similarly, we have cycle $\gamma_2(a,r,y)$ in C_1 covered by $(a,y) \in O_{\text{cut}} \subseteq O_1$. Thus, $(b,y) \in O_{\text{cl}} \subseteq O_2$ covering γ and contradicting our assumption.

Case
$\Pi_1 = \texttt{select}\, b!m[a].P_1 + S_1 \mid \texttt{select}\, c?m[x].P_2 + S_2 \mid \Pi$

Note that $\text{peer}_1 = \text{peer}_2$. The move involves moving an endpoint a from a process p to a process q over a channel consisting of endpoints b and c such that $b = \text{peer}(c)$, $\text{proc}(b) = p$, and $\text{proc}(c) = q$. Consider any cycle γ in C_2. We have three cases (1) endpoint a is not in γ, (2) a is in γ and c is in γ, and (3) a is in γ and c is not in γ. We prove the theorem for these 3 cases.
Case 1: a **is not in** γ. The only edge change between C_1 and C_2 is that we remove (a,p) and add (a,q). Since a is not on γ, γ exists in C_1 and is therefore covered by some obstruction in O_1 not involving a. The same obstruction is still present in O_2
Case 2: a **is in** γ **and** c **is in** γ Let γ be of the form $\gamma_1(a,q,c)(b,p,e)$, where γ_1 starts with $(\text{peer}_1(e),..)$. We have two sub cases. If γ_1 is

covered by an obstruction in C_2 we are done. Otherwise, γ_1 does not have an obstruction in C_2. We know that p cannot occur in γ_1, due to the fact that the cycle is primitive (recall that each process is traversed at most once in each cycle). Thus γ_1 is not covered by any obstructions in C_1. Thus, $(e,a) \in O_1$, since $\gamma_1,(a,p,e)$ is a cycle in C_1. Assume $m : \tau^t \to \sigma$. By the rewrite step we know that if $t = s$ then $(e,b) \in O_m \subseteq O_2$ covering γ. Otherwise $t = r$ and $(c,a) \in O_m \subseteq O_2$ covering γ.

Case 3: a is in γ and c is not in γ. Let γ be of the form $\gamma_1(f,q,a)$ in C_2, where γ_1 starts with $(\text{peer}_1(a),..)$. If sub-path γ_1 has an obstruction in C_2 we are done. Suppose γ_1 does not have any obstructions in C_2. Again we have two further sub cases. If p is not in the path γ_1, then we know that the sub-path γ_1 in C_1 does not have any obstructions either. However, since $\gamma_1(f,q,c)(b,p,a)$ is a cycle in C_1, we have that it should be covered by some obstruction. This obstruction cannot be (a,b) since that would preclude endpoint a from being sent over b. Thus, $(c,f) \in O_1$, and consequently, $(a,f) \in O_{\text{recvr}} \subseteq O_2$ and we are done. For the second sub case, we have that p is in the path γ_1. Then $\gamma_1 = \gamma_2(e,p,g)\gamma_3$, where $\text{proc}(e) = \text{proc}(g) = p$. Note that we therefore have cycles γ_2 and $(b,p,q)\gamma_3(f,q,c)$ in C_1 and therefore $(a,e) \in O_1$, since γ_2 is obstruction free. Then, we do another (final) case split. Either $(c,f) \in O_1$, which implies that $(a,f) \in O_{\text{recvr}} \subseteq O_2$ and we are done. Or, $(c,f) \notin O_1$, which implies that $(b,g) \in O_1$ (due to cycle $(b,p,g)\gamma_3(f,q,c)$ in C_1 and the fact that (b,g) is the only possible obstruction to cover it in C_1, since γ_3 has no obstructions due to the assumptions). Since $(b,g) \in O_1$ and $(a,e) \in O_1$ we have that $(g,e) \in O_{\text{sender}} \subseteq O_2$ in C_2, which contradicts our earlier assumption that γ_1 does not have any obstructions in C_2, and we are done. $\dashv$

A.2 Lemma 8 (Progress)

By inspection of the reduction rules and our classification from Section 3, we see that all processes must be of the form `select`... in order for the machine configuration to be globally stuck in a deadlock. Let $C = \langle\text{peer}; O; \Pi\rangle$ be such a configuration.

From the typing of `select`, we know that each process p_i exhibits a set S_i of selected endpoints that satisfy

$$\exists a \in S_i . a \notin O \vee \forall (a,b) \in O . b \in S_i$$

Let w_i be the existential witness a showing that S_i is a valid selection.

We prove the lemma in 2 steps. First we prove that there exist distinct processes p_i and p_j and an enabled channel (a, b), such that $\text{proc}(a) = p_i$ and $\text{proc}(b) = p_j$, and $\text{peer}(a) = b$. Second, we prove that p_i and p_j agree on a message to be exchanged.

We prove the first part by contradiction. Assume there is no enabled channel. Pick an arbitrary process p_0 and the witness w_0 of its valid selection S_0. Let $b_0 = \text{peer}(w_0)$ and $p_1 = \text{proc}(b_0)$. By channel construction (buffer processes), we know that $p_1 \neq p_0$ and by our assumption, we know that $b_0 \notin S_1$, otherwise we are done. Pick w_1, the witness of S_1 and continue this strategy. Since there are finitely many processes, we must at some point exhibit a cycle γ of processes and endpoints, where each segment has the form (b_i, p_i, w_i) and $w_i \in S_i$ but $b_i \notin S_i$. Since the configuration is cycle covered, we know that there exists a process p_j on γ and an obstruction (b_j, w_j), where b_j, w_j in γ. Since, w_j is the witness of the valid selection S_j, it must be the case that $b_j \in S_i$ which contradicts our assumption and exhibits an enabled channel $(b_j, \text{peer}(b_j))$, since $\text{peer}(b_j)$ was a witness and thus selected.

Let (a, b) be the enabled channel and $p = \text{proc}(a)$. The second part is trivial, since we know that a has type σ and b has type $\sim\sigma$ and both are part of their processes selection, it must be the case that one of them has type $?M$. Without loss of generality, assume $\sigma = ?M$. In that case, process p could only pass type checking if it is exhaustive w.r.t. M. Since the peer process also type checked, it must offer at least one message of m (rule T-OUT). Thus the reduction can take place. ⊣

Reachability and Boundedness in Time-Constrained MSC Graphs*

Paul Gastin[1], **Madhavan Mukund**[2] **and K Narayan Kumar**[2]
[1]*LSV, ENS Cachan, France*
Paul.Gastin@lsv.ens-cachan.fr
[2]*Chennai Mathematical Institute, Siruseri, India*
{madhavan,kumar}@cmi.ac.in

Abstract

Channel boundedness is a necessary condition for a message-passing system to exhibit regular, finite-state behaviour at the global level. For Message Sequence Graphs (MSGs), the most basic form of High-level Message Sequence Charts (HMSCs), channel boundedness can be characterized in terms of structural conditions on the underlying graph. We consider MSGs enriched with timing constraints between events. These constraints restrict the global behaviour and can impose channel boundedness even when it is not guaranteed by the graph structure of the MSG. We show that we can use MSGs with timing constraints to simulate computations of a two-counter machine. As a consequence, even the more fundamental problem of reachability, which is trivial for untimed MSGs, becomes undecidable when we add timing constraints. Different forms of channel boundedness also then turn out to be undecidable, using reductions from the reachability problem.

Keywords: Communicating systems, message sequence charts, timed specifications

1 Introduction

In a distributed system, several agents interact to generate a global behaviour. This interaction is usually specified in terms of scenarios, using message sequence charts (MSCs) [11]. A message sequence graph (MSG) is a finite directed graph with nodes labelled by MSCs. MSGs

*Partially supported by *Timed-DISCOVERI*, a project under the Indo-French Networking Programme.

are the most basic versions of High-level Message Sequence Charts (HMSCs) [12] and are a convenient mechanism for generating possibly infinite collections of MSCs.

Communicating finite-state machines (CFMs) are a natural implementation model for message-passing systems. In recent years, there has been a considerable body of work on the analysis of message-passing systems specified in terms of MSCs and communicating finite-state machines [4, 5, 8, 9, 15, 16]. A fruitful approach is to synthesize CFMs from MSC specifications and then use standard automata-theoretic techniques for formal verification.

One essential requirement for effective synthesis of CFMs from MSCs is channel boundedness. An MSC specification is said to be universally bounded if there is a uniform upper bound on the size of all channels along any execution consistent with the specification. A specification is existentially bounded if every computation can be scheduled in at least one way so that a uniform channel bound is maintained. Algorithms to synthesize CFMs from MSC specifications were originally obtained for universally bounded specifications [9] and later extended to the existentially bounded case [8].

MSG specifications are always existentially bounded. We also have a precise characterization of universal boundedness for MSGs [14]. Interestingly, the channel boundedness problem is known to be undecidable for CFMs [6], so the limited expressiveness of MSGs with respect to CFMs is responsible for making the problem decidable in the setting of MSGs.

In this paper, we consider the analysis of message-passing systems equipped with timing constraints. The basic MSC notation does not have any provision for describing explicit real-time constraints. On the other hand, timing is an important issue in practical specifications—for instance, how long should a server wait before deciding to drop an idle connection with a client?

Time-constrained MSCs (TC-MSCs) are an extension of the MSC notation in which we can specify timing constraints between pairs of events. If we label the nodes in an MSG with TC-MSCs, we obtain a time-constrained MSG (TC-MSG). We can regard each node in a TC-MSG as one phase of a protocol. To allow us to describe timing constraints in making the transition from one phase to another, we also permit time constraints along the edges between nodes in a TC-MSG.

On the automaton front, each component in a CFM can be replaced by a timed automaton [3] yielding a natural timed extension of the CFM model. Some progress has been made in extending the analysis of MSC specifications vis-a-vis CFMs to the timed setting [2, 7].

Our focus in this paper is the boundedness problem for TC-MSGs. Since the boundedness problem is already undecidable for untimed CFMs, it is clear that it is also undecidable for timed CFMs. Somewhat surprisingly, it turns out that channel boundedness is undecidable for timed CFMs even if there is no communication loop by which a sender gets feedback from the recipient of a message [10]. However, as in the untimed case, TC-MSGs are less expressive than timed CFMs, so these undecidability results cannot be transported directly to the TC-MSG setting.

Of course, if the underlying untimed MSG is universally bounded, so will any TC-MSG derived from it by adding timing constraints. However, it is also possible that timing constraints enforce universal boundedness even if the underlying untimed MSG does not satisfy the criterion described in [14].

On the other hand, even though untimed MSG specifications are always existentially bounded, it is not difficult to construct TC-MSGs in which the timing constraints do not guarantee existential boundedness. This is because timing constraints may prevent us from choosing the schedule required in the untimed case to guarantee a uniform bound.

Our main results are negative. We show that various variants of the boundedness problem are undecidable for TC-MSGs, even when we impose severe restrictions on the manner in which timing constraints can be used. The main technique that we use to demonstrate undecidability is a simulation of two counter machines [13] using TC-MSGs.

Our simulation makes crucial use of timing contraints across the edges of a TC-MSG. We believe that the boundedness problem is decidable for TC-MSG specifications without edge constraints. We have a sufficient condition for boundedness in this case, based on an analysis of a time-constrained producer-consumer system. However, decidability of boundedness in this case remains open.

The paper is organized as follows. We begin with some preliminaries about (timed) MSCs and MSGs. Section 3 formally describes the various versions of the boundedness problem that we look at in this paper. To show that boundedness is undecidable, in general, for TC-MSGs, we first establish that reachability is undecidable, in Section 4. We then show how to reduce reachability to boundedness. In Section 6, we strengthen our undecidability results to the setting where constraints can only be described using open intervals. In the next section, we show that we can obtain undecidability even with bounded channels. Finally, in Section 8 we show that we can restrict all edge constraints to refer to a single process and still establish undecidability. In Section 9 we obtain partial results concerning the decidability of boundedness for TC-MSGs without edge constraints. The paper concludes with a brief discussion.

2 Preliminaries

2.1 Message sequence charts

Let $\mathcal{P}$ be a finite set of processes that communicate using a finite set of message types $\mathcal{M}$ over reliable FIFO channels. For $p \in \mathcal{P}$, let $Act_p = \{p!q(m), p?q(m) \mid p \neq q \in \mathcal{P}, m \in \mathcal{M}\}$ be the set of communication actions for p. The actions $p!q(m)$ and $p?q(m)$ are read as *p sends m to q* and *p receives m from q*, respectively. Let $Act = \bigcup_{p \in \mathcal{P}} Act_p$.

Labelled posets An *Act*-labelled poset is a structure $M = (E, \leq, \lambda)$ where $(E, \leq)$ is a poset and $\lambda : E \to Act$ is a labelling function.

For $e \in E$, let $\downarrow e = \{e' \mid e' \leq e\}$. For $X \subseteq E$, $\downarrow X = \cup_{e \in X} \downarrow e$. For $p \in \mathcal{P}$ and $a \in Act$, we set $E_p = \{e \mid \lambda(e) \in Act_p\}$ and $E_a = \{e \mid \lambda(e) = a\}$, respectively.

Let $Ch = \{(p,q) \mid p \neq q\}$ denote the set of *channels*. For each $(p,q) \in Ch$, we define a relation $<_{pq}$ as follows, to capture the fact that channels are FIFO with respect to each message.

$$e <_{pq} e' \stackrel{\Delta}{=} \begin{array}{l} \lambda(e) = p!q(m), \\ \lambda(e') = q?p(m) \text{ and} \\ |\downarrow e \cap E_{p!q(m)}| = |\downarrow e' \cap E_{q?p(m)}| \end{array}$$

Finally, for each $p \in \mathcal{P}$, we define the relation $\leq_{pp} = (E_p \times E_p) \cap \leq$, with $<_{pp}$ standing for the largest irreflexive subset of $\leq_{pp}$.

Definition 1 *An MSC (over $\mathcal{P}$) is a finite Act-labelled poset $M = (E, \leq, \lambda)$ that satisfies the following conditions.*

1. *Each relation $\leq_{pp}$ is a linear order.*
2. *If $p \neq q$ then for each $m \in \mathcal{M}$, $|E_{p!q(m)}| = |E_{q?p(m)}|$.*
3. *If $e <_{pq} e'$, then $|\downarrow e \cap \left(\bigcup_{m \in \mathcal{M}} E_{p!q(m)}\right)| = |\downarrow e' \cap \left(\bigcup_{m \in \mathcal{M}} E_{q?p(m)}\right)|$.*
4. *The partial order $\leq$ is the reflexive, transitive closure of the relation $\bigcup_{p,q \in \mathcal{P}} <_{pq}$.*

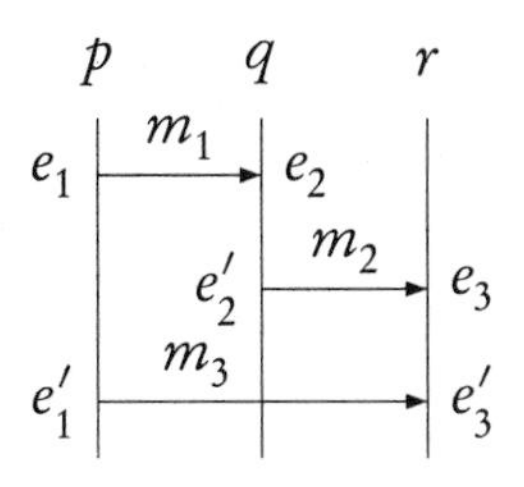

Figure 1: *An MSC*

The second condition ensures that every message sent along a channel is received. The third condition says that every channel is FIFO across all messages.

In diagrams, the events of an MSC are presented in *visual order.* The events of each process are arranged in a vertical line and messages are displayed as horizontal or downward-sloping directed edges. Fig. 1 shows an example with three processes $\{p, q, r\}$ and six events $\{e_1, e'_1, e_2, e'_2, e_3, e'_3\}$ corresponding to three messages—m_1 from p to q, m_2 from q to r and m_3 from p to r.

For an MSC $M = (E, \leq, \lambda)$, we let $\text{lin}(M) = \{\lambda(\pi) \mid \pi \text{ is a linearization of } (E, \leq)\}$. For instance, $p!q(m_1)q?p(m_1)q!r(m_2)p!r(m_3)$ $r?q(m_2)r?p(m_3)$ is one linearization of the MSC in Fig. 1.

MSC languages An *MSC language* is a set of MSCs. An MSC language $\mathscr{L}$ can also be seen as a word language L over *Act* corresponding to the linearizations of the MSCs in $\mathscr{L}$. For an MSC language $\mathscr{L}$, we set $\text{lin}(\mathscr{L}) = \bigcup\{\text{lin}(M) \mid M \in \mathscr{L}\}$.

Definition 2 *An MSC language $\mathscr{L}$ is said to be a* regular MSC language *if the word language* $\text{lin}(\mathscr{L})$ *is a regular language over Act.*

Let M be an MSC and $B \in \mathbb{N}$. We say that $w \in \text{lin}(M)$ is B-bounded if for every prefix v of w and for every channel $(p, q) \in Ch$, $\sum_{m \in \mathscr{M}} |\pi_{p!q(m)}(v)| - \sum_{m \in \mathscr{M}} |\pi_{q?p(m)}(v)| \leq B$, where $\pi_\Gamma(v)$ denotes the projection of v on $\Gamma \subseteq Act$. This means that along the execution of M described by w, no channel ever contains more than B-messages. We say that M is universally B-bounded if every $w \in \text{lin}(M)$ is B-bounded. An MSC language $\mathscr{L}$ is universally B-bounded if every $M \in \mathscr{L}$ is universally B-bounded. Finally, $\mathscr{L}$ is universally bounded if it is universally B-bounded for some B.

We then have the following result [9].

Theorem 3 *If an MSC language $\mathscr{L}$ is regular then it is universally bounded.*

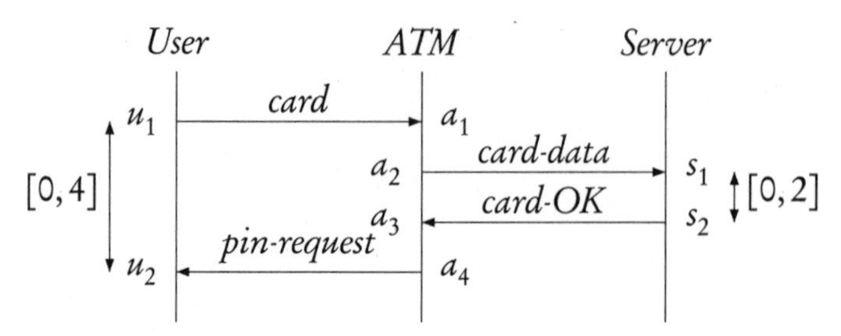

Figure 2: *A TC-MSC describing interaction with an ATM.*

A weaker notion of channel boundedness is existential boundedness. An MSC M is said to be existentially B-bounded if *some* $w \in \text{lin}(M)$ is B-bounded. Existential boundedness corresponds to choosing a good schedule for the events to ensure the channel bound B. An MSC language $\mathscr{L}$ is existentially B-bounded if every $M \in \mathscr{L}$ is existentially B-bounded. Finally, $\mathscr{L}$ is existentially bounded if it is existentially B-bounded for some B.

2.2 Time-constrained MSCs

A time-constrained MSC (denoted TC-MSC) is an MSC annotated with time intervals. For simplicity, we assume that the interval bounds are natural numbers. For $a, b \in \mathbb{N}$, we allow intervals that are open (a, b), closed $[a, b]$, half-open $(a, b]$, $[a, b)$, or unbounded $[a, \infty)$, (a, ∞). As usual, by (a, b), we mean $\{x \in \mathbb{R}_{\geq 0} \mid a < x < b\}$ and so on. Let $\mathscr{I}$ denote the set of all such intervals.

Definition 4 *Let $M = (E, \leq, \lambda)$ be an MSC. An* interval constraint *is a tuple $\langle (e_1, e_2), I\rangle$ where $e_1, e_2 \in E$ with $e_1 \leq_{pp} e_2$ for some $p \in \mathscr{P}$ or $e_1 <_{pq} e_2$ for some channel $(p, q) \in Ch$ and $I \in \mathscr{I}$.*

The restrictions on e_1 and e_2 ensure that an interval constraint is either local to a process or describes a bound on the delivery time of a single message. Fig. 2 shows a TC-MSC describing the interaction between a user, an ATM and a server. For instance, the constraint $[0, 2]$ on (s_1, s_2) specifies that the server is expected to respond to an authentication request within 2 time units.

Definition 5 *A* time-constrained MSC (TC-MSC) *is a pair $\mathscr{T} = (M, \mathscr{E}\mathscr{C})$ where $M = (E, \leq, \lambda)$ is an MSC and $\mathscr{E}\mathscr{C} \subseteq (E \times E) \times \mathscr{I}$ is a set of interval constraints such that each pair (e_1, e_2) is mapped to at most one interval.*

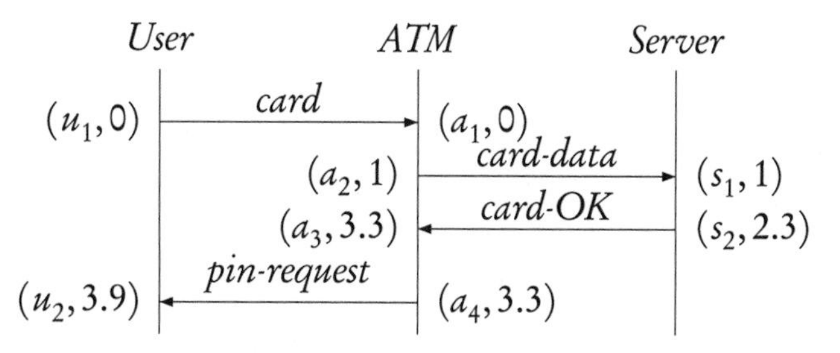

Figure 3: *A timed MSC describing interaction with an ATM.*

2.3 Timed MSCs

In a timed MSC, events are explicitly time-stamped so that the ordering on the time-stamps respects the partial order on the events.

Definition 6 *A* timed MSC *is pair* (M, τ) *where* $M = (E, \leq, \lambda)$ *is an MSC and* $\tau : E \to \mathbb{R}_{\geq 0}$ *assigns a nonnegative time-stamp to each event, such that for all* $e_1, e_2 \in E$*, if* $e_1 \leq e_2$ *then* $\tau(e_1) \leq \tau(e_2)$.

A timed MSC *realizes* a TC-MSC if the time-stamps assigned to events respect the interval constraints specified in the TC-MSC. Let $r \in \mathbb{R}_{\geq 0}$ and $I \in \mathcal{I}$. We write $r \models I$ to denote that r lies in the interval specified by I.

Definition 7 *Let* $M = (E, \leq, \lambda)$ *be an MSC,* $\mathcal{T} = (M, \mathcal{EC})$ *a TC-MSC and* $M_\tau = (M, \tau)$ *a timed MSC.* M_τ *is said to* realize $\mathcal{T}$ *if for each* $\langle(e_1, e_2), I\rangle \in \mathcal{EC}, \tau(e_2) - \tau(e_1) \models I$.

Fig. 3 shows a timed MSC that realizes the TC-MSC in Fig. 2.

We say that a TC-MSC is *realizable* if it is realized by at least one timed MSC. Realizability amounts to checking if the constraints in a TC-MSC are feasible. This can be checked by constructing a graph corresponding to the events with weighted, directed edges in which lower bounds are represented by negative weights and upper bounds by positive weights. We can then show that the constraints in the original TC-MSC are feasible if and only if this graph has no negative-weight cycles—see, for instance, [4].

2.4 Message sequence graphs

Message sequence graphs (MSGs) are finite directed graphs with designated initial and terminal vertices. Each vertex in an MSG is labelled by an MSC. The edges represent (asynchronous) MSC concatenation, in which one MSC is "pasted" below the other. Formally, MSC concatenation is defined as follows.

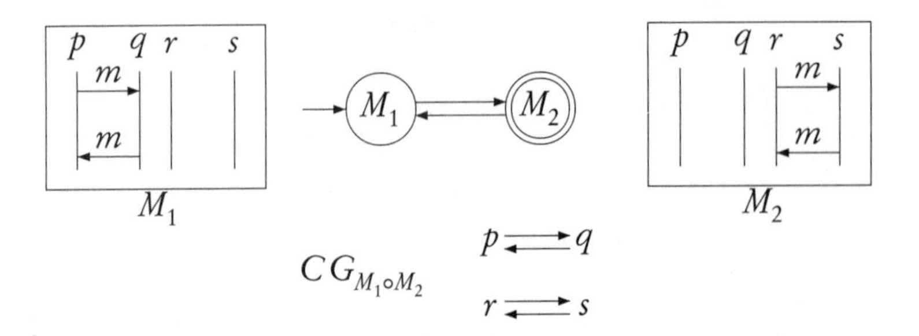

Figure 4: *A message sequence graph*

Let $M_1 = (E^1, \leq^1, \lambda_1)$ and $M_2 = (E^2, \leq^2, \lambda_2)$ be a pair of MSCs such that E^1 and E^2 are disjoint. The *(asynchronous) concatenation* of M_1 and M_2 yields the MSC $M_1 \circ M_2 = (E, \leq, \lambda)$ where $E = E^1 \cup E^2$, $\lambda(e) = \lambda_i(e)$ if $e \in E^i$, $i \in \{1,2\}$, and

$$\leq = \left(\leq^1 \cup \leq^2 \cup \bigcup_{p \in \mathscr{P}} E^1_p \times E^2_p \right)^* .$$

A *Message Sequence Graph* is a structure $G = (Q, \rightarrow, Q_{in}, Q_F, \Phi)$, where Q is a finite and nonempty set of states, $\rightarrow \subseteq Q \times Q$, $Q_{in} \subseteq Q$ is a set of initial states, $Q_F \subseteq Q$ is a set of final states and Φ labels each state with an MSC.

Let $\pi = q_0 \rightarrow q_1 \rightarrow \cdots \rightarrow q_n$ be a path through G. The MSC generated by π is $\Phi(\pi) = \Phi(q_0) \circ \Phi(q_1) \circ \cdots \circ \Phi(q_n)$. A path $\pi = q_0 q_1 \cdots q_n$ is a *run* if $q_0 \in Q_{in}$ and $q_n \in Q_F$. The language of MSCs accepted by G is $L(G) = \{\Phi(\pi) \mid \pi \text{ is a run through } G\}$.

An example of an MSG is depicted in Fig. 4. The initial state is marked $\rightarrow$ and the final state has a double circle. The language $\mathscr{L}$ defined by this MSG is *not* regular: $\mathscr{L}$ projected to $\{p!q(m), r!s(m)\}^*$ consists of $\sigma \in \{p!q(m), r!s(m)\}^*$ such that $|\pi_{p!q(m)}(\sigma)| = |\pi_{r!s(m)}(\sigma)| \geq 1$, which is not a regular string language.

In general, it is undecidable whether an MSG describes a regular MSC language [9]. However, in this paper, our main focus is not regularity but channel-boundedness—is it the case that the MSC language $\mathscr{L}$ defined by an MSG G is universally B-bounded?

It is easy to see that $\mathscr{L}$ is always existentially B-bounded. Since each MSC $M \in \mathscr{L}$ is generated by a path $\pi = q_0 \rightarrow q_1 \rightarrow \cdots \rightarrow q_n$, we can decompose $M = \Phi(\pi)$ as $\Phi(q_0) \circ \Phi(q_1) \circ \cdots \circ \Phi(q_n)$. Each individual $\Phi(q_i)$ is existentially B_i bounded for some bound B_i. By scheduling events so that $\Phi(q_i)$ is completed before we start $\Phi(q_{i+1})$, we observe that M is existentially B bounded for $B = \max_{i \in \{0,1,\ldots,n\}} B_i$. Thus, overall $\mathscr{L}$ must be existentially B_G bounded, where $B_G = \max_{q \in Q} B_q$ and B_q is the existential bound associated with $\Phi(q)$.

A necessary and sufficient condition for the MSC language of an MSG to be universally bounded is that the MSG be *locally strongly connected* [14]. To formalize this, we define the notion of a communication graph.

Communication graph For an MSC $M = (E, \leq, \lambda)$, let CG_M, *the communication graph of* M, be the directed graph $(\mathscr{P}, \mapsto)$ where:

- $\mathscr{P}$ is the set of processes of the system.
- $(p,q) \in \mapsto$ iff there exists an $e \in E$ with $\lambda(e) = p!q(m)$.

M is said to be *locally strongly connected* if every connected component of CG_M is strongly connected. An MSG G is said to be *locally strongly connected* if for each simple loop π in G, $\Phi(\pi)$ is locally strongly connected.

Notice that the MSC language defined by the MSG in Figure 4 is universally 1-bounded, though the language is not regular, and that the communication graph of the one simple loop in this MSG is in fact locally strongly connected.

2.5 Time-constrained MSGs

To describe infinite families of TC-MSCs, we label the nodes of an MSG with TC-MSCs instead of normal MSCs. We also permit process-wise timing constraints along the edges of the MSG. A constraint for process p along an edge $q \longrightarrow q'$ specifies a constraint between the final p-event of $\Phi(q)$ and the initial p-event of $\Phi(q')$, provided p actively participates in both these nodes. If p does not participate in either of these nodes, the constraint is ignored.

Definition 8 *A* time-constrained MSG (TC-MSG) *is a structure* $G = (Q, \rightarrow, Q_{in}, Q_F, \Phi, EdgeC)$, *where*

- *Q is a finite non-empty set of states with sets of initial and final states Q_{in} and Q_F, respectively, and $\rightarrow \subseteq Q \times Q$ is a transition relation, as in an MSG.*
- *Φ labels each node with a TC-MSC.*
- *$EdgeC \subseteq Q \times Q \times \mathscr{P} \times \mathscr{I}$ describes local constraints on the edges, with the restriction that $(q, q', p, I) \in EdgeC$ only if $q \longrightarrow q'$ and each triple (q, q', p) is mapped to at most one interval.*

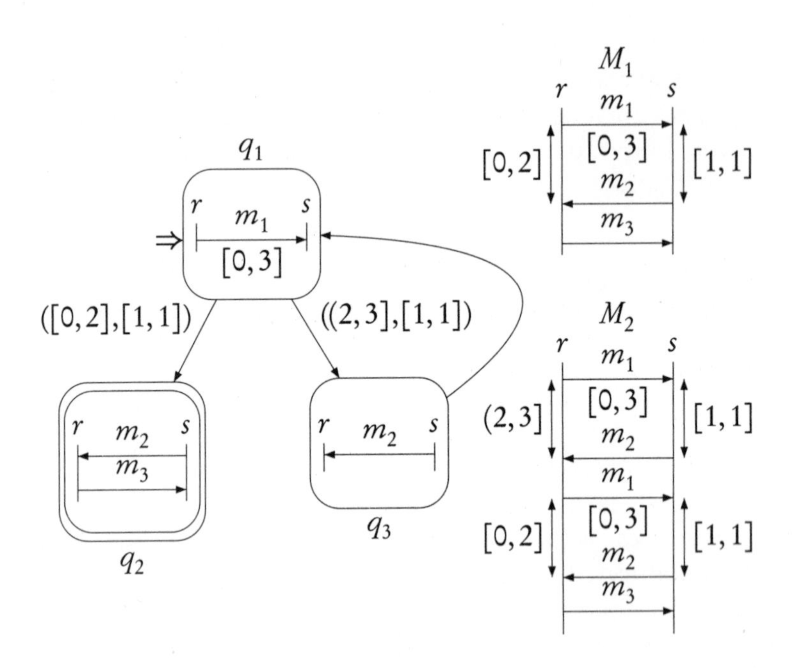

Figure 5: *A TC-MSG and some TC-MSCs that it generates*

For a path $\pi = q_0 q_1 \dots q_n$ through G, we define $\Phi(\pi)$, the TC-MSC generated by π as follows. We begin with the TC-MSC $\Phi(q_0) \circ \Phi(q_1) \circ \dots \circ \Phi(q_n)$. For each edge $q_i \longrightarrow q_{i+1}$, $0 \leq i < n$, if $(q_i, q_{i+1}, p, I) \in$ *EdgeC* we add a constraint I between the last p-event in $\Phi(q_i)$ and the first p-event in $\Phi(q_{i+1})$, provided p participates in both $\Phi(q_i)$ and $\Phi(q_{i+1})$. Fig. 5 shows a TC-MSG and some of the TC-MSCs that it generates.

3 Decision Problems for Time-Constrained MSGs

3.1 Channel-boundedness

The focus of this paper is to address the problem of channel-boundedness for time-constrained MSGs—that is, given a TC-MSG G, determine if it is universally or existentially bounded. Recall that the underlying untimed MSG is always existentially bounded and, if it is locally strongly connected, it is also universally bounded.

The situation for TC-MSGs is more complicated. It is possible that the timing constraints do not allow a TC-MSG to be existentially bounded. On the other hand, timing constraints may convert an MSG that is unbounded into one that is univerally bounded. We illustrate both scenarios with a simple TC-MSG modelling a producer-consumer system where one process keeps sending messages to the other, as shown in Figure 6.

In the untimed setting, this system is not universally bounded because in any MSC where k messages are sent, we can find a prefix in which all k messages are sent by P before the first message is received by C.

Proposition 9 *Consider a producer-consumer system with timing constraints as shown in Figure 6. The channel is universally bounded if and only if U is finite and either $l_P > 0$ or $l_C > 0$.*

Proof Suppose that U is finite and $l_P > 0$ and P sends a message every l_P time units starting at time 0, with each message delayed by U units, the maximum possible. We then have messages sent at times $0, l_P, 2l_P, \ldots$, which are received at times $U, U+l_P, U+2l_P, \ldots$ respectively. In this run, $\lceil \frac{U}{l_P} \rceil$ messages are sent from time 0 to time U. After this, with each new message inserted into the channel one old message is received, so the channel never grows beyond this bound.

On the other hand suppose that U is finite and $l_C > 0$ and there are B messages in the channel at time t. All these messages must be received by C before time $t+U$. However, at most $\lceil \frac{U}{l_C} \rceil$ messages can be received by C within the interval $[t, t+U]$, so $B \leq \lceil \frac{U}{l_C} \rceil$. Since t was arbitrary, the channel is universally bounded.

Conversely, if $U = \infty$ or $l_P = l_C = 0$, we can show that the channel is unbounded. Suppose $U = \infty$ and we propose a bound B. We can delay the receipt of the first message sent by P till $B+1$ messages have been inserted into the channel. On the other hand, if U is finite but $l_P = l_C = 0$, P can send $B+1$ (in fact, any number of messages we want) within U time units and all these messages can be received by C since $l_C = 0$.

Notice also that if $U = \infty$ and $l_C > u_P$, the language of this TC-MSG is not even existentially bounded. This is in sharp contrast to untimed MSGs, which are always existentially bounded.

Variants of the problem

Our basic problem is to check whether a TC-MSG is universally and/or existentially bounded. We can identify several ways to restrict the class of TC-MSGs under consideration.

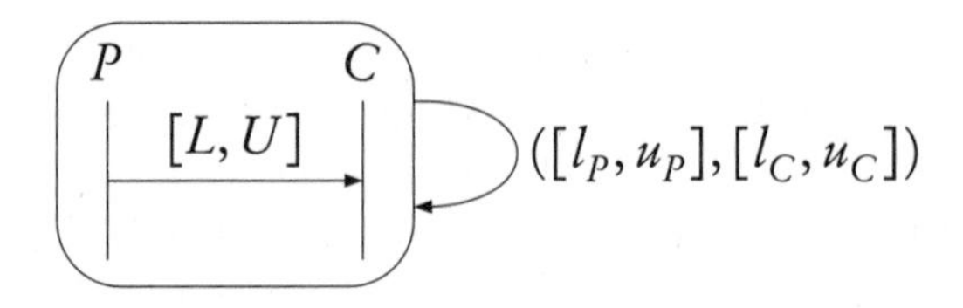

Figure 6: *Producer-Consumer with timing constraints*

Restrictions on constraints The first restriction is to do with edge constraints. In general, an edge in a TC-MSGs can have an independent constraint for each process that participates in both nodes connected by the edge. We consider two special cases:

- There are no edge constraints at all.
- Only one designated process p is permitted to have edge constraints (p is a fixed process for the entire TC-MSG).

We can also vary the type of constraints we consider. In general, as we have seen, the intervals we use in constraints can be closed, open or half open. In particular, we can have point intervals of the form $[a,a]$ which specify an exact delay. The special cases we can consider are:

- Both open and closed intervals are permitted, including point intervals.
- Only open intervals are permitted.

Restrictions on final states Normally, a TC-MSG is equipped with final states and we are only interested in paths from the initial state to one of the final states. We can drop the assumption that we have final states and consider all paths starting from an initial state.

Type of boundedness As we saw with the Producer-Consumer example, both universal and existential boundedness are nontrivial problems for TC-MSGs. The general question asks whether there exists a bound B such that the TC-MSG is existentially or universally B-bounded. We can also ask a weaker question: given a fixed bound B, is the TC-MSG existentially or universally bounded with respect to this fixed bound?

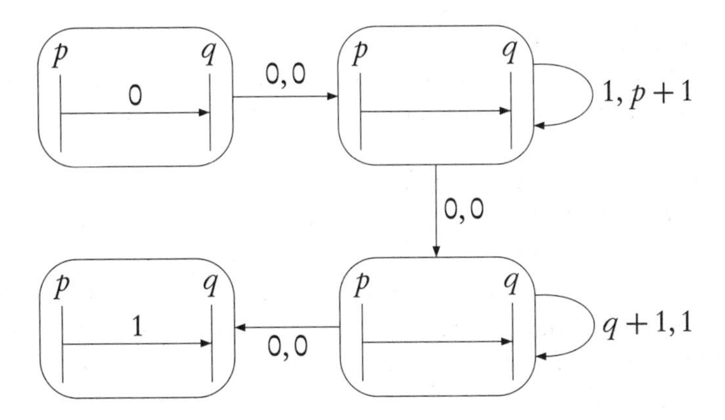

Figure 7: *Reachability in TC-MSGs is difficult*

3.2 Reachability in TC-MSGs

Let $G = (Q, \rightarrow, Q_{in}, Q_F, \Phi, EdgeC)$ be a TC-MSG. A state q is said to be *reachable* in G if there is a path $q_0 q_1 \dots q_n$ such that $q_0 \in Q_{in}$, $q_n = q$ and $\Phi(q_0) \circ \Phi(q_1) \circ \cdots \circ \Phi(q_n)$ is a realizable TC-MSC.

It turns out that even reachability is not trivial for TC-MSGs. Consider the example in Figure 7. (In this example and elsewhere, we denote point intervals $[a, a]$ by a single integer a.) To reach the last node we cannot take the shortest path and we need to iterate the first loop k times and the second loop ℓ times so that $kp - \ell q = 1$.

The negative results that we show for boundedness will, in fact, be derived from negative results for reachability—that is, we will reduce reachability to boundedness. When addressing the reachability problem, we will again consider restricted versions corresponding to the special cases on constraints, as discussed above in the context of boundedness.

4 Reachability With Edge Constraints is Undecidable

We first show that reachability is undecidable with unrestricted edge constraints. For this, we show that we can simulate the behaviour of a 2-counter machine [13]. As the name suggests, a 2-counter machine has two counters c_1 and c_2, each of which can hold a non-negative number. A program is a sequence of labelled instructions $\ell : I$, where I is one of the following:

- c_1++ or c_2++ which increments the value of the counter.

- `if` $c_1 \stackrel{?}{=} 0$ `goto` ℓ' `else` c_1−− which transfers control to the instruction labelled ℓ' if counter c_1 is zero, and otherwise it decrements c_1 and continues with the next instruction labelled $\ell + 1$. Indeed, we also have a similar instruction for c_2.

Observe, that 2-counter machines are deterministic. Thus, a given machine will either reach its final instruction and implicitly halt after a finite number of steps or perform an infinite computation.

To simulate a 2-counter machine, we construct a TC-MSG in which each node represents one labelled instruction $\ell : I$ of the 2-counter machine. We encode each counter by a pair of processes—the counter value is represented by the difference in time between the local clock values of the pair of processes associated with the counter.

Figure 8 shows a simple simulation for one counter c. Let t_p and t_q denote the time stamp of the last event on processes p and q, respectively. The value of counter c is encoded by $t_q - t_p$. We maintain, as an invariant, that $t_p \leq t_q$. Recall that a constraint of the form b denotes a point interval $[b, b]$. Notice that this simulation does not use any time constraints on messages, but that it does use point intervals.

The initial node synchronizes p and q, thereby setting $c = t_q - t_p$ to 0. Between two nodes, we always use edge constraints $(1, 1)$ enforcing that the time difference between the last event on process p (resp. q) in the previous node and the first event on process p (resp. q) in the next node is always 1. Therefore, the node *Freeze* preserves the value of the counter. The node labelled $c{+}{+}$ delays q, thereby incrementing c. Symmetrically, the node labelled $c{-}{-}$ delays p, which decrements c. Since the last message in node $c{-}{-}$ goes from p to q, we have $c = t_q - t_p \geq 0$ at the end. So this is realizable only if counter c was positive before entering node $c{-}{-}$. The node $c \stackrel{?}{=} 0$ checks if c is 0 by sending a message back from q to p. Let t_p and t_q be the time stamps of the last events on p and q in the previous node. The message in node $c \stackrel{?}{=} 0$ is sent at time $t_q + 1$ and received at time $t_p + 1 \geq t_q + 1$. Since our invariant demands that $t_p \leq t_q$, this is realizable precisely when $t_p = t_q$, which means $c = 0$. Note that the invariant is preserved.

Having shown how to encode counter values, it is a simple matter to construct a TC-MSG that simulates a given 2-counter machine. We use two pairs of processes (p_1, q_1) and (p_2, q_2) to encode the counters c_1 and c_2, respectively. By definition of the counter machine, exactly one counter is active in each instruction, for which we use the

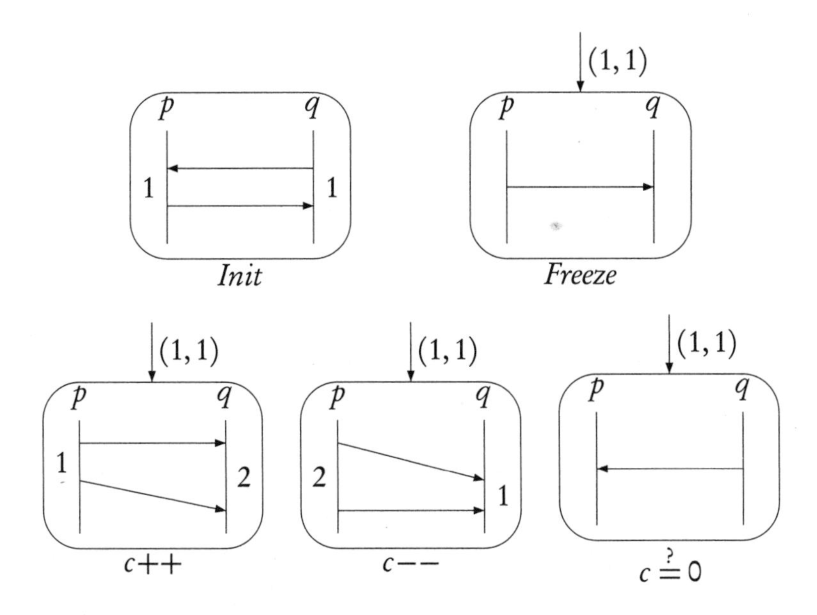

Figure 8: *Simulating a counter machine*

appropriate encoding described below. For the inactive counter, we preserve the value using the *Freeze* construction.

For each instruction $\ell : c{+}{+}$ ($c \in \{c_1, c_2\}$) of the 2-counter machine, we construct a state ℓ in the TC-MSG corresponding to $c{+}{+}$ and freezing the other counter. We connect state ℓ to the state(s) corresponding to instruction $\ell + 1$ in the TC-MSG to capture the implicit control flow in the counter machine.

For each instruction $\ell : \texttt{if}\ c \stackrel{?}{=} 0\ \texttt{goto}\ \ell'\ \texttt{else}\ c{-}{-}$ we create two states corresponding to $\ell_{c=0}$ and ℓ_{c--} which are labelled with the corresponding MSCs from Figure 8. Again, we freeze the inactive counter. We connect state $\ell_{c=0}$ to the state(s) corresponding to instruction ℓ' and state ℓ_{c--} to the state(s) corresponding to instruction $\ell + 1$.

Let $\ell_f : I_f$ be the final (halting) instruction of the 2-counter machine. Then, checking whether the corresponding node in the TC-MSG is reachable is equivalent to checking whether the given 2-counter machine halts. Since this is an undecidable problem for 2-counter machines, we have shown the following.

Theorem 10 *The reachability problem for TC-MSGs with arbitrary edge constraints is undecidable, even without constraints on message delays.*

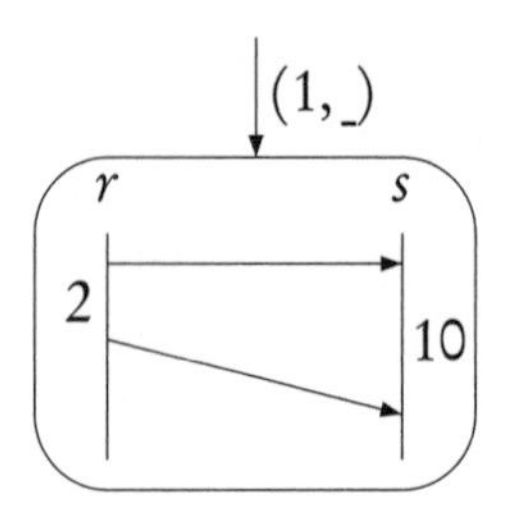

Figure 9: *Reducing reachability to boundedness*

5 Reducing Reachability to Boundedness

We add two new processes r and s to each node in our TC-MSG simulation of a 2-counter machine with two messages from r to s as shown in Figure 9. Events along r are tightly constrained by edge constraints. On the other hand, there are no edge constraints for s, events along s can be delayed arbitrarily.

There are two cases to consider.

- If the counter machine halts then the simulation is finite and channels are bounded by some B—we can calculate the bound from the length of the computation.
- If the counter machine does not halt then the simulation is infinite and r and s form a timed producer-consumer system in which the channel (r, s) is both existentially and universally unbounded, as analyzed in Section 3.1.

This reduction gives us the following result.

Theorem 11 *Both existential boundedness and universal boundedness are undecidable for TC-MSGs with arbitrary edge constraints, even without constraints on message delays.*

We note, in passing, that boundedness does not imply reachability. For instance, we can strengthen the notion of locally strongly connected TC-MSGs to obtain the class of *locally synchronized* TC-MSGs that have both bounded channels and regular behaviours [5, 9, 15]. However, reachability is still a nontrivial problem. In fact, it turns out that reachabililty is decidable for locally synchronized TC-MSGs, but this requires a somewhat sophisticated argument [2].

6 Simulating Counters With Open Intervals

We next consider the restriction where constraints can only be open intervals. Reachability remains undecidable even in this setting.

As before, we will model a counter by the difference in time across two processes p and q. However, the counter value is denoted not by $t_q - t_p$ but by $t_q - t_p - 1$. Thus, the counter is 0 when $t_q - t_p = 1$.

Our simulation of counters will use only open intervals. Instead of simulating p and q with timestamps that exactly capture the value of the counter, we use two pairs of processes (p_l, q_l) and (p_u, q_u) that serve as the *lower* and *upper* approximations of the value denoted by (p,q). We maintain as an invariant that $0 \leq t_{q_l} - t_{p_l} < t_q - t_p < t_{q_u} - t_{p_u}$.

The simulation is described in Figure 10. In the pictures, we show p and q, the processes with point intervals whose value we are trying to track, but this is only for reference. The actual simulation uses only the pairs (p_l, q_l) and (p_u, q_u)

The node *Init* sets up the invariant corresponding to $t_q - t_p = 1$, i.e., the counter value is 0.

In the exact simulation using p and q, each edge carries a constraint 1. Correponding to this, we compose nodes using edge constraints for the lower and upper approximations as shown in *Composition*. For the pair of nodes n and n' connected by such an edge, we use t to denote the times associated with the last events in n and t' to denote the times associated with the first events in n'. Then, we have:

$$\begin{array}{lcll}
t'_{q_l} - t'_{p_l} & < & t_{q_l} - t_{p_l} & \text{(by edge constraints),} \\
t_{q_l} - t_{p_l} & < & t_q - t_p & \text{(by assumption on } n\text{),} \\
t_q - t_p & = & t'_q - t'_p & \text{(exact delay of 1 on } p \text{ and } q\text{),} \\
t_q - t_p & < & t_{q_u} - t_{p_u} & \text{(by assumption on } n\text{),} \\
t_{q_u} - t_{p_u} & < & t'_{q_u} - t'_{p_u} & \text{(by edge constraints).}
\end{array}$$

From this, it follows that in n', we still have $0 \leq t'_{q_l} - t'_{p_l} < t'_q - t'_p < t'_{q_u} - t'_{p_u}$ as required.

The node c++ increments the counter. Once again using t' for the times of the second message and t for the times of the first one, we have

$$0 \leq t'_{q_l} - t'_{p_l} < t_{q_l} - t_{p_l} + 1 < t_q - t_p + 1 = t'_q - t'_p < t_{q_u} - t_{p_u} + 1 < t'_{q_u} - t'_{p_u},$$

so the lower and upper approximations correctly track c after the increment.

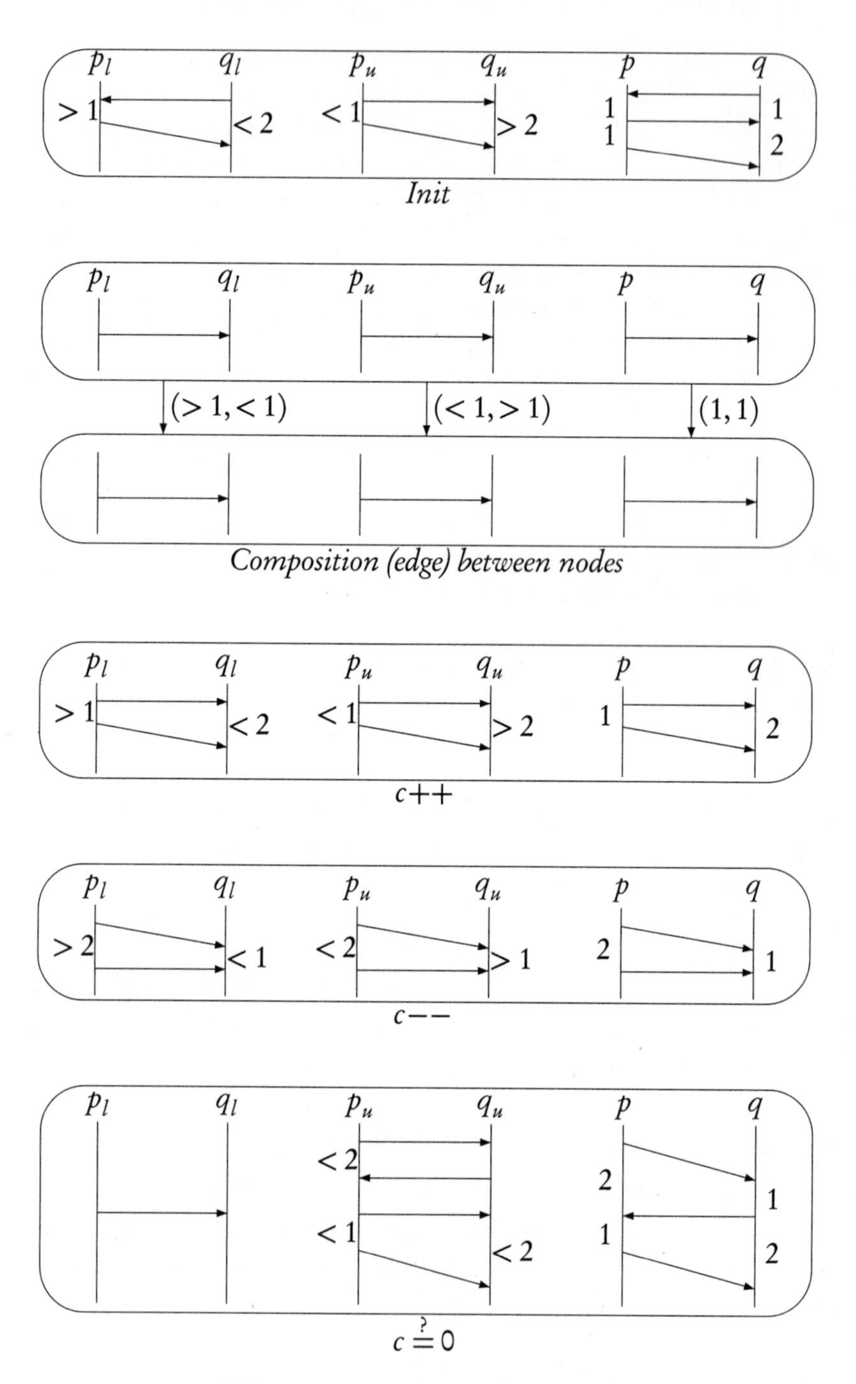

Figure 10: *Open interval simulation*

The node $c--$ decrements the counter. Once again using t' for the times of the second message and t for the times of the first one, we have

$$0 \le t'_{q_l} - t'_{p_l} < t_{q_l} - t_{p_l} - 1 < t_q - t_p - 1 = t'_q - t'_p < t_{q_u} - t_{p_u} - 1 < t'_{q_u} - t'_{p_u},$$

so the lower and approximations correctly track c after the increment.

As before, we *freeze* a value by sending a single message on both (p_l, q_l) and (p_u, q_u).

For $c \stackrel{?}{=} 0$, initially $1 \le t_q - t_p < t_{q_u} - t_{p_u}$. Once again using t' for the times of the second message and t for the times of the first one, we have $t_{q_u} \le t'_{q_u} \le t'_{p_u} < t_{p_u} + 2$. Hence, $t_{q_u} - t_{p_u} < 2$. From this, we deduce that $1 \le t_q - t_p < 2$, so $t_q - t_p = 1$ which means that $c = 0$.

Conversely, if $c = 0$, there is a timed MSC that realizes the path used so far with the property that $1 = t_q - t_p < t_{q_u} - t_{p_u} < 1 + \frac{1}{2}$. This means that the first two messages between p_u and q_u in the figure are realizable. The next two messages between p_u and q_u then "reset" the counter so that, at the end, $1 < t_{q_u} - t_{p_u}$.

We can now use two sets of lower and upper approximations to track two counters and set up a TC-MSG that simulates a 2-counter machine as in Section 4. It is a simple matter to add a suitably modified version of the MSC shown in Figure 9 to each node, so that we have the following results.

Theorem 12 *Reachability, existential boundedness and universal boundedness are all undecidable for TC-MSGs even if we restrict all constraints to open intervals and without constraints on message delays.*

7 Counter Simulation With Bounded Channels

We can use a more sophisticated construction to simulate a counter in which all channels are 1-bounded. The main ingredients are shown in Figure 11. The counter value c is encoded as the difference $t_q - t_p$. This quantity is manipulated by sending messages on the channels (p, p') and (q, q'), with timing constraints on the send and receive events along processes (p, p') and (q, q').

The main complication is that the construction for $c--$ does not prevent the value in c from going below 0. Note however that the TC-MSC for $c \stackrel{?}{=} 0$ can be realized only when $t_q - t_p = 0$, i.e., it accurately checks that $c = 0$ without assuming that the counter is non-negative. We use this fact to implement a decrementation which "terminates"

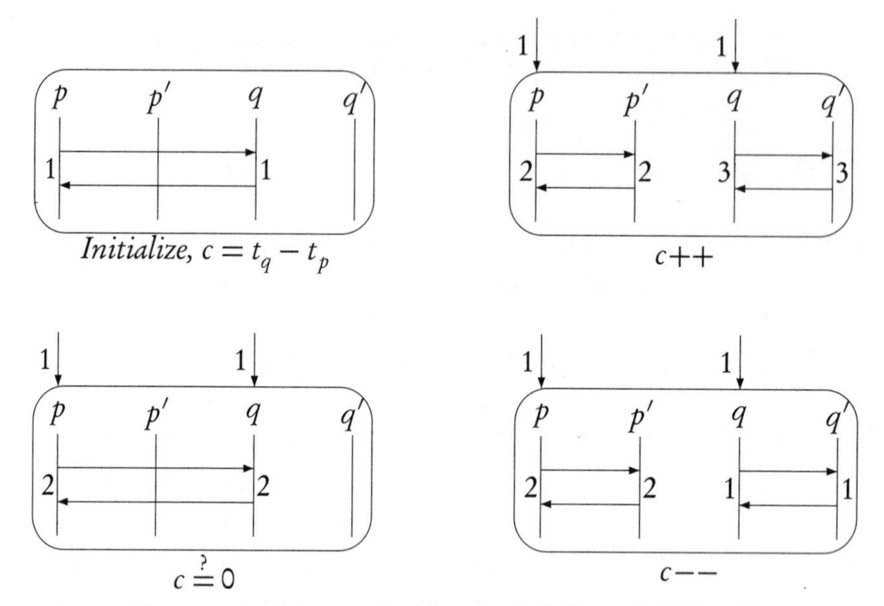

Figure 11: *Counter simulation with bounded channels*

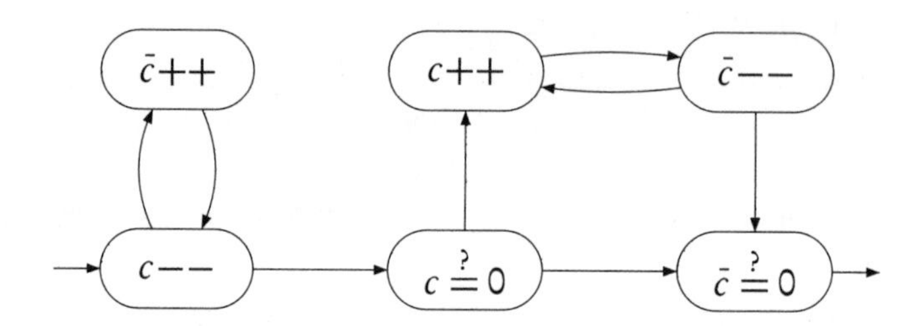

Figure 12: *Decrementation without going below* 0

only if the counter stay above 0. We copy the contents of c to another counter $\bar{c}$ by repeatedly decrementing c and incrementing $\bar{c}$, allowing the loop to terminate when $c \stackrel{?}{=} 0$. In this case, we transfer the contents back from $\bar{c}$ to c using the same trick. The precise construction is depicted in Figure 12. If $c < 0$ after the original decrement, we have a livelock in which we cycle through the first loop an infinite number of times, thus effectively blocking the simulation. So the simulation (may) terminates only if the counter is non-negative after the initial decrement.

For *Freeze*, we exchange a pair of messages between p, p' and between q, q' with the same delay along each process. So the node is similar to that of c++ where we replace 3 by 2.

With these ingredients in place, we can once again simulate a 2-counter machine with a TC-MSG using six pairs of processes,

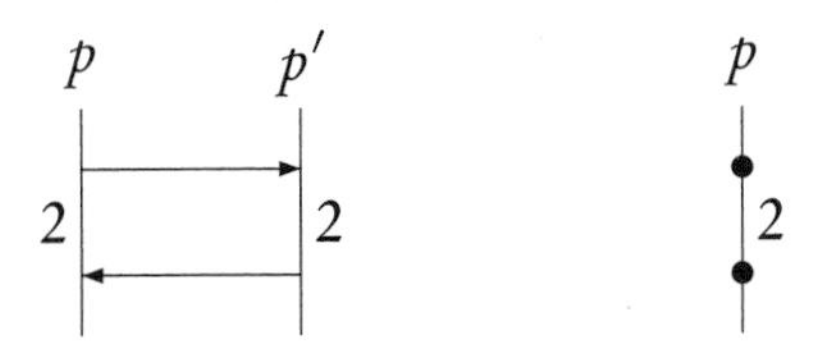

Figure 13: *Messages can be eliminated*

$\{(p_1, p_1'), (q_1, q_1'), (p_2, p_2'), (q_2, q_2'), (p, p'), (q, q')\}$, to encode the three counters c_1, c_2 and $\bar{c}$. The main difference with our earlier construction is that even when the run of the 2-counter machine is infinite, the corresponding TC-MSC traced out by our TC-MSG is 1-bounded.

The fact that this simulation of a 2-counter machine has 1-bounded channels allows us to sharpen our undecidability result for boundedness. In the earlier reduction from reachability to boundedness, we had included the MSC in Figure 9 within each node of the 2-counter simulation. Here, instead, we create a single separate node containing this TC-MSC, with a self-loop. We then add an edge from the node in our TC-MSG corresponding to the final instruction $\ell_f : I_f$ of the 2-counter machine to this new node.

Clearly, the new node with the TC-MSC from Figure 9 is reachable if and only if the 2-counter machine that we are simulating terminates at $\ell_f : I_f$. Once we enter this new node, we start generating a TC-MSC that is not existentially or universally bounded for *any* choice of B. On the other hand, if the 2-counter machine computation never terminates, we always remain within the 1-bounded portion of the simulation. Thus, if we can decide whether this TC-MSG is B-bounded, either existentially or universally, for any $B \geq 1$, we can also decide whether the 2-counter machine halts. Hence, we have the following result.

Theorem 13 *Checking whether a TC-MSG with arbitrary edge contraints is existentially or universally bounded with respect to a fixed bound B is undecidable for every $B \geq 1$.*

In fact, the construction can be further simplified if we permit internal events along processes. We can then replace each exchange of a pair of messages along the channel (p, p') by two internal events on p with the same interval constraint, as shown in Figure 13. But we still need to exchange messages between p and q to initialize the counter or to check $c \stackrel{?}{=} 0$. Thus, we can divide the number of processes by 2 in this simulation.

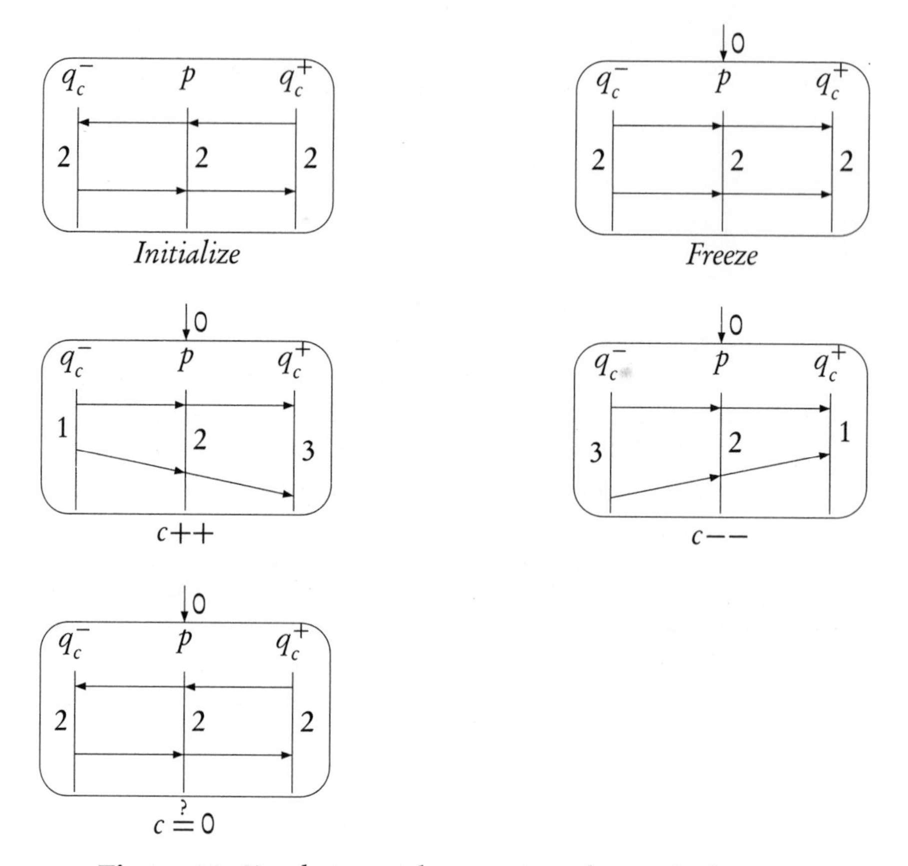

Figure 14: *Simulation with constraints along a single process*

8 Edge Constraints Along a Single Process

The last case we consider is when edge constraints are only permitted for a fixed process. This is a natural restriction — for instance this process could be a controller that coordinates between the different phases of a protocol. Let us denote this special process p. To simulate a counter c, we add two processes q_c^+ and q_c^-. Let t_c^+, t_c^- and t denote the time of the last event along processes q_c^+, q_c^- and p, respectively. We then maintain the value of c in terms of an upper approximation $c^+ = t_c^+ - t$ and a lower approximation $c^- = t - t_c^-$, so that we always have $c^+ \geq c \geq c^- \geq 0$.

Figure 14 shows the basic constructs needed for the simulation. As usual, the node labelled *Init* ensures that $t_c^+ = t = t_c^-$, so initially we have $c^+ = c^- = 0$. The node labelled *Freeze* preserves the current value of c. The nodes labelled $c{+}{+}$ and $c{-}{-}$ correspond to incrementing and decrementing the value of c, respectively. The node labelled $c \stackrel{?}{=} 0$ check if the value of c is 0.

Any sequence $\sigma = s_1 s_2 \ldots s_n$ of states where each s_j is one among $c{+}{+}$, $c{-}{-}$, *Freeze* and $c \stackrel{?}{=} 0$ yields a TC-MSC M_σ. We say that σ describes a *valid* computation if when executing the corresponding sequence of instructions the counter c never goes below 0 and whenever a zero-test is performed then the value of c is indeed 0. Note that, if σ describes a valid computation then M_σ is realizable by a T-MSC such that after each instruction we have $c^+ = c = c^- \geq 0$.

Conversely, to justify that c^+ and c^- track the value of c accurately, we show that any T-MSC that realizes M_σ maintain the invariant $c^+ \geq c \geq c^- \geq 0$. Let us denote $|\sigma|_a$ the number of occurrences of the letter a in the sequence σ so that $c = |\sigma|_{c++} - |\sigma|_{c--}$. We verify that the invariant holds with the following calculation.

$$\begin{array}{rcl} c^+ & = & t_c^+ - t \\ & \geq & 3|\sigma|_{c++} + 2|\sigma|_{c \stackrel{?}{=} 0} + 2|\sigma|_{Freeze} + |\sigma|_{c--} - \\ & & 2(|\sigma|_{c++} + |\sigma|_{c \stackrel{?}{=} 0} + |\sigma|_{Freeze} + |\sigma|_{c--}) \\ & = & |\sigma|_{c++} - |\sigma|_{c--} = c \end{array}$$

$$\begin{array}{rcl} c^- & = & t - t_c^- \\ & \leq & 2(|\sigma|_{c++} + |\sigma|_{c \stackrel{?}{=} 0} + |\sigma|_{Freeze} + |\sigma|_{c--}) - \\ & & -(3|\sigma|_{c--} + 2|\sigma|_{c \stackrel{?}{=} 0} + 2|\sigma|_{Freeze} + |\sigma|_{c++}) \\ & = & |\sigma|_{c++} - |\sigma|_{c--} = c \end{array}$$

Note that we always have $c^- = t - t_c^- \geq 0$ since the last message of each node always goes from q_c^- to p. Hence, if M_σ is realizable then $c \geq c^- \geq 0$ and the counter never goes below 0 during the computation.

Moreover, the realization of the TC-MSC in $c \stackrel{?}{=} 0$ implies $t_c^+ = t = t_c^-$ at the end, i.e., $c^+ = c^- = 0$. Using the invariant, we deduce that the counter must be 0 whenever we use state $c \stackrel{?}{=} 0$ in the sequence σ.

The simulation of a 2-counter machine with a TC-MSG follows along the usual lines. In each MSC of our simulation, we add a copy of the TC-MSC from Figure 9. Since edge-constraints are permitted only on p, we use p to play the role of r and introduce s as a fresh process. Then, from our argument, the channel (p, s) will be bounded if and only if the 2-counter machine that we are simulating has a halting computation. This yields the following result.

Theorem 14 *Reachability, existential boundedness and universal boundedness are all undecidable for TC-MSGs even if all edge constraints in the TC-MSG are restricted to a fixed process and without constraints on message delays.*

9 Decidability Without Edge Constraints

If we have no edge constraints in a TC-MSG, reachability is decidable and the status of boundedness is still open.

9.1 Reachability

Since there are no edge constraints, for a path $\pi = q_0 q_1 \dots q_n$, we can always choose to execute the events in $\Phi(\pi) = \Phi(q_0) \circ \Phi(q_1) \circ \dots \circ \Phi(q_n)$ one node at a time, as in an untimed MSG. The only difficulty that can arise is that a TC-MSC labelling a node is not realizable—that is, the constraints within it are not feasible. We can check this easily, as explained in Section 2.3, and delete nodes that are labelled by unrealizable TC-MSCs. Reachabilty then amounts to finding a path in the reduced graph after eliminating infeasible nodes.

9.2 Boundedness

When there are no constraints on edges, one strategy is to extend the producer-consumer analysis in Section 3 and check universal and existential boundedness. Recall that an MSG in which every node is labelled by a nonempty MSC is universally bounded if for every simple loop, the communication graph of the MSC described by the loop is locally strongly connected. Given this, it suffices to concentrate on simple loops where the communication graph of the underlying MC is not locally strongly connected.

Fix a simple loop π whose underlying MSC is $M = \Phi(\pi)$ such that CG_M is not locally strongly connected. Let (p, q) be a pair of processes such that there is an edge from p to q in CG_M, but no path back from q to p. If we iterate M, p and q play the role of P and C from the producer-consumer system described in Section 3. This system can be completely analyzed if we know the constraints on messages from p to q as well as between successive send events on p and receive events on q. The only difficulty is that the constraints along p and q may arise due to the transitive closure of dependencies through other pairs of events.

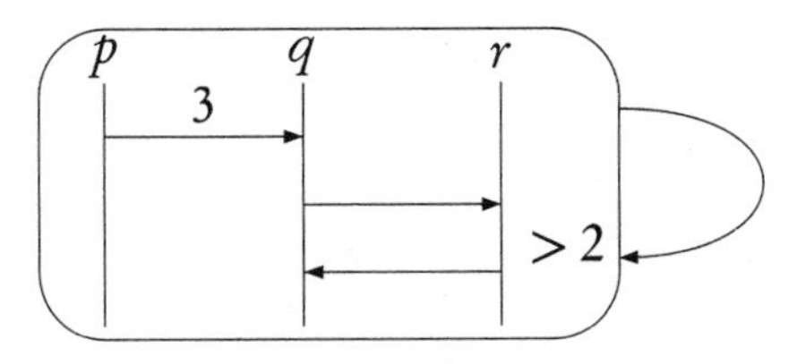

Figure 15: *Transitive dependency*

For example, in the system shown in Figure 15, the channel (p, q) is apparently unbounded since there is no explicit lower bound on consecutive sends by p or consecutive receives by q, though there is an upper bound on message delays along (p, q). However, the lower bound between the two events along r implicitly enforces a lower bound between the two corresponding events along q and hence imposes a lower bound between the receipt of consecutive messages by q along the channel (p, q). Thus, this loop satisifies the condition specified in Proposition 9 for the channel (p, q) to be universally bounded.

Unfolding the loop π several times may "reveal" timing constraints between two occurrences of the same send event $p!q(m)$ in the loop, or two occurrences of the same receive event $q?p(m)$. These timing constraints may be induced by constraints in the SCC of p or q respectively. Doing so, we may obtain for some message from p to q induced constraints $[l_p, u_p]$ and $[l_q, u_q]$ for successive send and receive events, respectively. Considering in addition the bounds $[L, U]$ carried by this message, we may apply Proposition 9 to prove that this loop is universally bounded if U is finite and $l_p > 0$ or $l_q > 0$.

Unfortunately, we do not obtain a necessary and sufficient condition following this approach. Consider the system in Figure 16. There is a single loop whose communication graph is not strongly connected. If we focus on the channel (p, r), we can regard this as a producer consumer system. In this system, we cannot derive an upper bound for the message from p to r sent in the third node. Nevertheless, a careful analysis shows that the channel (p, r) is indeed bounded by $\frac{2+U_1+U_2}{\ell}$.

Thus, we have a sufficient condition to ensure that a TC-MSG without edge constraints is bounded, but not a necessary one.

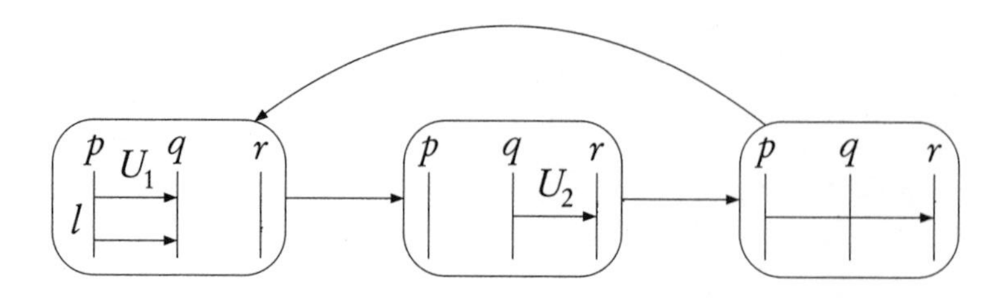

Figure 16: *Boundedness without edge constraints*

10 Discussion

We have shown that reachability and boundedness are, in general, undecidable for TC-MSGs. This is in contrast to the positive results for both questions in the untimed setting, confirming the suspicion that adding real-time constraints to specifications of distributed systems is tricky. As we have shown, even the simplest form of timing constraints across edges of a TC-MSG induce undecidability. Without edge constraints, reachability becomes decidable for TC-MSGs. We conjecture that boundedness is also decidable for this class of TC-MSGs.

The main reason why timed distributed systems are difficult to analyze is that global time acts as a covert channel to convey information across processes. This observation has also been made in the context of timed CFMs [10]. One way to get around this problem is to allow each component's clock to drift with respect to other components. The difficulty is to describe a model that permits this without making the notion of time across components meaningless. A model of timed CFMs with independently evolving clocks has been introduced in [1]. It remains to be seen if similar ideas can be incorporated into the TC-MSG framework.

References

[1] S. Akshay, B. Bollig, P. Gastin, M. Mukund and K. Narayan Kumar: Distributed Timed Automata with Independently Evolving Clocks. *Proc. CONCUR 2008*, Springer LNCS **5201** (2008) 82–97.

[2] S. Akshay, M. Mukund and K. Narayan Kumar: Checking Coverage for Infinite Collections of Timed Scenarios *Proc. CONCUR 2007*, Sringer LNCS **4703** (2007) 181–196.

[3] R. Alur and D. Dill: A Theory of Timed Automata. *Theor. Comput. Sci.*, **126** (1994) 183–225.

[4] R. Alur, G. Holzmann and D. Peled: An analyzer for message sequence charts. *Software Concepts and Tools*, **17(2)** (1996) 70–77.

[5] R. Alur and M. Yannakakis: Model checking of message sequence charts. *Proc. CONCUR'99*, Springer Lecture Notes in Computer Science **1664**, (1999) 114–129

[6] D. Brand and P. Zafiropulo: On communicating finite-state machines. *J. ACM*, **30(2)** (1983) 323–342.
[7] P. Chandrasekaran and M. Mukund: Matching Scenarios with Timing Constraints. *Proc. FORMATS 2006*, Springer LNCS 4202 (2006) 98–112.
[8] B. Genest, D. Kuske and A. Muscholl: A Kleene theorem and model checking algorithms for existentially bounded communicating automata. *Inf. Comput.* **204(6)** (2006) 920–956.
[9] J.G. Henriksen, M. Mukund, K. Narayan Kumar, M. Sohoni and P.S. Thiagarajan: A Theory of Regular MSC Languages. *Inf. Comput.*, **202(1)** (2005) 1–38.
[10] P. Krcál and W. Yi: Communicating Timed Automata: The More Synchronous, the More Difficult to Verify. *Proc. CAV 2006*, Springer LNCS 4144 (2006) 249–262.
[11] ITU-T Recommendation Z.120: *Message Sequence Chart (MSC)*. ITU, Geneva (1999).
[12] S. Mauw and M.A. Reniers: High-level message sequence charts. *Proc. SDL'97*, Elsevier (1997) 291–306.
[13] M. Minsky: *Computation: finite and infinite machines*. Prentice-Hall (1967).
[14] R. Morin: On Regular Message Sequence Chart Languages and Relationships to Mazurkiewicz Trace Theory. *FoSSaCS 2001*, Springer LNCS **2030** (2001) 332–346.
[15] A. Muscholl and D. Peled: Message sequence graphs and decision problems on Mazurkiewicz traces. *Proc. MFCS'99*, Springer Lecture Notes in Computer Science **1672**, (1999) 81–91
[16] A. Muscholl, D. Peled, and Z. Su: Deciding properties for message sequence charts. *Proc. FOSSACS'98*, LNCS **1378**, Springer-Verlag (1998) 226–242.

Test Generation from Integrated System Models Capturing State-based and MSC-based Notations

Ankit Goel, Abhik Roychoudhury
Department of Computer Science
National University of Singapore
{ankit,abhik}@comp.nus.edu.sg

Abstract

Model-based testing methods for reactive systems use a formal behavioral model of the system to generate test cases which are tried on the software implementation. Traditionally, such testing techniques rely on state diagram like behavioral models which emphasize intra-process control flow over inter-process communication. In this paper, we use system models which combine intra-process control flow (as expressed in State Diagrams) with inter-process communication (as described in Message Sequence Charts or MSCs). We develop algorithms to generate tests cases from our integrated system model, corresponding to a given test case specification. We employ our techniques on a large real-life case study involving an automotive infotainment system.

Keywords: Model-based testing, message sequence charts, communicating systems

1 Introduction

Model-based testing is a well-known software development activity. The key idea in model-based testing is to develop an explicit behavioral model of the software from informal requirements. This forms a precise specification of the software's intended behaviors. The behavioral model is searched to generate a test-suite or a set of test cases. These test cases are tried on the software (which might have been constructed manually or semi-automatically) to check the software's behaviors and match them with the intended behaviors as described

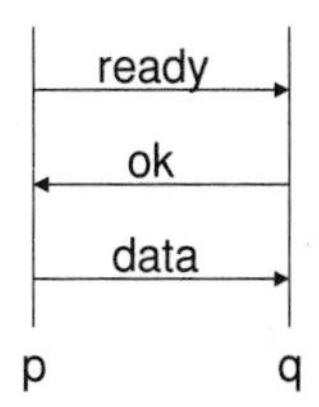

Figure 1: *An example MSC*

by the model. A collection of contributions in this area appears in the book [2].

Researchers have studied the methodological and technical issues in model-based test generation, focusing on how a given model can be best exploited for constructing test suites of real applications. In these works, different kinds of state machine like executable models have been used. Each process of the system-under-test is modeled as a (finite) state machine or I/O automata, or as a process algebraic expression (which can be compiled into a state machine). A common thread linking all these works is that they rely on *intra-process behavioral models*.

However, there has been little effort in utilizing *inter-process behavioral models*, such as the ones using Message Sequence Charts (MSCs) [26] based notations, for the purpose of test generation. Visually, a Message Sequence Chart depicts communicating processes as vertical lines; communication between processes are shown by horizontal or downward sloping message arrows between these vertical lines. Thus, MSCs emphasize *inter-process communication*. For illustration, the MSC shown in Figure 1 represents an interaction scenario between two processes p and q. In this paper, we investigate the use of *MSC-based executable models* [7, 21, 20, 9, 19] for test case generation. In particular, we focus on the notation of *Interacting Process Classes* (IPC) [9] and study test generation from it.

In the IPC notation, we describe each process-class (*i.e.* a collection of processes) in the system-under-test as a labeled transition system. However, each action label in a labeled transition system denotes a guarded Message Sequence Chart (MSC) instead of an atomic action, such as a message send or receive. Thus, if a MSC γ involves processes p and q, γ will appear in the action labels in the transition systems for process p *and* process q. *A global system execution trace in this model is a sequence of MSCs.* Further, we can symbolically execute such models without maintaining the states of individual processes in a process class.

One of the main advantages of using such MSC-based models over State-based models for test generation is that, we can represent test cases and test case specifications at a higher level than conventional model-based test generation methods. A test case specification is a user input used for guiding the test generation process [22]. Given an IPC system model where the set of all MSCs appearing in the model is Γ, we present a test case as a finite sequence of MSCs drawn from Γ. A test case specification is also a sequence of MSCs drawn from Γ where any sequence of charts containing the test case specification as a subsequence is said to satisfy the test case specification. This high level view of a test case can be helpful in providing a quick and intuitive understanding of its behavior. The user can get a fairly good idea of a test case behavior by looking at the sequence of MSCs it contains, rather than having to examine the low level details involving message send/receive events or local computations.

Also, since our IPC models can be executed symbolically, we can efficiently generate symbolic test cases for process classes with many (active) objects. Such interacting process classes are common in many application domains — telecommunication (phone and switch classes), avionics (class of aircrafts being controlled) and automotive infotainment (class of multimedia devices operating inside a car). We note that, our notion of symbolic test case groups together various concrete test-cases with similar behavior, differing only in the identities of objects of various classes participating in a test case. This makes our technique particularly suitable for testing control systems with many similar interacting processes.

Last but not the least, we note that in many application domains involving distributed control systems, the software requirements specify possible system scenarios as Message Sequence Charts (MSCs). Examples of such MSC-based software requirements are easy to find e.g. see [5, 6] for two real-life examples of distributed control systems from avionics and automotive domain. Since our behavioral model itself is MSC-based, we are able to exploit the (fragments of) MSCs present in the requirements as MSCs in the behavioral model. This establishes a tighter connection between informal software requirements, formal software models, and the test cases generated from such models. Such relationships enable easy traceability of the test execution results back to the original requirements [10].

Finally, we would like to mention that the focus of our current work is on the *test case generation* in the form of MSCs from the MSC-based executable models of process classes. There are existing works (*e.g.* [16]) covering test case execution and assigning test verdicts for the test cases derived in the form of MSCs.

Before giving the technical description of our method, we discuss related work (Section 2) and the scenario-based modeling language on which our method is based (Section 3).

2 Related Work

Model-based testing has been conventionally pursued using state diagram like behavioral models. There exists a rich theory of testing for labeled transition systems (*e.g.* see [1, 14]). The focus is on developing testing-based formal conformance or pre-order relations between implementations and models. These conformance relations can then be used to drive test generation for certifying conforming implementation as correct via testing. The TGV [13] and the STG [3] tools embody test generation work in this direction. In recent years, Pretschner et. al. [17, 18] have used state machine descriptions and symbolic execution of such descriptions (via Constraint Logic Programming) for generating tests. *Symstra* [25], a framework for generating unit tests for class level testing, also uses symbolic execution for efficient generation of test cases. We note that the symbolic test generation methods in these works are different from ours. Existing methods typically represent collections of data values within processes in the system-under-test symbolically using constraints. Each process in the system-under-test is still represented concretely, whereas we symbolically represent processes within a process class.

Message Sequence Charts (MSCs) have been historically used in software testing. The well-known Testing and Test Control Notation (popularly known as TTCN [12]) is an industrially adopted standardized test language based on MSCs. However, the conventional usage of MSCs is in describing test case specifications — the "constraints" that the generated tests should satisfy.* Typically MSCs do not appear in the behavioral models from which the tests are generated. In our work, we directly use MSCs as the *building block* for defining executable behavioral models (from which the tests are generated) as well as test case specifications (which the tests should satisfy).

We are aware that test generation from HMSCs has been studied (*e.g.* in UBET [23]). This is an inherently easier problem than ours, since it generates sequences of nodes from a graph where each node is an MSC. Moreover, HMSC specifications are not executable and due to the presence of "implied scenarios" [24] it is not even always

*The test case specification is given as a MSC (or a partial order of events) and the generated test-cases are given as a total order of events which respect this partial order.

possible to get a per-process implementation exactly corresponding to a HMSC description.

We note that there have been efforts to use temporal logics interpreted over finite sequences and model checking for those properties to generate witness traces (*e.g.* see [11]). This is similar to our problem of generating witness tests corresponding to a given test specification. However, [11] simply uses depth-first search which may produce unnecessarily long witness traces. We have investigated search heuristics which can generate short witness traces with reasonable efficiency. Our heuristics are based on A^* search [15], and appear in Section 4.

3 System Modeling

In the following, we first present a case-study which is used as a running example throughout the paper. Then, we briefly describe the modeling language and outline its symbolic execution semantics (see [9] for details).

3.1 Case Study – MOST

The MOST (Media Oriented Systems Transport) [5] is a networking standard that has been designed for interconnecting various classes of multimedia components in automobiles. It is currently maintained by the "MOST Cooperation", an umbrella organization consisting of various automotive companies and component manufacturers like BMW, Daimler-Chrysler and Audi. The MOST network employs a ring topology, supporting easy plug and play for addition and removal of any new devices. It has been designed to suit applications that need to network multimedia information along with data and control functions.

A node (or the device) in the MOST network is a physical unit connected to the network via a Network Interface Controller (NIC). A MOST system may consist of up to 64 such nodes with identical NICs. A network device generally consists of various functional blocks, such as a tuner, an amplifier, CD player etc. Each such functional block provides a number of functionalities which may be directly available to user via human-machine interface or for use by other devices in the network; for example, a CD player provides Play, Stop and Eject functions. A special function block called the NetBlock is present in all the devices and provides functions related to the entire device. Various specification documents for MOST are available at `http://www.mostcooperation.com/publications/` One of the specifications, namely the 'MOST Dynamic Specification', presents the general description of the dynamic behavior in a MOST network, encompassing: a) Network Management, b) Connection Management, and c) Power Management. Network management ensures secure communication between applications over the MOST network by maintaining and providing most recent information about various nodes, whereas Connection Management deals with the protocols for establishing communication channels between nodes in the network. Network wake-up and shutdown are handled by the Power Management component.

For our experiments, we modeled only the Network management part of MOST. From the network management perspective the system consists of: (i) Network Master (**NM**), a specific node which maintains the *central registry* containing network address information about various devices and their functional blocks in the network, and (ii) Network Slaves (**NS**), which are the remaining nodes in the network. The NM has two main functions: (a) maintaining the *central registry* with most up-to date information, and (b) providing this information to various nodes when requested. The requirements document describes the NM using two parallel processes: one requests the configuration from NS when required, and other receives the registration information sent by them. The configuration information received from slaves is checked for various errors, such as invalid or duplicate functional block addresses etc. The validity of the central registry is reflected in the NM variable 'System State' which is set to 'OK' when registry is valid and 'NotOK' otherwise. At system startup the state is always 'NotOK', and subsequently becomes 'OK' once NM is able to complete the network scanning and update the central registry without any errors. Also, some network slave may enter/leave the network resulting in 'Network Change Event' (NCE), which causes the NM to re-scan the network and communicate the changes in the registry, if any.

3.2 Interacting Process Classes (IPC)

We model a reactive system as a network of interacting process classes where processes with similar functionalities are grouped into a class. For each class, a labeled transition system is used to capture the common sequences of the computational and communication actions that the objects belonging to the class can go through. While a computational action represents a local computation within an object, a communication action represents an interaction among objects of various process classes. Each communication action names a transaction and a role played by an object of the class in the transaction. *Message Sequence Charts* are used to represent transactions.

Guarded Message Sequence Charts as Transactions

We fix a set of **transactions** Γ with γ ranging over Γ. A transaction $\gamma = (I : Ch)$ consists of a *guard* I and a Message Sequence Chart Ch. Each **lifeline**[†] in Ch is annotated with a pair (p, ρ) where p is the name of a class -from which an object playing this lifeline is to be drawn- and ρ is the chart role to be played by the lifeline ("sender", "receiver" etc.) in the interaction specified by Ch. Note that a chart Ch can have two lifelines (p, ρ_1) and (p, ρ_2), that is, *objects of the same process class may play multiple lifelines in the execution of an MSC.* For a transaction $\gamma = (I : Ch)$, the guard I consists of a conjunction of guards, one for each lifeline of Ch. In a transaction, the guard $I_r = (\Psi, \Lambda)$ associated with lifeline $r = (p, \rho)$ specifies the conditions that must be satisfied by an object O_r belonging to the class p in order for it to be eligible to play the lifeline r. These conditions consists of two components: (i) a *propositional formula* Ψ built from boolean predicates regarding the values of the (instantiated) variables owned by O_r, and (ii) a *regular expression* Λ that should be satisfied by the sequence of actions that O_r has so far gone through.

High-level LTS of a process class

We fix a set of process classes $\mathscr{P}$ with p, q ranging over $\mathscr{P}$. The transition system describing the common control flow of all the objects belonging to the class p is denoted as TS_p and is a structure of the form $TS_p = \langle S_p, Act_p, \rightarrow_p, init_p, V_p, v_{init_p} \rangle$. Here, S_p is the set of local states, $init_p \in S_p$ is the initial state and $\rightarrow_p \subseteq S_p \times Act_p \times S_p$ is the transition relation. The set of actions Act_p are the set of lifelines/roles that the p-objects can play in the transactions in Γ. Accordingly, a member of Act_p will be a triple of the form (γ, p, ρ) with $\gamma \in \Gamma$, $\gamma = (I : Ch)$ and (p, ρ) is a lifeline of chart Ch. We write (γ, p, ρ) as γ_ρ when p is clear from the context. Effect of computational steps performed by an object is described with the help of the set of variables V_p associated with p. The above description for each process class constitutes a specification in the **Interacting Process Class (IPC)** model.

[†]Each vertical line in a MSC Ch denotes a communicating process and is called a lifeline (eg. see Fig. 1).

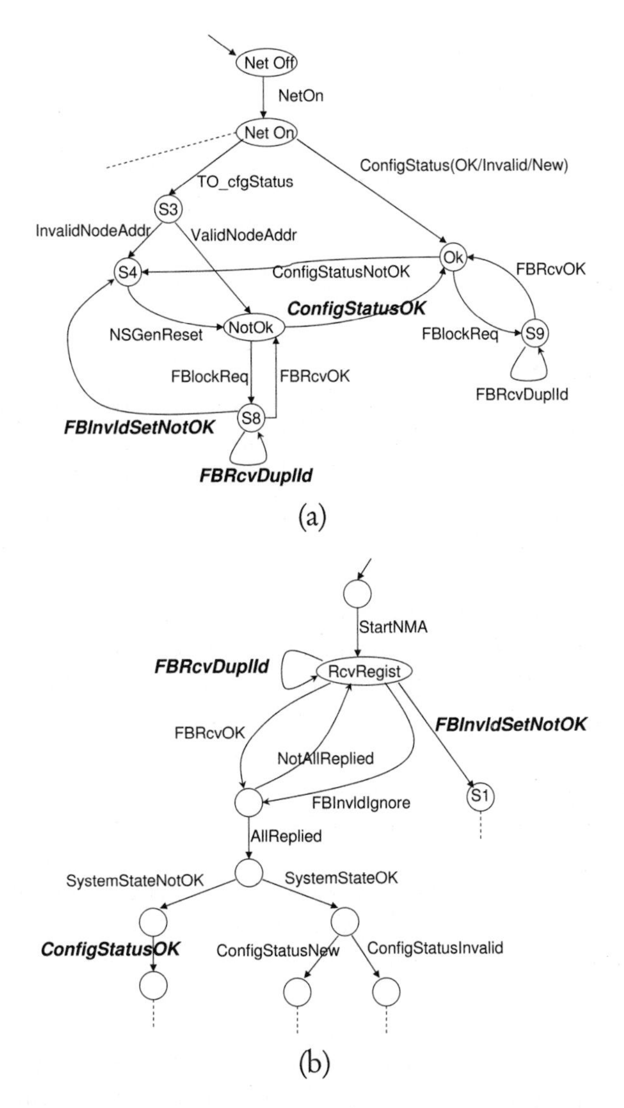

Figure 2: *Transition system fragments: (a) Network Slaves: TS_{NS}, and (b) Network Master process responsible for receiving configuration from slaves: $TS_{\text{NM-B}}$. Transaction roles are not shown in action labels to reduce visual clutter.*

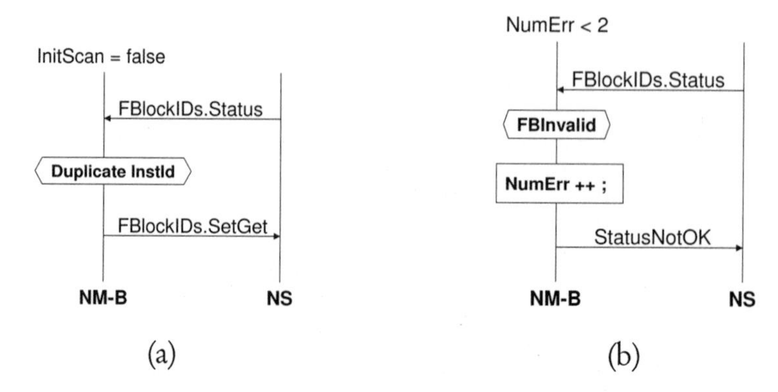

Figure 3: *MSCs (a) FBRcvDuplId, (b) FBInvldSetNotOK.*

Modeling MOST

The behavior description in the requirements document for MOST is given in the form of "high-level MSCs from the view-point of a particular process". Interestingly, the "high-level MSCs" in the requirements document are actually not HMSCs. In reality, they are rather *mistakenly* called as "high-level MSCs", and they reflect -at a very high level- the control flow *within* a process. In addition, several "scenario MSCs" in system execution are provided in the requirements document. The "scenario MSCs" of the requirements document correspond to a sequence of MSCs in our modeling, that is, a scenario in system execution. Using these scenarios, we elaborated on the very high-level per-process control flow given by the "high-level MSCs" of the document. This gave us the labeled transition system for each process class and the MSCs corresponding to the action labels of these transition systems. There was however some additional work involved in our modeling since the requirements document used both message passing as well as shared variables for inter-process communication, whereas our modeling language does not support shared variables.

As described earlier (Section 3.1), the MOST system consists of a network-master NM and several network-slaves (NS). We modeled NM using two process classes: 'NM-A', for requesting configuration from slaves and 'NM-B' for receiving responses from slaves. Also, environment is modeled as a separate process class, which takes part in network on/off events and causes the network change event (NCE) whenever a node leaves or joins the network. All network slaves are modeled as a single process class 'NS'; the number of slaves can be varied easily by changing the number of objects of this class.

Fragments of the transition system descriptions of the NS class, and the NM-B (process responsible for receiving the configuration information from slaves), are shown in Figure 2. The edges in these transition systems are labeled with transactions or guarded MSCs. Recall that the edges in the high-level transition systems of a process class in our IPC model are labeled with (transaction, role) pairs; here we have not shown the roles in each transaction to reduce visual clutter. The transactions corresponding to various transitions were easily obtained from the descriptions given in the requirements document. However obtaining the control flow for the per-process transition systems was not so straightforward. The "scenario MSCs" in the requirements document (which correspond to sequences of MSCs describing overall system behavior) were helpful in modeling the per-process control flow correctly.

For illustration, two transactions– *FBRcvDuplId* and *FBInvldSetNotOK*, from the MOST example are shown in Figure 3. Transaction *FBRcvDuplId* (Fig. 3 (a)) represents a scenario in which the address information provided by a network slave clashes with that of another slave which has replied earlier. Note that, in this transaction lifeline NM-B has a boolean guard *InitScan =false*. Thus, in order to execute this transaction, NM-B's local variable *InitScan* must be *false*. In case of transaction *FBInvldSetNotOK* (Fig. 3 (b)), a network slave replies with an invalid address information. This represents an error situation, such that NM-B keeps track of the error count using variable *NumErr*. If *NumErr* < 2, then this transaction is executed setting the system state to 'NotOK'.

Overall, the IPC specification for MOST comprises of *53 transactions* (or guarded MSCs) with process classes NS, NM-A and NM-B containing 18, 6 and 21 control states respectively.

3.3 IPC Execution Semantics

We now describe the symbolic execution semantics of our IPC model. The *symbolic* nature of our execution semantics (where the state of each individual object of a process class is not maintained) is ingrained in the definition of a configuration. The operational semantics defines how to move from one configuration to another by executing an MSC or transaction.

The definition of configuration uses the notion of "behavioral partition", which is central for defining the grouping of objects. A *behavioral partition* captures the state of an object during execution. For a process class $p \in \mathcal{P}$ in a given IPC specification (see Section 3.2) and a set of transactions Γ, we define the least set of minimal DFAs H_p as follows. A DFA $\mathcal{A}$ is in H_p if and only if there exists a transaction

$\gamma = (I : Ch) \in \Gamma$, such that $r = (p, \rho)$ is a lifeline of chart Ch, (Ψ, Λ) is the guard of lifeline r and $\mathscr{A}$ is the minimal DFA corresponding to regular expression Λ.

Definition 1 Behavioral Partition
Let $\{TS_p = (S_p, Act_p, \rightarrow_p, init_p, V_p, v_{init_p})\}_{p \in \mathscr{P}}$ *be an IPC description, and* $H_p = \{\mathscr{A}_1, \ldots, \mathscr{A}_k\}$ *be the set of minimal DFAs defined for class p. Then a* behavioral partition beh_p *of class p is a tuple* $(s, q_1, \ldots, q_k, v)$, *where*

$$s \in S_p, q_1 \in Q_1, \ldots, q_k \in Q_k, v \in Val(V_p).$$

Q_i *is the set of states of automaton* $\mathscr{A}_i$ *and* $Val(V_p)$ *is the set of all possible valuations of variables* V_p*. We use* BEH_p *to denote the set of all behavioral partitions of class* p.

Thus, two p-objects O_1 and O_2 of process class p are in the same *behavioral partition* if and only if the following conditions hold.

- O_1 and O_2 are currently in the same state of TS_p,
- They have the same valuation of local variables V_p, and
- Their current histories lead to the same state for the minimal DFAs accepting the regular expression guards appearing in TS_p.

This implies that the computation trees of two objects in the same behavioral partition are isomorphic. Since, several objects of a process class can be in the same state during execution (i.e. they map to the same behavioral partition), our *abstract execution semantics* simply keep track of the number of objects in various behavioral partitions at runtime and do not maintain their individual states and identities. To explain how abstract execution takes place, we now define the notion of an "abstract configuration".

Definition 2 Abstract Configuration
Let $\{TS_p\}_{p \in \mathscr{P}}$ *be an IPC specification* $\mathscr{S}$ *such that each process class p contains* N_p *objects. An abstract configuration of the IPC is defined as follows.*

$$\mathsf{cfg_a} = \{count_p\}_{p \in \mathscr{P}}$$

- $count_p : BEH_p \rightarrow \mathbb{N}$ *is a mapping s.t.*

$$\Sigma_{b \in BEH_p} count_p(b) = N_p$$

- BEH_p *is the set of all behavioral partitions of class* p,

So $count_p(b)$ *is the number of objects in partition* b. *The set of all* abstract *configurations of an IPC* $\mathcal{S}$ *is denoted as* $\mathscr{C}_{\mathcal{S}}$.

To keep our discussion simple, here we only consider the case where the number of objects N_p of a process class p is a given positive integer constant. In general, our IPC model allows N_p to be specified as ω, standing for an unbounded number of p objects. For more details, interested readers may refer to [9].

The system moves from one abstract configuration to another by executing a transaction $\gamma \in \Gamma$. For each lifeline $r = (p, \rho)$ of γ, our abstract execution semantics determines a *witness* behavioral partition $b_p = (s, q_1, \ldots, q_k, v) \in BEH_p$, from which a p-object can be chosen to execute lifeline r. Let (Ψ, Λ) be the guard of lifeline r and $\mathrm{H}_p = \{\mathscr{A}_1, \ldots, \mathscr{A}_k\}$ be the set of minimal DFAs defined for class p. Then, partition b_p must satisfy the following conditions–

1. $s \xrightarrow{(\gamma_\rho)} s'$ is a transition in TS_p
2. $v \in Val(V_p)$ satisfies the propositional guard Ψ.
3. For all $1 \leq i \leq k$, if $\mathscr{A}_i$ is the DFA corresponding to the regular expression of Λ, then q_i is an accepting state of $\mathscr{A}_i$.

Further, if partition b_p is assigned as *witness* partition of n roles in γ, then $count_p(b) \geq n$. This ensures that n distinct p-objects are chosen to execute these n lifelines in γ. Consequently, execution of transaction γ proceeds by selecting a distinct object from each *witness* partition to execute the corresponding lifeline. An object chosen from partition $b = (s, q_1, \ldots, q_k, v)$ and executing lifeline $r = (p, \rho)$ in γ, moves to the corresponding *destination* partition $b' = (s', q_1', \ldots, q_k', v')$ after executing r, such that–

- $s \xrightarrow{(\gamma_\rho)} s'$ is a transition in TS_p.
- for all $1 \leq i \leq k, q_i \xrightarrow{(\gamma_\rho)} q_i'$ is a transition in DFA $\mathscr{A}_i$.
- $v' \in Val(V_p)$ is the effect of executing γ_ρ on v.

Finally, the new abstract configuration reached after execution of transaction γ is determined by updating the object counts of various witness and destination behavioral partitions.

Example

For illustration, consider process class NS with 50 network-slaves in a MOST network. Assume that currently all slaves are residing in the control state *S8* of its transition system TS_{NS} in Figure 2 (a), while process NM-B is in the control state *RcvRegist* of its transition system $TS_{\mathrm{NM-B}}$ (Fig. 2 (b)). Let the values of NM-B's local variables *InitScan* and *NumErr* be *false* and *0* respectively. Further, assume that there are no local variables in the NS class and no regular expression guard for any transaction. Then, this configuration corresponds to

$$c = \{\langle S8 \rightarrow 50\rangle, \langle (RcvRegist, InitScan = false, Numerr = 0) \rightarrow 1\rangle\}.$$

For process class NS, its behavioral partition is described using only the control state *S8* from TS_{NS}. For process class NM-B, a behavioral partition consists of control state *RcvRegist* from $TS_{\mathrm{NM-B}}$ and the values of its local variables *InitScan* and *NumErr*. Note that we have only shown behavioral partitions with non-zero objects, which is how we keep track of them during simulation. Also, we have omitted NM-A's state for the purpose of illustration. We can now execute the transaction *FBInvldSetNotOK* (transition $S8 \rightarrow S4$ in Figure 2(a) and $RcvRegist \rightarrow S1$ in Figure 2(b)) which is shown in Fig. 3 (b), with any one slave object executing the lifeline marked NS. The resulting configuration is

$$c' = \{\langle S4 \rightarrow 1, S8 \rightarrow 49\rangle, \langle (S1, InitScan = false, Numerr = 1) \rightarrow 1\rangle\}.$$

Thus, the slave object participating in *FBInvldSetNotOK* moves to control state *S4* in TS_{NS}, while NM-B moves to control state *S1* in $TS_{\mathrm{NM-B}}$ with the value of its variable *NumErr* getting updated to 1. This captures the basic idea in our symbolic execution semantics (note that we did not maintain the local states of 50 slave objects separately).

4 Meeting Test Specifications

In this section we describe the automatic generation of test cases from an IPC model based on a user-provided test case specification. The user gives *a sequence of transactions*, as a test specification. The test generation procedure makes use of guided search to generate a transaction sequence containing the user-provided test specification sequence as a *subsequence*. Note that it is possible that there is no execution sequence that can satisfy a given test specification.

4.1 Problem Formulation

The user gives a sequence of transactions $\tau_1, \tau_2, \ldots, \tau_n$ as the test specification. The test-case generation procedure aims at producing one or more test sequences of the form $\tau_1^1, \ldots, \tau_1^{i1}, \tau_1, \tau_2^1, \ldots, \tau_2^{i2}, \tau_2, \ldots, \tau_n^{in}, \tau_n$.

This problem can be viewed as finding a witness trace in the IPC model satisfying the Linear-time Temporal Logic (LTL) [4] property $\mathbf{F}(\tau_1 \wedge \mathbf{F}(\tau_2 \wedge (\ldots(\mathbf{F}\tau_n)\ldots)))$. We always generate only finite witness traces (*i.e.*, a finite sequence of transactions) such that any infinite trace obtained by extending our finite witness trace will satisfy the above-mentioned LTL property. This can be accomplished by standard search strategies like breadth-first or depth-first search. Breadth-first search produces shortest-possible test traces (the sequence of MSCs generated), but it is expensive in time and memory. On the other hand, *depth-first* search can help us find test-cases efficiently, but the generated sequence of MSCs may not be optimal in length. Hence, we investigate intelligent search heuristics for this problem.

4.2 A^* Search

Various well-known heuristic search strategies such as best-first, and A^* (pronounced A-star) [15] have been shown to be useful in *test-case generation* [18], and *model checking* [8]. The heuristics mainly differ in the evaluation function used by them, which gives the "desirability" of expanding a node in the state graph. The search proceeds by expanding the graph choosing the most desirable node first. The evaluation function $f(s)$ used in A^*, which evaluates the state s during state-space exploration (lower score is better), consists of two parts– (i) $g(s)$, giving the shortest generating path length for state s, i.e. length of the shortest path from start-state to s, and (ii) $h(s)$, the *estimated* cost of the cheapest path to the goal state from s. The evaluation function $f(s) = g(s) + h(s)$ gives the estimate of cheapest path from start state to the goal state, passing through s. If $h^*(s)$ is the actual cost of the cheapest path from a state s to the goal state and $h(s) \leq h^*(s)$, for all states s, then the heuristic is said to be *admissible* and guarantees to find the shortest path to goal state, if one exists.

We adapt and modify the A^* algorithm to guide our test-selection process. While searching for witness traces, we break the search into steps, such that each step aims at generating a transaction sequence up to the next uncovered transaction in the test specification. So if the test specification is $\tau_1, \tau_2, \ldots, \tau_n$ and so far the search has produced a witness for $\tau_1, \ldots, \tau_{i-1}$ (where $i \leq n$), the "goal" is the next transaction in the test specification — τ_i. The search for this "goal" will of course start from a state appearing at the end of the witness trace found for $\tau_1, \ldots, \tau_{i-1}$. Then, for a state s visited while searching for a path to τ_i, the value of function $g(s)$ in A^*'s evaluation function determines the length of the shortest path to s seen so far from the initial state, and covering the test specification transactions that have already executed (in the order of their occurrence). While, for computing the h function, we only focus on the next goal (τ_i) instead of the whole remaining sequence to be covered ($\tau_i, \ldots, \tau_n$).

During the search we maintain a global search tree $\mathcal{T}$ capturing the states visited so far. A **global state** and the **successor** of a global state, appearing in $\mathcal{T}$ during the test generation are defined as follows.

Definition 3 Global state
Given an IPC system model and a test case specification $\tau_1, \ldots, \tau_n$, a **global state** *$s = (c, i)$ consists of: (i) an abstract system* **configuration** *c in the IPC model (see Section 3.3) and (ii) i, the index of the next* **goal transaction** *τ_i (while searching for the test case) in the test case specification, where $1 \leq i \leq n$.*

Definition 4 Successor state
Given a global state *$s = (c, i)$ as defined above, a global state $s' = (c', i')$ is a successor state of s if and only if: (i) c' is obtained from c in our symbolic execution semantics by executing some transaction τ', and (ii) $i' = i + 1$ if $\tau' = \tau_i$ (i.e. the next goal transaction), and $i' = i$ otherwise. We use* **Succ(s)** *to denote the set of all possible successors of s.*

In the search tree $\mathcal{T}$, each edge from a state s to its successors is labeled with the transaction name that was executed at s leading to that successor. Also for each state s in $\mathcal{T}$ we maintain the values $g(s)$ and $h(s)$.

Computation of heuristic function h. To compute $h(s)$ for a given a state $s = (c, i)$, we consider the process classes involved in the transaction τ_i, that is, the process classes whose objects should appear as the lifelines of the MSC in τ_i (the next goal transaction). Let this set of classes be $classes(\tau_i)$. For each process class $p \in classes(\tau_i)$, we determine the length of the shortest path in TS_p from current control state(s)[‡] of p-objects to the source state(s) of the transition(s) which appear as a lifeline in τ_i. Note that, we pre-compute the shortest paths between all state-pairs in the Labeled Transition Systems (LTSs) of different process-classes once and for all, at the beginning of the test generation. Clearly, different objects of a process class p can be in different control states — so we need to consider all the control states in which any object of a process class p is currently in. Let the

[‡] TS_p is the transition system describing process class p.

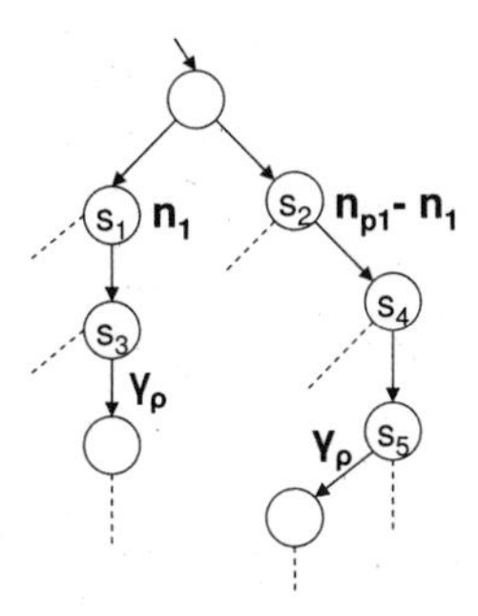

Figure 4: *Transition system fragment for process class p_1*

shortest distance computed in this way for process class p be denoted as $d_p^{\tau_i}$.

For illustration, consider a fragment of transition system describing a process class p_1 as shown in Figure 4. Assume that p_1 has n_{p1} objects such that n_1 (> 0) p_1-objects are currently in control state s_1, while remaining $n_{p1} - n_1$ objects are in control state s_2 (see Fig. 4). Let the next goal transaction be γ and (p_1, ρ) be the only lifeline involving process class p_1 in transaction γ (*i.e.* $p_1 \in classes(\gamma)$). Now, in the transition system of p_1 (fragment of which is shown in Fig. 4), the shortest paths to the control state(s) with an outgoing transition γ_ρ from states s_1 and s_2 respectively are— $s_1.s_3$ and $s_2.s_4.s_5$, having path lengths *one* and *two* (see Fig. 4). Hence, we get $d_{p1}^{\gamma} = 1$.

Finally, we define

$$h(s) \stackrel{def}{=} max_{p \in classes(\tau_i)} d_p^{\tau_i} \text{ and } f(s) = g(s) + h(s) \qquad (1)$$

Intuitively h gives a lower bound on the trace length required to execute τ_i from s, and therefore, gives us an *admissible* heuristic.

4.3 Test Generation Algorithm

We now discuss the overall test generation procedure described using two functions, one top level or driver function **genTest** presented in Algorithm 1 which makes use of the second function **genTrace** described in Algorithm 2 (appears in Appendix A). In particular, the function *genTest* (Algorithm 1) takes as input three parameters: a set S of global states, i is the index of the current goal transaction and n is the length of the test-case specification $\tau_1, \ldots, \tau_n$. During test generation, states $s = (c, i)$ from the set S are evaluated (i.e. $f(s)$ is computed using Equation 1), and the one with the minimum value of function f is chosen and explored (i.e. all its successors are generated). The initial call to this procedure is `genTest({s_init},1,n)` where $\mathtt{s_{init}} = (c_{init}, 1)$, c_{init} is the initial *configuration* in our IPC system model. The search tree $\mathcal{T}$ is also initialized, having a single root node s_{init}.

Algorithm 1 genTest(S,i,n)

1: **while** $S \neq \phi$ **do**
2: $\langle GoalStates, UnExplored \rangle \leftarrow$ **genTrace**(S,τ_i)
3: **if** *GoalStates* $= \emptyset$ **then**
4: **if** $i = 1$ **then**
5: report Failure and Exit
6: **else**
7: **return**
8: **end if**
9: **else if** $i < n$ **then**
10: **genTest**($GoalStates$,$(i+1)$,n)
11: $S \leftarrow$ *UnExplored*
12: **else**
13: output *witness(GoalStates)* and Exit $/{*}i = n{*}/$
14: **end if**
15: **end while**

For finding a trace from states in the set S to the next goal transaction τ_i, *genTest* calls `genTrace(`S`,`τ_i`)` (line 2 of Algorithm 1). The function *genTrace* described in Algorithm 2 then tries to find a path leading to the execution of transaction τ_i from states in S (see Appendix A). It returns: (a) *GoalSet*, the set containing the states reached after successfully finding a path from a state in S to the transaction τ_i, and (b) the set *UnExplored*, which contains states that were reached but not explored when searching for the trace. If *GoalSet* is empty, this means that *genTrace* failed to generate a trace from S reaching τ_i. In this case, if $i = 1$, the *genTest* algorithm reports failure and exits, since it cannot find any trace up to the first transaction in the test specification; otherwise, the *genTest* algorithm backtracks (line 7 of Algorithm 1) and tries to find another trace for the previous goal transaction using the remaining unexplored states from that step. This occurs by assigning the set *Unexplored* to S when a failed recursive call returns (line 11 of Algorithm 1). The new S is explored further due to outermost **while** loop. If *GoalSet* is not empty, *genTest* is called recursively using the set *GoalSet* as the starting states for finding the next transaction (τ_{i+1}) in the test specification. In this way if we have encountered τ_n and *GoalSet* is still not empty, a witness trace satisfying the test specification has been found. The *genTest* algorithm then outputs the witness trace and exits.

5 Experimental Results

Given our modeling of the MOST network-management, we performed experiments for (a) generating test cases corresponding to various test specifications, (b) comparing test case generation for symbolic vs concrete model execution for a given test specification, and (c) testing an implementation of the MOST protocol (derived independently as C++ code) using test cases derived.

Witness trace generation

We derived a number of test specifications based on various scenario MSCs given in the MOST Dynamic specification. These test specifications represent different use cases. Test cases were successfully derived for these test specifications using the A^*-based heuristic approach. Experimental results showing the witness trace lengths and test-generation times corresponding to four test specifications are shown in Table 1 in the columns under 'RESULTS A'. Recall that each witness trace contains the corresponding test specification as a subsequence. All experiments were performed on a machine with 3GHz CPU and 1 GB of memory.

We also generated witness traces for the same test specifications using breadth-first (BFS) and depth-first strategies (DFS). In most of the cases A^* based approach generated optimal length traces, i.e. same as those generated using BFS[§], taking only a fraction of time. On the other hand, test cases generated using DFS were up to 3 times longer than those generated using A^* heuristic.

For illustration, let us consider the following test specification (corresponding to the first test specification in Table 1): *FBRcvDuplId, ConfigStatusOk*. As discussed in Section 4, for a given test specification (given as a sequence of transactions), test generation procedure attempts to find a witness trace (another sequence of transactions) containing the test specification as a subsequence. The transaction *FBRcvDuplId* (transition $S8 \rightarrow S8$ in Figure 2(a) and *RcvRegist* $\rightarrow$ *RcvRegist* in Figure 2(b)) corresponds to the scenario shown in Figure 3 (a)– which takes place when the configuration information sent

---

[§]Given a test specification $\tau_1, \ldots, \tau_n$, we employ A^* based search to find a smallest path to τ_1, search from that occurrence of τ_1 to find a smallest path to τ_2 and so on. Clearly, adding up these smallest paths may not produce the minimal path containing $\tau_1, \ldots, \tau_n$ as a subsequence.

Example	Test Spec.		RESULTS A		RESULTS B				
			Witn. trace	Test gen.	Expl.	# Witnesses		Exec. Times(sec)	
	#	Length	length	Time(sec)	Depth	S	C	S	C
MOST (3 slaves)	1	2	28	67	30	16	–	272	> 10 min.
	2	2	19	4.5	19	7	21	28	480
	3	3	32	312	32	4	–	460	> 10 min.
	4	2	34	445	34	–	–	> 10 min.	> 10 min.

Table 1: *RESULTS A: Witness test generation– test lengths & generation times. RESULTS B: Comparing Symbolic/Concrete test generation, S ≡ Symbolic, C ≡ Concrete.*

by slave to master clashes with the configuration information for some other slave node already registered with the master. In response, when it is not the first time that the system is being scanned after the network was powered on (as indicated by the guard 'InitScan = false' of lifeline NM-B in Figure 3 (a)), network master assigns a new id for this slave and sends this value for acceptance to the slave via message *FBlockIDs.SetGet*. This new value may be accepted or rejected by the slave node. In case it is accepted, then the new value is entered in the central registry by NM-B. The transaction *ConfigStatusOk* (whose MSC is not shown here) corresponds to the communication of 'SystemState = OK' by network master to the slave nodes, once all the nodes have responded correctly. As shown in Table 1, a test-case MSC sequence of length 28 was obtained within 67 seconds for this test specification. In this manner, the test specifications, though small in length, can be used to derive test cases representing complex system behaviors.

Symbolic nature of our tests

Recall that our IPC models inherently support *symbolic* execution of process classes (Section 3.3) and hence generation of symbolic tests where processes are grouped together in terms of behavior. To evaluate the advantage of symbolic test generation, we generated all possible witness traces corresponding to the given test specifications using both *symbolic* and *concrete*[¶] execution semantics. This was done by exploring the MOST system model up to a given depth bound. For a given test specification, in order to guarantee the generation of at least one witness trace during exploration, the length of its witness trace derived from A^*-based test generation was used as a cut-off for the depth bound.

The results comparing the test-suite size and test generation times appear in Table 1 in the columns under 'RESULTS B'. As we can

[¶]The state of each object is maintained separately in concrete execution semantics.

observe, using symbolic execution, all the witness traces could be generated (up to the given depth bound) for the first 3 test specifications. Whereas, for concrete execution, we were only able to generate the results for the second test specification. Moreover, in this case, 21 concrete test cases were grouped into 7 symbolic test cases with much lower generation time for the symbolic case. Note that, we used the time limit of 10 minutes for these set of experiments.

Use of our tests
We note that the tests generated from our IPC model can be executed on a distributed system implementation, that is, executable code in a programming language. This is the case where the system implementation is generated manually using the informal system requirements as a guide. In this case, the tests increase our confidence in system implementation. Alternatively, it is conceivable that the system implementation is generated from the system model, and so are the test cases. In this case, the tests can be used to increase our confidence in the system model itself (which is generated manually from the informal requirements). For more detailed discussion on the use of model-based tests in model-driven software development, the reader is referred to [10].

6 Discussion

In this paper, we have studied test generation for reactive systems using models which combine state-based intra process control flow with scenario-based inter process communication. The key novelty of our work is the uniform usage of scenarios in the system model, in test specification or test purpose as well as in generated test cases. Our hypothesis is that the scenarios are widely present in system requirements documents for expressing system behavior. So, a scenario-based system modeling and testing methodology can lead to wider and more direct exploitation of the requirements documents for reliable software design.

References

[1] E. Brinksma. A theory for the derivation of tests. In *Proc. 8th Int. Conf. Protocol Specification, Testing and Verification*, pages 63–74. North-Holland, 1988.

[2] M. Broy, B. Jonsson, J-P. Katoen, M. Leucker, and A. Prestchner, editors. *Model-Based Testing of Reactive Systems*, volume 3472 of *Lecture Notes in Computer Science*. Springer, 2005.

[3] Duncan Clarke, Thierry Jéron, Vlad Rusu, and Elena Zinovieva. Stg: a tool for generating symbolic test programs and oracles from operational specifications. volume 26, pages 301–302. ACM, 2001.
[4] E.M. Clarke, O. Grumberg, and D. Peled. *Model Checking*. MIT Press, 1999.
[5] MOST cooperation. Media oriented system transport. `http://www.mostcooperation.com/`.
[6] CTAS. Center TRACON automation system. `http://www.ctas.arc.nasa.gov`.
[7] W. Damm and D. Harel. LSCs: Breathing life into message sequence charts. *Formal Methods in System Design*, 19(1):45–80, 2001.
[8] Stefan Edelkamp, Alberto Lluch Lafuente, and Stefan Leue. Directed explicit model checking with HSF-SPIN. In *SPIN workshop on Model checking of software*, volume 2057, pages 57–79. Springer-Verlag, 2001.
[9] Ankit Goel, Sun Meng, Abhik Roychoudhury, and P. S. Thiagarajan. Interacting process classes. In *ICSE*, pages 302–311. ACM, 2006.
[10] Ankit Goel and Abhik Roychoudhury. Synthesis and traceability of scenario-based executable models. In *Invited Paper, ISoLA*, pages 347–354. IEEE Computer Society, 2006.
[11] Elsa L. Gunter and Doron Peled. Temporal debugging for concurrent systems. In *TACAS*, volume 2280, pages 431–444. Springer-Verlag, 2002.
[12] Telelogic Inc. Testing and test control notation (TTCN-3). `http://www.telelogic.com/corp/standards/ttcn-3.cfm`.
[13] Claude Jard and Thierry Jéron. Tgv: theory, principles and algorithms: A tool for the automatic synthesis of conformance test cases for non-deterministic reactive systems. *Int. J. Softw. Tools Technol. Transf.*, 7(4):297–315, 2005.
[14] Rocco De Nicola and Matthew Hennessy. Testing equivalence for processes. In *Intl. Colloqium on Automata Languages and Programming (ICALP)*, volume 154 of *Lecture Notes In Computer Science*, pages 548–560. Springer-Verlag, 1983.
[15] N.J. Nilsson. *Principles of Artificial Intelligence*. Morgan Kaufmann, 1993.
[16] S. Pickin, C. Jard, T. Jeron, Jean-Marc Jezequel, and Y. Le Traon. Test Synthesis from UML Models of Distributed Software. *IEEE Trans. Softw. Eng.*, 33(4):252–269, 2007.
[17] A. Pretschner et al. One evaluation of model-based testing and its automation. In *ICSE*, pages 392–401. ACM, 2005.
[18] Alexander Pretschner. Classical search strategies for test case generation with constraint logic programming. In *In Proc. Formal Approaches to Testing of Software*, pages 47–60. BRICS, 2001.
[19] A. Roychoudhury, A. Goel, and B. Sengupta. Symbolic Message Sequence Charts. In *ESEC-FSE*, pages 275–284. ACM, 2007.
[20] A. Roychoudhury and P.S. Thiagarajan. Communicating transaction processes. In *Intl. Conf. on Application of Concurrency to System Design (ACSD)*, page 157. IEEE Computer Society, 2003.
[21] B. Sengupta and R. Cleaveland. Triggered message sequence charts. In *FSE*, pages 167–176. ACM, 2002.
[22] J. Tretmans and E. Brinksma. Côte de Resyste – Automated Model Based Testing. In *Progress 2002 – 3rd Workshop on Embedded Systems*, pages 246–255. STW Technology Foundation, 2002.
[23] UBET. Ubet, 1999. `http://cm.bell-labs.com/cm/cs/what/ubet/`.
[24] S. Uchitel, J. Kramer, and J. Magee. Detcting implied scenarios in message sequence chart specifications. In *ESEC-FSE*, pages 74–82. ACM, 2001.
[25] Tao Xie, Darko Marinov, Wolfram Schulte, and David Notkin. Symstra: A framework for generating object-oriented unit tests using symbolic execution. In *TACAS*, volume 2280 of *Lecture Notes In Computer Science*, pages 365–381, 2005.
[26] Z.120. Message Sequence Charts (MSC'96), 1996.

Appendix

A Test generation Algorithm *genTrace*

The function *genTrace* described in Algorithm 2 takes as input two parameters: a set S of states to be explored, and the current goal transaction τ. It then tries to generate a trace to the current goal transaction τ from a state in S. It maintains the set *Open* containing the states yet to be explored and the set *GoalSet* to store the state(s) reached after executing the *goal* transaction τ. The set *GoalSet* is initially $\emptyset$. At the end of each iteration of the **while** loop (lines 2–35, Algorithm 2) if *GoalSet* is not empty, then we have found a trace up to the current goal transaction. We return the set *GoalSet* and also the set *Open* containing the unexplored states (lines 32–33, Algorithm 2).

In each iteration of the **while** loop, a state s with minimum value of $f(s)$ $(= g(s) + h(s))$ is chosen and removed from *Open* (line 3, Algorithm 2). All successors of s are generated and the search tree capturing states visited so far is updated. Note that in the search tree capturing the states visited so far, we maintain *one* of the predecessors of each visited state s as the "parent" of s. This is done to remember the shortest path from the start state to state s, such that it includes all previous goal transactions (i.e. transactions appearing in the test specification that have already been executed) in the order of their occurrence.

Thus, the **parent pointer** of each state s is set to a state x such that (a) s is a successor of x and (b) the current shortest path from start state to s consists of the current shortest path from start state to x (covering the previous goal transactions in the order executed) and the edge $x \rightarrow s$. For each successor m of state s, the following updates need to be performed.

Case 1: (line 7, Algorithm 2) If m has *not* been reached earlier, it is added either to *GoalSet* or *Open* (lines 9 and 12, Algorithm 2), depending on whether m is reached by executing the current goal transaction τ or not. The value of h function (appearing in A^*'s evaluation function) is also computed accordingly. Finally, we compute the value of g function of A^*'s evaluation function, and set the node s as m's parent in the search tree capturing visited states.

Case 2: (line 18, Algorithm 2) Otherwise, we check if $m \in Open$ and if so, we know that m has been reached earlier via an alternate path. In this case if the new path to m is shorter, we accordingly update the function g for m. Note that the function h remains unchanged since

Algorithm 2 genTrace(S, τ): adapted from A^*

```
1: Open ← S; GoalSet ← ∅
2: while Open ≠ ∅ do
3:    s ← getMin_f(Open); Open ← Open − {s}
4:    for all m ∈ successors(s) do
5:       setSuccessor(m,s) /* Add m as a successor of s*/
6:       if notVisited(m) then
7:          /* Case 1: m not visited yet */
8:          if (m is a destination state of a τ transaction) then
9:             GoalSet ← GoalSet ∪ {m}
10:            h(m) ← distance from m to next goal transaction (after
               τ), use Equation (1)
11:         else
12:            Open ← Open ∪ {m}
13:            h(m) ← distance from m to τ, use Equation (1)
14:         end if
15:         g(m) ← g(s) + 1 /* length of generating path */
16:         setParent(m,s)
17:      else if m ∈ Open then
18:         /* Case 2: m reached earlier, successors not explored */
19:         if (g(s) + 1) < g(m) then
20:            /* we have found a shorter path to m via s */
21:            g(m) ← g(s) + 1; setParent(m,s)
22:         end if
23:      else
24:         /* Case 3: m reached earlier and successors explored */
25:         if (g(s) + 1) < g(m) then
26:            /* we have found a shorter path to m via s */
27:            g(m) ← g(s) + 1; setParent(m,s);
28:            updateAll_g(m) /*Update value of function g and par-
               ent of nodes from m downwards*/
29:         end if
30:      end if
31:   end for
32:   if GoalSet ≠ ∅ then
33:      return (GoalSet,Open)
34:   end if
35: end while
36: return (∅,∅)
```

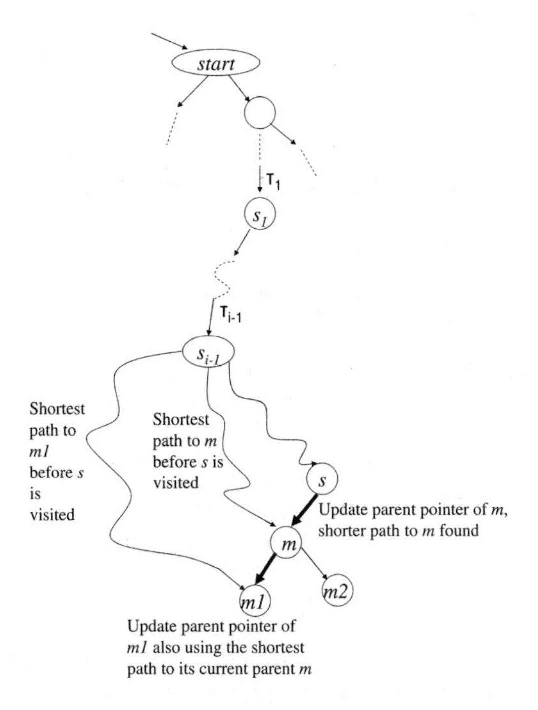

Figure 5: *Updating parent pointers in search tree — line 28 of* genTrace *(Alg. 2)*

it estimates the length of the path from m to τ, which is not affected by the path through which m is reached from the start state.
Case 3: (line 24, Algorithm 2) In case m is not present in *Open* also, then it was reached *and* its successors were explored earlier. Again, we check if the new path to m is shorter. If it is so, then its g value is updated and parent pointer set to s. Further, in this case we check the children of m in the search tree capturing the visited states. For all the successors already pointing to m as their parent, their g value is updated. For all other successors, if the new path to them via m is shorter, then m is set as their parent, also updating their g value. This process is repeated for successors of these successors and so on (see Figure 5 for a pictorial explanation).

How Hard is Smart Play-Out? On the Complexity of Verification-Driven Execution *

David Harel[1] **, Hillel Kugler**[2] **, Shahar Maoz**[1] **, Itai Segall**[1]
[1]*The Weizmann Institute of Science, Israel*
{dharel,shahar.maoz,itai.segall}@weizmann.ac.il
[2]*Microsoft Research, Cambridge, UK*
hkugler@microsoft.com

Abstract

Smart play-out is a method for executing declarative scenario-based requirements, which utilizes powerful model-checking or planning algorithms to run the scenarios and avoid some of the violations that can be caused by naïve execution. In this paper, we investigate the complexity of smart play-out. Specifically, we use a reduction from QBF in order to show that smart play-out for a most basic subset of the scenario-based language of LSC is PSPACE-hard. The main advantage of our proof compared to a previous one by Bontemps and Schobbens is that ours is explicit, and takes advantage of the visual features of the LSC language. We also show that for a subset of the language, in which no multiple running copies are allowed, the problem is NP-hard.

Keywords: Scenario based specifications, live sequence charts, model-checking, complexity

1 Introduction

Live sequence charts (LSCs) [2] constitute a visual formalism for inter-object scenario-based specification and programming of reactive systems. The language extends classical *message sequence charts* (MSC) [10], mainly by adding universal and existential modalities. Thus, LSCs

*The research was supported in part by The John von Neumann Minerva Center for the Development of Reactive Systems at the Weizmann Institute of Science and by a Grant from the G.I.F., the German-Israeli Foundation for Scientific Research and Development.

distinguish between behaviors that may happen in the system (existential, cold) and those that must happen (universal, hot). A universal chart contains a *prechart*, which specifies the scenario which, if satisfied, forces the system to satisfy also the scenario given in the actual chart body.

An operational semantics was defined for the language in [8]. The result of this semantics is an execution technique for LSCs, called *play-out*, in which an LSC specification consisting of a set of LSCs is executed directly. However, the execution engine presented in [8] is not enough, due to its naïve nature. At each point in time, it chooses one step that is legal at that time, without considering the consequences of one choice or the other. Thus, it may choose steps that eventually lead to a violation even though other steps could have avoided it. The notion of *smart play-out* is therefore introduced in [7]. Two implementations have been suggested to date for smart play-out [7, 9].

In this paper, we analyze the theoretical bounds on the complexity of smart play-out, and prove the problem to be PSPACE-hard. We also show an interesting subset of the problem, in which no multiple running copies are allowed, to be NP-hard.

Despite the high theoretical worst case complexity, in practice, there are interesting LSC specifications for which smart play-out can be applied successfully. We believe that since LSC specifications are man-made, they will inherently have some "logical" structure, which can be exploited for optimizing smart play-out techniques, causing them to be practical. This issue is further discussed in Section 5.4. Another reason that smart play-out may be effective despite its worst case complexity is rooted in the significant advances made over the past few years in verification methods and tools for analyzing large complex systems.

The paper is organized as follows. Section 2 presents preliminary material on LSCs. Section 3 defines the problem of smart play-out. In Section 4 we prove the complexity results for the problem. In Section 5 we discuss some issues related to the smart play-out problem and its complexity. Finally, Section 6 covers some related and future work. The formal proofs omitted from the body of the paper are given in Appendix A.

2 Preliminaries

LSCs inherit the syntactic structure and visual representation from MSCs. An LSC contains vertical lines, termed *lifelines*, which denote objects, and *events*, involving one or more lifelines, thus inducing a partial order. The most basic construct of the language are messages: a message is denoted by an arrow between two lifelines (or from a lifeline to itself), representing the event of the source object sending a message to the target object. Objects are partitioned into environment-controlled and system-controlled ones, and messages are said to be controlled by the environment or system according to the sending object. A universal LSC, the kind used to drive the execution, is divided into two parts, the prechart and the main chart. The intended semantics is that whenever the prechart is satisfied, the main chart must also be satisfied. LSCs are multi-modal; almost any construct in the language can be either cold (usually denoted by the color blue) or hot (denoted by red), with a semantics of "may happen" or "must happen", respectively. If a cold element is violated (say a condition that is not true when reached), this is considered a legal behavior and some appropriate action is taken. Violation of a hot element, however, is considered a violation of the specification and must be avoided.

An LSC may also contain conditions, *if-then-else* constructs, etc., as well as more advanced constructs such as symbolic instances (representing classes instead of individual objects), and forbidden elements. Using conditions, one can also express *anti-scenarios*, i.e., scenarios that are forbidden, by introducing an LSC with the entire forbidden scenario as a prechart, and a hot FALSE condition as the main chart.

An example of an LSC can be seen in Figure 1(a). The prechart is denoted by the dashed blue hexagon, and the main chart by the solid black rectangle. The prechart contains a single message, in which the object $Y\{i-1\}T$ sends `Choose_Next` to itself. The main chart states that XiT should send `m` to XiF, then YiT should send `m` to YiF, etc. Note the `SYNC` construct, dictating an order between these two messages, which would have otherwise been unordered.

An operational semantics was defined for the language in [8]. The result of this semantics is an execution technique for LSCs, called *play-out*, in which an LSC specification consisting of a set of LSCs is executed directly. In this semantics, the notion of *cut* is used. The cut includes one location on each lifeline, separating the set of messages that have already occurred from those that have not. Whenever a message that is minimal in the prechart is sent, a *running copy* of this chart is created, and the cut starts to progress. A runtime *configuration* of an LSC specification is a set of running copies and their corresponding cuts. Messages appearing immediately after the cut are termed *enabled*. If a message that appears in a chart is sent while not being enabled in it, the chart is violated (either causing a violation of the specification or a graceful exit of the chart, according to the modalities/temperatures).

Note that if a message appears twice in a chart, once as minimal and again as non-minimal, then this chart may have multiple running copies at runtime: if the second appearance is enabled, and the message occurs, then this copy must advance its cut to after the message and another copy opens, advancing its own cut to be directly after the minimal event. The complexity for the case where multiple running copies are not allowed is discussed in Section 4.2.

3 The Smart Play-Out Problem

Given an initial configuration, we define a *superstep* to be a finite set of steps executed by the system and ending in a state in which the system has no more obligations. This means that after performing the superstep all the main charts † have completed successfully. We also assume that all events in the superstep were taken as a result of appearing explicitly in the main chart of an LSC and being enabled; that is, we do not consider the case of taking events 'spontaneously', which may be in general helpful by allowing violation of precharts, amounting to making less commitments to satisfy main charts.

The problem of *smart play-out* is defined as the problem of finding a legal superstep from a given initial configuration if such a superstep exists, or deciding and reporting that no such superstep exists otherwise. This problem is proven to be PSPACE-hard in [1]. In the present paper, we give an alternative proof, by a reduction from QBF, the canonical PSPACE-complete problem [3].

We prove this complexity for the subset of LSCs containing only messages. As a first step, we allow also synchronization constructs and anti-scenarios (charts for which the main chart is merely a FALSE condition, causing the completion of the prechart to be an immediate violation of the specification), and then explain how these can be removed.

Note that we assume nothing about the structure of the charts themselves. For example, a message may appear several times in a chart. In particular, a message that is minimal in the prechart can also appear again in the same chart. This feature allows multiple copies of the same chart to be active simultaneously at runtime, each at a different cut. Although our main focus is on the most general case, the

† To simplify the discussion we assume that the main charts contain only system events.

case in which multiple running copies are disallowed is also discussed in Section 4.2.

4 How Hard is Smart Play-Out?

4.1 The general problem

Theorem 1 *Any smart play-out mechanism supporting messages, synchronization constructs, and anti-scenarios is PSPACE-hard.*

Proof We prove this by a reduction from QBF. Given a QBF formula $\varphi = \exists x_1 \forall y_1 \exists x_2 \forall y_2 \cdots \exists x_n \forall y_n(\psi)$, where ψ is a CNF formula over variables $\{x_1, y_1, \ldots, x_n, y_n\}$, we build a system and an LSC specification. The specification is built such that a superstep exists (from a specific initial configuration) if and only if the formula is true.

Intuitively, the superstep will backtrack over the variables x_i, y_i, where it will non-deterministically choose a value for the x_i's and check both options for the y_i's. For each such assignment to all variables, it checks that the CNF formula ψ holds.

We now describe the system and specification generated, first intuitively and then more formally. The objects in the system are $v_i T, v_i F$ for each variable $v_i, v \in \{x, y\}$. The assignment of a TRUE value to a variable v_i is denoted by a message `m` passing from $v_i T$ to $v_i F$ and then back. A FALSE value is denoted by the same two messages, in the reverse order (i.e., first from $v_i F$ to $v_i T$).

Some of these objects may also send various messages to themselves, as follows.

- A message `Choose_Next` sent from an object $y_i T$ to itself denotes the fact that x_i and y_i have both received values and the execution can proceed to the next pair.
- A message `Done` sent from an object $y_i T$ to itself indicates a backtrack from y_i — if y_i was previously TRUE, then the FALSE value should now be checked, and if it was FALSE the execution will backtrack to $i - 1$.
- A message `Check` sent from either $x_i T$ or $y_i T$ to itself denotes the fact that all variables are assigned values. The execution will now "resend" these values and check that the assignment is legal (i.e., that ψ holds).
- A message `Done_Check` sent from either $x_i T$ or $y_i T$ to itself denotes the fact that this variable was checked. Once all variables

x_i, y_i are checked without the specification being violated, we know that the current assignment is legal (i.e., ψ holds).

The specification consists of three groups of LSCs, with the following roles:

- Type 1 LSCs take care of the backtracking stage. They backtrack over the variables x_i and y_i, while non-deterministically choosing values for the x_i's and checking both options for the y_i's.
- Type 2 LSCs act as the memory of the system; they keep the most recent assignment to each variable, and once all variables are assigned values they "resend" these values, so that ψ can be checked.
  If a backtracking iteration changes only some of the values, these LSCs are the ones in charge of remembering the values of the rest of the variables.
- Type 3 LSCs check that ψ holds in the current assignment. Each LSC checks one clause, and if a violated clause is detected, it causes the specification to be violated.

Formally, the system consists of $4n+1$ objects, labeled:

$$x_1T, x_1F, y_1T, y_1F, \ldots, x_nT, x_nF, y_nT, y_nF, \text{ and } y_0T$$

and of $12n+m+1$ LSCs (where m is the number of clauses in ψ and n is half the number of variables appearing in ψ), as follows.

- Type 1 LSCs:
  - *Type1A* charts: Figure 1(a) shows n different charts (for $i=1,\ldots,n$). Each such chart, upon seeing a `Choose_Next` message from y_{i-1} to itself, sends `m` from x_iT to x_iF, then from y_iT to y_iF and from y_iF to y_iT, and, finally, the message `Choose_Next` from y_iT to itself.
  - *Type1B* charts: Figure 1(b) shows n charts, similar to the Type 1A charts, only the first main chart message is from x_iF to x_iT (as opposed to the other way around in Type 1A).
  - *Type 1C* charts: Figure 1(c) shows n charts, stating that whenever `m` is sent from y_iT to y_iF and back, and then y_iT sends `Done` to itself, m should be sent from y_iF to y_iT and back, and, finally, y_iT should send `Choose_Next` to itself.
  - *Type 1D* charts: Figure 1(d) shows n charts, stating that whenever `m` is sent from y_iF to y_iT and back, and then

$y_i T$ sends Done to itself, $y_{i-1} T$ should also send Done to itself.
  - *Type 1E* charts: Figure 1(e) shows a chart, specifying that once $y_n T$ sends Choose_Next to itself, all objects $v_i T, v \in \{x, y\}, i \in \{1, \ldots, n\}$ should send themselves Check and Done_Check. After they all do so, $y_n T$ should send Done to itself.

- Type 2 LSCs:

  - *Type 2A* charts: Figure 2(a) shows $2n$ different charts (for $v \in \{x_i, y_i\}$, $i = 1, \ldots, n$). Each such chart states, for a specific object pair vT, vF, that if vT sends m to vF, vF sends m to vT, and then vT sends Check to itself, vT should once again send m to vF, vF should send m to vT, and, finally, vT should send Done_Check to itself.
  - *Type 2B* charts: Figure 2(b) shows $2n$ charts, similar to Type 2A charts, except that the order between the vT sending m and the vF sending m is switched, both in the prechart and in the main chart.
  - *Type 2C*, *Type 2D* charts: Figures 2(c) and 2(d) show $2n$ charts each, similar to Type 2A and Type 2B charts, respectively, with the addition of object vT sending Done_Check to itself before the message Check in the prechart.

- Type 3 LSCs:
  Let $\psi = C_1 \wedge C_2 \wedge \cdots \wedge C_m$. We introduce a single chart for each clause $C_j, j = 1, \ldots, m$. An example of this chart, corresponding to the clause $x_1 \vee y_1 \vee \neg y_2$, is given in Figure 3. The example states that the following scenario is forbidden: the scenario contains three independent sequences, with no explicit order between them. The first contains object $x_1 T$ sending Check to itself, $x_1 F$ sending m to $x_1 T$, $x_1 T$ sending m to $x_1 F$ and $x_1 T$ sending Done_Check to itself. The second sequence is similar, for objects $y_1 T$ and $y_1 F$. Finally, the third sequence is for objects $y_2 T$ and $y_2 F$, except that the order of the two m messages is switched.
  It is easy to see how this example generalizes to any clause C. All literals v_k in C are represented similarly to x_1 and y_1 in the example, and all literals $\neg v_k$ are represented similarly to y_2.

Note that the size of the specification is polynomial in the size of the input: it consists of $12n + m + 1$ LSCs, with sizes as follows. The Type 1E LSC is of size polynomial in n (it contains $2n$ objects, each sending a constant number of messages). Type 3 LSCs are of size polynomial in the size of the clauses of ψ (and all together, polynomial in the size of ψ itself). All other LSCs are of constant size.

We also introduce another set, called Type 1C′ LSCs; see Figure 4. These are similar to Type 1C charts, with an extra `Choose_Next` message sent from $y_i T$ to itself in the prechart (before the `Done` message). These LSCs are not part of the final specification, but are used as an intermediate step in the proof below.

Throughout the rest of the proof, we denote the event of object o_1 sending message `msg` to object o_2 by $o_1 \xrightarrow{\texttt{msg}} o_2$.

We now explain intuitively why the specification constructed here has a legal superstep, triggered by an external $y_0 T \xrightarrow{\texttt{Choose\_Next}} y_0 T$ event, if and only if φ is true. The formal proof for this claim is given in Appendix A.

We explain this by incrementally adding the chart types to the specification. First, consider a specification containing only charts of types 1A, 1B, 1C′, 1D and 1E. These represent the backtracking "engine". This engine backtracks over the x_is and y_is, where in backtracking level i, $x_i T$ and $x_i F$ send each other the message `m` in a non-deterministic order, then $y_i T \xrightarrow{\texttt{m}} y_i F$ and $y_i F \xrightarrow{\texttt{m}} y_i T$ are sent in this order, and the following backtracking level is called (chart types 1A, 1B). Upon return of this call, $y_i F \xrightarrow{\texttt{m}} y_i T$ and $y_i T \xrightarrow{\texttt{m}} y_i F$ are sent in this order (which is to be contrasted to the eariler case), and the following backtracking level is called again (chart Type 1C'). Upon return of this second call, this level returns as well (chart Type 1D). At the halting condition, i.e., at level n, all $x_i T$ and $y_i T$ objects send themselves `Check` and `Done_Check` in this order (chart 1E). The idea is that for variable v_i, the order between $v_i T \xrightarrow{\texttt{m}} v_i F$ and $v_i F \xrightarrow{\texttt{m}} v_i T$ represents the value given to it. Thus, for x variables, the order is chosen non-deterministically and for y variables both options are checked. The halting condition is the point where the CNF formula ψ will be checked against the assigned values.

Type 2A and Type 2B charts act as memory – whenever a variable v_i is assigned some value (i.e., $v_i T \xrightarrow{\texttt{m}} v_i F$ and $v_i F \xrightarrow{\texttt{m}} v_i T$ are sent in a certain order), the appropriate 2A or 2B chart will advance past the first two prechart messages. Then, following a $vT \xrightarrow{\texttt{Check}} vT$ message, the two m messages will be resent in the same order. 2C and 2D are similar, and will resend the messages if a $vT \xrightarrow{\texttt{Done\_Check}} vT$ is sent before the $vT \xrightarrow{\texttt{Check}} vT$ (which happens whenever this

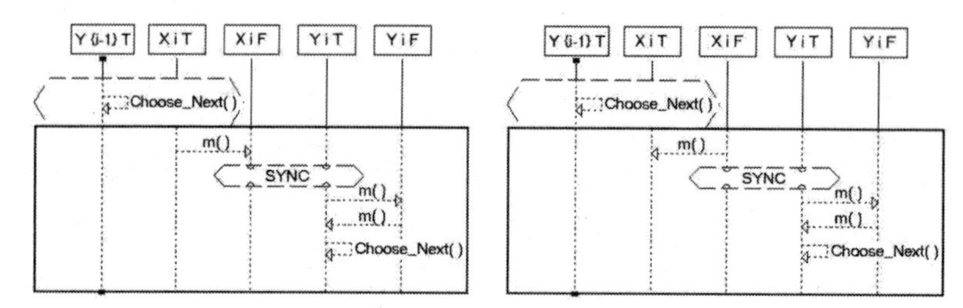

(a) Type 1A charts: n charts, which, together with Type 1B charts, nondeterministically choose an x_i, and then choose $y_i = T$.

(b) Type 1B charts: n charts, which, together with Type 1A charts, nondeterministically choose an x_i, and then choose $y_i = T$.

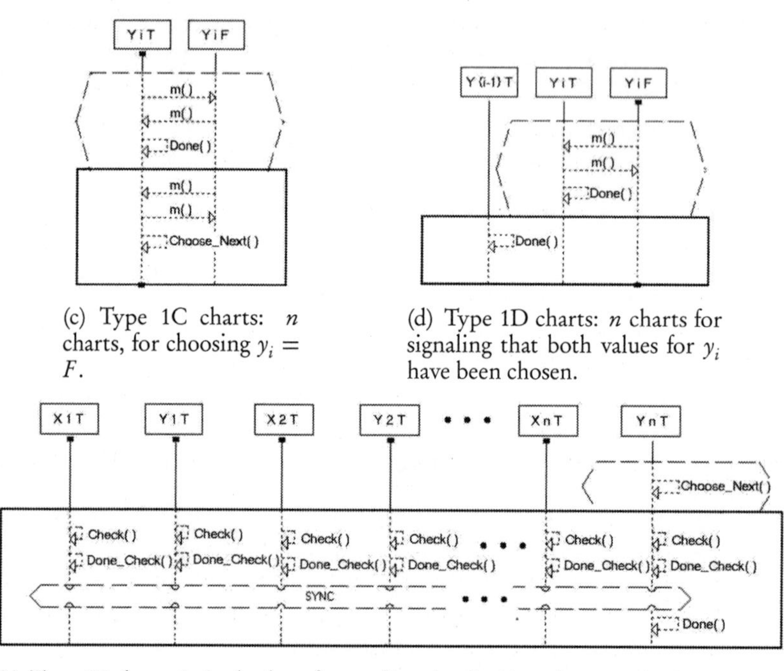

(c) Type 1C charts: n charts, for choosing $y_i = F$.

(d) Type 1D charts: n charts for signaling that both values for y_i have been chosen.

(e) Type 1E chart: A single chart for sending the **Check** and **Done_Check** messages.

Figure 1: *LSCs of Type 1, forcing the smart play-out mechanism to check all values for y variables and choose a value for each x variable.*

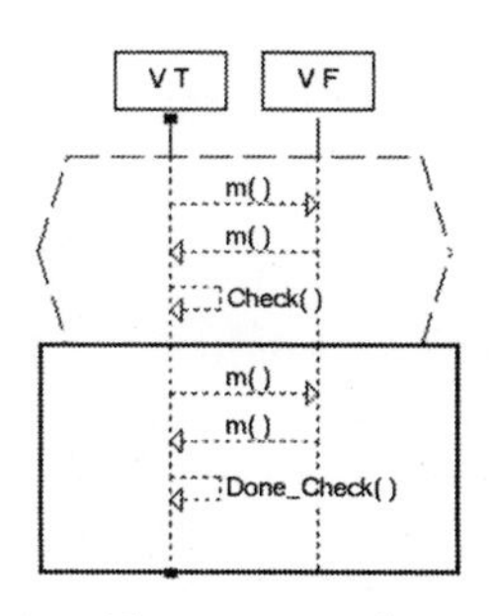

(a) Type 2A charts: for "resending" the last value, if true, of each variable v after a $v_t \xrightarrow{\textsf{Check}} v_t$ message.

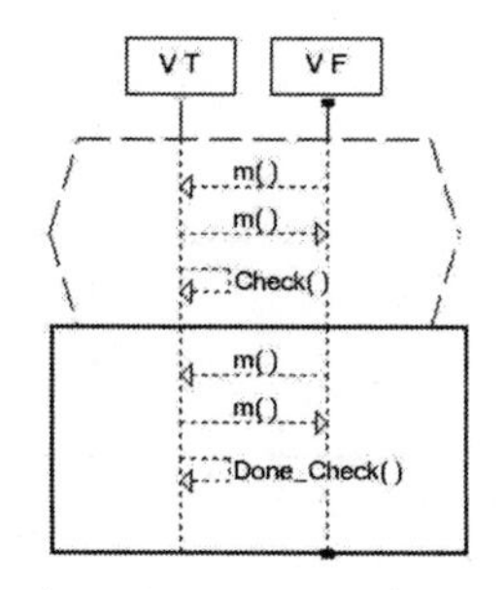

(b) Type 2B charts: for "resending" the last value, if false, of each variable v after a $v_t \xrightarrow{\textsf{Check}} v_t$ message.

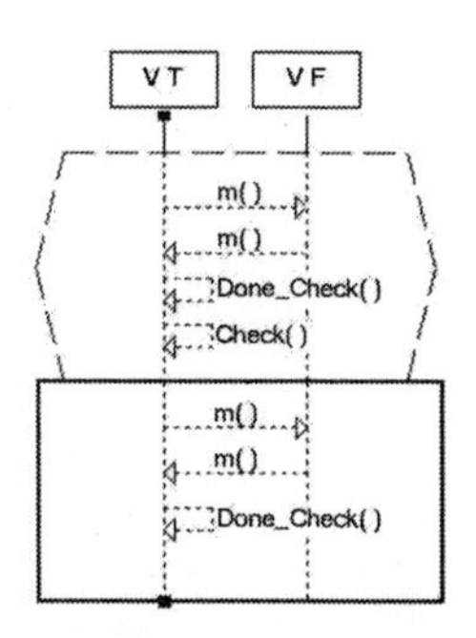

(c) Type 2C charts: for "resending" the last value, if true, of each variable v after a $v_t \xrightarrow{\textsf{Check}} v_t$ message, if it was not changed since the previous Check message.

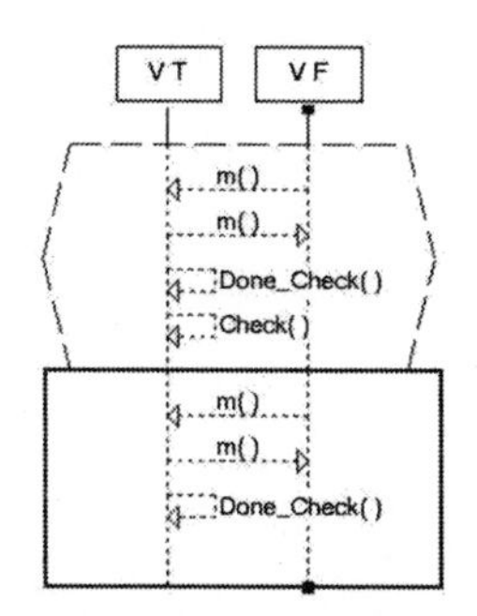

(d) Type 2D charts: for "resending" the last value, if false, of each variable v after a $v_t \xrightarrow{\textsf{Check}} v_t$ message, if it was not changed since the previous Check message.

Figure 2: *LSCs of Type 2, causing the last value of each variable to be "resent" after each* Check *message.*

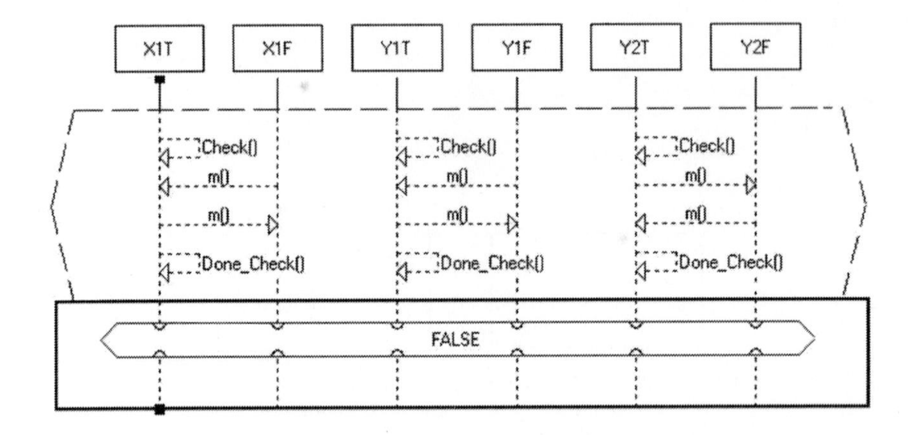

Figure 3: *An example for a Type 3 LSC, corresponding to the clause* $x_1 \vee y_1 \vee \neg y_2$. *The chart causes the case where* $x_1 = F, y_1 = F, y_2 = T$ *to be forbidden.*

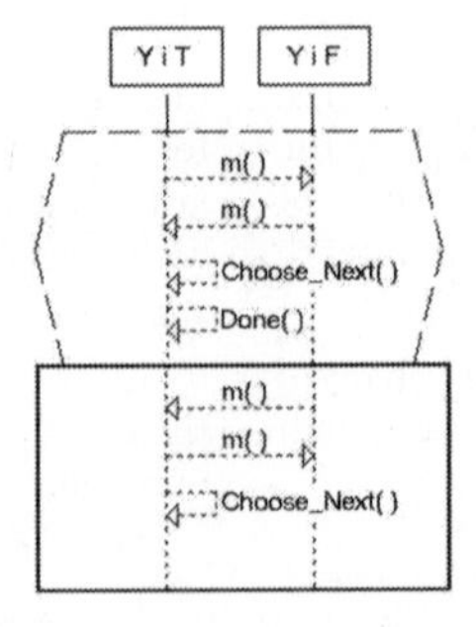

Figure 4: *Type 1C′ charts: similar to Type 1C but with an extra* $v_iT \xrightarrow{\texttt{Choose\_Next}} v_iT$ *message in the prechart.*

variables has not been reassigned a value since the last time the halting condition was reached). Since $vT \xrightarrow{\texttt{Check}} vT$ is sent for all x_is and y_is at the halting condition of the backtracking algorithm (of the Type 1 charts), augmenting the Type 1 charts with Type 2 charts causes all variables to resend their last $v_iT \xrightarrow{\texttt{m}} v_iF$ and $v_iF \xrightarrow{\texttt{m}} v_iT$ messages (in the same order they were last sent) whenever the halting condition is reached.

Type 3 charts wait for this resend at the halting condition. Each chart of Type 3 checks a single clause in ψ. Its prechart will be satisfied if and only if the clause does not hold, i.e., all the variables relevant to it were assigned values that are negations of those in the clause. Since Type 3 charts are anti-scenarios (i.e., with a hot FALSE condition as the main chart), if the prechart of a Type 3 chart is ever satisfied, the specification is violated. Thus, augmenting the specification with Type 3 charts assures that whenever the halting condition in the backtracking algorithm is reached, all clauses are satisfied, and ψ holds.

Therefore, the full specification has a legal superstep, following an external $y_0T \xrightarrow{\texttt{Choose\_Next}} y_0T$ event, if and only if the backtracking algorithm manages to assign values to the x_is and y_is such that ψ always holds, which can happen if and only if φ is true.

⊣

We are now ready to move on to the main theorem of the paper — the PSPACE-hardness of smart play-out mechanisms:

Theorem 2 *Any smart play-out mechanism supporting messages is PSPACE-hard.*

Proof It is enough to show how the synchronization constructs and anti-scenarios of the above construction can be reduced to messages. There are three types of constructs that need to be removed, as follows.

1. The SYNC constructs on two objects, appearing in Type 1A and Type 1B charts, can easily be replaced by a newly introduced message between the two lifelines (a different message for each chart). Each of these messages appears only once in the specification, therefore their only effect is in synchronizing the two lifelines, much like the original synchronization construct.
2. The SYNC construct in the Type 1E chart can be replaced by a series of new messages from each lifeline to the following one (i.e., x_1T to y_1T, and then y_1T to x_2T, etc.). This will ensure

that the $y_n T \xrightarrow{\texttt{Done}} y_n T$ message will be executed only after all `Done_Check` messages were sent, which is the purpose of the SYNC.

3. Anti-scenarios in Type 3 charts can be removed as follows. First, we duplicate the chart. In one copy, we use two newly introduced messages `msg1` and `msg2`, and specify that they must appear in a certain order. In the second, we require them to appear in the opposite order. Clearly, no superstep can satisfy both, so that the new pair of LSCs acts, in combination, as an anti-scenario that prevents the play-out mechanism from satisfying the prechart.

Clearly, after making these changes, the specification is still polynomial in the size of the input. ⊣

4.2 Multiple running copies

In general, the semantics of LSC and of play-out allows multiple copies of the same chart to be simultaneously active. This may happen if messages that are minimal in the chart are allowed to reappear in it. A similar issue was mentioned in [11] as one that requires special attention — forbidding events from appearing more than once in the same chart allowed a more succinct translation of the LSC language into temporal logic. It remains an open question as to whether this is a necessary condition for such a succinct translation to exist.

Both known PSPACE-hardness proofs — the one in [1] and ours — use this feature in order to "implement memory". For example, in our proof this is heavily relied on in the Type 2 LSCs: the prechart "remembers" the last value assigned to each variable, and the main chart "resends" it. It is crucial that when the main chart resends this value, a new copy opens with its prechart "remembering" this value.

We do not know whether the complexity of the problem drops when one forbids simultaneous multiple running copies of the same chart. What we can prove, however, is that the smart play-out problem, even without multiple running copies, is NP-hard. While this does not answer the question completely, it does show that the problem is still hard enough even in this constrained case. Note that even without multiple running copies, the superstep may still be of length exponential in the size of the specification. This is discussed further in Section 5.3.

Theorem 3 *Smart play-out without multiple running copies is NP-hard.*

Proof We prove this by a reduction from SAT. Similarly to the proof of the main theorem above, given a CNF formula $\psi = C_1 \land C_2 \land \cdots \land C_m$ over variables $\{x_1, \ldots, x_n\}$, we build a system and an LSC specification. The specification is built such that a superstep exists (from a specific initial configuration) if and only if the formula is satisfiable.

The system consists of $2n+1$ objects, named $x_1T, x_1F, x_2T, x_2F, \ldots, x_nT, x_nF$ and O. The truth assignment to $x_1, \ldots x_n$ is represented similarly to that in the proof Theorem 1, i.e., x_i is assigned TRUE if x_iT sends `m` to x_iF and then x_iF sends `m` to x_iT, and is assigned FALSE if the same two messages are sent in reverse order. The object O can send the message `Start` to itself, which is the event that will trigger the superstep.

The specification contains $2n + m$ LSCs, of two types, as follows.

- Type 1 charts: these will be in charge of assigning values to the variables:
  - Type 1A charts: n charts, where chart i states that whenever O sends `Start` to itself, then x_iT must send `m` to x_iF.
  - Type 1B charts: similar to Type 1A charts, except that in the main chart x_iF sends `m` to x_iT.
- Type 2 charts: These are similar to the Type 3 charts in the earlier construction, and make sure that each clause is satisfied by the assignment. Given a clause C, let X_C^T be the literals appearing positive in C, and X_c^F be those appearing negative. The LSC corresponding to C is an anti-scenario, forbidding the scenario in which $O \xrightarrow{\texttt{Start}} O$ is sent, followed by the events $x_iF \xrightarrow{\texttt{m}} x_iT; x_iT \xrightarrow{\texttt{m}} x_iF$ for all $x \in X_C^T$ and the events $x_iT \xrightarrow{\texttt{m}} x_iF; x_iF \xrightarrow{\texttt{m}} x_iT$ for all $x \in X_C^F$.

The specification is clearly of size polynomial in that of the input. Moreover, since the minimal event in all LSCs is unique and never appears again in the chart, there will never be any multiple running copies of the same chart.

It is left to show that following an event $O \xrightarrow{\texttt{Start}} O$, a superstep exists if and only if the formula is satisfiable.

- Assume a superstep exists. Clearly, from the Type 1 charts, the superstep contains each object x_iT sending one `m` message to

$x_i F$ and vice versa (in some order between the two). Since no Type 2 charts are violated, it follows that for each clause C, either $\exists x \in X_C^T$ s.t. $x_i T \xrightarrow{\mathtt{m}} x_i F$ was sent first, or $\exists x \in X_C^F$ s.t. $x_i F \xrightarrow{\mathtt{m}} x_i T$ was sent first. If we assign TRUE to x_i if and only if $x_i T \xrightarrow{\mathtt{m}} x_i F$ was sent first, it follows that for this assignment, each clause is satisfied, and therefore ψ is satisfied.

- Assume an assignment to $x_1, \ldots, x_n$ exists s.t. ψ is satisfied. Consider a superstep in which $\forall x_i$ assigned TRUE, the events $x_i T \xrightarrow{\mathtt{m}} x_i F; x_i F \xrightarrow{\mathtt{m}} x_i T$ are sent, and $\forall x_i$ assigned FALSE, the events $x_i F \xrightarrow{\mathtt{m}} x_i T; x_i T \xrightarrow{\mathtt{m}} x_i F$ are sent. Similarly to the above argument, the superstep satisfies all Type 1 charts, and does not satisfy the prechart of any anti-scenario in the Type 2 charts, and is therefore a legal superstep for the specification.

⊣

5 Discussion

5.1 Advanced constructs

LSCs is a rich language, which in addition to messages and synchronization constructs that we have used in our proofs, contains also conditions, object properties, assignments, messages with parameters, and control flow constructs like if-then-else and loops. The original smart play-out indeed handles these more advanced features of the language (see [7]). Thus, it is of interest to understand whether allowing the use of these constructs increases the complexity of smart play-out.

The addition of bounded loops does not affect smart play-out complexity, as such loops can be unravelled to represent all the iterations explicitly. However, the effect of using condition evaluation and assignments over object properties and variables depends on their domain and on the operators allowed. Thus, these may add no complexity but might render smart play-out undecidable in the case of infinite domains.

We note that unbounded loop constructs (i.e., star) may introduce additional complexity, as they allow the user to add another level of non-determinism to the LSC program: one may create a specification where the next cut is not uniquely determined by the current cut and the next event. It is of interest to check how these special cases may affect the complexity of smart play-out.

5.2 Upper bound

To date, there are two implementations of smart play-out: one based on model-checking [7] and one based on AI-style planning (termed planned play-out) [9]. However, neither support multiple running copies (see 4.2). Thus, although both use a reduction into known PSPACE problems, they cannot be used as a proof for an upper bound for the complexity of the general problem. A tight upper bound for the problem is yet unknown.

5.3 Superstep length

We note that the length of the superstep may be exponential in the size of the specification. For example, in the specification used in the proof of section 4.1, a legal superstep will traverse all possible assignments to the y_i's, and will therefore be of exponential length.

Interestingly, this observation still holds even when multiple running copies are not allowed. For example, consider a one-object system, with a specification consisting of k LSCs, $S_k = \{L_1, L_2, \ldots, L_k\}$, over an alphabet $\Sigma_k = \{m_1, m_2, \ldots, m_{k+1}\}$, such that for all $1 \le i \le k$, L_i states that whenever m_i is sent, eventually m_{i+1} needs to be sent twice. Following m_1, the only legal superstep for S_k is of length exponential in k (for example, for S_2, the superstep is $m_1, m_2, m_3, m_3, m_2, m_3, m_3$).

Thus, one may consider the problem of *bounded smart play-out*: given a specification, an initial configuration, and an integer k, return a superstep of length up to k if and only if one exists. This problem may be of interest in practice. We leave its analysis to future work.

5.4 Accelerations

Although the worst case complexity of smart play-out is PSPACE-hard, in practice we expect LSC specifications to be less complex. The main purpose of smart play-out is to resolve dependencies between LSCs. Though there may be many such dependencies in the worst case, our experience shows that in many interesting cases they are limited. This fact gives rise to the hope that appropriate heuristics could render the problem feasible for many practical specifications.

In other work in our group [6], and inspired by standard compiler and model checking optimization techniques, we suggest an algorithm that uses various acceleration techniques for smart play-out. The accelerations are based on approximating the set of LSCs that may participate in the current superstep, and on separating the elements that may cause dependencies between the LSCs in the specification (in the context of a given configuration) from the elements that may not do so. While the former require smart play-out, the latter can be handled efficiently in a more naïve fashion. All this is aimed at reducing the size of the model without affecting the soundness and completeness of finding a correct superstep. Clearly, such accelerations are heuristic in nature, and do not reduce the complexity of the problem in the worst case.

5.5 Is smart play-out good enough?

Smart play-out addresses the limitations of naïve play-out and finds a legal superstep if one exists. However, looking only one superstep (or a finite number of supersteps) ahead is still quite limited. Intuitively, if the LSC specification is "too deep", smart play-out may not be able to distinguish between a superstep that allows the system to continue playing (forever) from one that allows the environment to eventually force the execution into a violation.

Indeed, in [4], it is shown that smart play-out, however often repeated, is strictly weaker than full synthesis from LSCs, as was defined in [5]. On the other hand, [4] also shows that for a given LSC specification, there exists a k such that smart play-out that looks k supersteps ahead is as good as full synthesis.

6 Related and Future Work

Bontemps and Schobbens [1] examine the complexity of various problems related to LSCs, including reachability, language inclusion, and variants of synthesis. Specifically, they use a reduction from the halting problem of a PSPACE-bounded Turing machine to show that for LSC, reachability, inclusion and satisfiability are PSPACE-hard. Finally, the authors of [1] claim that their results can be adapted to show that determining whether a finite superstep exists is PSPACE-complete, but an explicit proof is not given. In contrast, our proof uses a reduction from QBF and is explicit.

A number of questions related to the complexity of smart play-out remain open. As mentioned earlier, these include explicit investigation of additional langauge constructs, the question of whether allowing multiple copies really affects the difficulty of the problem, and establishing a tight upper bound for the general problem. In addition, following [9], we consider the complexity of finding all supersteps (from a given configuration) to be an interesting question.

References

[1] Y. Bontemps and P. Y. Schobbens. The complexity of live sequence charts. In V. Sassone, editor, *Proc. 8th Int. Conf. Foundations of Software Science and Computational Structures (FoSSaCS'05)*, volume 3441 of *Lecture Notes in Computer Science*, pages 364–378. Springer, 2005.

[2] W. Damm and D. Harel. LSCs: Breathing Life into Message Sequence Charts. *J. on Formal Methods in System Design*, 19(1):45–80, 2001. Preliminary version in Proc. 3rd IFIP Int. Conf. on Formal Methods for Open Object-Based Distributed Systems (FMOODS'99), (P. Ciancarini, A. Fantechi and R. Gorrieri, eds.), Kluwer Academic Publishers, 1999, pp. 293-312.

[3] M. R. Garey and D. S. Johnson. *Computers and Intractability: A Guide to the Theory of NP-Completeness*. W. H. Freeman, 1979.

[4] D. Harel, A. Kantor, and S. Maoz. On the Power of Play-Out for Scenario-Based Programs. In D. Dams, U. Hannemann, and M. Steffen, editors, *Concurrency, Compositionality, and Correctness, Festschrift in Honor of Willem-Paul de Roever*. Springer, 2009. To appear.

[5] D. Harel and H. Kugler. Synthesizing State-Based Object Systems from LSC Specifications. *Int. J. of Foundations of Computer Science*, 13(1):5–51, February 2002. (Also in *Proc. 5th Int. Conf. on Implementation and Application of Automata* (CIAA 2000), Springer-Verlag, pp. 1–33. Preliminary version appeared as technical report MCS99-20, Weizmann Institute of Science, 1999.).

[6] D. Harel, H. Kugler, S. Maoz, and I. Segall. Accelerating Smart Play-Out of Scenario-Based Specifications. *Submitted*, 2009.

[7] D. Harel, H. Kugler, R. Marelly, and A. Pnueli. Smart play-out of behavioral requirements. In M. Aagaard and J. W. O'Leary, editors, *Proc. 4th Int. Conf. on Formal Methods in Computer-Aided Design (FMCAD '02)*, pages 378–398. Springer-Verlag, 2002.

[8] D. Harel and R. Marelly. *Come , Let's Play: Scenario-Based Programming Using LSCs and the Play-Engine*. Springer-Verlag, 2003.

[9] D. Harel and I. Segall. Planned and traversable play-out: A flexible method for executing scenario-based programs. In O. Grumberg and M. Huth, editors, *Proc. 13th Int. Conf. on Tools and Algorithms for the Construction and Analysis of Systems (TACAS'07)*, volume 4424 of *Lecture Notes in Computer Science*, pages 485–499. Springer, 2007.

[10] ITU. International Telecommunication Union Recommendation Z.120: Message Sequence Charts. Technical report, 1996.

[11] H. Kugler, D. Harel, A. Pnueli, Y. Lu, and Y. Bontemps. Temporal Logic for Scenario-Based Specifications. In N. Halbwachs and L. D. Zuck, editors, *Proc. 11th Int. Conf. on Tools and Algorithms for the Construction and Analysis of Systems (TACAS '05)*, volume 3440 of *Lecture Notes in Computer Science*, pages 445–460. Springer, 2005.

Appendix

A Completion of Proofs

In proving Theorem 1 we omitted the formal proof that a superstep exists in the specification, following an external $y_0 T \xrightarrow{\texttt{Choose\_Next}} y_0 T$ event, if and only if φ is true. This is proved in the following series of claims.

Claim 1 *Consider a specification consisting only of the Type 1 charts, where Type 1C charts are replaced by Type 1C′ charts. Any legal execution of it, triggered by an external event $y_0 T \xrightarrow{\texttt{Choose\_Next}} y_0 T$, can be described by the following backtracking pseudo-code:*

1: **procedure** CHOOSE(i)
2: **if** $i = n+1$ **then**
3: **for all** $v_j \in \{x_j, y_j : j = 1, \ldots, n\}$ **do**
4: $v_j T \xrightarrow{\texttt{Check}} v_j T$
5: $v_j T \xrightarrow{\texttt{Done\_Check}} v_j T$
6: **end for**
7: $y_n T \xrightarrow{\texttt{Done}} y_n T$
8: Return
9: **end if**
10: **nondeterministic switch**
11: **case**
12: $x_i T \xrightarrow{\texttt{m}} x_i F$
13: $x_i F \xrightarrow{\texttt{m}} x_i T$
14: **case**
15: $x_i F \xrightarrow{\texttt{m}} x_i T$
16: $x_i T \xrightarrow{\texttt{m}} x_i F$
17:
18: **end switch**
19: $y_i T \xrightarrow{\texttt{m}} y_i F$
20: $y_i F \xrightarrow{\texttt{m}} y_i T$
21: Async call to Choose(i+1)
22: Wait until $y_i T \xrightarrow{\texttt{Done}} y_i T$ is sent
23: $y_i F \xrightarrow{\texttt{m}} y_i T$

24: $y_i T \xrightarrow{\texttt{m}} y_i F$
25: Async call to Choose(i+1)
26: Wait until $y_i T \xrightarrow{\texttt{Done}} y_i T$ is sent
27: $y_{i-1} T \xrightarrow{\texttt{Done}} y_{i-1} T$
28: **end procedure**

Proof For each $i = 1, \ldots, n$, the four LSC instantiations of Type 1A, Type 1B, Type 1C′, and Type 1D charts implement the `Choose` procedure with parameter i, as follows.

The Type 1A and Type 1B charts begin the execution. Lines 10–21 are the exact description of their main charts (where the non-deterministic choice comes from choosing which one progresses first), according to LSC's operational semantics.

Whenever a Type 1A and Type 1B chart pair completes (note that this always happens simultaneously), the cut of the corresponding Type 1C′ chart is right before the last prechart message ($y_i T \xrightarrow{\texttt{Done}} y_i T$). Note that this state will not be violated, since neither of the corresponding Type 1A, Type 1B or Type 1C′ charts can be active at that time. Once this message ($y_i T \xrightarrow{\texttt{Done}} y_i T$) is sent, the main chart will be activated. This corresponds to line 22 in the pseudo-code. The main chart clearly implements lines 23 - 25.

Similarly, the Type 1D chart will be activated by the next $y_i T \xrightarrow{\texttt{Done}} y_i T$ message (thus implementing line 26), and its main chart implements line 27.

The Type 1E chart clearly implements the halting condition in lines 2 - 9. (Note that the `for all` command in Line 3 is interpreted as all j's executed in any order, not necessarily sequential, and with any possible interleaving between the messages corresponding to different values of j.)

⊣

Claim 2 *Consider a specification consisting of all Type 1 and Type 2 charts (with the original Type 1C charts, not Type 1C′ as used in Claim 1). Any legal execution of it, triggered by an external* $y_0 T \xrightarrow{\texttt{Choose\_Next}} y_0 T$ *event, can be described by the following pseudo-code (changes from claim 1 underlined see line 5):*

1: **procedure** CHOOSE(i)
2: **if** $i = n + 1$ **then**
3: **for all** $v_j \in \{x_j, y_j : j = 1, \ldots, n\}$ **do**

4: $v_j T \xrightarrow{\texttt{Check}} v_j T$
5: $\underline{\text{Repeat the last message } \texttt{m} \text{ sent back and forth between}}$
$\underline{v_j T \text{ and } v_j F \text{ (or vice versa)}}$
6: $v_j T \xrightarrow{\texttt{Done\_Check}} v_j T$
7: **end for**
8: $y_n T \xrightarrow{\texttt{Done}} y_n T$
9: Return
10: **end if**
11: **nondeterministic switch**
12: **case**
13: $x_i T \xrightarrow{\texttt{m}} x_i F$
14: $x_i F \xrightarrow{\texttt{m}} x_i T$
15: **case**
16: $x_i F \xrightarrow{\texttt{m}} x_i T$
17: $x_i T \xrightarrow{\texttt{m}} x_i F$
18:
19: **end switch**
20: $y_i T \xrightarrow{\texttt{m}} y_i F$
21: $y_i F \xrightarrow{\texttt{m}} y_i T$
22: Async call to Choose(i+1)
23: Wait until $y_i T \xrightarrow{\texttt{Done}} y_i T$ is sent
24: $y_i F \xrightarrow{\texttt{m}} y_i T$
25: $y_i T \xrightarrow{\texttt{m}} y_i F$
26: Async call to Choose(i+1)
27: Wait until $y_i T \xrightarrow{\texttt{Done}} y_i T$ is sent
28: $y_{i-1} T \xrightarrow{\texttt{Done}} y_{i-1} T$
29: **end procedure**

Proof This proof has two parts. Fact 1: Augmenting the model by the Type 2 charts, while replacing Type 1C′ charts with Type 1C charts, does not affect the backtracking part of the pseudo-code. Fact 2: The Type 2 LSCs implement line 5. We start by proving the second of these.

Fact 2: Assuming Fact 1 holds, whenever the pseudo-code reaches line 4 (in which the message $v_j T \xrightarrow{\texttt{Check}} v_j T$ is sent), for each variable v_j, exactly one of the corresponding Type 2 charts has its cut right before the $v_j T \xrightarrow{\texttt{Check}} v_j T$ message in the prechart, and no main chart of Type 2 charts is active. This is true, since for each v_j, both $v_j T \xrightarrow{\texttt{m}} v_j F$ and $v_j F \xrightarrow{\texttt{m}} v_j T$ were necessarily sent. If $v_j T \xrightarrow{\texttt{Done\_Check}} v_j T$ was sent afterwards, then either the Type 2C or Type 2D chart has a cut in that location, otherwise this holds for the Type 2A or Type 2B chart.

The $v_j T \xrightarrow{\texttt{Check}} v_j T$ message in line 4, therefore, necessarily activates exactly one of the Type 2 charts for each variable v_j, which corresponds to the correct order in which the last $v_j T \xrightarrow{\texttt{m}} v_j F$ and $v_j F \xrightarrow{\texttt{m}} v_j T$ messages were sent. This will cause them to be sent again. Also note that the message $v_j T \xrightarrow{\texttt{Done\_Check}} v_j T$ appears now in two active main charts — the Type 1E chart, and the active Type 2 chart. According to LSC semantics, this message may not be sent until enabled in both. This causes the Type 1E chart to block until all the $v_j T \xrightarrow{\texttt{m}} v_j F$ and $v_j F \xrightarrow{\texttt{m}} v_j T$ messages are resent, therefore postponing the execution of line 6 until after line 5 completes.

Fact 1: First introduce the Type 1C charts and Type 2 charts into the specification from Claim 1 (without removing the Type 1C′ ones yet). Clearly, Type 2 LSCs are activated by the $v_j T \xrightarrow{\texttt{Check}} v_j T$ message, which can be sent only by the Type 1E chart. Whenever this chart is active, no other Type 1 charts are active. Moreover, for each $i = 1, \ldots, n$, either the corresponding Type 1C′ chart, or the Type 1D chart has a cut right before the last prechart message (the Type 1C precharts were violated by the $y_i T \xrightarrow{\texttt{Choose\_Next}} y_i T$ message). Now, whenever a Type 2 chart resends the last $v_j T \xrightarrow{\texttt{m}} v_j F$ and $v_j F \xrightarrow{\texttt{m}} v_j T$ messages, this prechart will be violated, but another copy of it will start. For Type 1D charts, it will reach the exact same location (right before the last prechart message). Both Type 1C and Type 1C′ precharts will open, both with a cut following the two `m` messages. Therefore, when the $v_j T \xrightarrow{\texttt{Done\_Check}} v_j T$ message is sent (and the Type 2 charts all finish), the backtracking algorithm continues from the exact same state as in Claim 1: Type 1C and Type 1D charts are activated (where the main chart of Type 1C is identical to that of Type 1C′), and Type 1C′ precharts are violated and closed.

Since Type 1C′ charts never become active in this run, we can now remove them from the specification without changing the execution. ⊣

Claim 3 *A legal superstep in the specification defined above, triggered by an external* $y_0T \xrightarrow{\texttt{Choose\_Next}} y_0T$ *event, exists if and only if* φ *is true.*

Proof First note that adding the Type 3 charts to the specification does not change the result of the above claim (as long as no Type 3 chart causes a violation), since Type 3 charts are only anti-scenarios, and not LSCs with "real" main charts that can drive the execution. Also note that for Type 1 and Type 2 charts alone, any choice from among the non-deterministic possibilities reflects a legal superstep. Therefore, a superstep exists for the entire specification if and only if there exist choices for each of the non-deterministic possibilities such that Type 3 charts are never violated.

The minimal messages in the precharts of Type 3 charts are `Check` messages from the relevant vT objects to themselves. These are sent in the halting condition of the pseudo-code above (line 4). Therefore, these precharts will become active whenever a halting condition is reached. For each object vT, following the `Check` message, the corresponding vT and vF objects resend their last exchange of the message `m`.

Assume a superstep exists. Consider, for each variable v, the exchange of messages `m` replayed by the Type 2 charts (i.e., in line 5). Let $v = \text{TRUE}$ if the order is $vT \xrightarrow{\texttt{m}} vF; vF \xrightarrow{\texttt{m}} vT$, and $v = \text{FALSE}$ otherwise. Now consider the set of all values that all variables take in all the executions of line 5. It is clear that a Type 3 anti-scenario holds (i.e., it violates the specification) if and only if none of the corresponding literals is satisfied. Since a superstep exists for each such set of values, no Type 3 anti-scenario holds; i.e., for each clause c_i, at least one of its literals is satisfied, and therefore the whole CNF formula ψ holds. From the flow of the backtracking algorithm, it is clear that it finds a value for each $\exists$ term and checks all $\forall$ terms, such that ψ holds, and therefore φ is true.

Now assume a superstep does not exist, and by contradiction that φ is true. Thus, there exist decisions for the backtracking algorithm such that whenever the halting condition is reached, ψ holds. The algorithm, along with these decisions, induce a legal superstep, in contradiction to the assumption.

⊣

Symmetry in Petri Nets

Jonathan Hayman **Glynn Winskel**
Computer Laboratory, University of Cambridge, England
{jonathan.hayman,glynn.winskel}@cl.cam.ac.uk

Abstract

An algebraic treatment of symmetry in Petri nets is proposed. The standard definition of Petri net is that it has precisely one initial marking. Motivated by work on defining symmetry across models for concurrency, we extend the definitions of forms of net to allow them to have multiple initial markings. Existing coreflections between event structures and occurrence nets and between occurrence nets and P/T nets are generalized, and from them coreflections between categories of models with symmetry are obtained.

Keywords: Petri nets, event structures, unfoldings, category theory

1 Introduction

Petri nets are a widely used model for concurrency. They play a fundamental role analogous to that of transition systems, but, by capturing the effect of events on *local* components of state, it becomes possible to describe how events might occur concurrently, how they might conflict with each other and how they might causally depend on each other. Here an algebraic treatment of symmetry on Petri nets is proposed.

Without doubt symmetry is important and plays a role, at least informally, in many models, and often in the analysis of processes. It is, for instance, present in security protocols due to the repetition of essentially similar sessions [2, 5, 1], can be exploited to increase efficiency in model checking [17], and is present whenever abstract names are involved [6].

Of course, there are undoubtedly several ways to adjoin symmetry to Petri nets. The method we use has some history. It was motivated through the need to extend the expressive power of event structures and the maps between them [23, 24]. One important reason to extend the treatment of symmetry beyond event structures to Petri nets is the potentially more compact and algorithmically-amenable representation nets afford. Another reason for extending the method of adjoining symmetry is to obtain a characterisation of the unfolding of general nets *up to symmetry* [7]: There is an implicit symmetry in a Petri net where a place can be marked more than once. That symmetry is inherited by its unfolding, and, if not made explicit there, will spoil the uniqueness required by its universal characterisation. Once symmetry is added, a coreflection *up to symmetry* between occurrence nets and general nets is obtained.

Roughly, a symmetry in a Petri net is described as a relation between its runs as causal nets, the relation specifying when one run is similar to another up to symmetry; of course, if runs are to be similar, they should have similar futures, as well as pasts. More technically and generally, a relation of symmetry is expressed as a span of open maps which form a pseudo equivalence — it is said to form an equivalence when the span of maps is jointly monic. One motivation for the work in [7] is to apply this general algebraic method to adjoin symmetry to a model, to the instance of Petri nets, and obtain a universal charactersiation of the unfolding of nets. But another motivation is that Petri nets provide a good testing ground for the method of adjoining symmetries.

In our work it became apparent that all but one of the general issues we encountered in considering general nets also arose in considering just safe or P/T nets. The usual categories of Petri nets attach to each net an initial marking to represent the state in which process represented by the net initially lies. The initial marking is essential to understanding the behaviour of the net. Singly-marked nets are, however, unable to express very natural symmetries on nets. Applying the scheme in [23, 24] to obtain a way of defining symmetry on nets, nets with a single initial marking do not even allow the symmetry of the two places in the net $\odot\odot$ to be expressed. This phenomenon, which appears even for occurrence nets, is frustrating since it only occurs at the initial marking; such symmetry arising, for example, in the postconditions of events can be expressed. While even for safe nets the introduction of symmetry on nets leads us to drop the requirement that a symmetry be a joint monic relation — joint monicity was imposed in [23, 24], but if we were to insist on joint monicity we simply could not express some reasonable symmetries — see the Conclusion.

The work tests a method of adjoining symmetries and provides a rationale for a certain, probably rather innocent, extension of Petri nets. It argues for the enlargement of nets to include multiple initial markings. This extension does however force us to review the existing adjunctions between nets, and in particular the unfoldings of safe and P/T nets into occurrence nets and event structures. In summary, in this paper, we generalize the definition of nets to allow them to have a *set* of initial markings. We extend the existing coreflections* between categories of event structures and occurrence nets and between categories of P/T nets and occurrence nets to this new setting. The coreflections are shown to extend to categories with symmetry adjoined.

Notation

In what follows, it will be necessary to use a little notation when dealing with multisets and sets (a full, formal treatment of the use of multisets in the setting of Petri nets can be found in the appendix of [22]). We write $R \cdot X$ for the result of applying the (multi)relation R to the (multi)set X, $+$ for the union of multisets and $-$ for the partial operation of subtraction of multisets. A partial function f from X to Y will be written $f : X \rightarrow_* Y$. We write $f(x) = *$ if f is undefined at x. The image under f of the (multi)set Z comprising elements of X is denoted fZ. We write R^+ for the transitive closure of a relation R and R^* for the reflexive, transitive closure of R.

2 Symmetry in Concurrency

In [23, 24], a symmetry in model X, an object in a category of models $\mathcal{C}$, is a span

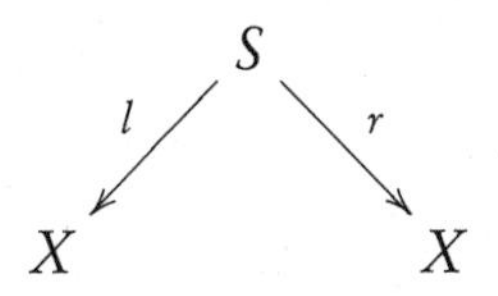

which we write $(X; l, r : S \rightarrow X)$. The span should represent a pseudo equivalence, so it is required to satisfy the standard axioms of reflexivity, symmetry and transitivity presented in Appendix A. For simplicity, we assume that the category $\mathcal{C}$ has pullbacks. The morphisms l and r of the symmetry are required to be *open* morphisms [9].

A morphism of $\mathcal{C}$ is open with respect to a path category $\mathbb{P}$ that is a subcategory of $\mathcal{C}$. That is, for a path category $\mathbb{P} \hookrightarrow \mathcal{C}$, a morphism

*A coreflection is an adjunction of which the left adjoint is full and faithful. Equivalently [10], a coreflection is an adjunction of which the unit is a natural isomorphism.

$f : X \to Y$ is said to be $\mathbb{P}$-open if, for any morphism $m : P \to P'$ in $\mathbb{P}$ and morphisms $p : P \to X$ and $p' : P' \to Y$ in $\mathscr{C}$, whenever the diagram

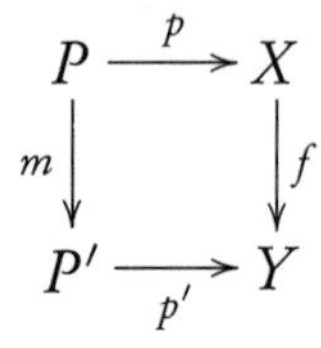

$$\begin{array}{ccc} P & \xrightarrow{p} & X \\ {\scriptstyle m}\downarrow & & \downarrow{\scriptstyle f} \\ P' & \xrightarrow[p']{} & Y \end{array}$$

commutes, *i.e.* $fp = p'm$, there is a morphism $h : P' \to X$ such that the two triangles in the following diagram commute

$$\begin{array}{ccc} P & \xrightarrow{p} & X \\ {\scriptstyle m}\downarrow & \nearrow{\scriptstyle h} & \downarrow{\scriptstyle f} \\ P' & \xrightarrow[p']{} & Y \end{array}$$

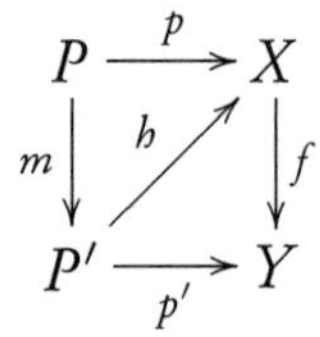

i.e. $hm = p$ and $fh = p'$. Open maps, with respect to suitable path categories, can give rise to well-known forms of bisimulation. For instance, with respect to a category of labelled sequences, spans of open maps of the category of labelled transition systems give rise to Milner and Park's definition of bisimilarity [9]. Spans of open maps of (labelled) event structures and nets, taking paths to be (labelled) partially ordered multisets of events, exhibit a strengthened version [9, 13] of history preserving bisimulation as introduced in [14, 18].

Given a category $\mathscr{C}$ with pullbacks, we will form a category with symmetry $\mathscr{S}_{\mathbb{P}}\mathscr{C}$. The objects of $\mathscr{S}_{\mathbb{P}}\mathscr{C}$ are constructed as above, being tuples $(X; l, r)$ where $l, r : S \to X$ are $\mathbb{P}$-open morphisms in $\mathscr{C}$ that satisfy the axioms of reflexivity, symmetry and transitivity presented in Appendix A. The morphisms of the category $\mathscr{S}_{\mathbb{P}}\mathscr{C}$ are morphisms of $\mathscr{C}$ that preserve symmetry in the sense that a morphism $f : (X; l, r) \to (X'; l', r')$ in $\mathscr{S}_{\mathbb{P}}\mathscr{C}$ is a morphism $f : X \to X'$ in $\mathscr{C}$ for which there exists a morphism $h : S \to S'$ in $\mathscr{C}$ making the two squares in the following diagram commute, where S is the domain of l and r and S' is the domain of l' and r':

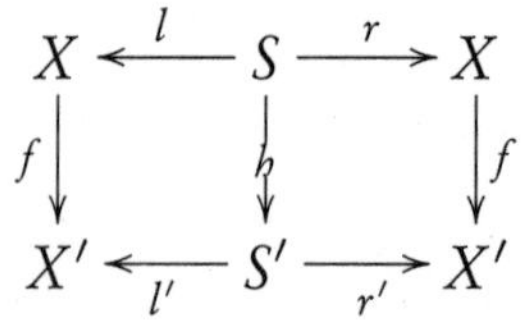

$$\begin{array}{ccccc} X & \xleftarrow{l} & S & \xrightarrow{r} & X \\ {\scriptstyle f}\downarrow & & \downarrow{\scriptstyle h} & & \downarrow{\scriptstyle f} \\ X' & \xleftarrow[l']{} & S' & \xrightarrow[r']{} & X' \end{array}$$

The definition of symmetry presented here differs from that in [23] in that it is a *pseudo* equivalence rather than an equivalence. In particular, there an object with symmetry $(X; l, r)$ requires the morphisms l and r to be jointly monic. This relaxation has turned out to be necessary in other situations [19, 7] and is discussed in the conclusion.

2.1 Symmetry, functors and adjunctions

Let $\mathscr{C}$ and $\mathscr{D}$ be categories upon which symmetry can be placed, *i.e.* with pullbacks and subcategories $\mathbb{P}$ and $\mathbb{Q}$, respectively, of paths from which open maps can be drawn. We obtain the categories with symmetry $\mathscr{S}_{\mathbb{P}}\mathscr{C}$ and $\mathscr{S}_{\mathbb{Q}}\mathscr{D}$.

Say that a functor $F : \mathscr{C} \to \mathscr{D}$ preserves open maps if, for any $\mathbb{P}$-open map $f : X \to X'$ of $\mathscr{C}$, the morphism $F(f) : F(X) \to F(X')$ is $\mathbb{Q}$-open in $\mathscr{D}$. Say that F preserves pullbacks of $\mathbb{P}$-open morphisms if, for any two $\mathbb{P}$-open morphisms $f : X \to Y$ and $f' : X' \to Y$ that have a pullback P with pullback morphisms $p : P \to X$ and $p' : P \to X'$ in $\mathscr{C}$, then the object $F(P)$ with pullback morphisms $F(p)$ and $F(p')$ is a pullback of $F(f)$ and $F(f')$ in $\mathscr{D}$.

Proposition 1 *A functor $F : \mathscr{C} \to \mathscr{D}$ between categories described above yields a functor $\mathscr{S}F : \mathscr{S}_{\mathbb{P}}\mathscr{C} \to \mathscr{S}_{\mathbb{Q}}\mathscr{D}$ defined on objects $(X; l, r)$ of $\mathscr{S}_{\mathbb{P}}\mathscr{C}$ as $\mathscr{S}F(X; l, r) = (FX; Fl, Fr)$ and on morphisms $f : (X; l, r) \to (X'; l', r')$ as $\mathscr{S}F(f) = F(f)$ if F preserves open maps and preserves pullbacks of $\mathbb{P}$-open maps.*

Proof It is easy to see, given that F preserves pullbacks, that $(FX; Fl, Fr)$ satisfies the requirements to be an element of $\mathscr{S}_{\mathbb{Q}}\mathscr{D}$. It is also easy to show that $\mathscr{S}F(f)$ is a map preserving symmetry as a consequence of f being a map preserving symmetry. $\dashv$

Any adjunction

$$\mathscr{C} \underset{G}{\overset{F}{\rightleftarrows}} \mathscr{D} \quad (F \dashv G)$$

in which the functors F and G satisfy the constraints above of preserving open maps and preserving pullbacks of open maps (noting that the functor G automatically preserves all pullbacks as a consequence of it being a right adjoint) gives rise to an adjunction between the categories enriched with symmetry.

Proposition 2 *Let $\mathscr{C}$ and $\mathscr{D}$ be categories with pullbacks equipped with subcategories $\mathbb{P}$ and $\mathbb{Q}$, respectively, of path objects with respect to which open maps are defined. Suppose that the functors $F : \mathscr{C} \to \mathscr{D}$ and $G : \mathscr{D} \to \mathscr{C}$ both preserve open maps, that F preserves pullbacks of $\mathbb{P}$-open morphisms, that G preserves pullbacks of $\mathbb{Q}$-open morphisms, and furthermore that $F \dashv G$, i.e. F is left adjoint to G. The functor $\mathscr{S}F : \mathscr{S}_{\mathbb{P}}\mathscr{C} \to \mathscr{S}_{\mathbb{Q}}\mathscr{D}$ defined in Proposition 1 is left adjoint to the functor $\mathscr{S}G : \mathscr{S}_{\mathbb{Q}}\mathscr{D} \to \mathscr{S}_{\mathbb{P}}\mathscr{C}$, i.e.*

$$\mathscr{S}_{\mathbb{P}}\mathscr{C} \underset{\mathscr{S}G}{\overset{\mathscr{S}F}{\rightleftarrows}} \mathscr{S}_{\mathbb{Q}}\mathscr{D} \quad (\bot).$$

3 Petri Nets

Petri nets were introduced by Petri in 1962 and are an important model of concurrent computation. We now proceed to define the variants of net that we shall consider, referring the reader to [15, 25] for a fuller introduction to net theory. As discussed in the introduction, we generalize nets so that they might possess more than one initial marking. As we do so, a guiding intuition in forming the definitions, particularly in the extended definition of occurrence net, shall be that each initial marking can be thought of as being given rise to by some special, hidden event that is in conflict with all the other events giving rise to the other initial markings.

The key reason in the present setting for extending the definition of nets in this way is that nets with just a single initial marking do not allow obvious symmetries on nets to be expressed. The reader unfamiliar with categories of Petri nets may wish to return to the following account later, skipping directly to Section 3.1.

Consider the span of open morphisms in Figure 1(a) representing the symmetry of the two conditions p and p': the symmetry relates the condition p to p' through the condition (p, p') and through (p', p), so it is symmetric as required. Such a symmetry might arise from unfolding a general net with an event that places two tokens in a single condition [7]. If we were to remove the event from the net, we would still wish to be able to represent the symmetry of p and p'. However, a span satisfying the requirements for being a pseudo equivalence can only be obtained by allowing the span to be from a net with more than one initial marking, as seen in Figure 1(b). We are therefore obliged to consider categories of nets with multiple initial markings.

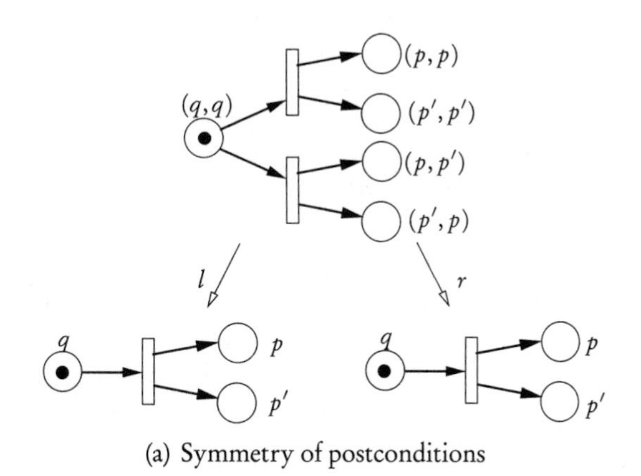

(a) Symmetry of postconditions

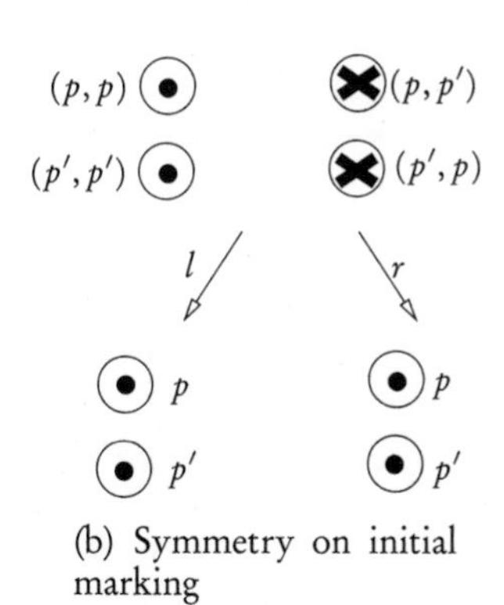

(b) Symmetry on initial marking

For each span, the morphism l on conditions projects to the first element of the pair and the morphism r projects to the second.

Figure 1: *Net symmetry as spans*

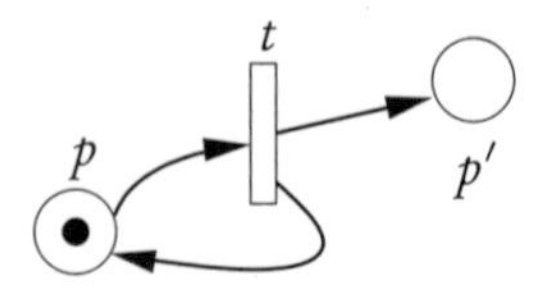

Figure 2: *Example P/T net N*

3.1 P/T nets and safe nets

A P/T net comprises sets of *conditions* (or *places*) and *events* (or *transitions*). Conditions are normally depicted as circles and events as rectangles. There are arcs from conditions to events and arcs from events to conditions, yielding a *flow relation* on the net. In Figure 2, the P/T net N has one event, t, and two conditions, p and p'. The flow relation indicates that p flows into t and t flows into both p and p'. A relatively standard requirement on nets, that simplifies the presentation of Section 5, is to require every event to have at least one condition flowing into it. It is also standard to require there to be no isolated conditions, where a condition is said to be isolated if neither does it occur as a pre- or a postcondition of some event nor does it occur in some initial marking. Other than these, we shall make no further assumptions about P/T nets.

The state of a P/T net is represented by the number of *tokens*, drawn as dots, that each condition contains. This can be considered as a multiset of conditions called the *marking* of the net. In the net N, the condition p holds one token. As discussed, a net is defined with a *set* of *initial markings* representing the set of initial states in which the net could be. Any initial marking is required to contain at most one token in any condition, so each initial marking is itself appropriately described as a set. Formally:

Definition 3 *A P/T net is a 4-tuple*

$$(P, T, F, \mathbb{M})$$

where

- *P is the set of conditions (or places),*
- *T is the set of events (or transitions), disjoint from P,*
- *$F \subseteq (P \times T) \cup (T \times P)$ is the flow relation, and*
- *$\mathbb{M} \subseteq \mathscr{P}ow(P)$ is the set of initial markings, each of which is a set.*

The net must contain no isolated places and, for each event $t \in T$, there must exist $p \in P$ such that $p\ F\ t$.

We shall call a net *singly-marked* if it follows the standard definition, having only one initial marking. A singly-marked net is a tuple (P,T,F,M) of which M is the single initial marking. We shall sometimes, when necessary to explicitly disambiguate the the old singly-marked nets from the nets with sets of initial markings introduced here, call the new nets *multiply-marked*. Be aware, however, that the set of markings of a multiply-marked net could be a singleton set or even empty.

The marking of a P/T net changes according to the occurrence of events: an event t can occur if every condition p that flows into t holds at least one token, *i.e.* each condition that flows into t occurs at least once in the marking. In this situation, the event is said to have *concession*. The resulting marking is obtained by taking a single token from each condition that flows into t and adding a single token to each condition that flows from t. This is called the *token game* for P/T nets. For instance, the token game for the net N in Figure 2 allows the event t to occur in the initial marking drawn, removing a token from condition p and placing tokens in conditions p and p' to yield a marking in which both conditions contain exactly one token.

For any event t and condition p, we adopt the notations:

$$\begin{array}{rclcrcl} {}^\bullet t &=& \{p \mid p\,F\,t\} & \quad & {}^\bullet p &=& \{t \mid t\,F\,p\} \\ t^\bullet &=& \{p \mid t\,F\,p\} & & p^\bullet &=& \{t \mid p\,F\,t\} \end{array}$$

We call the set ${}^\bullet t$ the *preconditions* of t and $t^\bullet$ the *postconditions* of t. The token game gives rise to a transition relation on markings,

$$M \xrightarrow{t} M' \stackrel{\Delta}{\iff} {}^\bullet t \le M \ \&\ M' = M - {}^\bullet t + t^\bullet.$$

Note that we apply the operations of multiset union $+$ and multiset subtraction $-$ to the sets ${}^\bullet t$ and $t^\bullet$, regarding these sets as multisets. A marking that can be obtained through a sequence of events from some initial marking is said to be *reachable* from that initial marking. Observe that although each initial marking of a P/T net is a set, it need not in general be the case that every reachable marking is itself a set. For instance, the net in Figure 2 has the following sequence of event occurrences:

$$\left\{ \begin{array}{c} p \mapsto 1 \\ p' \mapsto 0 \end{array} \right\} \xrightarrow{t} \left\{ \begin{array}{c} p \mapsto 1 \\ p' \mapsto 1 \end{array} \right\} \xrightarrow{t} \left\{ \begin{array}{c} p \mapsto 1 \\ p' \mapsto 2 \end{array} \right\}$$

After two occurrences of the event t, there are two tokens in the condition p'.

We say that a net is *safe* if all its reachable markings are sets.

3.2 Occurrence nets

Occurrence nets were introduced in [12] as a class of net suited to giving the semantics of more general kinds of net in a way that directly represents the causal dependencies of elements of the net, for example that a particular event must have occurred at some earlier stage for a particular condition to become marked, and how the occurrence of elements of the net might conflict with each other. Technically, they can be thought-of as safe nets with acyclic flow relations such that every condition occurs as a postcondition of at most one event, for every condition there is a reachable marking containing that condition, and for every event there is a reachable marking in which the event can occur. We extend their original definition to account for the generalization to having a set of initial markings.

Definition 4 *An* occurrence net $O = (B, E, F, \mathbb{M})$ *is a safe net satisfying the following restrictions:*

1. $\forall M \in \mathbb{M} : \forall b \in M : ({}^{\bullet}b = \emptyset)$
2. $\forall b' \in B : \exists M \in \mathbb{M} : \exists b \in M : (b\ F^{*}\ b')$
3. $\forall b \in B : (|{}^{\bullet}b| \leq 1)$
4. F^{+} *is irreflexive and, for all* $e \in E$, *the set* $\{e' \mid e'\ F^{*}\ e\}$ *is finite*
5. $\#$ *is irreflexive, where*

$$e \#_{\mathrm{m}} e' \overset{\Delta}{\iff} e \in E \ \&\ e' \in E \ \&\ e \neq e' \ \&\ {}^{\bullet}e \cap {}^{\bullet}e' \neq \emptyset$$

$$b \#_{\mathrm{m}} b' \overset{\Delta}{\iff} \exists M, M' \in \mathbb{M} : (M \neq M' \ \&\ b \in M \ \&\ b' \in M')$$

$$x \# x' \overset{\Delta}{\iff} \exists y, y' \in E \cup B : y \#_{\mathrm{m}} y' \ \&\ y\ F^{*}\ x \ \&\ y'\ F^{*}\ x'$$

The flow relation F of an occurrence net O indicates how occurrences of events and conditions causally depend on each other and the relation $\#$ indicates how they conflict with each other, with $\#_{\mathrm{m}}$ representing immediate conflict. Two events are in immediate conflict if they share a common precondition, so that the occurrence of one would mean that the other could not occur in any subsequent marking. Two conditions are in immediate conflict if they occur in different initial markings, so if one occurs in a reachable marking there is no subsequent reachable marking in which the other occurs. This corresponds to the intuition at the beginning of this section, that the hidden events giving rise to each initial marking should be in conflict with each other.

The *concurrency* relation $\mathrm{co}_O \subseteq (B \cup E) \times (B \cup E)$ of an occurrence net O may be defined as follows:

$$x \ \mathrm{co}_O \ y \overset{\Delta}{\iff} \neg(x \# y \text{ or } x\ F^+\ y \text{ or } y\ F^+\ x)$$

We often drop the subscript O when we write co_O if the net O is obvious from the context. The concurrency relation is extended to sets of conditions A in the following manner:

$$\begin{aligned}\mathrm{co}\,A \overset{\Delta}{\iff} \ & (\forall b, b' \in A:\ b \ \mathrm{co} \ b') \text{ and} \\ & \{e \in E \mid \exists b \in A. e\ F^*\ b\} \text{ is finite}\end{aligned}$$

Proposition 5 *Let $O = (B, E, F, \mathbb{M})$ be an occurrence net. Any subset $A \subseteq B$ satisfies* co A *iff there exists a reachable marking M of O such that $A \subseteq M$.*

The events and conditions of an occurrence net $(B, E, F, \mathbb{M})$ can only occur from a unique initial marking. For $x \in B \cup E$ and $M \in \mathbb{M}$, write $M\ F^*\ x$ if there exists $b \in M$ such that $b\ F^*\ x$. It is easy to see that M is unique: for any $M, M' \in \mathbb{M}$, if $M\ F^*\ x$ and $M'\ F^*\ x$ then $M = M'$.

Proposition 6 *Let $O = (B, E, F, \mathbb{M})$ be an occurrence net. For any $b \in B$ and $M \in \mathbb{M}$, if $M\ F^*\ b$ then there exists M' such that $b \in M'$ and M' is reachable from M. For any $e \in E$ and $M \in \mathbb{M}$, if $M\ F^*\ e$ then there exists M' such that e has concession in M' and M' is reachable from M.*

An occurrence net O gives rise to a set of singly-marked occurrence nets obtained by splitting the net O at each marking.

Definition 7 *Let $O = (B, E, F, \mathbb{M})$ be an occurrence net. The* marking decomposition *of O is a family of singly-marked occurrence nets $(O_M)_{M \in \mathbb{M}}$ in which the net O_M has conditions B_M and events E_M defined as*

$$B_M = \{b \in B \mid M\ F^*\ b\} \qquad E_M = \{e \in E \mid M\ F^*\ e\},$$

each net O_M inherits the flow relation of O and has initial marking M.

Morphisms of nets are introduced in the following section. It will then be possible to say that an occurrence net can be recovered, up to isomorphism, by placing the elements of its marking decompositions side-by-side, each element with its own initial marking. This will amount to taking the coproduct of the nets in its marking decomposition. Consequently, a multiply marked occurrence net can be partitioned into a family of singly-marked occurrence nets in such a way that the flow relation does not cross the partitions.

3.3 Morphisms and categorical constructions

Morphisms on Petri nets, apart from the slight generalization to multiple initial markings introduced here, were first presented in [20]. They embed the structure of one net into another in a way that preserves the behaviour (the token game) of the original net.

Definition 8 *Let $N = (B,E,F,\mathbb{M})$ and $N' = (B',E',F',\mathbb{M}')$ be P/T nets. A morphism $(\eta,\beta) : N \to N'$ comprises a partial function $\eta : E \to_* E'$ and a relation $\beta \subseteq B \times B'$ satisfying the following criteria:*

- $\forall M \in \mathbb{M}\ \exists M' \in \mathbb{M}' : \beta M \subseteq M' \quad \& \quad \forall b' \in M'\ \exists! b \in M : \beta(b,b')$,
- $\forall e \in E : \beta{}^\bullet e \subseteq {}^\bullet\eta(e) \quad \& \quad \forall b' \in {}^\bullet\eta(e)\ \exists! b \in {}^\bullet e : \beta(b,b')$, *and*
- $\forall e \in E : \beta e^\bullet \subseteq \eta(e)^\bullet \quad \& \quad \forall b' \in \eta(e)^\bullet\ \exists! b \in e^\bullet : \beta(b,b')$.

Note that we regard ${}^\bullet\eta(e) = \eta(e)^\bullet = \emptyset$ if $\eta(e) = *$. Regarding these sets as multisets, using multiset notation we might equivalently have written:

$$\begin{array}{l} \forall M \in \mathbb{M} : \beta \cdot M \in \mathbb{M}' \\ \forall e \in E : \beta \cdot {}^\bullet e = {}^\bullet\eta(e) \\ \forall e \in E : \beta \cdot e^\bullet = \eta(e)^\bullet \end{array}$$

We write $\mathbf{PT}^\sharp$ for the category of P/T nets, $\mathbf{Safe}^\sharp$ for the category of safe nets and $\mathbf{Occ}^\sharp$ for the category of occurrence nets. In each case, we omit the superscript $^\sharp$ to indicate the old categories of singly-marked nets, for example writing **Occ** for the category of singly-marked occurrence nets.

We shall say that a net morphism (η,β) is *synchronous* if η is a total function on events. We add the subscript $_\mathrm{s}$ to denote categories with only synchronous morphisms, for example writing $\mathbf{Occ}^\sharp_\mathrm{s}$ for the category of multiply-marked occurrence nets with synchronous morphisms between them. A morphism (η,β) is a *folding* morphism if it is synchronous and the relation β is also a (total) function. We add the subscript $_\mathrm{f}$ to denote categories with only folding morphisms, for example writing $\mathbf{PT}^\sharp_\mathrm{f}$ for the category of multiply-marked P/T nets with folding morphisms between them.

A morphism $(\eta,\beta) : N \to N'$ respects the token game for nets in the sense that

$$\text{if } M \xrightarrow{e} M' \text{ in } N \text{ then } \beta \cdot M \xrightarrow{e} \beta \cdot M' \text{ in } N'.$$

It will be useful later to point out that, if the nets N and N' are occurrence nets, the morphism preserves markings giving rise to elements of the occurrence net, it reflects conflict and it reflects the F relation in the following sense:

Proposition 9 *Let $O_1 = (B_1, E_1, F_1, \mathbb{M}_1)$ and $O_2 = (B_2, E_2, F_2, \mathbb{M}_2)$ be occurrence nets. For events $e_1, e'_1 \in E_1$, write $e_1 \asymp_1 e'_1$ iff either $e_1 = e'_1$ or $e_1 \# e'_1$. Define $\asymp_2$ similarly for events in E_2. For any morphism $(\eta, \beta) : O_1 \to O_2$ in $\mathbf{Occ}^\#$:*

- *for any $b_1 \in B_1$ and $M \in \mathbb{M}_1$, if $M\ F_1^*\ b_1$ and $\beta(b_1, b_2)$ then $\beta \cdot M\ F_2^*\ b_2$*
- *for any $e_1 \in E_1$, if $\eta(e_1)$ defined and $M\ F_1^*\ e_1$ then $\beta \cdot M\ F_2^*\ \eta(e_1)$*
- *for any $e_1, e'_1 \in E_1$ and $e_2, e'_2 \in E_2$:*

$$\eta(e_1) = e_2\ \&\ \eta(e'_1) = e'_2\ \&\ e_2 \asymp_2 e'_2 \Longrightarrow e_1 \asymp_1 e'_1$$

- *for any $b_1, b'_1 \in B_1$ and $b_2, b'_2 \in B_2$:*

$$\beta(b_1, b_2)\ \&\ \beta(b'_1, b'_2)\ \&\ b_2 \asymp_2 b'_2 \Longrightarrow b_1 \asymp_1 b'_1$$

- *for any $e_2 \in E_2$, $b_1 \in B_1$ and $b_2 \in B_2$:*

$$e_2\ F_2\ b_2\ \&\ \beta(b_1, b_2) \Longrightarrow \exists! e_1 \in E_1 :\ e_1\ F_1\ b_1\ \&\ \eta(e_1) = e_2$$

- *for any $e_1 \in E_1$, $e_2 \in E_2$ and $b_2 \in B_2$:*

$$\eta(e_1) = e_2\ \&\ b_2\ F_2\ e_2 \Longrightarrow \exists! b_1 \in B_1 :\ b_1\ F_1\ e_1\ \&\ \beta(b_1, b_2)$$

It follows that morphisms in the category $\mathbf{Occ}^\#$ also preserve the concurrency relation on both events and conditions.

Coproducts in the categories of singly-marked safe nets and singly-marked occurrence nets were studied in [20]. There, the construction of $N_1 + N_2$ essentially involves 'gluing' the nets N_1 and N_2 together at their initial markings. The generalization to allow multiple initial markings allows a somewhat simpler construction in the categories $\mathbf{Occ}^\#$ and $\mathbf{PT}^\#$, where the nets are forced to operate on disjoint sets of conditions.

Proposition 10 *Let $(N_i)_{i \in I}$ be a family of P/T nets where $N_i = (P_i, T_i, F_i, \mathbb{M}_i)$ for each $i \in I$. The net $\sum_{i \in I} N_i = (P, T, F, \mathbb{M})$ defined as*

$$\begin{aligned} P &= \{\mathsf{in}_i p \mid i \in I\ \&\ p \in P_i\} \\ T &= \{\mathsf{in}_i t \mid i \in I\ \&\ t \in T_i\} \end{aligned}$$

$$(\mathsf{in}_i x)\, F\, (\mathsf{in}_j y) \iff i = j \ \&\ x\, F_i\, y$$
$$\mathbb{M} = \{\{\mathsf{in}_i p \mid p \in M\} \mid i \in I \ \&\ M \in \mathbb{M}_i\}$$

is a coproduct in the category $\mathbf{PT}^\sharp$ *with coproduct injections* $\mathsf{in}_i : N_i \to \sum_{j\in I} N_j$.

Furthermore, the construction gives coproducts in the categories $\mathbf{Occ}^\sharp$ *and* $\mathbf{Safe}^\sharp$ *and the categories with synchronous morphisms.*

As mentioned at the end of the previous section, a multiply-marked occurrence net can be recovered (up to isomorphism) by taking the coproduct of the nets arising from each initial marking.

Proposition 11 *Let* O *be an occurrence net and* $(O_M)_{M\in\mathbb{M}}$ *be its marking decomposition. Then* $O \cong \sum_{M\in\mathbb{M}} O_M$ *through an isomorphism natural in* O.

We conclude this section by noting that the category $\mathbf{PT}^\sharp$, in addition to having coproducts, also has pullbacks. This result, a mild generalization of [4] where it was shown that the category of safe nets has pullbacks, is important in being able to enrich P/T nets and occurrence nets with symmetry.

Proposition 12 *The category* $\mathbf{PT}^\sharp$ *has pullbacks.*

A consequence of the coreflection between $\mathbf{Occ}^\sharp$ and $\mathbf{PT}^\sharp$ will be that the category $\mathbf{Occ}^\sharp$ also has pullbacks.

4 Event Structures

Event structures [12, 21] represent a computational process as a set of event occurrences, recording how these event occurrences *causally depend* on each other. An event structure also records how the occurrence of an event indicates that the process has taken a particular branch. For the variant of event structure that we shall consider, called *prime* event structures, this amounts to recording how event occurrences *conflict* with each other.

Definition 13 *A (prime) event structure is a 3-tuple*

$$ES = (E, \leq \#),$$

where

- E *is the set of* events *(more precisely, event occurrences),*
- $\leq \subseteq E \times E$ *is the partial order of* causal dependency, *and*
- $\# \subseteq E \times E$ *is the irreflexive, symmetric binary relation of* conflict.

An event structure must satisfy the following axioms:

1. *each event causally depends on only finitely many other events, i.e.* $\{e' \mid e' \leq e\}$ *is finite for all* $e \in E$, *and*
2. *if* $e_1 \# e_2$ *and* $e_1 \leq e_1'$ *then* $e_1' \# e_2$.

The intuition is that if we have $e \leq e'$ for two events e and e', then the event e must have occurred prior to any occurrence of e'. If we have $e\#e'$, then the occurrence of e precludes the occurrence of event e' at any later stage. An event structure is said to be *elementary* if the conflict relation is empty. The first axiom ensures that an event structure only consists of event occurrences that can eventually take place, not relying on an infinite number of prior event occurrences. The second axiom asserts that if the occurrence of an event e_2 precludes the occurrence of an event e_1 upon which the event e_1' causally depends, then the event e_2 precludes the occurrence of the event e_1'. We say that two events e_1 and e_2 are *concurrent*, written e_1 co e_2, if there is no causal dependency between them and they do not conflict, *i.e.* $e_1 \text{ co } e_2 \iff \neg(e_1 \# e_2 \text{ or } e_1 \leq e_2 \text{ or } e_2 \leq e_1)$. We write $e < e'$ if $e \leq e'$ but $e \neq e'$.

The computational states of an event structure, called its *configurations*, are represented by the sets of events that have occurred. Every configuration must be consistent with the relations of conflict and causal dependency. Formally, $x \subseteq E$ is a configuration of an event structure $(E, \leq, \#)$ if it satisfies the following two properties:

- Conflict-freedom: $\forall e, e' \in x : \neg(e\#e')$
- Downwards-closure: $\forall e, e' \in E : e \leq e' \;\&\; e' \in x \Longrightarrow e \in x$.

We write $\mathscr{D}(ES)$ for the set of configurations of ES and write $\mathscr{D}^0(ES)$ for the set of finite configurations of ES. We write $\lceil e \rceil$ for $\{e' \mid e' \leq e\}$, the least configuration containing the event e.

4.1 Morphisms

We now introduce morphisms of event structures. A morphism $\eta : ES \rightarrow ES'$ is a function from the events of ES to the events of ES' that expresses how the behaviour of ES embeds into ES' in the sense that the function preserves the the configurations of the event structure and also preserves the atomicity of events.

Definition 14 *Let* $ES = (E, \leq, \#)$ *and* $ES' = (E', \leq', \#')$ *be event structures. A morphism* $\eta : ES \rightarrow ES'$ *consists of a partial function*

$\eta : E \rightarrow_* E'$ *such that for all* $x \in \mathscr{D}(ES)$*:*

$$\begin{aligned} & \eta x \in \mathscr{D}(ES) \\ \& \quad & \forall e, e' \in x : \eta(e), \eta(e') \text{ defined } \& \ \eta(e) = \eta(e') \Longrightarrow e = e' \end{aligned}$$

A morphism is said to be synchronous *if it is a total function on events.*

In fact, it is only necessary to consider finite configurations $x \in \mathscr{D}^0(ES)$ in the requirement on morphisms above. It is easy to see that if $x \xrightarrow{e} x'$ then $\eta x \xrightarrow{\eta(e)} \eta x'$.

We obtain a category **ES** of (prime) event structures with event structures as objects and morphisms as described above. The identity morphism on an event structure is the identity function on its underlying set of events, and composition of morphisms occurs as composition of functions. We also obtain a category $\mathbf{ES}_s$ of event structures with synchronous morphisms between then. We write **Elem** for the category of elementary event structures (event structures with no conflict) and $\mathbf{Elem}_s$ for the category of elementary event structures with synchronous morphisms between them. Elementary event structures can be thought of as paths of event structures and nets, and these categories will later be used to define open maps of event structures and nets.

Before moving on to consider their relationship with Petri nets, we note that the categories **ES** and $\mathbf{ES}_s$ have coproducts obtained by forming the disjoint union of their events using injections in_i, placing two events in conflict if they occur in different components of the coproduct.

Proposition 15 *Let* $(ES)_{i \in I}$ *be a family of event structures indexed by* I*, where* $ES_i = (E_i, \leq_i, \#_i)$*. A coproduct of these event structures in the category* **ES** *and also in the category* $\mathbf{ES}_s$ *is the event structure* $\sum_{i \in I} ES_i = (E, \leq, \#)$ *with events* $E = \{\mathsf{in}_i e \mid i \in I \ \& \ e \in E_i\}$ *and relations*

$$\begin{aligned} \mathsf{in}_i e \leq \mathsf{in}_j e' & \iff i = j \ \& \ e \leq_i e' \\ \mathsf{in}_i e \ \# \ \mathsf{in}_j e' & \iff i \neq j \text{ or } (i = j \ \& \ e \#_i e') \end{aligned}$$

For each $j \in I$*, the function* $\mathsf{in}_j : ES_j \rightarrow \sum_{i \in I} ES_i$ *defined as* $\mathsf{in}_i(e) = \mathsf{in}_i e$ *is the associated injection into the coproduct.*

The category of event structures also has pullbacks. Their construction is given in Appendix C of [23]. Their direct construction is hard, being most easily seen in the category of stable families, so we shall not present it here.

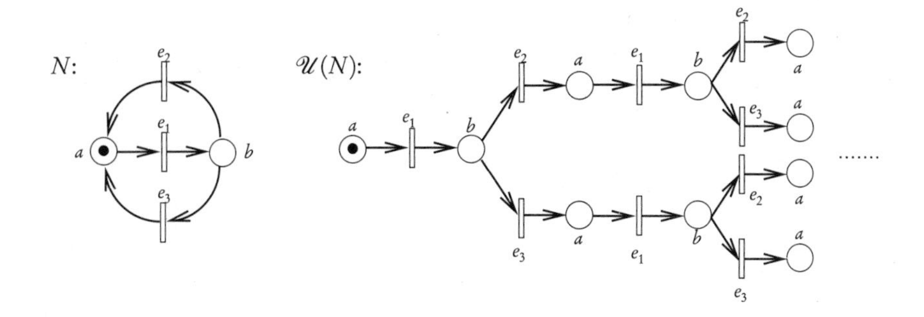

Figure 3: *The unfolding of a safe net*

5 Results on Singly-Marked Nets

We briefly recount some results on singly-marked nets. We shall first describe the coreflection between singly-marked safe nets and occurrence nets and shall then give the coreflection between singly-marked occurrence nets and event structures. The constructions used in the second coreflection shall be used in the description of the coreflection between multiply-marked occurrence nets and event structures.

5.1 Singly-marked occurrence nets and safe nets

The category **Occ** is a coreflective subcategory of **Safe**:

$$\mathbf{Occ} \underset{\mathcal{U}}{\overset{\hookrightarrow}{\perp}} \mathbf{Safe}$$

The left adjoint is the inclusion of the category of occurrence nets into the category of safe nets. The right adjoint *unfolds* the safe net to an occurrence net. An example is presented in Figure 3, with events and conditions in the unfolding labelled by the corresponding events and conditions in the original net. The occurrence net unfolding of a safe net, first defined in [12], describes how the event occurrences of the safe net causally depend on and conflict with each other due to their effect on the holding of conditions. We shall not describe the unfolding concretely here, though its definition occurs in Section 7 in the more general setting of multiply-marked P/T nets.

The unfolding $\mathcal{U}(N)$ of a safe net N is equipped with a morphism $\epsilon_N : \mathcal{U}(N) \to N$ in the category **Safe** relating the unfolding back to the original net. The adjunction between safe nets and occurrence nets [20, 21] is proved by showing that $\mathcal{U}(N)$ and ϵ_N are *cofree* over

N. That is, for any safe net N and morphism of nets $(\theta, \alpha) : O \to N$ from an occurrence net, there is a unique morphism of occurrence nets $(\pi, \gamma) : O \to \mathcal{U}(N)$ such that the following diagram commutes:

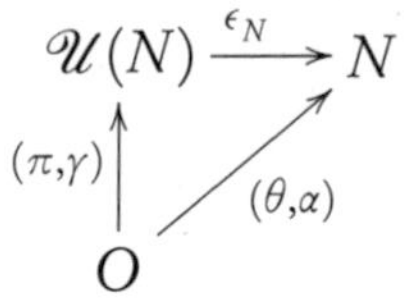

$$\begin{array}{ccc} \mathcal{U}(N) & \xrightarrow{\epsilon_N} & N \\ {\scriptstyle (\pi,\gamma)} \uparrow & \nearrow {\scriptstyle (\theta,\alpha)} & \\ O & & \end{array}$$

It is a standard result of category theory that this is sufficient to give the adjunction [10]. The adjunction is a coreflection because the inclusion **Occ** $\hookrightarrow$ **Safe** is full and faithful. The coreflection also goes through for synchronous morphisms to show that $\mathbf{Occ}_s$ is a coreflective subcategory of $\mathbf{Safe}_s$.

5.2 Event structures and singly-marked occurrence nets

There is a coreflection that embeds the category of event structures into the category of singly-marked occurrence nets.

$$\mathbf{ES} \underset{\mathcal{E}}{\overset{\mathcal{N}}{\rightleftarrows}} \mathbf{Occ} \quad (\bot)$$

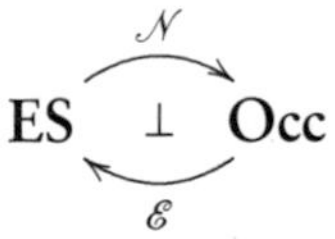

The functor $\mathcal{N}$ constructs an occurrence net from an event structure, saturating the events of the event structure with as many conditions as possible that are consistent with the the relations of causal dependency and conflict in the original event structure. The functor $\mathcal{E}$ strips away the conditions from the occurrence net to reveal the underlying causal dependency and conflict relations on events. Since we shall use the constructions in forming a coreflection between event structures and multiply-marked occurrence nets, we now give the constructions $\mathcal{E}$ and $\mathcal{N}$ concretely. A coreflection can also be obtained via the category of asynchronous transition systems as in [25].

The functor $\mathcal{E} : \mathbf{Occ} \to \mathbf{ES}$

The functor $\mathcal{E}$ takes an occurrence net to an event structure by interpreting causal dependency on the events of the occurrence net as the transitive closure of the flow relation and obtaining the conflict relation as in Definition 4.

Definition 16 *Let $O = (B, E, F, \mathbb{M})$ be an occurrence net. The event structure $\mathscr{E}(O) = (E, \leq, \#)$ has the same events as O, inherits conflict from O as in Definition 4 and has $e \leq e'$ iff $e\, F^*\, e'$.*

It is an immediate consequence of the definitions that $\mathscr{E}(O)$ is an event structure for any occurrence net O. Recalling that a morphism between occurrence nets O and O' is a pair (η, β) of which $\eta : E \rightarrow_* E'$ is a partial function on their underlying sets of events, we obtain the operation of the functor on morphisms.

Proposition 17 *Let $(\eta, \beta) : O \rightarrow O'$ be a morphism in* **Occ**. *Then $\eta : \mathscr{E}(O) \rightarrow \mathscr{E}(O')$ is a morphism in* **ES**.

It is straightforward to see that defining $\mathscr{E}(\eta, \beta) = \eta$ yields an operation that preserves identities and composition, so $\mathscr{E}$: **Occ** $\rightarrow$ **ES** is a functor. This is easily seen to restrict to categories with synchronous morphisms, so also $\mathscr{E} : \mathbf{Occ}_s \rightarrow \mathbf{ES}_s$.

The functor $\mathscr{N}$: ES $\rightarrow$ Occ

We now consider how to form an occurrence net from an event structure. As stated earlier, the essential idea is to form an occurrence net with the same events as the original event structure, adding as many conditions as possible that are consistent with the causal dependency and conflict relations of the original event structure.

Definition 18 *Let $ES = (E, \leq, \#)$ be a event structure. The net $\mathscr{N}(ES)$ is defined as $(B, E, F, \{M\})$, where*

$$
\begin{aligned}
M &= \{(\emptyset, A) \mid A \subseteq E \;\&\; (\forall a, a' \in A : a \asymp a')\} \\
B &= M \cup \begin{array}{l}\{(e, A) \mid e \in E \;\&\; A \subseteq E \;\&\; (\forall a, a' \in A : a \asymp a') \\ \qquad\qquad \& \; (\forall a \in A : e < a)\}\end{array} \\
F &= \{(e, (e, A)) \mid (e, A) \in B\} \\
&\quad \cup \{((x, A), e) \mid (x, A) \in B \;\&\; e \in A\}
\end{aligned}
$$

The net is formed with conditions (e, A) indicating that all the events in A are in conflict with each other and all causally depend on e. There are conditions $(\emptyset, A)$ to indicate just that the events in A are in conflict with each other but might not causally depend on some other event. The net formed is *condition-extensional* in the sense that any two conditions with precisely the same beginning- and end-events are identified. The occurrence net of an example event structure is presented in Figure 4.

Proposition 19 *The net $\mathscr{N}(ES)$ is an occurrence net. Furthermore, for any event structure ES we have $\mathscr{E}(\mathscr{N}(ES)) = ES$.*

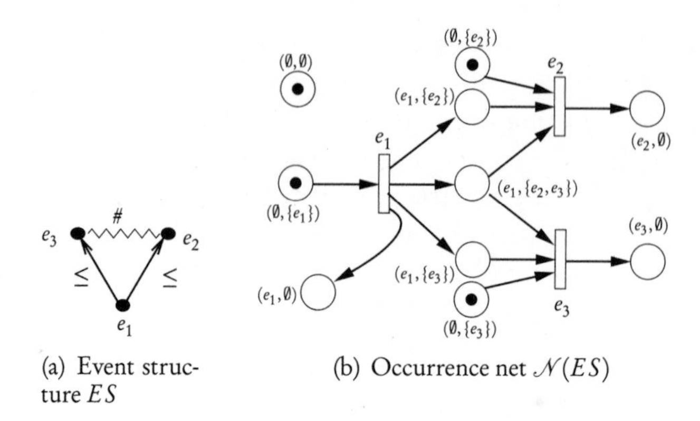

(a) Event structure ES

(b) Occurrence net $\mathcal{N}(ES)$

Figure 4: *An event structure with its associated occurrence net*

Freeness and morphisms

In order to obtain a coreflection, this time it is easier to show a *freeness* result. This is sufficient, also, to show how the operation $\mathcal{N}$ extends to a functor [10].

Proposition 20 *For any event structure in* **ES**, *the net* $\mathcal{N}(ES)$ *and morphism* $\mathsf{id}_{ES} : ES \to \mathcal{E}(\mathcal{N}(ES))$ *is free over* ES *with respect to* $\mathcal{E}$. *That is, for any occurrence net* O *and morphism* $\eta : ES \to \mathcal{E}(O)$ *in* **ES** *there is a unique morphism* $(\pi, \gamma) : \mathcal{N}(ES) \to O$ *in* **Occ** *such that the triangle in the following diagram commutes:*

$$\begin{array}{ccccc} \mathcal{N}(ES) & \qquad & \mathcal{E}(\mathcal{N}(ES)) & \xleftarrow{\mathsf{id}_{ES}} & ES \\ \Big\downarrow{\scriptstyle (\pi,\gamma)} & & \Big\downarrow{\scriptstyle \mathcal{E}(\pi,\gamma)=\pi} & \swarrow{\scriptstyle \eta} & \\ O & & \mathcal{E}(O) & & \end{array}$$

Hence the functor $\mathcal{N}$: **ES** $\to$ **Occ** is left-adjoint to the functor $\mathcal{E}$: **Occ** $\to$ **ES**. Since the unit of the adjunction is a natural isomorphism (in this case, the identity), the adjunction is a coreflection.

Proposition 20 also applies using the categories $\mathbf{ES_s}$ and $\mathbf{Occ_s}$ in place of **ES** and **Occ**, so a coreflection is obtained for the categories with synchronous morphisms. We shall to use the same symbols to represent the functors $\mathcal{N} : \mathbf{ES_s} \to \mathbf{Occ_s}$ and $\mathcal{E} : \mathbf{Occ_s} \to \mathbf{ES_s}$.

6 Relating Event Structures and Occurrence Nets

We now progress to consider a coreflection between event structures and the multiply-marked occurrence nets presented in Section 3.2. To obtain an adjunction, it is necessary to restrict attention to categories of occurrence nets and event structures with synchronous morphisms (morphisms that are total on events). We shall briefly mention how partiality could be recovered at the end of this section.

We first define how a multiply-marked occurrence net forms an event structure, giving rise to a functor $\mathcal{E}^{\sharp}_{s} : \mathbf{Occ}^{\sharp}_{s} \to \mathbf{ES}_{s}$.

The construction of the event structure $\mathcal{E}^{\sharp}_{s}(O)$ from a multiply-marked occurrence net O is similar to that presented in Section 5.2. The events of $\mathcal{E}^{\sharp}_{s}(O)$ are simply the events of O; causal dependency of events is obtained from the flow relation F; and the conflict relation on events is obtained from the conflict relation of O. Recall that the conflict relation on the occurrence net places two events in conflict if they are given rise to by different initial markings.

Definition 21 *Let $O = (B, E, F, \mathbb{M})$ be an occurrence net. The event structure $\mathcal{E}^{\sharp}_{s}(N)$ is $(E, \leq, \#)$ where*

$$e \leq e' \iff e\, F^{*}\, e'$$

and # is the conflict relation on the occurrence net O in Definition 4.

The operation $\mathcal{E}^{\sharp}_{s}$ extends to a functor $\mathcal{E}^{\sharp}_{s} : \mathbf{Occ}^{\sharp}_{s} \to \mathbf{ES}_{s}$ by taking a morphism of occurrence nets $(\eta, \beta) : O \to O'$ to

$$\mathcal{E}^{\sharp}_{s}(\eta, \beta) = \eta.$$

It is relatively straightforward to show that $\eta : \mathcal{E}^{\sharp}_{s}(O) \to \mathcal{E}^{\sharp}_{s}(O')$ is a morphism of event structures and that $\mathcal{E}^{\sharp}_{s}$ satisfies the requirements for being a functor.

The specification of a functor from event structures to occurrence nets with multiple initial markings is less straightforward. The generalization of occurrence nets to allow them to possess more than one initial marking gives rise to two distinct ways in which their events may be in conflict.

'Early' conflict Any event in an occurrence net can occur in a marking reachable from precisely one initial marking. The events may conflict if they arise from distinct initial markings.

'Late' conflict As with singly-marked occurrence nets, two events e_1 and e_2 might be in conflict because they either share a

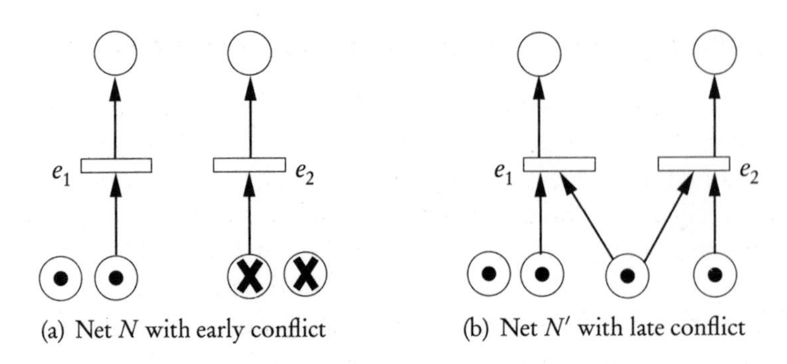

(a) Net N with early conflict

(b) Net N' with late conflict

Figure 5: *Nets with early conflict and late conflict*

precondition or there might exist events e_1' and e_2' that share a common precondition for which e_1 causally depends on e_1' and e_2 causally depends on e_2'.

Quite clearly, all conflict in singly-marked occurrence nets is late conflict. The old functor $\mathcal{N} : \mathbf{ES}_s \to \mathbf{Occ}_s$ from event structures to singly-marked occurrence nets therefore uses late conflict to represent conflict in the event structure.

In the category $\mathbf{Occ}_s^\sharp$, early conflict embeds into late conflict. Consider, for instance, the nets in Figure 5. There is a morphism preserving events from N to N'. Late conflict, however, does not embed into early conflict; there is no morphism preserving events from N' to N.

The old functor $\mathcal{N}$ can be seen, as a result, not to be left adjoint to the new functor $\mathcal{E}_s^\sharp$. If it were, there would have to be a morphism preserving events of the form $\mathcal{N}(\mathcal{E}_s^\sharp(N)) \to N$. It is easy to see that the event structures $\mathcal{E}_s^\sharp(N)$ and $\mathcal{E}_s^\sharp(N')$ are equal, comprising two events e_1 and e_2 that are in conflict. The net $\mathcal{N}(\mathcal{E}_s^\sharp(N))$ is isomorphic to N', so the required morphism does not exist. The problem is that in constructing the net $\mathcal{N}(\mathcal{E}_s^\sharp(N))$, early conflict is replaced by late conflict. We must therefore define a new functor $\mathcal{N}_s^\sharp : \mathbf{ES}_s \to \mathbf{Occ}_s^\sharp$ that ensures that if two events of a net N are in early conflict then they remain in early conflict in the net $\mathcal{E}_s^\sharp(\mathcal{N}_s^\sharp(N))$.

The functor $\mathcal{N}_s^\sharp$ will involve the *compatibility* relation to distinguish which pairs of events in $\mathcal{E}_s^\sharp(N)$ could possibly have been in early conflict in the net N. Let $ES = (E, \leq, \#)$ be an event structure. The compatibility relation $\frown \subseteq E \times E$ is defined as:

$$e \frown e' \overset{\Delta}{\iff} \neg(e\#e')$$

Two events are compatible if there exists a configuration containing them both. The compatibility relation is symmetric and reflexive, so its transitive closure $\frown^+$ is an equivalence relation. The event structure ES can be partitioned into a family $(ES_c)_{c\in C}$ of $\frown^+$-equivalence classes. Each $\frown^+$-equivalence class ES_c is an event structure with conflict and causal dependency inherited from ES. Any event of ES_c is in conflict with every event of ES_d in the event structure ES if $c \neq d$.

Lemma 22 *Let ES be an event structure and let the $\frown^+$-equivalence classes contained in ES form the family $(ES_c)_{c\in C}$ for some indexing set C. Each equivalence class ES_c is an event structure and $ES \cong \sum_{c\in C} ES_c$ through an isomorphism natural in ES, taking the coproduct in the category* $\mathbf{ES}_s$ *defined in Proposition 15.*

Two events of the event structure $\mathcal{E}_s^{\sharp}(N)$ cannot be in early conflict if they are $\frown^+$-related. Given an event structure ES, we are now able to define the occurrence net $\mathcal{N}_s^{\sharp}(ES)$.

Definition 23 *Let ES be an event structure generated by the family of $\frown^+$-equivalence classes $(ES_c)_{c\in C}$. Define*

$$\mathcal{N}_s^{\sharp}(ES) \triangleq \sum_{c\in C} \mathcal{N}(ES_c).$$

It is clear that $\mathcal{N}_s^{\sharp}(ES)$ is an occurrence net since $\mathcal{N}(ES_c)$ is known to be a (singly-marked) occurrence net for each $c \in C$ and the coproduct of occurrence nets is itself an occurrence net.

An important observation when considering the coreflection between event structures and multiply-marked occurrence nets will be that morphisms of event structures preserve the relation $\frown^+$.

Lemma 24 *Let $ES_1 = (E_1, \leq_1, \#_1)$ and $ES_2 = (E_2, \leq_2, \#_2)$ be event structures with compatibility relations $\frown_1$ and $\frown_2$ respectively. Let $\eta : ES_1 \to ES_2$ be a morphism in* $\mathbf{ES}_s$*. For any $e, e' \in E_1$, if $e \frown_1^+ e'$ then $\eta(e) \frown_2^+ \eta(e')$.*

The previous lemma relies on the fact that morphisms are synchronous, *i.e.* total on events. It need not be the case that if $e_1 \frown_1^+ e_2$ and $\eta(e_1)$ and $\eta(e_2)$ are defined for some non-synchronous morphism η then $\eta(e_1) \frown_2^+ \eta(e_2)$.

Let the $\frown_1^+$-equivalence classes of ES_1 be the family $(ES_c)_{c\in C}$ and the $\frown_2^+$-equivalence classes of ES_2 be the family $(ES_d)_{d\in D}$, and suppose that there is a synchronous morphism $\eta : ES_1 \to ES_2$ in $\mathbf{ES_s}$. As a consequence of the previous lemma, for all $c \in C$ and $d \in D$:

$$\begin{aligned} & \exists e \in E_c : E_d = \{e_2 \mid \eta(e) \subset_2^+ e_2\} \\ \iff \quad & \forall e \in E_c : E_d = \{e_2 \mid \eta(e) \subset_2^+ e_2\}. \end{aligned}$$

We may therefore define a function $\hat{\eta} : C \to D$ as $\hat{\eta}(c) = d$ iff $\exists e \in E_c : E_d = \{e_2 \mid \eta(e) \subset_2^+ e_2\}$, which informs that the event structure ES_c within ES_1 is taken by η to $ES_{\hat{\eta}(c)}$ in ES_2. The morphism $\eta : ES_1 \to ES_2$ therefore restricts to a morphism $\eta_c : ES_c \to ES_{\hat{\eta}(c)}$. Applying the old functor $\mathcal{N} : \mathbf{ES_s} \to \mathbf{Occ_s}$ from event structures to singly-marked occurrence nets, we obtain a morphism $\mathcal{N}(\eta_c) : \mathcal{N}(ES_c) \to \mathcal{N}(ES_d)$. We therefore have a morphism

$$\mathsf{in}_c \circ \mathcal{N}(\eta_c) : \mathcal{N}(ES_c) \to \mathcal{N}^\sharp(ES_2)$$

for each $c \in C$. Since $\mathcal{N}^\sharp(ES_1) = \sum_{c \in C} \mathcal{N}(ES_c)$ is a coproduct, we have a morphism

$$\mathcal{N}^\sharp(\eta) : \mathcal{N}^\sharp(ES_1) \to \mathcal{N}^\sharp(ES_2).$$

This construction is straightforwardly shown to form a functor $\mathcal{N}^\sharp : \mathbf{ES_s} \to \mathbf{Occ}^\sharp_\mathbf{s}$.

We now proceed to show that $\mathcal{N}^\sharp_s$ is left adjoint to the functor $\mathcal{E}^\sharp_s$ giving the coreflection

$$\mathbf{ES_s} \underset{\mathcal{E}^\sharp_s}{\overset{\mathcal{N}^\sharp_s}{\rightleftarrows}} \mathbf{Occ}^\sharp_\mathbf{s} \quad (\bot)$$

This will also specify the action of the operation $\mathcal{N}^\sharp_s$ on morphisms of event structures, yielding a functor $\mathcal{N}^\sharp_s : \mathbf{ES_s} \to \mathbf{Occ}^\sharp_\mathbf{s}$.

Theorem 25 *The functors $\mathcal{E}^\sharp_s$ and $\mathcal{N}^\sharp_s$ form a coreflection: There is an isomorphism of hom-sets*

$$\phi_{ES,N} : \mathbf{ES_s}(ES, \mathcal{E}^\sharp_s(N)) \cong \mathbf{Occ}^\sharp_\mathbf{s}(\mathcal{N}^\sharp_s(ES), N),$$

natural in ES and N and, furthermore, the functor $\mathcal{N}^\sharp_s$ is full and faithful.

Proof Let the occurrence net $O = (B, E, F, \mathbb{M})$ have marking decomposition $(O_M)_{M \in \mathbb{M}}$. Suppose that the $\frown^+$-decomposition of the event structure ES is $(ES_c)_{c \in C}$. We have the following chain of isomorphisms.

$$
\begin{aligned}
\mathbf{ES}_s(ES, \mathscr{E}^\sharp_s(O)) &\cong \mathbf{ES}_s(\textstyle\sum_{c\in C} E_c, \sum_{M\in\mathbb{M}} \mathscr{E}(O_M)) && (1)\\
&\cong \textstyle\prod_{c\in C} \mathbf{ES}_s(ES_c, \sum_{M\in\mathbb{M}} \mathscr{E}(O_M)) && (2)\\
&\cong \textstyle\prod_{c\in C} \sum_{M\in\mathbb{M}} \mathbf{ES}_s(ES_c, \mathscr{E}(O_M)) && (3)\\
&\cong \textstyle\prod_{c\in C} \sum_{M\in\mathbb{M}} \mathbf{Occ}_s(\mathscr{N}(ES_c), O_M) && (4)\\
&\cong \textstyle\prod_{c\in C} \mathbf{Occ}^\sharp_s(\mathscr{N}(ES_c), \sum_{M\in\mathbb{M}} O_M) && (5)\\
&\cong \mathbf{Occ}^\sharp_s(\textstyle\sum_{c\in C} \mathscr{N}(ES_c), \sum_{M\in\mathbb{M}} O_M) && (6)\\
&\cong \mathbf{Occ}^\sharp_s(\mathscr{N}^\sharp_s(ES), O) && (7)
\end{aligned}
$$

Isomorphism (1) is an immediate consequence of Lemma 22 and the definition of $\mathscr{E}^\sharp_s(O)$. Isomorphisms (2) and (6) are from the universal characterization of coproduct. Isomorphism (3) is a straightforward consequence of Lemma 24. Isomorphism (4) arises from the old adjunction between singly-marked occurrence nets and event structures. Isomorphism (5) is a straightforward consequence of Proposition 9. Finally, isomorphism (7) follows from Proposition 11 and the definition of $\mathscr{N}^\sharp_s(ES)$. We omit the proofs that the isomorphisms are natural.

The functor $\mathscr{N}^\sharp_s$ is easily seen to be full and faithful as a consequence of the functor $\mathscr{N}$ being full and faithful, due to the existing coreflection between singly-marked occurrence nets and event structures. ⊣

6.1 Symmetry

We now wish to show that the coreflection above between event structures and occurrence nets with synchronous morphisms extends to the categories with symmetry. To attach symmetry to these categories, we must choose a path category such that open maps are preserved by the adjunctions. The appropriate paths of event structures in this situation are elementary event structures and the appropriate paths of occurrence nets are the images under the functor $\mathscr{N}^\sharp_s : \mathbf{ES}_s \to \mathbf{Occ}^\sharp_s$ of elementary event structures. Write $\mathscr{N}^\sharp_s\mathbf{Elem}_s$ for the subcategory of $\mathbf{Occ}^\sharp_s$ formed as the image of $\mathbf{Elem}_s$ under the functor $\mathscr{N}^\sharp_s$. Open maps of safe nets with elementary event structures as paths were studied in [13].

A general result about open maps presented in [9] shows that the functors $\mathscr{E}^\sharp_s$ and $\mathscr{N}^\sharp_s$ preserve open maps as defined here. The remaining lemma in proving the required coreflection is:

Lemma 26 *The functor $\mathscr{N}^\sharp_s$ preserves pullbacks of* $\mathbf{Elem}_s$*-open maps.*

Therefore, since right adjoints preserve limits, we obtain the desired coreflection.

Theorem 27 *The functor $\mathscr{S}\mathscr{N}^\sharp_s$ is left adjoint to the functor $\mathscr{S}\mathscr{E}^\sharp_s$.*

$$\mathscr{S}_{\mathbf{Elem}_s}\mathbf{ES}_s \underset{\mathscr{S}\mathscr{E}^\sharp_s}{\overset{\mathscr{S}\mathscr{N}^\sharp_s}{\rightleftarrows}} \mathscr{S}_{\mathscr{N}^\sharp\mathbf{Elem}_s}\mathbf{Occ}^\sharp_s \quad (\perp)$$

Furthermore, the adjunction is a coreflection.

The account so far has been restricted to categories of nets and event structures with synchronous morphisms. To lift this restriction and still obtain an adjunction, event structures may presumably be extended to record information on early conflict, essentially by considering families of event structures. We shall not, however, go further into the precise definition of the structure of "event structures with early conflict" in the present paper.

7 Relating Occurrence Nets and P/T Nets

In this section, we turn to considering an adjunction between the category of P/T nets and its full subcategory of occurrence nets.

$$\mathbf{Occ}^\sharp \underset{\mathscr{U}}{\overset{\hookrightarrow}{\rightleftarrows}} \mathbf{PT}^\sharp \quad (\perp)$$

The left adjoint is the inclusion arising from every occurrence net being a P/T net; the right adjoint is the functor $\mathscr{U}$ which unfolds a P/T net to an occurrence net. The adjunction is a coreflection, with $\mathbf{Occ}^\sharp$ being a coreflective subcategory of $\mathbf{PT}^\sharp$. We shall work within the broader categories of occurrence and P/T nets with morphisms that can be partial on events, although the coreflection cuts down to yield a coreflection between the subcategories of occurrence nets and P/T nets with synchronous morphisms. The proof of the coreflection follows that in [20] for safe nets and occurrence nets, with a little generalization to account for multiple markings.

We present the coreflection between multiply-marked nets rather than singly-marked nets because, as we shall see, it restricts to a coreflection between singly-marked occurrence and P/T nets. The adjunction between the categories of multiply-marked nets is not obtained algebraically from the adjunction between singly-marked nets in an analogous manner to the adjunction in the previous section since, in general, a multiply-marked P/T net cannot be expressed as the coproduct of singly-marked P/T nets.

An important fact when dealing with occurrence nets is that any element of an occurrence net occurs at a unique *depth*. For a condition b and an event e of an occurrence net $O = (B, E, F, \mathbb{M})$, depth is defined as:

$$\begin{array}{rcll} \text{depth}(b) & = & 0 & \text{if } \exists M \in \mathbb{M} : (b \in M) \\ \text{depth}(b) & = & \text{depth}(e) & \text{if } e\, F\, b \end{array}$$

$$\text{depth}(e) = 1 + \max\{\text{depth}(b) \mid b \in B \;\&\; b\, F\, e\}$$

The occurrence net unfolding of a P/T net N is defined inductively: The unfolding to depth zero $\mathcal{U}_0(N)$ is defined first and then the unfolding $\mathcal{U}_n(N)$ to depth n is used to define the unfolding $\mathcal{U}_{n+1}(N)$ to depth $n+1$. The unfolding $\mathcal{U}(N)$ is obtained as a colimit of this sequence. We shall not present details here, but shall characterize the unfolding uniquely.

Theorem 28 *Let $N = (P, T, F_N, \mathbb{M}_N)$ be a P/T net. The unfolding $\mathcal{U}(N) = (B, E, F, \mathbb{M})$ is the unique occurrence net to satisfy*

$$\begin{array}{rcl} B & = & \{(M, p) \mid M \in \mathbb{M}_N \;\&\; p \in M\} \\ & & \cup\{(\{e\}, p) \mid e \in E \;\&\; p \in \eta(e)^\bullet\} \\ E & = & \{(A, t) \mid A \subseteq B \;\&\; t \in T \;\&\; \text{co}\, A \;\&\; \beta A = {}^\bullet\eta(t)\} \\ & & \quad (X, p)\; F\; (A, t) \iff (X, p) \in A \\ & & \quad (A, t)\; F\; (X, p) \iff X = \{(A, t)\} \\ \mathbb{M} & = & \{\{(M, p) \mid p \in M\} \mid M \in \mathbb{M}_N\} \end{array}$$

where co *is the concurrency relation on $\mathcal{U}(N)$ and*

$$\eta(A, t) = t \qquad \beta((X, p), p') \iff p = p'.$$

Furthermore, $\epsilon_N = (\eta, \beta) : \mathcal{U}(N) \to N$ is a morphism in the category $\mathbf{PT}^\sharp$.

Proof Existence of the net and morphism follows from the inductive definition. Uniqueness follows from the fact that any element of an occurrence net occurs at unique depth. ⊣

7.1 A coreflection

Now that we have presented the operation of unfolding a P/T net to form an occurrence net, we characterize the operation as being right adjoint to the inclusion of the category of occurrence nets into the category of P/T nets. To do so, we show that the construction is cofree with respect to N.

Theorem 29 *For any P/T net N and occurrence net O, if there is a morphism* $(\theta, \alpha) : O \to N$ *in the category* $\mathbf{PT}^\sharp$ *then there is a unique morphism* $(\pi, \gamma) : O \to \mathcal{U}(N)$ *in the category* $\mathbf{Occ}^\sharp$ *such that the following diagram commutes:*

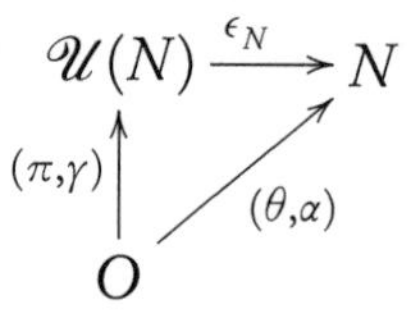

$$\begin{array}{ccc} \mathcal{U}(N) & \xrightarrow{\epsilon_N} & N \\ {\scriptstyle (\pi,\gamma)}\uparrow & \nearrow {\scriptstyle (\theta,\alpha)} & \\ O & & \end{array}$$

As standard [10], this cofreeness result implies that the operation $\mathcal{U}(N)$ extends to being a functor $\mathcal{U} : \mathbf{PT}^\sharp \to \mathbf{Occ}^\sharp$ which is right adjoint to the forgetful functor $\mathbf{Occ}^\sharp \hookrightarrow \mathbf{PT}^\sharp$. Since $\mathbf{Occ}^\sharp$ is a full and faithful subcategory of $\mathbf{PT}^\sharp$, it follows that the adjunction is a coreflection.

The coreflection is readily seen to restrict to give a coreflection between the categories with synchronous morphisms $\mathbf{Occ}^\sharp_s$ and $\mathbf{PT}^\sharp_s$ and to categories of singly-marked nets **Occ** and **PT**.

7.2 Symmetry

An important class of net, *causal nets*, is often encountered in describing paths of nets. They shall form a path category from which open maps of nets may be obtained.

Definition 30 *A causal net is an occurrence net with empty conflict relation and with at most one initial marking.*

We denote the category of causal nets with net morphisms **Caus**. We shall not go into detail here, but the requirements for a net morphism to be **Caus**-open are stronger than those required for it to be $\mathcal{N}^\sharp$**Elem**-open, requiring that the morphism is a bijection when restricted to any reachable marking, in addition to the path lifting condition required for $\mathcal{N}^\sharp$**Elem**-bisimilarity.

We wish to show that the coreflection above extends to give an adjunction between the categories enriched with symmetry

$$\mathscr{S}_{\mathbf{Caus}}\mathbf{Occ}^{\#} \underset{\mathscr{S}\mathscr{U}}{\overset{\hookrightarrow}{\underset{\longleftarrow}{\perp}}} \mathscr{S}_{\mathbf{Caus}}\mathbf{PT}^{\#}$$

The requirement that the inclusion $\mathbf{Occ}^{\#} \hookrightarrow \mathbf{PT}^{\#}$ and the functor $\mathscr{U} : \mathbf{PT}^{\#} \to \mathbf{Occ}^{\#}$ preserve open maps follows from the earlier coreflection and results on preservation of open maps through coreflections [9]. The functor $\mathscr{U}$ preserves limits since it is a right adjoint, so all that remains is to show that the inclusion $\mathbf{Occ}^{\#} \hookrightarrow \mathbf{PT}^{\#}$ preserves pullbacks of **Caus**-open maps.

It can be proved that all **Caus**-open morphisms of the category $\mathbf{Occ}^{\#}$ are folding morphisms (morphisms that are functions on conditions). That is, any **Caus**-open morphism in $\mathbf{Occ}^{\#}$ occurs in the category $\mathbf{Occ}^{\#}_{\mathrm{f}}$. The category $\mathbf{Occ}^{\#}_{\mathrm{f}}$ gives us a handle on pullbacks of open maps, since pullbacks in the category $\mathbf{Occ}^{\#}_{\mathrm{f}}$ are known to exist and are relatively straightforwardly characterized; they coincide with the pullbacks taken in $\mathbf{PT}^{\#}$.

Lemma 31 *The inclusions*

$$\mathbf{Occ}^{\#}_{\mathrm{f}} \hookrightarrow \mathbf{Occ}^{\#}$$
$$\mathbf{Occ}^{\#}_{\mathrm{f}} \hookrightarrow \mathbf{PT}^{\#}_{\mathrm{f}}$$
$$\mathbf{PT}^{\#}_{\mathrm{f}} \hookrightarrow \mathbf{PT}^{\#}$$

preserve pullbacks.

Consequently, the inclusion $\mathbf{Occ}^{\#} \hookrightarrow \mathbf{PT}^{\#}$ preserves pullbacks of **Caus**-open morphisms, despite the fact that the inclusion does not preserve pullbacks of *all* morphisms.

As an immediate consequence of Lemma 31 and Proposition 2, a coreflection between categories enriched with symmetry can now be demonstrated.

Theorem 32 *The unfolding functor* $\mathscr{S}\mathscr{U} : \mathscr{S}_{\mathbf{Caus}}\mathbf{PT}^{\#} \to \mathscr{S}_{\mathbf{Caus}}\mathbf{Occ}^{\#}$ *is right adjoint to the inclusion* $\mathscr{S}_{\mathbf{Caus}}\mathbf{Occ}^{\#} \hookrightarrow \mathscr{S}_{\mathbf{Caus}}\mathbf{PT}^{\#}$. *Furthermore, the adjunction is a coreflection.*

8 Conclusion

In this paper, we have shown how symmetry can be applied to two forms of Petri net, P/T nets and occurrence nets, and that this necessitates extending the definition of nets to allow them to have multiple initial markings. A coreflection connecting these two categories was given and this was shown to extend to a coreflection between the categories with symmetry. A coreflection between categories of event structures and occurrence nets was given, also extending to categories with symmetry.

In defining symmetry in occurrence nets and P/T nets, we made the most natural choice when specifying open maps by using the path category formed from causal nets. The coreflection between nets with symmetry is therefore based on spans of maps that are open with respect to the category of causal nets. The most natural choice of path for event structures, however, is the category of elementary event structures. The coreflection between event structures with symmetry and occurrence nets with symmetry is therefore based on spans of maps that are open with respect to the category of elementary event structures. We leave it for future work to consider more fully how the two coreflections are connected.

An alternative, explicit notion of symmetry on Petri nets has been used in [16]. The use of coreflections to connect models for concurrency is described in [25]. The unfolding of safe nets was defined in [12]. The operation was extended to unfolding P/T nets by Engelfriet in [3], where the unfolding was characterized as a limit within a lattice of partial unfoldings. In [11] a coreflection between occurrence nets and (singly-marked) *semi-weighted* nets is given. Semi-weighted nets generalize P/T nets net in that they allow conditions to occur with multiplicity greater than one as preconditions to events, though the postconditions of any event must form a set as must the initial marking of the net. Presumably the coreflection extends to one between multiply-marked nets, but we have chosen not to study it since the morphisms of semi-weighted nets are slightly more complicated than those here because they are multirelations on conditions rather than relations.

The definition of symmetry in Section 2 is different from that in [23, 24] since it requires a symmetry to be a span that is a *pseudo* equivalence rather than an equivalence. As such, the maps l and r of an object with symmetry need not be jointly monic according to the definition here. In Figure 6, we give a symmetry that happens to be jointly monic in $\mathbf{Occ}^\sharp$. When the morphisms l and r are considered in the category $\mathbf{PT}^\sharp$ (or in $\mathbf{Safe}^\sharp$), however, the morphisms are

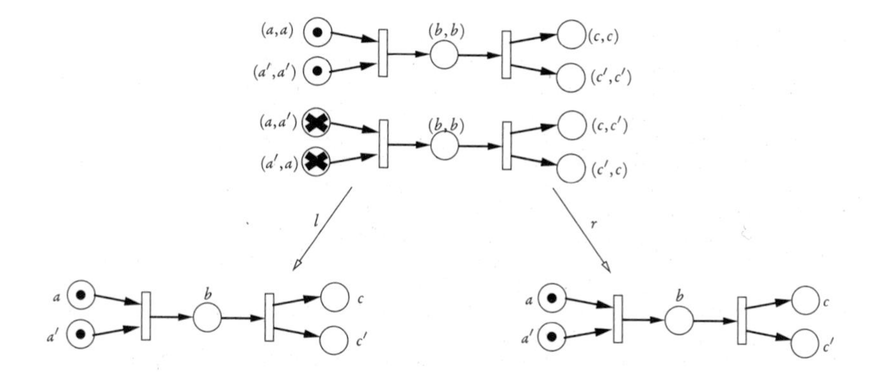

Figure 6: *A symmetry* $(N; l, r)$ *with (folding) morphisms* $l(x, y) = x$ *and* $r(x, y) = y$.

not jointly monic. With the restriction to jointly monic maps, a symmetry of occurrence nets would not be a symmetry of P/T nets by virtue of the fact that any occurrence net is a P/T net. Let the symmetry in Figure 6 be denoted $(N; l, r)$ spanning from the net S. In fact, it can be seen that there is no corresponding jointly monic symmetry in the category $\mathbf{PT}^\sharp$ since the image of the net S in $N \times N$, taking the product in $\mathbf{PT}^\sharp$, has more behaviour than S. Consequently, we would fail to obtain a coreflection if we were to restrict attention to jointly monic maps.

More generally, we note that the addition of symmetry seems to answer a call from net theory for more general event structures with which to understand the unfolding of nets [8], though here, to be completely compelling, we would need also to address the collective token game — work for the future.

References

[1] F. Crazzolara and G. Winskel. Events in security protocols. In *ACM Conference on Computer and Communications Security*, 2001.

[2] S. Doghmi, J. Guttman, and F. Thayer. Searching for shapes in cryptographic protocols. In *Proc. TACAS'07*, 2007.

[3] J. Engelfriet. Branching processes of Petri nets. *Acta Informatica*, 28:575–591, 1991.

[4] E. Fabré. On the construction of pullbacks for safe Petri nets. In *Proc. ICATPN '06*, volume 4024 of *Lecture Notes in Computer Science*, 2006.

[5] F. Fabrega, J. Herzog, and J. Guttman. Strand spaces: Why is a security protocol correct. In *Proc. IEEE Symposium on Security and Privacy*. IEEE Computer Society Press, May 1998.

[6] M. J. Gabbay and A. M. Pitts. A new approach to abstract syntax with variable binding. *Formal Aspects of Computing*, 13:341–363, 2001.

[7] J. Hayman and G. Winskel. The unfolding of general Petri nets. To appear at FSTTCS '08.

[8] P. Hoogers, H. Kleijn, and P. Thiagarajan. An event structure semantics for general Petri nets. *Theoretical Computer Science*, 153:129–170, 1996.
[9] A. Joyal, M. Nielsen, and G. Winskel. Bisimulation from open maps. In *Proc. LICS '93*, volume 127(2) of *Information and Computation*, 1995.
[10] S. MacLane. *Categories for the Working Mathematician*. Springer, 1971.
[11] J. Meseguer, U. Montanari, and V. Sassone. On the semantics of Place/Transition Petri nets. *Mathematical Structures in Computer Science*, 7:359–397, 1996.
[12] M. Nielsen, G. Plotkin, and G. Winskel. Petri nets, event structures and domains, Part 1. *Theoretical Computer Science*, 13:85–108, 1981.
[13] M. Nielsen and G. Winskel. Petri nets and bisimulation. *Theoretical Computer Science*, 1–2(153):211–244, Janurary 1996.
[14] A. M. Rabinovich and B. A. Trakhtenbrot. Behaviour structures and nets. *Fundamenta Informaticae*, 11(4), 1988.
[15] W. Reisig. *Petri Nets*. EATCS Monographs on Theoretical Computer Science. Springer-Verlag, 1985.
[16] V. Sassone. An axiomatization of the category of petri net computations. *Mathematical Structures in Computer Science*, 8(2):117–151, 1998.
[17] A. P. Sistla. Employing symmetry reductions in model checking. *Computer Languages, Systems and Structures*, 30(3–4):99–137, 2004.
[18] R. J. van Glabbeek and U. Goltz. Equivalence notions for concurrent systems and refinement of actions. In *Proc. MFCS '89*, volume 379 of *Lecture Notes in Computer Science*, 1989.
[19] G. Winskel. The symmetry of stability. Forthcoming.
[20] G. Winskel. A new definition of morphism on Petri nets. In *Proc. STACS '84*, volume 166 of *Lecture Notes in Computer Science*, 1984.
[21] G. Winskel. Event structures. In *Advances in Petri Nets, Part II*, volume 255 of *Lecture Notes in Computer Science*. Springer, 1986.
[22] G. Winskel. Petri nets, algebras, morphisms and compositionality. *Information and Computation*, 72(3):197–238, 1987.
[23] G. Winskel. Event structures with symmetry. *Electronic Notes in Theoretical Computer Science*, 172, 2007.
[24] G. Winskel. Symmetry and concurrency. In *Proc. CALCO '07*, May 2007. Invited talk.
[25] G. Winskel and M. Nielsen. Models for concurrency. In *Handbook of Logic and the Foundations of Computer Science*, volume 4, pages 1–148. Oxford University Press, 1995. BRICS report series RS-94-12.

Appendix

A Equivalences

Assume a category $\mathscr{C}$ with pullbacks. Let $l, r : S \to G$ be a pair of morphisms in $\mathscr{C}$. They form a *pseudo equivalence* iff they satisfy:

Reflexivity there is a map ρ such that

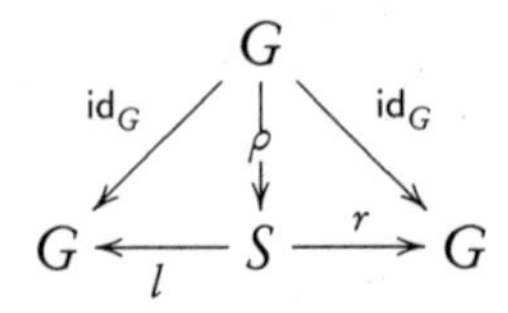

commutes;

Symmetry there is a map σ such that

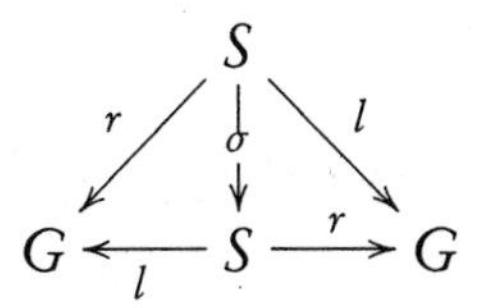

commutes; and

Transitivity there is a map τ such that

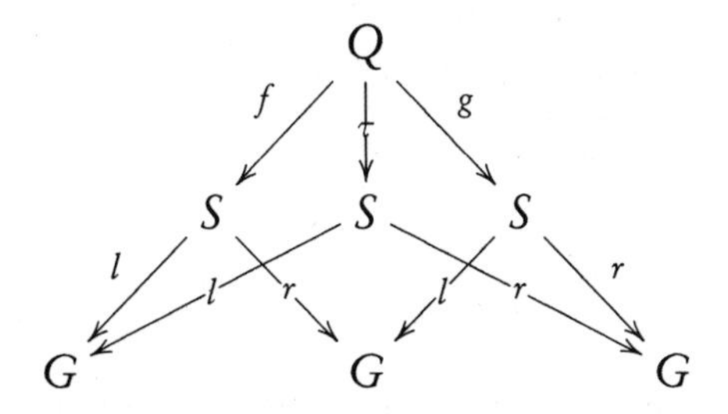

commutes, where Q, f, g is a pullback of r, l.

If, furthermore, the maps l and r are jointly monic, then they form an *equivalence*.

Steps and Coverability in Inhibitor Nets

Jetty Kleijn[1] **and Maciej Koutny**[2]
[1] *LIACS, Leiden University*
P.O.Box 9512, NL-2300 RA Leiden, The Netherlands
kleijn@liacs.nl
[2] *School of Computing Science, Newcastle University*
Newcastle upon Tyne NE1 7RU, U.K.
maciej.koutny@ncl.ac.uk

Abstract

For Petri nets with inhibitor arcs, properties like reachability and boundedness are undecidable and hence constructing a coverability tree is not feasible. Here it is investigated to what extent the coverability tree construction might be adapted for Petri nets with inhibitor arcs. Emphasis is given to the (a priori) step sequence semantics which cannot always be simulated by firing sequences. All this leads to the notion of a step coverability tree which may be of use for the analysis of the step behaviour of certain subclasses of Petri nets with inhibitor arcs.

Keywords: Petri nets, inhibitor arcs, step semantics, step coverability tree, boundedness, reachability, decidability

1 Introduction

Petri nets [20] are a generic, formal approach to concurrent computation based on notions of local states, local actions, and their relationships (together defining the underlying structure or 'net'). Whether or not a local action ('transition') can occur and its effect when it does, depend only on the local states ('places') to which it relates. Among a variety of net models introduced and investigated over the past few decades [24, 23], Elementary Net Systems (EN-systems) [25, 22] and Place/Transition systems (PT-systems) [5] are two basic classes. EN-systems are the more fundamental of the two, whereas PT-nets support a more convenient modelling and use of (potentially unboundedly many) resources. The switch from EN-system to PT-nets is in essence an extension from sets (Booleans) to multisets (natural numbers).

Generally speaking, the dynamics of a Petri net model is defined through a specific 'firing rule', describing enabledness (to occur at a global state or 'marking') and the effect (on the marking) of the occurrence of a single transition. In addition, there usually is also a step firing rule (for sets or multisets of simultaneously occurring transitions) as a derived notion relating to independence or concurrency (or rather simultaneity) of transition occurrences. This results in a natural way in operational semantics with a behavioural description in terms of firing sequences or step sequences. Moreover, reachability graphs (labelled transition systems) combine (step) firing sequences and state information. These essentially sequential semantical representations are the most straight-forward approaches which proved to be very useful as they allow behavioural analysis and verification (including model checking [27]). (Alternative semantics of EN-systems and PT-nets, providing more and explicit information on concurrency, causality, and conflict relations between occurrences of actions/transitions can be based on occurrence nets (processes), partial orders, traces, event structures and related temporal logics approaches [26, 18].)
Since state spaces may be infinite, for verification purposes an important (and often assumed or guaranteed by construction) property is 'boundedness' of the Petri net which amounts to saying that its state space is finite. A standard tool to decide this for PT-nets is the 'coverability tree' introduced in [12] and then investigated, among others, in [7, 19]. What is more, coverability trees can also provide a tool for deciding many other relevant behaviour problems, such as mutual exclusion, even in the case of infinite state spaces. Recently, ideas derived from the Karp and Miller construction have been applied in [4] to obtain a non-trivial decidability result in the area of quasi-static scheduling policies.

In net models like EN-systems and PT-nets it is not possible to test for the absence of resources (zero-testing), and so, quite early on an additional kind of relationship between places and transitions has been considered in the form of inhibitor arcs [1, 9, 19]. An inhibitor arc gives the possibility of testing rather than producing and consuming resources. If a place is connected to a transition by an inhibitor arc, then this transition cannot occur if that place is not empty. The extension with inhibitor arcs gives the resulting model of PTI-nets the expressive power of Turing machines and thus has its price. Net languages become recursively enumerable rather than recursive, and decidability for certain important behavioural properties, such as reachability, is lost [8, 1, 6, 21]. In our own investigations aimed at the development of a process semantics for PTI-nets with the view of verifying them through model checking, we combined techniques first proposed for EN-systems with inhibitor arcs [11] (to deal with the difference between concurrency and simultaneity in the context of inhibitor nets) and PT-nets [10] (in which concurrency relations between transitions may depend on preceding history). This has led in [13, 14] to the proposal of two constructions, one for general (possibly unbounded) PTI-nets and the other for PTI-nets with complemented (hence bounded) inhibitor places. Hence in the latter case it was possible to adopt an approach as developed for EN-systems with inhibitor arcs, resulting in a simple yet fully satisfactory solution.

In this paper we return to the basic questions concerning the boundedness of PTI-nets. Since boundedness (like reachability) is undecidable for general PTI-nets [8], we have however to be satisfied with partial solutions to our questions. An important line of attack here is the construction of a coverability tree which in finite time should provide information useful for a behavioural analysis of the net. Since inhibitor arcs destroy the monotonicity in the behaviour (having more resources available in a PTI-net may imply loss of behaviour), the coverability tree construction has to be modified. In particular to guarantee termination of the construction, the full class of PTI-nets has to be restricted. That is exactly why in [2], 'primitive' PTI-nets have been introduced, a subclass of PTI-nets which includes the ordinary PT-nets and still has more expressive power. However, the results in [2], as well as those by others on decidability issues for PTI-nets, are derived for the firing sequence semantics (or a step sequence semantics with the same reachability properties). We are, however, primarily interested in PTI-nets operating under the *a priori* step sequence semantics for which reachability is a richer concept than for the firing sequence semantics. Therefore, in this paper we set out to investigate issues relating to coverability and the a priori step sequence semantics in PTI-nets.

After a preliminary section, we introduce first the basic notations and concepts relating to PTI-nets, and discuss their operational semantics. We compare the purely sequential firing sequence semantics and the a priori step sequence semantics in view of their reachable markings (the state spaces they define). Next, we reconsider boundedness and reachability for PTI-nets. Using a result from [21], it can be shown that in case of no more than one inhibitor place, at least firing sequence reachability is a decidable property. We are mostly interested in the most general 'weighted' variant of PTI-nets, but as we demonstrate in that section, for reachability and boundedness it is sufficient to consider only the unweighted or 'simple' PTI-nets. In Section 5, the standard coverability tree construction for PT-nets is revisited. We recall the properties which make coverability trees useful and these serve later as guidelines when we discuss similar constructions for PTI-nets. Then we try to adapt the construction for PTI-nets (with the firing sequence semantics). Though the resulting tree reflects properly the unboundedness of places it is not adequate. Even if it terminates (for the subclass of PTI-nets with one inhibitor place), the new construction provides no more than a semi-algorithm for boundedness. In the main Section 6, we investigate the coverability tree construction for the a priori step sequence semantics. First of all we have to extend the labelling of its edges from single transitions to steps, because steps cannot always be simulated by firing sequences. Then it turns out, that it is not only the non-monotonicity which spoils the algorithm, but also the potential unboundedness of the steps. To properly capture this aspect of concurrency, the concept of a 'covering' or 'extended' step is introduced, which we see as a main contribution of this paper. Combining covering steps and the property of primitivity as in the coverability tree construction in [2] leads to the construction of a step coverability tree for PTI-nets. We show that the algorithm always terminates and that the resulting tree can indeed be used to decide whether a primitive PTI-net working under the a priori step sequence semantics is bounded. Moreover, similar to the coverability tree of PT-nets, the step coverability tree may be useful also to decide other properties. We give the example of executability of steps (comparable to the usefulness of transitions in PT-nets). In a way, the step coverability tree may be a new tool which could be useful also for other kinds of Petri nets, including PT-nets (when step executability is considered), as argued in the concluding section, or for Petri nets operating under the maximal parallelism execution semantics. To enhance the readability we have moved a few rather technical proofs to the Appendix.

2 Preliminaries

We use standard mathematical notation, in particular, $\uplus$ denotes disjoint set union, $\mathbb{N} = \{0, 1, 2, \ldots\}$ the set of natural numbers, and ω the first infinite ordinal. We assume that $\omega + \omega = \omega$, $\omega - \omega = \omega$, $n < \omega$, $n - \omega = 0$, $0 \cdot \omega = 0$ and $\omega + n = \omega - n = k \cdot \omega = \omega$, where n is any natural number and k any positive natural number.

A multiset (over a set X) is a function $\mu : X \to \mathbb{N}$, and an *extended multiset* (over X) is a function $\mu : X \to \mathbb{N} \cup \{\omega\}$. In this paper, X

will always be a finite set. We denote $x \in \mu$ if $\mu(x) > 0$, and call the set of all such x the *carrier* of μ. For two extended multisets μ and μ' over X, we denote $\mu \leq \mu'$ if $\mu(x) \leq \mu'(x)$ for all $x \in X$. We then also say that μ' *covers* μ. As usual, $\mu(x) < \mu'(x)$ if $\mu(x) \leq \mu'(x)$ and $\mu(x) \neq \mu'(x)$. Any subset of X may be viewed through its characteristic function as a multiset over X, and a multiset may always be considered as an extended multiset. The multiset $\mathbf{0}$ and the extended multiset Ω are given respectively by $\mathbf{0}(x) \stackrel{\mathrm{df}}{=} 0$ and $\Omega(x) \stackrel{\mathrm{df}}{=} \omega$ for all x.

In the examples, we will use notations like $\{w^2yz^\omega\}$ to denote an extended multiset μ such that $\mu(w) = 2$, $\mu(y) = 1$, $\mu(z) = \omega$ and $\mu(x) = 0$, for all $x \in X \setminus \{w, y, z\}$. (For the examples, this kind of notation will not lead to confusion with sets consisting of a single sequence.)

The sum of two extended multisets is given by $(\mu+\mu')(x) \stackrel{\mathrm{df}}{=} \mu(x)+\mu'(x)$, the difference by $(\mu - \mu')(x) \stackrel{\mathrm{df}}{=} \max\{0, \mu(x) - \mu'(x)\}$, and the multiplication of an extended multiset by a natural number by $(n \cdot \mu)(x) \stackrel{\mathrm{df}}{=} n \cdot \mu(x)$. The cardinality of μ is defined as $|\mu| \stackrel{\mathrm{df}}{=} \sum_{x \in X} \mu(x)$. We write μ_ω for the set of all x such that $\mu(x) = \omega$, and $\mu_{\omega \mapsto k}$ is a multiset such, for all x, $\mu_{\omega \mapsto k}(x) = k$ if $x \in \mu_\omega$, and $\mu_{\omega \mapsto k}(x) = \mu(x)$ otherwise.

If μ is a multiset, μ' an extended multiset over the same set and $k \geq 0$, then we say that μ is a *k-approximation* of μ' if, for all x, $\mu(x) = \mu'(x)$ if $\mu'(x) < \omega$, and otherwise $\mu(x) > k$. We denote this by $\mu \Subset_k \mu'$.

In some of the proofs we will be referring to Dickson's Lemma which states that every infinite sequence of extended multisets (over a common set) contains an infinite non-decreasing subsequence. Another important technical tool is *König's Lemma* by which every infinite, finitely branching tree has an infinite path starting from the root.

3 PT-nets with Inhibitor Arcs

This section introduces the notation and terminology for Place/Transition nets (PT-nets, for short) and PT-nets with inhibitor arcs (PTI-nets) and discusses their operational semantics. We first define their underlying structures.

A *net* is a triple $\mathcal{N} = (P, T, W)$ such that P and T are disjoint finite sets of *places* and *transitions*, respectively, and $W : (T \times P) \cup (P \times T) \to \mathbb{N}$ is the *weight function* of $\mathcal{N}$. In diagrams, places are drawn as circles and transitions as rectangles. If $W(x,y) \geq 1$ for some $(x,y) \in (T \times P) \cup (P \times T)$, then (x,y) is an *arc* leading from x to y. As usual, arcs are annotated with their weight if this is 2 or more. A double headed arrow between p and t indicates that $W(p,t) = W(t,p) = 1$. We assume that, for every $t \in T$, there is a place p such that $W(p,t) \geq 1$ or $W(t,p) \geq 1$ (i.e., transitions are never isolated).

An *inhibitor net* is a net together with a (possibly empty) set of *weighted inhibitor arcs* leading from places to transitions. An inhibitor net $\mathcal{N}$ is specified as a tuple (P,T,W,I) such that (P,T,W) is a net (the underlying net of $\mathcal{N}$) and I — the *inhibitor mapping* — is an extended multiset over $P \times T$. If $I(p,t) = k \in \mathbb{N}$, then p is an *inhibitor place* of t meaning intuitively that t can only be executed if p does not contain more than k tokens (defined below); in particular, if $k = 0$ then p must be empty. $I(p,t) = \omega$ means that t is not inhibited by the presence of tokens in p. If I always returns 0 or ω, then we are dealing with *unweighted* inhibitor arcs which can only be used to test whether a place is empty or not. A net (P,T,W), without inhibitor arcs, can be considered as a special instance of an inhibitor net by identifying it with the inhibitor net (P,T,W,Ω). In diagrams, inhibitor arcs have small circles as arrowheads. As for the standard Petri net arcs, inhibitor arcs are annotated by their weights. In this case, the weight 0 is not shown, and if $I(p,t) = \omega$, then there is no inhibitor arc at all between p and t.

Given a transition t of an inhibitor net $\mathcal{N} = (P,T,W,I)$, we denote by $t^\bullet$ the multiset of places given by $t^\bullet(p) \stackrel{\text{df}}{=} W(t,p)$, by ${}^\bullet t$ the multiset of places given by ${}^\bullet t(p) \stackrel{\text{df}}{=} W(p,t)$, and by ${}^\circ t$ the extended multiset of places given by ${}^\circ t(p) \stackrel{\text{df}}{=} I(p,t)$. These notations extend to finite multisets U of transitions in the following way: $U^\bullet \stackrel{\text{df}}{=} \sum_{t \in U} U(t) \cdot t^\bullet$ and ${}^\bullet U \stackrel{\text{df}}{=} \sum_{t \in U} U(t) \cdot {}^\bullet t$ are multisets of places, while ${}^\circ U$ defined by ${}^\circ U(p) \stackrel{\text{df}}{=} \min(\{\omega\} \cup \{{}^\circ t(p) \mid t \in U\})$, is an extended multiset of places. For a place p, we denote by ${}^\bullet p$ and $p^\bullet$ the multisets of transitions given by $p^\bullet(t) \stackrel{\text{df}}{=} W(p,t)$ and ${}^\bullet p(t) \stackrel{\text{df}}{=} W(t,p)$, respectively.

The states of an inhibitor net $\mathcal{N} = (P,T,W,I)$ are given in the form of markings. A *marking* of $\mathcal{N}$ is a multiset of places. Following the standard terminology, given a marking M of $\mathcal{N}$ and a place $p \in P$, we say that p is marked (under M) if $M(p) \geq 1$ and that $M(p)$ is the number of tokens in p. In diagrams, every token in a place is drawn as a small black dot. Also, if the set of places of $\mathcal{N}$ is implicitly ordered, $P = \{1,\ldots,n\}$, then we will represent any marking M of $\mathcal{N}$ as the n-tuple $(M(1),\ldots,M(n))$ of natural numbers.

Transitions represent actions which may occur at a given marking and then lead to a new marking. First, we discuss the *sequential semantics* of inhibitor nets based on the standard (and non-controversial) definition for the occurrence of single transitions.

A transition t of $\mathcal{N} = (P, T, W, I)$ can occur at a marking M of $\mathcal{N}$ if for each place p, the number of tokens $M(p)$ is at least $W(p,t)$, the number of tokens that t needs as input from that place according to the weight function. In addition, each inhibitor place p of t should not contain more than $I(p,t)$ tokens. Formally, t is *enabled* at M, denoted by $M[t\rangle$, if ${}^\bullet t \leq M \leq {}^\circ t$. If t is enabled at M, then it can be *executed* (or *fired*) leading to the marking $M' \stackrel{\mathrm{df}}{=} M - {}^\bullet t + t^\bullet$, denoted by $M[t\rangle M'$. Thus M' is obtained from M by deleting $W(p,t)$ tokens 'consumed' by t from each place p and adding $W(t,p)$ tokens to each place p as output 'produced' by t. A *firing sequence* from a marking M to marking M' in $\mathcal{N}$ is a possibly empty sequence of transitions $\sigma = t_1 \ldots t_n$ such that

$$M = M_0 [t_1\rangle M_1 [t_2\rangle M_2 \cdots M_{n-1} [t_n\rangle M_n = M' ,$$

for some markings $M_1, \ldots, M_{n-1}$ of $\mathcal{N}$. If σ is a firing sequence from M to M', then we write $M [\sigma\rangle_{fs} M'$ and call M' *fs–reachable* from M (in $\mathcal{N}$). Note that every marking is *fs*–reachable from itself by the empty firing sequence.

Figure 1 shows two inhibitor nets each with a marking $(1,1,0,0)$. The first of them, $\mathcal{N}_1$, has three non-empty firing sequences starting from $(1,1,0,0)$: $\sigma_1 = t$, $\sigma_2 = u$ and $\sigma_3 = ut$. However, the other one, $\mathcal{N}_2$, allows only the first two, σ_1 and σ_2. Moreover, the set of markings *fs*–reachable from the marking $(1,1,0,0)$ for $\mathcal{N}_1$ comprises $(1,1,0,0)$, $(1,0,0,1)$, $(0,1,1,0)$ and $(0,0,1,1)$, whereas for $\mathcal{N}_2$ it comprises only $(1,1,0,0)$, $(1,0,0,1)$ and $(0,1,1,0)$.

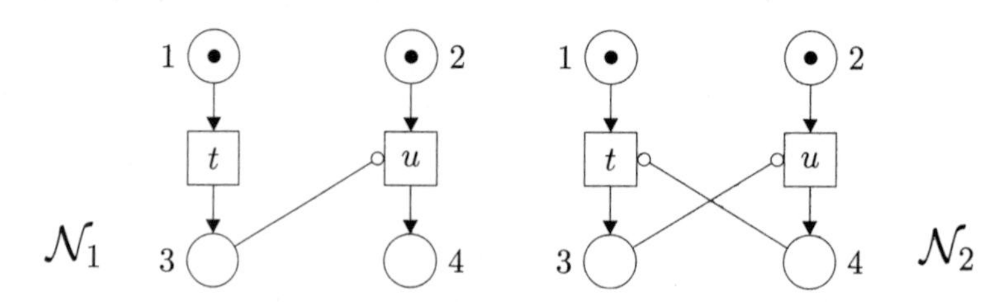

Figure 1: *Two marked inhibitor nets.*

Next we define a semantics of inhibitor nets in terms of concurrently occurring transitions. A *step* of an inhibitor net $\mathcal{N} = (P, T, W, I)$ is a finite multiset of transitions, $U : T \to \mathbb{N}$. The enabledness of steps is not defined in a unique way in the literature. Following [11, 13, 14, 17], we consider here the operationally defined *a priori step sequence semantics* which is based on a direct generalization of the enabledness of single transitions to multisets of transitions. A step U is *a priori enabled* or simply *enabled*, at a marking M of $\mathcal{N}$ if ${}^{\bullet}U \leq M \leq {}^{\circ}U$. Thus, in order for U to be enabled at M, for each place p, the number of tokens in p under M should at least be equal to the accumulated number of tokens that are needed as input to each of the transitions in U, respecting their multiplicities in U. By the second inequality, each place p which is an inhibitor place of some transition t occurring in U, should contain no more than $I(p,t)$ tokens. If U is a priori enabled at M, then it can be *executed* leading to the marking $M' \stackrel{\mathrm{df}}{=} M - {}^{\bullet}U + U^{\bullet}$, denoted $M[U\rangle M'$. Thus the effect of executing U is the accumulated effect of executing each of its transitions (taking into account their multiplicities in U). Note that the empty step $\mathbf{0}$ is enabled at every marking of $\mathcal{N}$, and that its execution has no effect, i.e., $M' = M$. An (*a priori*) *step sequence* from a marking M to marking M' in $\mathcal{N}$ is a possibly empty sequence $\tau = U_1 \ldots U_n$ of non-empty steps U_i such that

$$M = M_0\,[U_1\rangle\, M_1\,[U_2\rangle M_2 \cdots M_{n-1}\,[U_n\rangle\, M_n = M' ,$$

for some markings $M_1, \ldots, M_{n-1}$ of $\mathcal{N}$. If τ is a step sequence from M to M' we write $M\,[\tau\rangle\, M'$ and M' is said to be *a priori reachable* or simply *reachable* from M (in $\mathcal{N}$). Note that every marking is reachable from itself by the empty step sequence.

A *Place/Transition net with inhibitor arcs* (or PTI-net) is an inhibitor net equipped with an initial marking. It is specified as a tuple $\mathcal{N} = (P,T,W,I,M_0)$, where $\mathcal{N}' = (P,T,W,I)$ is its underlying inhibitor net, and M_0 is a marking of $\mathcal{N}'$. If $I = \Omega$, then $\mathcal{N}$ is a *Place/Transition net* (or PT-net) which may also be specified as (P,T,W,M_0). All terminology and notation with respect to enabling, firing, and steps are carried over from $\mathcal{N}'$ to $\mathcal{N}$.

A *step sequence* of $\mathcal{N}$ is an (a priori) step sequence starting from its initial marking M_0. The set of all its step sequences is $steps(\mathcal{N}) \stackrel{\mathrm{df}}{=} \{\tau \mid \exists M : M_0[\tau\rangle M\}$. The set of *reachable* markings of $\mathcal{N}$ is given by $[M_0\rangle \stackrel{\mathrm{df}}{=} \{M \mid \exists \tau : M_0[\tau\rangle M\}$. Similarly, the set of all firing sequences of $\mathcal{N}$ is $fs(\mathcal{N}) \stackrel{\mathrm{df}}{=} \{\sigma \mid \exists M : M_0[\sigma\rangle_{fs} M\}$, and the set of *fs–reachable* markings of $\mathcal{N}$ is $[M_0\rangle_{fs} \stackrel{\mathrm{df}}{=} \{M \mid \exists \sigma : M_0[\sigma\rangle_{fs} M\}$.

Coming back to Figure 1, we observe that $\mathcal{N}_1$ has four non-empty step sequences: $\tau_1 = \{t\}$, $\tau_2 = \{u\}$, $\tau_3 = \{u\}\{t\}$ and $\tau_4 = \{tu\}$, while $\mathcal{N}_2$, on the other hand, generates τ_1, τ_2 and τ_4. As a result, the set of reachable markings for $\mathcal{N}_2$ comprises $(1,1,0,0)$, $(1,0,0,1)$, $(0,1,1,0)$ and $(0,0,1,1)$, which is different from its set of *fs*–reachable markings which does not include $(0,0,1,1)$. Thus, in contrast to, e.g., PT-nets, the a priori enabled steps of PTI-nets cannot always be sequentialised to a firing sequence (with the same number of occurrences of each transition. Moreover, this example has as an important implication that

> *for PTI-nets executed under the a priori step sequence semantics, marking reachability cannot be reduced to marking fs–reachability.*

This observation is a main motivation for the investigation reported in the rest of this paper. In more formal terms, we can characterise situations where steps cannot be sequentialised in the following way.

A non-singleton step U of transitions of $\mathcal{N}$ is (structurally) *non-split* if there is a (multiplicity respecting) enumeration of its elements, $t_1, \ldots, t_n$, such that there is no place p such that $t_n^\bullet(p) > 0$ and ${}^\circ t_1(p) \in \mathbb{N}$, nor p such that $t_i^\bullet(p) > 0$ and ${}^\circ t_{i+1}(p) \in \mathbb{N}$ for some $i < n$.

Proposition 1 *Let U be a step enabled at a marking M of $\mathcal{N}$ such that there is no (multiplicity respecting) enumeration $t_1, \ldots, t_n$ of its elements for which $\{t_1\} \ldots \{t_n\}$ is a step sequence enabled at M. Then U contains a non-split sub-step W.*

Proof Let $U = \{u_1, \ldots, u_n\}$ and G be a directed graph with the nodes $v_1, \ldots, v_n$, where each v_i is labelled by $l(v_i) = u_i$, and there is an arc from v_i to v_j if $l(v_i)^\bullet \cap {}^\circ l(v_j) \neq \varnothing$. From the non-existence of a sequentialisation of U it follows that G must have a cycle. The labels of the nodes of such a cycle define a non-split sub-step of U. $\dashv$

That is, non-split steps cannot always be fully sequentialised (this may depend on the weights) and, as a consequence, if we execute $\mathcal{N}$ in a sequential semantics then some of the markings reachable in the a priori semantics may be unreachable.

A term often used in connection with steps — for any form of step semantics — is *auto-concurrency* (of a transition). This means that there is an enabled step U such that $U(t) \geq 2$ for at least one transition t. Furthermore, a PTI-net exhibits *unbounded auto-concurrency* if there is a transition t such that for every integer n one can find a reachable marking which enables a step U such that $U(t) \geq n$.

As a preview of the later construction of coverability trees, we now briefly mention the concepts of *extended markings* and *extended steps* generalising the finite multisets of respectively places and transitions defining the execution semantics of PTI-nets. It should be stressed that the ω-components in these extended multisets do not represent actual tokens or fired transitions, rather, they indicate that the number of tokens or simultaneous firings of transitions can be arbitrarily high. The transition and step enabling and firing, as well as the result of executing transitions(s), are defined in the same way as for the finite case. Recall that we postulated $\omega - \omega = \omega$, and so an ω marked place remains ω marked even after the execution of a step which 'removes' from it ω tokens.

3.1 Alternative semantics of PTI-nets

As we already mentioned, the a priori step sequence semantics for PTI-nets is not the only one to be found in the literature. An alternative is provided by the a posteriori semantics (used in [3, 2] for the case of unweighted inhibitor arcs) in which a step U is *a posteriori enabled* at a marking M if ${}^\bullet U \leq M$ and $M + (U - \{t\})^\bullet \leq {}^\circ t$ for each transition t with $U(t) \geq 1$. Thus the difference with the a priori approach lies in the second inequality which states that, for each transition occurring in U, there is no combination of the other transition occurrences that will produce an inhibiting amount of tokens in any of its inhibitor places. Switching to the a posteriori interpretation can have a dramatic effect on marking reachability. Taking again Figure 1 and the PTI-net $\mathcal{N}_2$, we observe that it has only two non-empty a posteriori step sequences: $\tau_1 = \{t\}$ and $\tau_2 = \{u\}$. As a result, the set of a posteriori reachable markings for $\mathcal{N}_2$ comprises only $(1,1,0,0)$, $(1,0,0,1)$ and $(0,1,1,0)$. It is interesting to observe that on the one hand, as for the a priori approach, the a posteriori enabledness of a singleton multiset coincides with the enabledness of its only transition. On the other hand however, the treatment of multisets in the a posteriori approach is not a direct lifting of the enabledness of single transitions to multisets of transitions. Finally, note that as a consequence of the check for the effect of the firing of the transitions, every a posteriori enabled step can be sequentialised and a posteriori step reachability coincides with *fs*-reachability. Actually, all transition occurrences in an a posteriori enabled step can be executed in any order, as a firing sequence from the current marking.

Another — intermediate — variation of the step sequence semantics is provided in [28] where it is assumed that a step of transitions is enabled if it is a priori enabled and, in addition, it is possible to find at least one sequential way of executing its members. For example, the step $\{t\,u\}$ would be enabled in the sense at the initial marking of $\mathcal{N}_1$ in Figure 1, but not at the initial marking of $\mathcal{N}_2$. (Note that $\{t\,u\}$ would be rejected by the a posteriori semantics in both cases.)

4 Boundedness and Reachability

A place p of a PTI-net $\mathcal{N} = (P, T, W, I, M_0)$ is *bounded* if there is $n \in \mathbb{N}$ such that $M(p) \leq n$ for every marking M reachable from M_0; otherwise it is *unbounded*. $\mathcal{N}$ itself is *bounded* if all its places are bounded. In addition, in the sequential semantics where we consider only those markings of $\mathcal{N}$ which are *fs*–reachable, we may use corresponding terminology adding the prefix *fs*– leading to: *fs–bounded* and *fs–unbounded*. Considering the example PTI-net shown in Figure 2, one can easily see that the inhibitor place is unbounded, and the other place bounded, under both sequential and a priori step semantics.

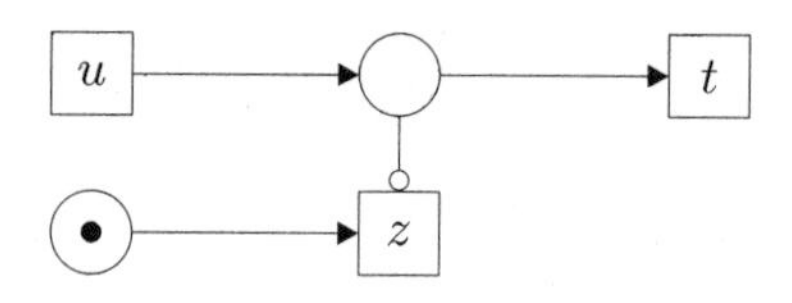

Figure 2: *A PTI-net.*

The *place (fs–)boundedness* problem for PTI-nets is to decide whether a given place of a PTI-net is (*fs*–)bounded; the *(fs–)boundedness* problem is to decide whether all places in a given PTI-net are (*fs*–)bounded. The *(fs–)reachability* problem is concerned with deciding whether a given marking is (*fs*–)reachable from the initial one.

It is well-known that the reachability problem for PT-nets is decidable [16, 15]. Also for PTI-nets with no more than one unweighted inhibitor arc, the *fs*–reachability problem is decidable [21] and thus also the reachability problem. To see that reachability can indeed be reduced to *fs*–reachability in the case of a unique unweighted inhibitor arc between transition t and place p_{inh}, we observe first that it follows from Proposition 1 that reachability and *fs*–reachability are the same if p_{inh} is not an output place of t. Otherwise one can simulate multiple occurrences of t in a step using the following construction.
We may assume that there is no arc from p_{inh} to t since otherwise t is never enabled and can simply be deleted. It suffices to add fresh places, p_{mutex} (marked initially with single token), p'_{mutex} and p''_{mutex} (initially empty), and p' for every original place p other than p_{mutex} (initially empty), together with transitions, t', u, w and t_p for every original place p other than p_{mutex}. Then one adds a number of arcs (unweighted, unless stated otherwise), as follows. First, each original transition other than t is connected with p_{mutex} using a pair of arcs pointing in opposite directions; moreover, we add an arrow from p_{mutex} to t, and from t to p'_{mutex}. Transition t' acquires the original

(weighted) incoming connectivity from t and the outgoing (weighted) connectivity to p_{inh} but not the inhibitor arc; moreover, the outgoing (weighted) connectivity to each place p other than p_{inh} is redirected to p', and t' is connected with p'_{mutex} using a pair of arcs pointing in opposite directions. We then add arcs from p'_{mutex} to u, from u to p''_{mutex}, from p''_{mutex} to w, and from w to p_{mutex}. For each place p other than p_{mutex} we add arcs from p' to t_p and from t_p to p. Finally, each t_p is connected with p''_{mutex} using a pair of arcs pointing in opposite directions. Thus simultaneous firings of t are now simulated by one (first) occurrence of t followed by the appropriate number of occurrences of t'. The places p_{mutex}, p'_{mutex}, and p''_{mutex} sequentialise the behaviour and prevent that in the meantime the rest of the net is already affected by these occurrences of t and t'. Now it is not difficult to see the direct correspondence between the reachable markings of the original net and the *fs*-reachable markings in the simulating net.

Both *fs*-reachability and reachability are however undecidable for PTI-nets with two or more inhibitor places [1, 9].

We now provide a construction to simulate the inhibitor arcs connected to a single inhibitor place by one unweighted inhibitor arc. This makes it possible to extend the decidability result from [21].

Theorem 2 *The fs-reachability problem for PTI-nets with one inhibitor place reduces to the fs-reachability problem for PTI-nets with one unweighted inhibitor arc.*

Proof Assume that $\mathcal{N}$ is a PTI-net with exactly one inhibitor place p_{inh} and let M be a marking of $\mathcal{N}$. Our aim is to reduce the problem of checking whether M is reachable from the initial marking M_0 to that of the *fs*-reachability of a related marking $\widetilde{M}$ in a newly created PTI-net $\mathcal{N}'$ with a single unweighted inhibitor arc.

Let $t_1, \ldots, t_n$ be the transitions inhibited by p_{inh}, with weights k_1, $\ldots$, k_n, respectively. Moreover, let m_i be the weight of the ordinary arc from p_{inh} to t_i for $i = 1, \ldots, n$. Without loss of generality, we may assume that $k_i \geq m_i$, for every i, since otherwise t_i is never enabled and can simply be deleted.

Consider the transformation which first removes the transitions $t_1, \ldots, t_n$, and then adds new places: p_{mutex}, p_{test}, r_{test}, p_i, r_i, l_i, v_{ij} (for $i = 1, \ldots, n$ and $j = 0, \ldots, k_i - m_i$), as well as transitions: τ, u_{ij}, w_{ij}, t'_i, t''_i (for $i = 1, \ldots, n$ and $j = 0, \ldots, k_i - m_i$). Their connections and initial marking are described in Figure 3. In addition, ${}^\bullet t'_i = {}^\bullet t_i$ and $t''^\bullet_i = t_i^\bullet$ for $i = 1, \ldots, n$. Let $\mathcal{N}'$ be the resulting net.

One can see that if M is *fs*-reachable in $\mathcal{N}$ then the marking $\widetilde{M}$ of $\mathcal{N}'$ such that, for all places p: $\widetilde{M}(p_{mutex}) = 1$, $\widetilde{M}(p) = M(p)$ if $p \in P$,

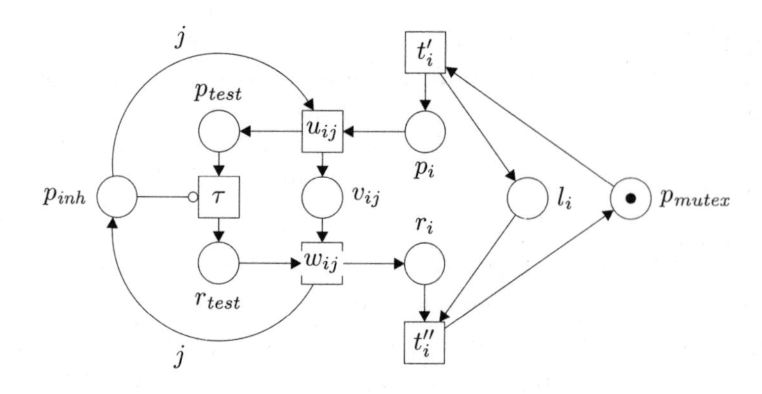

Figure 3: *Transformation from one inhibitor place to one inhibitor arc (fragment).*

and otherwise $\widetilde{M}(p)=0$; is *fs*–reachable in $\mathcal{N}'$ and vice versa.
The basic idea is that for any firing sequence σ of $\mathcal{N}$, the firing of t_i in $\mathcal{N}$ can be simulated by the firing of $t'_i u_{ij} \tau w_{ij} t''_i$ in $\mathcal{N}'$ resulting in a firing sequence $\widetilde{\sigma}$ leading to $\widetilde{M}$ Conversely, since $\widetilde{M}(p_{mutex})=1$, any firing sequence σ' of $\mathcal{N}'$ leading to $\widetilde{M}$ can be rearranged to a firing sequence σ'' such that there exists a firing sequence σ of $\mathcal{N}$ leading to M and such that $\sigma''=\widetilde{\sigma}$. Since $\mathcal{N}'$ has one unweighted inhibitor arc, we are done. ⊣

It thus follows that

Corollary 3 *The fs–reachability problem for PTI-nets with one inhibitor place is decidable.*

The transformation in the proof above can however not be used for the a priori step sequence semantics, as a consequence of the role of the place p_{mutex} which sequentialises the firings of the simulating transitions.

A PTI-net is said to be *simple* if it has only unweighted inhibitor arcs (its inhibitor mapping returns always 0 or ω) and, moreover, also only unweighted ordinary arcs (its weight function always returns 0 or 1). With each PTI-net we can associate a simple PTI-net with equivalent marking (*fs*–)reachability and (*fs*–)boundedness problems as follows.

Let $\mathcal{N}=(P,T,W,I,M_0)$ be a PTI-net. Without loss of generality, we assume that, for every place p, there is at most one transition t connected to it by an inhibitor arc, i.e., such that $I(p,t)\in\mathbb{N}$. (We can always make enough copies of a place retaining the standard connectivity and distribute the inhibitor arcs among them). For each inhibitor place p, we let inh_p be the weight of the only inhibitor arc attached to it. Each inhibitor place p is now provided with two 'assistants' p_1 (initially inh_p tokens) and p_2 (initially empty) connected to p via two new transitions w_p and u_p using unweighted arcs such that ${}^\bullet w_p = \{p, p_1\}$, $w_p{}^\bullet = \{p_2\}$ and ${}^\bullet u_p = \{p_2\}$, $u_p{}^\bullet = \{p, p_1\}$. All inhibitor arc weights are changed into weight 0. Observe that p_2 acts as a 'store' of tokens in p with p_1 as a bound on the remaining capacity. Place p can be successfully tested for emptiness in the new net if it can be emptied using at most inh_p executions of w_p moving tokens from p to p_2. These tokens are returned to p by executing transition u_p.

Next we apply a construction to the PTI-net obtained above which should yield a simple PTI-net with equivalent boundedness and reachability problem. It is essentially the transformation given in [7, 19] to switch from a general PT-net to an equivalent unweighted PT-net. Let $\max_p = \max(\{W(p,t) \mid t \in T\} \cup \{W(t,p) \mid t \in T\})$, for all original places p, and $\max_{p_1} = \max_{p_2} = \max_p$, for all new assistant places. The idea is now that the tokens in every place p can be distributed over a ring of $\max_p$ places (arranged in a circular fashion with transitions connecting neighbouring places; together these places and transitions induce an unweighted directed cycle) and that each weighted arc connecting p with a transition t can be represented by the corresponding number of unweighted arcs connecting t with individual places from the conglomerate. Note however that it might be that some of the tokens marking an original inhibitor place p are in 'store' in p_2. For this reason, every transition t taking tokens from a place p with an inhibitor arc will be represented by several copies in the new inhibitor net: for each pair of natural numbers k, m such that $k + m = W(p,t)$ and $m \leq inh_p$ there will be a representant of t taking k tokens from the ring of places representing p and m tokens from the ring representing p_2 (with unweighted arcs pointing to this new transition); and this copy of t adds m tokens to the ring representing p_1. If t has more than one inhibitor place as input place, then each of its representatives corresponds with a combination of choices of such k, m for each of these inhibitor places. If p is an output place of t, then there will be $W(t,p)$ unweighted arcs from the representants of t to the ring of p in the new net. Every (unweighted) inhibitor arc from a place p to transition t is replaced by inhibitor arcs from every place in the ring of p to each representant of t.

By construction, the resulting PTI-net is simple and every firing sequence of the original net has an obvious translation into a firing sequence in the newly constructed net and vice versa. Moreover, for every a priori step sequence of each net, there is a corresponding one in the other net. Also the markings of both nets are directly related as sketched above and corresponding (firing, step) sequences lead to corresponding markings. Thus we can conclude that the decidability status of the boundedness problem and that of the reachability problem of the original PTI-net can be derived in the new simple PTI-net.

To avoid complicated proofs, in the last part of the paper, we consider simplicity as a normal form of PTI-nets.

5 Coverability Tree for the Sequential Semantics

In this section, we first recall the construction of coverability trees for PT-nets [12, 7, 19], and then investigate a possible way of extending this construction to PTI-nets.

5.1 Coverability tree construction for PT-nets

A coverability tree $CT = (V, A, \mu, v_0)$ for a PT-net $\mathcal{N} = (P, T, W, M_0)$ has a set of nodes V, a root node v_0, and a set of directed labelled arcs A. Each node v is labelled by an extended marking $\mu(v)$ of $\mathcal{N}$. An α-labelled arc from v to w will be denoted as $v \xrightarrow{\alpha} w$. We write $v \leadsto_A^{\sigma} w$ (or simply $v \leadsto_A w$) to indicate that node w can be reached from another node v with σ as the sequence of labels along the path from v to w. The algorithm for the construction of coverability trees which is given in Table 1, assumes the sequential semantics for PT-nets (recall that this doesn't affect reachability because enabled steps can always be sequentialized).

A coverability tree is a finite representation of the reachable markings of a PT-net. Initially, it has one node corresponding to the initial marking. A node labelled with an (extended) marking that already occurs as a label of a processed node is terminal and doesn't need to be processed since its successors already appear as successors of this earlier node. (Strictly speaking the algorithm is not deterministic, but with this interpretation the defined reachability structure is unique; see also Fact 2.) For each transition enabled at the marking of a node that is being processed, a new node and an arc labelled with that transition between these two nodes is added. The label of the new node is the extended marking reached by executing that transition. A key aspect of the algorithm in Table 1 is the condition which allows one to replace some of the integer components of an extended marking

Table 1: *Algorithm generating a coverability tree of a PT-net*

$CT = (V, A, \mu, v_0)$ where $V = \{v_0\}, A = \varnothing$ and $\mu(v_0) = M_0$
unprocessed $= \{v_0\}$
`while` *unprocessed* $\neq \varnothing$
  `let` $v \in$ *unprocessed*
  `if` $\mu(v) \notin \mu(V \setminus unprocessed)$ `then`
  `for every` $\mu(v)[t\rangle M$
    $V = V \uplus \{w\}$ and $A = A \cup \{v \xrightarrow{t} w\}$
      and *unprocessed* $=$ *unprocessed* $\cup \{w\}$
    `if` there is u such that $u \rightsquigarrow_A v$ and $\mu(u) < M$
    `then` $\mu(w)(p) = ($`if` $\mu(u)(p) < M(p)$ `then` ω `else` $M(p))$
    `else` $\mu(w) = M$
  unprocessed $=$ *unprocessed* $\setminus \{v\}$

by ω's. Suppose that at some point of the operation of the algorithm, we generated through the firing of a transition an extended marking M. Then, provided that there is an ancestor node of the current node labelled by marking M' such that $M' < M$, we replace each $M(p)$ by ω whenever $M'(p) < M(p)$. The intuition behind such decision is that the sequence of transitions labelling the path from the ancestor node to the newly generated one can be repeated indefinitely, implying the unboundedness of any place p for which $M'(p) < M(p)$.

The following are well-known facts about the algorithm in Table 1 and its result (see, e.g., [2]). They demonstrate that the algorithm in Table 1 always terminates and, moreover, that in the coverability tree obtained, all firing sequences of the PT-net are represented; each reachable marking of the PT-net is covered by an extended marking; and each ω-component corresponds exactly with an unbounded number of tokens in that place.

Let CT be the coverability tree generated for PT-net $\mathcal{N}$ by a run of the algorithm in Table 1. The first result is that CT is finite, or in other words, the algorithm always terminates.

Fact 1 *CT is finite.*

The next result is fairly technical but it has a clear interpretation, namely, it states that any firing sequence of the PT-net can be re-traced in the coverability tree although sometimes one needs to 'jump' from one node to another provided that the two nodes are labelled by the same extended marking.

Fact 2 *For each firing sequence $M_0[t_1\rangle M_1 \ldots M_{n-1}[t_n\rangle M_n$ of $\mathcal{N}$, there are arcs $v_0 \xrightarrow{t_1} w_1, v_1 \xrightarrow{t_2} w_2, \ldots, v_{n-1} \xrightarrow{t_n} w_n$ in CT such that:*

$$\mu(w_i) = \mu(v_i) \text{ for } i = 1, \ldots, n-1.$$
$$M_i \leq \mu(v_i) \text{ (for } i = 0, \ldots, n-1\text{) and } M_n \leq \mu(w_n).$$

Note, however, that the converse of Fact 2 does not, in general, hold. That is, there may be traversals of a coverability tree which do not correspond to valid firing sequences of the PT-net. This highlights the difference between coverability trees and reachability graphs as in the latter a converse of Fact 2 does hold (but the counterpart of Fact 1 does not!)

Fact 3 *For every node v of CT and $k \geq 0$, there is a reachable marking M of $\mathcal{N}$ which is a k-approximation of $\mu(v)$, i.e., $M \Subset_k \mu(v)$.*

This result validates the meaning of extended markings appearing in the coverability tree, by showing they are in some sense minimal (note that the all-ω extended marking $\boldsymbol{\Omega}$ covers any marking of the PT-net, but is usually too rough to be a useful approximation). More precisely, Fact 3 shows that the ω-components in an extended marking appearing in CT indicate that there are reachable markings of $\mathcal{N}$ which simultaneously grow arbitrarily large on all places with an ω and are exactly the same on all the remaining places. A straightforward application of this is that coverability trees can be used to decide the boundedness of places.

Fact 4 *A place p of $\mathcal{N}$ is bounded* iff $\mu(v)(p) \neq \omega$ *for every node v of CT.*

5.2 Adapting the coverability tree construction to inhibitor arcs

When trying to extend the construction of coverability trees as given in Table 1 to PTI-nets, the main problem one encounters, is the *non-monotonicity* of nets with inhibitor arcs: given two markings $M' < M$ and a firing sequence σ which can be fired from M', one cannot be sure that σ can also be fired from M. As a consequence, the condition for generating ω-components may be too weak. It can be strengthened by making sure that no inhibitor arc features were used along the path from u to v for those places in which the number of tokens has grown. We thus modify the construction in Table 1 by replacing the line

> `if` there is u such that $u \rightsquigarrow_A v$ and $\mu(u) < M$

by

> `if` there is u such that $u \rightsquigarrow_A^{\sigma} v$ and $\mu(u) < M$ and such that $\mu(u)(p) < M(p)$ implies that ${}^{\circ}t'(p) = \omega$, for all transitions t' in σt

From here we will refer to this modification of the algorithm in Table 1 as the *modified CTC*. In what follows, $\mathcal{N} = (P, T, W, I, M_0)$ is a PTI-net and CT an object (optimistically referred to as a coverability tree) generated for $\mathcal{N}$ by the modified CTC. Note that this algorithm only considers the firing of single transitions and not of steps. Hence, if it works correctly, it provides us with a coverability tree for PTI-nets under the sequential semantics.

First we demonstrate that the construction terminates at least for PTI-nets with one inhibitor place. (In case of no inhibitor places, the PTI-net is a PT-net and the modification would be void.)

Theorem 4 *If $\mathcal{N}$ has exactly one inhibitor place then the modified CTC always terminates.*

Proof Suppose that the algorithm generates an infinite CT, and that p is the only inhibitor place of $\mathcal{N}$. Since T is finite, CT is finitely branching. Hence there exists, by König's Lemma, an infinite path ξ from the root. Furthermore, the ω-components in markings are never changed to integers when moving to a child node, and we thus may assume that there is a node starting from which all markings labelling the nodes of ξ have ω-components at exactly the same positions. Let ϕ be a sequence of such nodes. Since the markings labelling nodes in ξ are all different (by the definition of the algorithm) and there are only finitely many places, it follows from Dickson's lemma that there is a subsequence $v_1 v_2 \ldots$ of ϕ such that $\mu(v_1) < \mu(v_2) < \ldots$. We then observe that $\mu(v_i)(p) \neq \mu(v_{i+1})(p)$, for every i, since otherwise we would have $\mu(v_i)_\omega \neq \mu(v_{i+1})_\omega$. In particular, this means that $\mu(v_i)(p) \neq \omega$, for every i, since ω-components are persistent along the arcs in CT.

We have therefore shown that $\mu(v_1)(p) < \mu(v_2)(p) < \ldots < \omega$. Hence between each pair v_i and v_{i+1}, there is at least one arc labelled by a transition which is inhibited by p. Thus there is an infinite sequence $i_1 < i_2 < \ldots$ such that $\mu(v_{i_j})(p) = l$, for every j, where l is an integer less or equal to the highest of the weights of inhibitor arcs adjacent to p. Thus, again by the fact that there are only finitely many places and Dickson's lemma, there is an infinite sequence $m_1 < m_2 < \ldots$ such that $\mu(v_{i_{m_1}}) < \mu(v_{i_{m_2}}) < \ldots$. This, however, means that the algorithm would need to generate infinitely many ω-components, a contradiction. $\dashv$

We do not know at the moment whether Theorem 4 holds for all PTI-nets with two inhibitor places. There is however a PTI-net with three inhibitor places for which the modified coverability tree construction does not terminate.

Proposition 5 *There are PTI-nets for which the modified CTC will never terminate.*

Proof Consider the PTI-net $\mathcal{N}$ in Figure 4.

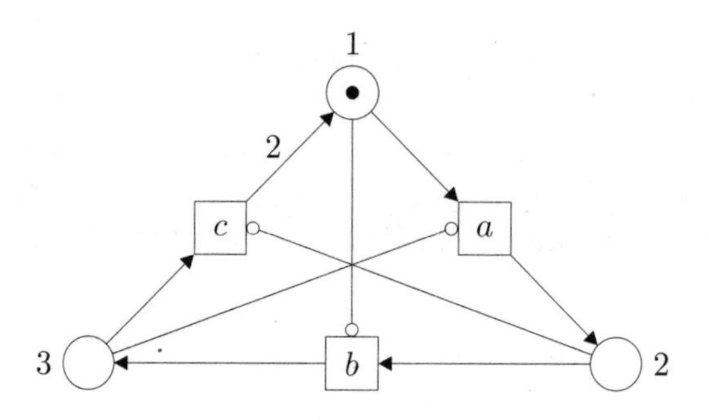

Figure 4: *A PTI-net for which the modified CTC does not terminate.*

$\mathcal{N}$ can execute exactly one infinite sequence of transitions $\sigma = \sigma_1\sigma_2\sigma_3\ldots$, where

$$\sigma_i = \underbrace{aa\ldots a}_{i\ times}\ \underbrace{bb\ldots b}_{i\ times}\ \underbrace{cc\ldots c}_{i\ times}$$

for every $i \geq 1$. (Note that there is no choice offered at any stage.) The sequence σ has the following properties:

- No subsequence of σ involving all three transitions can ever be repeated.
- No markings reachable through two different finite prefixes of σ are the same.
- Between any two $<$-comparable markings reachable through two different finite prefixes of σ, each of the three transitions a, b and c must be fired at least once.

Hence the modified CTC will never generate any ω components, and so never terminates when applied to the PTI-net in Figure 4. ⊣

The next result shows that *CT* encodes in a sound way the unboundedness of places even if the algorithm does not terminate, and so we obtain a counterpart of Fact 3 which holds for all PTI-nets.

Theorem 6 *For every node v of CT and $k \geq 0$, there is a reachable marking M of $\mathcal{N}$ which is a k-approximation of $\mu(v)$, i.e., $M \Subset_k \mu(v)$.*

Proof (sketch) We proceed by induction on the distance from the root. In the base case, $v = v_0$ is the root of the tree and so $\mu(v) = M_0$. Suppose that the result holds for a node w, $w \xrightarrow{t} v$ and $\mu(w)[t\rangle M'$.

By the induction hypothesis, there is $M_1 \in [M_0\rangle$ such that $M_1 \Subset_{k+|^\bullet t|} \mu(w)$. We have $M_1[t\rangle M_2$ and, for every $p \in P$, $\mu(v)(p) < \omega$ implies $M_2(p) = M'(p) = \mu(v)(p)$.

Thus M_2 satisfies the required condition for all p such that $\mu(v)(p) < \omega$ or $\mu(w)(p) = \omega$. Therefore, the required condition may not be satisfied only if there are places p such that $\mu(v)(p) = \omega$ and $\mu(w)(p) < \omega$. In such a case, by the construction, we have that there is a node u and a path $u = w_1 \xrightarrow{t_1} w_2 \ldots w_n \xrightarrow{t_n} w_{n+1} = v$ (i.e., $w_n = w$ and $t_n = t$) in the tree *CT* such that $\mu(u) < M'$. Let $k' = k + |^\bullet t| + k \cdot \sum_{i=1}^{n} |^\bullet t_i|$. By the induction hypothesis, there is $M_3 \in [M_0\rangle$ such that $M_3 \Subset_{k'} \mu(w)$. One can then show that $\sigma = t(t_1 \ldots t_n) \ldots (t_1 \ldots t_n)$ (with k times $(t_1 \ldots t_n)$) is a firing sequence enabled at the marking M_3 leading to a marking which is a k-approximation of $\mu(v)$. ⊣

The above result cannot, in general, be reversed in the sense that not all finite coverability trees provide full information about the unbounded places of a PTI-net (not even if it has only a single inhibitor arc).

Proposition 7 *There is a PTI-net with one inhibitor arc and an unbounded place p such that the modified CTC does not yield a CT with a node label with an ω-component corresponding to p.*

Proof Consider the PTI-net $\mathcal{N}$ in Figure 5 together with its finite coverability tree *CT* which is unique up to isomorphism.

It may be observed that place 3 which is unbounded because of the infinite firing sequence

$$abc\,abc\,abc\,abc\;\ldots$$

is not detected as such by the modified CTC. ⊣

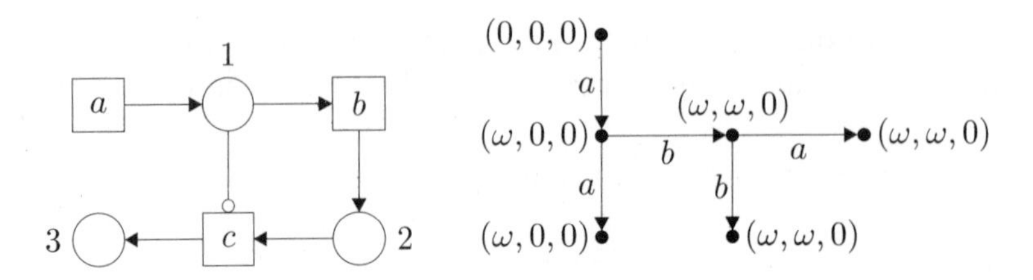

Figure 5: *A PTI-net for which the modified CTC in Table 1 does not detect all unbounded places.*

Thus to conclude this section, for the sequential semantics of PTI-nets we only have the modified CTC which provides some information on possible (simultaneous) unboundedness of places, but supports no more than a semi-algorithm for the (place) *fs*–boundedness problem.

6 Coverability Tree and Step Semantics

We now turn to the a priori step sequence semantics of PTI-nets. We start by re-considering the very concept of a coverability tree in the context of a semantics based on steps rather than single transitions.

To start with, we observe that if a PTI-net does not exhibit unbounded auto-concurrency and we know the bound k on the size of steps enabled at the reachable markings, then the boundedness problem can be reduced to the interleaving case. Simply, one can add (finitely many) transitions representing all potential steps. I.e., for each step U of transitions satisfying $U(t) \leq k$ for each transition t of the net, we add a fresh transition t_U such that ${}^\bullet t_U = {}^\bullet U$, $t_U{}^\bullet = U^\bullet$ and ${}^\circ t_U = {}^\circ U$. It is easy to see that a place is bounded in the resulting PTI-net under the sequential semantics if and only if it is bounded in the original PTI-net. Note also that such a transformation does not create additional inhibitor places.

As we have already seen, non-monotonicity in the executions of inhibitor nets is the reason why the standard definition of a coverability tree is too weak to detect unbounded places. The inhibitor net in Figure 5 illustrates one of the ways in which this actually happens. Intuitively, the example combines non-monotonicity with the persistence of ω-components in the extended markings labelling the nodes of a coverability tree. The mechanism is quite simple: since such components are never replaced by integer values when generating descendant nodes, one can miss the chance of detecting the situation that a place can be emptied at some point in the future and de-activating a current inhibitor constraint. But it would be wrong to think that this is the only way in which non-monotonicity can spoil the construction.

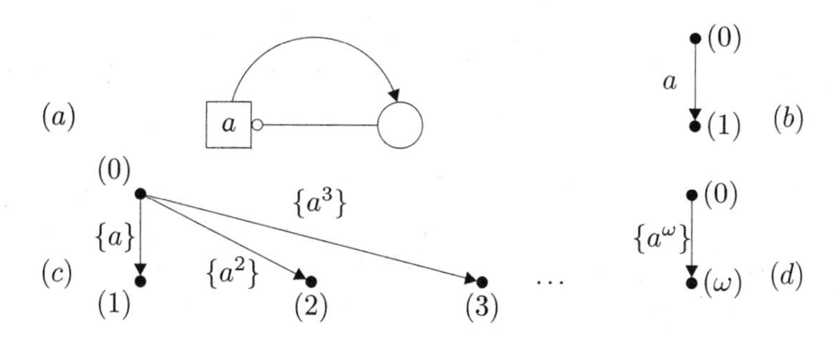

Figure 6: *PTI-net with associated reachability and coverability trees.*

Consider, for example, the PTI-net in Figure 6(a). Its interleaving coverability graph, shown in Figure 6(b), is fully satisfactory as in this case no place is unbounded. The situation changes radically when we start generating a coverability tree for the a priori step sequence semantics, using steps instead of single transitions and a natural adaptation of the CTC shown in Table 1. The reason is that in such a case there are no ω-markings at all, but the generated CT is infinite, as shown in Figure 6(c). This is unsatisfactory since, intuitively, one should be able to handle unboundedness in simple cases like this. Intuitively, the example inhibitor net exhibits unboundedness 'in breadth' which cannot be replaced by unboundedness 'in depth'. This never happens in the case of PT-nets and the difference is caused by the non-monotonicity in the behaviours of inhibitor nets.

To address the problem, we propose to adapt the coverability tree construction by incorporating not only ordinary steps, but also 'infinite' steps. For our example this leads to the *step coverability tree* shown in Figure 6(d), where the infinite step $\{a^\omega\}$ covers infinitely many steps $\{a^i\}$, and leads to an extended marking which implies the unboundedness of the only place.*

Table 2 shows a generic algorithm for constructing a step coverability tree for a given PTI-net. It is similar to the construction described in Table 1 but uses extended steps rather than single transitions to label edges. Since the set of steps enabled at a marking can

*Note that if one required that each transition had both at least one input place and at least one output place (rather than just being non-isolated), we would still maintain the same approach. Simply, in our example, we would add a new input place to transition a and introduce a transition with a marked loop to fill this input place with an arbitrary large number of tokens.

Table 2: *Algorithm generating a step coverability tree of a PTI-net; select and $\sqsubset$ need to be specified separately, depending on the subclass of PTI-nets under consideration*

$SCT = (V, A, \mu, v_0)$ where $V = \{v_0\}, A = \varnothing$ and $\mu(v_0) = M_0$
unprocessed $= \{v_0\}$
`while` *unprocessed* $\neq \varnothing$
  `let` $v \in$ *unprocessed*
  `if` $\mu(v) \notin \mu(V \setminus unprocessed)$ `then`
  `for every` $\mu(v)[U\rangle M$ with $U \in select(\mu(v))$
    $V = V \uplus \{w\}$ and $A = A \cup \{v \xrightarrow{U} w\}$
      and *unprocessed* $=$ *unprocessed* $\cup \{w\}$
    `if` there is u such that $u \rightsquigarrow_A v$ and $\mu(u) \sqsubset M$
    `then` $\mu(w)(p) =$ (`if` $\mu(u)(p) < M(p)$ `then` ω `else` $M(p)$)
    `else` $\mu(w) = M$
  unprocessed $=$ *unprocessed* $\setminus \{v\}$

be infinite, the `for`-loop is executed for steps from a *finite* yet sufficiently representative subset *select*(.) of extended steps enabled at the extended marking under consideration. Another difference in comparison with the original construction is the use of the relation $\sqsubset$ to compare extended markings rather than $<$. In the next subsection, we will instantiate this algorithm for a subclass of PTI-nets.

6.1 Primitive PTI-nets

Here we concentrate on the class of PTI-nets introduced in [2] which enjoy the property that once an inhibitor place contains more than a certain threshold of tokens (its emptiness limit), no transition which tests it for emptiness can occur anymore.

A PTI-net $\mathcal{N} = (P, T, W, I, M_0)$ is *primitive* (or is a PPTI-net) if there is an integer EL (the 'emptiness limit') such that for every reachable marking M and every inhibitor place p, if $M(p) > EL$ then for every marking M' reachable from M and transition t enabled at M', it is the case that $M'(p) > {}^{\circ}t(p)$. For example, the PTI-net in Figure 6 is trivially primitive with $EL = 0$ since its only place is an inhibitor place and output place of the only transition. In general, if no inhibitor place has an outgoing (ordinary) arc, we may set $EL = 0$, and if there are no inhibitor places (i.e., if $\mathcal{N}$ is a PT-net), we may set $EL = -1$. In [2] the threshold value EL is chosen separately for each individual inhibitor place. Since the number of places in $\mathcal{N}$ is finite, ours is an equivalent definition (though less efficient algorithmically).

In what follows we consider a fixed simple PPTI-net $\mathcal{N} = (P, T, W, I, M_0)$ with a fixed emptiness limit EL. (Note that primitivity is preserved by the construction described at the end of Section 4.)

For PPTI-nets, the algorithm in Table 2 is instantiated as follows:

- *select*$(\mu(v))$ is the set of all extended steps of transitions U enabled at $\mu(v)$ such that $U(t) \in \{0, 1, \ldots, EL, \omega\}$, for each transition t such that $\{t^\omega\}$ is enabled at $\mu(v)$.
- For any two extended markings, M and M', we have $M \sqsubseteq M'$ if $M(p) \leq M'(p)$, for all places p, and $M(p) = M'(p)$ for all inhibitor places p, whenever $M(p) \leq EL$.
  Moreover, $M \sqsubset M'$ if $M \sqsubseteq M'$ and $M \neq M'$.

We refer to the algorithm resulting from this instantiation as the SCTC (step coverability tree construction).

Intuitively, *select*$(\mu(v))$ is defined in such a way that if a non-selected extended step enabled at $\mu(v)$ inserts some tokens into an inhibitor place p, then it necessarily inserts at least $EL+1$ tokens, making from this point on the inhibiting features of p void. And the step itself will be covered by at least one step in *select*$(\mu(v))$.

The ordering $\sqsubset$ was introduced in [2] and is intended to ensure that inhibitor places are treated as such (and their marking not being replaced by ω's) until the threshold value EL has been passed. It is easy to show (see [2]) that Dickson's lemma also holds for this ordering, i.e., every infinite sequence of extended markings contains an infinite subsequence ordered w.r.t. $\sqsubseteq$.

Figure 7 shows how the algorithm works for three PTI-net where inhibitor arcs influence the execution semantics.

Let SCT be the step coverability tree generated for $\mathcal{N}$ by a run of the SCTC. Our immediate aim is to re-establish the main properties of the CTC defined for PT-nets. First, we show that SCT is finite and SCTC always terminates (cf. Fact 1).

Theorem 8 *SCT is finite.*

Proof Suppose that SCT is not finite. We first observe that, since *select*$(\mu(v))$ is always a finite set which follows directly from the definition and T being finite, SCT is finitely branching. Hence, by König's Lemma, there is an infinite path $v_0 v_1 \ldots$ from the root. By Dickson's Lemma for $\sqsubseteq$ and the definition of SCTC, there is an

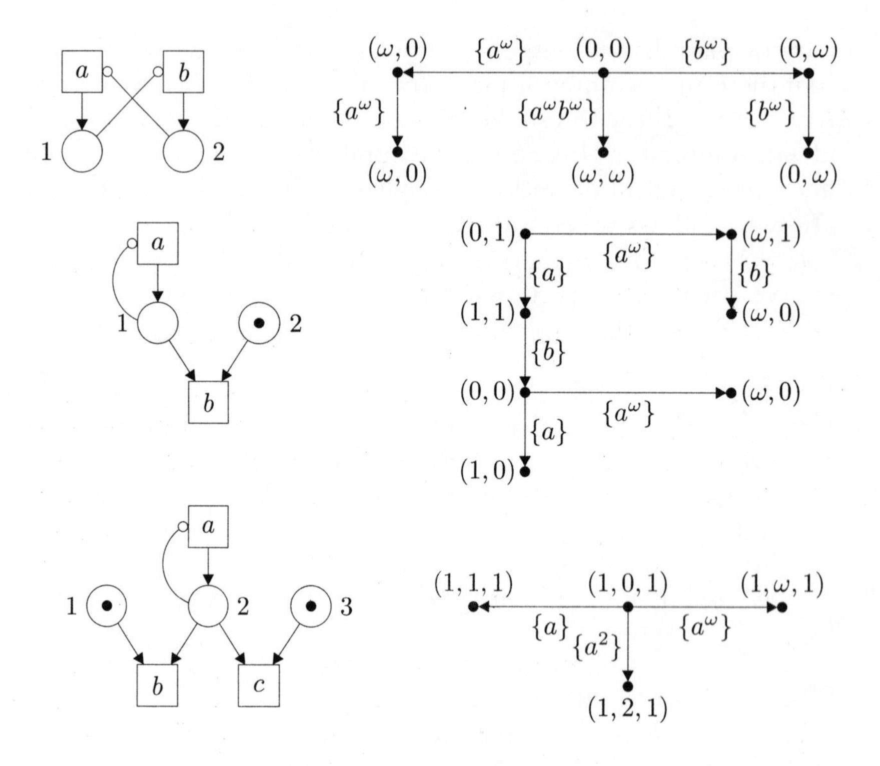

Figure 7: *PPTI-nets with EL* $= 0$ *(top), EL* $= 1$ *(middle) and EL* $= 2$ *(bottom), and their coverability trees derived according to the algorithm in Table 2. Note that for the last tree we only show the root and its three child nodes.*

infinite sequence of indices $i_1 < i_2 < \ldots$ such that $\mu(v_{i_1}) \sqsubseteq \mu(v_{i_2}) \sqsubseteq \ldots$ is a sequence of distinct markings. Hence $\mu(v_{i_1}) \sqsubset \mu(v_{i_2}) \sqsubset \ldots$. For every $j > 1$, by $\mu(v_{i_j}) \sqsubset \mu(v_{i_{j+1}})$ and the construction, there is p such that $\mu(v_{i_j})(p) < \omega = \mu(v_{i_{j+1}})(p)$. Hence the number of ω-components in $\mu(v_{i_{|P|+1}})$ is greater than the total number of places, a contradiction. $\dashv$

The next result is similar to Fact 2; we show that every step sequence of the PTI-net can be retraced in the SCTC if not exactly, then at least through a step sequence covering it.

Proposition 9 *For each step sequence* $M_0[U_1\rangle \ldots [U_n\rangle M_n$ *of* $\mathcal{N}$, *there are arcs* $v_0 \xrightarrow{V_1} w_1, v_1 \xrightarrow{V_2} w_2, \ldots, v_{n-1} \xrightarrow{V_n} w_n$ *in SCT such that:*

- $U_i \leq V_i$ *for* $i = 1, \ldots, n$.

- $\mu(w_i) = \mu(v_i)$ *for* $i = 1, \ldots, n-1$.
- $M_i \leq \mu(v_i)$ *(for* $i = 0, \ldots, n-1$*) and* $M_n \leq \mu(w_n)$.
- *For all places* p, $\mu(v_i)(p) = \omega$ *whenever* $M_i(p) \neq \mu(v_i)(p)$ *for* $i = 0, \ldots, n-1$, *and* $\mu(w_n)(p) = \omega$ *whenever* $M_n(p) \neq \mu(w_n)(p)$.

Proof This is an immediate consequence of Proposition 19 in the Appendix. ⊣

Moreover, the extended markings appearing in *SCT* cover in a minimal way the markings of $\mathcal{N}$ and their ω-components indicate the components which simultaneously grow arbitrarily large. In other words, we obtain a counterpart of Fact 3.

Proposition 10 *For every node* v *of SCT and* $k \geq 0$, *there is a reachable marking* M *of* $\mathcal{N}$ *which is a* k*-approximation of* $\mu(v)$, *i.e.,* $M \Subset_k \mu(v)$.
Proof See Appendix. ⊣

As an immediate consequence of the last two results, we now can formulate a central result of this paper that SCTC can be used to decide the boundedness of PPTI-nets working under the a priori step sequence semantics.

Theorem 11 *A place* p *of* $\mathcal{N}$ *is bounded* iff *there is no node* v *in the coverability tree constructed by the algorithm in Table 2 such that* $\mu(v)(p) = \omega$. *Consequently,* $\mathcal{N}$ *is bounded* iff *no extended marking annotating a node of its step coverability tree contains an* ω*-component.*
Proof Follows from Propositions 9 and 10. ⊣

We finally observe that SCTC can also be used to decide marking coverability.

Theorem 12 *For a marking* M *of* $\mathcal{N}$, *there is a reachable marking* M' *of* $\mathcal{N}$ *such that* $M \leq M'$ iff *there is a node* v *in SCT such that* $M \leq \mu(v)$.
Proof Follows from Proposition 9. ⊣

6.2 Deciding step executability

The coverability tree constructed as in Table 2 has arcs labelled by extended steps, and so it is a valid question to ask whether such a tree could be used to investigate issues related to the executability of steps. In other words, there may now be an opportunity to investigate concurrency aspects with the help of coverability trees. In what follows, the *step executability* problem for PTI-nets is to decide whether a step of a PTI-net can be executed at some of its reachable markings. As it turns out, this problem is indeed decidable for the subclass of PPTI-nets. We start by providing two auxiliary results.

Again, $\mathcal{N} = (P, T, W, I, M_0)$ is a PPTI-net (not necessarily simple) and SCT is any step coverability tree for $\mathcal{N}$ generated by the SCTC.

Proposition 13 *If step U is enabled at a reachable marking M of $\mathcal{N}$, then there is an arc $v \xrightarrow{W} w$ in SCT such that $M \leq \mu(v)$ and $U \leq W$.*
Proof Follows from Proposition 9. ⊣

Proposition 14 *For every $k \geq 0$ and every W labelling an arc in SCT, there is a step U enabled at a reachable marking of $\mathcal{N}$ satisfying $U \Subset_k W$.*

Proof Let $U = W_{\omega \mapsto k+1}$ and $k' = |{}^{\bullet}U|$. Moreover, let $v \xrightarrow{W} w$ be an arc in SCT. From W being enabled at $\mu(v)$, it follows that $({}^{\bullet}W)_{\omega} \subseteq \mu(v)_{\omega}$. By Proposition 10, there is $M \in [M_0\rangle$ such that $M \Subset_{k'} \mu(v)$ and so U is enabled at M. This and $U \Subset_k W$ completes the proof. ⊣

We then obtain a result which, together with Theorem 8, implies that the step executability problem for PPTI-nets is decidable.

Theorem 15 *A step U is enabled at some reachable marking of $\mathcal{N}$* iff *there is an arc in SCT labelled by W such that $U \leq W$.*
Proof ($\Longrightarrow$) Follows from Proposition 13.
($\Longleftarrow$) Follows from Proposition 14 and the observation that if a step U' is enabled at a marking of a PTI-net and $U \leq U'$, then U is also enabled. ⊣

Corollary 16 *A transition t of $\mathcal{N}$ is dead* iff *there is no arc in SCT labelled by a step containing t.*

In order to improve the efficiency of the algorithm in Table 2 one could try to reduce the size of the set *select*$(\mu(v)$. A natural possibility would be, as in [2], to define the values $EL(p)$ individually for each inhibitor place to be as small as possible. Another would be to require that only those steps be selected which cannot be replaced by the sequential execution of their elements.

7 Concluding Remarks

The step coverability tree construction can be used to decide boundedness and other properties of primitive PTI-nets working under the a priori step sequence semantics. It must be noted however that primitivity itself is an undecidable property even when only firing sequences are considered (see [2]). Still, as argued in [2], primitive PTI-nets are an interesting subclass and often primitivity is guaranteed by construction. In particular, PTI-nets with bounded and complemented inhibitor places satisfy primitivity and, as noted in Section 6.1, PT-nets can be considered as PPTI-nets with $EL = -1$. Hence the results we obtained here are directly applicable to these classes of PTI-nets. Note that for PT-nets, $\sqsubset$ becomes $<$, and $U \in select(\mu(v))$ if U is enabled at $\mu(v)$ and $U(t) = \omega$, for each transition t such that $\{t^{\omega}\}$ is enabled at $\mu(v)$.

It might not seem to be prudent to use SCTC for the investigation of properties of PT-nets, as a sequential construction would in general exhibit a much lower degree of branching and therefore yield smaller trees. However, the situation changes if we move to a problem which could be seen as a counterpart of the marking reachability, but this time involving steps. First, directly from Theorem 15, we obtain

Corollary 17 *Let $\mathcal{N}$ be a PT-net and CT any coverability tree for $\mathcal{N}$ generated by SCTC. Then a step U is enabled at some reachable marking of $\mathcal{N}$* iff *there is an arc in CT labelled by W such that $U \leq W$.*

What is more, the standard CTC cannot be used to decide step executability. A counterexample is provided by the two PT-nets in Figure 8(a, b) for which the algorithm in Table 1 generates the same coverability tree shown in Figure 8(c). Yet, clearly, the first one enables arbitrarily large steps at all reachable markings whereas the latter enables only singleton steps. Note, again, that if one required that each transition had both at least one input place and at least one output place (rather than just being non-isolated), we would could still produce a pair of PT-nets as in Figure 8. Simply, in our example, we would give both transitions fresh input places and new transitions with marked loop filling these input place with an arbitrary large number of tokens.

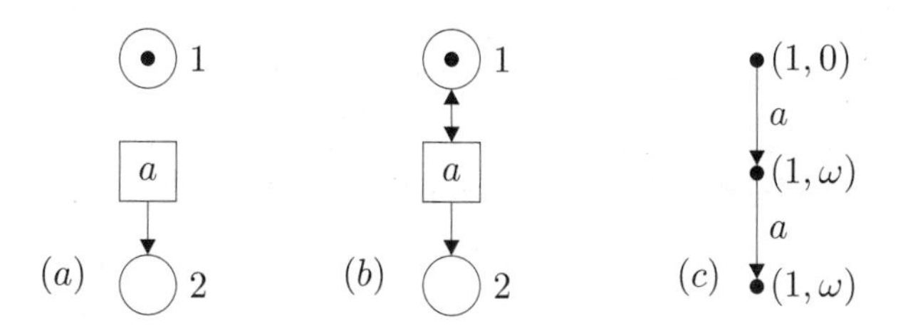

Figure 8: *Two PT-nets and their sequential coverability tree.*

Acknowledgement

We would like to thank the anonymous reviewers for their very careful reading of the manuscript and constructive comments.

References

[1] T. AGERWALA. *A Complete Model for Representing the Coordination of Asynchronous Processes*, Hopkins Computer Research Report 32, Johns Hopkins University (1974)

[2] N. BUSI. *Analysis Issues in Petri Nets with Inhibitor Arcs*, Theoretical Computer Science 275 (2002) 127-177

[3] N. BUSI AND G.M. PINNA. *Process Semantics for Place/Transition Nets with Inhibitor and Read Arcs*, Fundamenta Informaticae 40 (1999) 165-197

[4] PH. DARONDEAU, B. GENEST, P.S. THIAGARAJAN AND S. YANG. *Quasi-Static Scheduling of Communicating Tasks*, Proc. of CONCUR'08, Lecture Notes in Computer Science 5201 (2008) 310-324

[5] J. DESEL AND W. REISIG. *Place/Transition Petri Nets*, In: [23] 122-173

[6] J. ESPARZA AND M. NIELSEN. *Decidability Issues for Petri Nets*, Petri Net Newsletter 47 (1994) 5-23

[7] M. HACK. *Decision Problems for Petri Nets and Vector Addition Systems*, Technical Memo 59, Project MAC, MIT (1975)

[8] M. HACK. *Petri Net Languages*, Technical Report 159, MIT (1976)

[9] M. HACK. *Decidability Questions for Petri Nets*, PhD Thesis, MIT (1976)

[10] P.W. HOOGERS, H.C.M. KLEIJN AND P.S. THIAGARAJAN. *A Trace Semantics for Petri Nets*, Information and Computation 117 (1995) 98-114

[11] R. JANICKI AND M. KOUTNY. *Semantics of Inhibitor Nets*, Information and Computation 123 (1995) 1-16

[12] R.M. KARP AND R.E. MILLER. *Parallel Program Schemata*, J. Comput. Syst. Sci. 3 (1969) 147-195

[13] H.C.M. KLEIJN AND M. KOUTNY. *Process Semantics of General Inhibitor Nets*, Information and Computation 190 (2004) 18-69

[14] H.C.M. KLEIJN AND M. KOUTNY. *Infinite Process Semantics of Inhibitor Nets*, Proc. of ICATPN'06, Lecture Notes in Computer Science 4024 (2006) 282-301

[15] S.R. KOSARAJU. *Decidability of Reachability in Vector Addition Systems*, Proc. of STOC'82, ACM (1982) 267-281

[16] E.W. MAYR. *An Algorithm for the General Petri Net Reachability Problem*, SIAM J. Comput. 13 (1984) 441-460

[17] M. PIETKIEWICZ-KOUTNY. *Synthesis of ENI-systems Using Minimal Regions*, Proc. of CONCUR'98, Lecture Notes in Computer Science 1466 (1998) 565-580

[18] M. MUKUND AND P.S. THIAGARAJAN. *Linear Time Temporal Logics over Mazurkiewicz Traces*, Proc. of MFCS'96, Lecture Notes in Computer Science 1113 (1996) 62-92

[19] J.L. PETERSON. *Petri Net Theory and the Modeling of Systems*, Prentice Hall (1981)

[20] C.A. PETRI. *Fundamentals of a Theory of Asynchronous Information Flow*, Proc. of IFIP Congress'62, North Holland, Amsterdam (1962) 386-390

[21] K. REINHARDT. *Reachability in Petri Nets with Inhibitor Arcs*, Technical Report WSI-96-30, Wilhelm Schickard Institut für Informatik, Universität Tubingen (1996)

[22] G. ROZENBERG AND J. ENGELFRIET. *Elementary Net Systems*, In: [23] 12-121

[23] W. REISIG AND G. ROZENBERG. *Lectures on Petri Nets I: Basic Models, Advances in Petri Nets*, Lecture Notes in Computer Science 1491 (1998)

[24] G. ROZENBERG AND P.S. THIAGARAJAN. *Petri Nets: Basic Notions, Structure, Behaviour*, Current Trends in Concurrency, Lecture Notes in Computer Science 224 (1986) 585-668

[25] P.S. THIAGARAJAN. *Elementary Net Systems*, Advances in Petri Nets 1986, Lecture Notes in Computer Science 254 (1987) 26-59
[26] P.S. THIAGARAJAN. *Some Behavioural Aspects of Net Theory*, Theoretical Computer Science 71 (1990) 133-153
[27] A. VALMARI. *The State Explosion Problem*, In: [23] 429-528
[28] W. VOGLER. *Partial Order Semantics and Read Arcs*, Proc. of MFCS'97, Lecture Notes in Computer Science1295 (1997) 508-517

Appendix

A Proofs of various results

As in subsection 6.1, we consider here a simple PPTI-net $\mathcal{N} = (P, T, W, I, M_0)$ with emptiness limit EL. For every marking M, $Z(M)$ denotes the set of inhibitor places which are *active* as such in markings reachable from M, i.e., $p \in Z(M)$ if there is M' reachable from M and t enabled at M' such that $p \in {}^\circ t$.
An extended marking M' *Z-covers* a marking M if, for every place p,

- $M(p) \leq M'(p)$. (C1)
- $p \in Z(M)$ implies $M(p) = M'(p)$. (C2)
- $M(p) \neq M'(p)$ implies $M'(p) = \omega$. (C3)

We denote this by $M \trianglelefteq M'$.

Proposition 18 *If $M \trianglelefteq M'$ and $M[U\rangle\widehat{M}$ then $M'[U\rangle\widehat{M}'$ and $\widehat{M} \trianglelefteq \widehat{M}'$.*
Proof That $M'[U\rangle\widehat{M}'$ for some $\widehat{M}'$ follows from (C1) and (C2) for M and M'. Clearly, (C1) holds for $\widehat{M}$ and $\widehat{M}'$. Moreover, since $Z(\widehat{M}) \subseteq Z(M)$ and (C2) holds for M and M', it also holds for $\widehat{M}$ and $\widehat{M}'$. Similarly, if $\widehat{M}(p) \neq \widehat{M}'(p)$ then $M(p) \neq M'(p)$ and so, by (C2) for M and M', $M'(p) = \omega$. Hence $\widehat{M}'(p) = \omega$, and we conclude that $\widehat{M} \trianglelefteq \widehat{M}'$. ⊣

Proposition 19 *For each step sequence $M_0[U_1\rangle \ldots [U_n\rangle M_n$ of $\mathcal{N}$, there are arcs $v_0 \xrightarrow{V_1} w_1, v_1 \xrightarrow{V_2} w_2, \ldots, v_{n-1} \xrightarrow{V_n} w_n$ in SCT such that:*

- *$U_i \leq V_i$ for $i = 1, \ldots, n$.*
- *$\mu(w_i) = \mu(v_i)$ for $i = 1, \ldots, n-1$.*
- *$M_i \trianglelefteq \mu(v_i)$ (for $i = 0, \ldots, n-1$) and $M_n \trianglelefteq \mu(w_n)$.*

Proof We proceed by induction on n. Clearly, the base case for $n = 0$ holds. Assume that the result holds for n and consider $M_n[U_{n+1}\rangle M_{n+1}$.

Let v_n be the first generated node such that $\mu(v_n) = \mu(w_n)$. As $M_n \trianglelefteq \mu(v_n)$ and $M_n[U_{n+1}\rangle M_{n+1}$, it follows from Proposition 18 that there exists M such that $\mu(v_n)[U_{n+1}\rangle M$ and $M_{n+1} \trianglelefteq M$. Let V_{n+1} be the $\sqsubset$-smallest marking in *select*($\mu(v_n)$) satisfying $U_{n+1} \leq V_{n+1}$ (such a step always exists). Moreover, let M' be such that $\mu(v_n)[V_{n+1}\rangle M'$.

Consider now any place p such that $M(p) \neq M'(p)$. Then there is t such that $p \in {}^\bullet t \cup t^\bullet$ and $U_{n+1}(t) \neq V_{n+1}(t)$. From $U_{n+1} \leq V_{n+1}$, the $\leq$-minimality of V_{n+1} and the definition of *select*(), it then follows that $U_{n+1}(t) > EL$ and $V_{n+1}(t) = \omega$. Consequently, if $p \in {}^\bullet t$ then $\mu(v_n)(p) = \omega = M(p) = M'(p)$, a contradiction. So $p \in t^\bullet$ and we have $M(p) \geq U_{n+1}(t) > EL$ and $M'(p) = \omega$. Hence we have shown that:

$$M(p) \neq M'(p) \implies EL < M(p) < \omega = M'(p). \quad (\dagger)$$

Taking M_{n+1} and M', clearly (C1) and (C3) hold due to $M_{n+1} \trianglelefteq M$ and (†). Suppose now that $p \in Z(M_{n+1})$ and $M_{n+1}(p) \neq M'(p)$. By $M_{n+1} \trianglelefteq M$, we have $M_{n+1}(p) = M(p)$ and so $M(p) \neq M'(p)$. Hence, by (†), $M(p) > EL$. This, $M_{n+1}(p) = M(p)$ and the fact that $\mathcal{N}$ is primitive, mean that $p \notin Z(M_{n+1})$, contradiction. Hence (C2) also holds and so $M_{n+1} \trianglelefteq M'$.

Now, during the processing of v_n an arc $v_n \xrightarrow{V_{n+1}} w_{n+1}$ is created such that $M_{n+1} \trianglelefteq \mu(w_{n+1})$. The latter follows from the following, for any place p:

- (C1) follows from $M_{n+1}(p) \leq M'(p)$ and the fact that $M'(p) \neq \mu(w_{n+1})(p)$ implies $\mu(w_{n+1})(p) = \omega$.
- To show (C2), suppose that $p \in Z(M_{n+1})$. Then, by $M_{n+1} \trianglelefteq M'$, we have $M_{n+1}(p) = M'(p) \neq \omega$. If $\mu(w_{n+1})(p) = M'(p)$ then $\mu(w_{n+1})(p) = M_{n+1}(p)$. Otherwise, there is a node v such that $\mu(v) \sqsubset M'$ and $\mu(v)(p) < M'(p)$. The latter and $\mu(v) \sqsubset M'$ implies $\mu(v)(p) > EL$. We have $M_{n+1}(p) = M'(p) > \mu(v)(p) > EL$ and, by $\mathcal{N}$ being primitive, $p \notin Z(M_{n+1})$, a contradiction.
- To show (C3), suppose that $M_{n+1}(p) \neq \mu(w_{n+1})(p) \neq \omega$ Then, by construction, $\mu(w_{n+1})(p) = M'(p)$ and so $M_{n+1}(p) \neq M'(p)$. Hence, by $M_{n+1} \trianglelefteq M'$, we have $M'(p) = \omega$, a contradiction.

This completes the proof of the induction step. ⊣

Proof of Proposition 10

Without loss of generality we may assume $k > EL$ and proceed by induction on the distance from the root of the nodes of the tree.

We denote $maxcons \stackrel{\text{df}}{=} \sum_{(w,U,v)\in A} |^\bullet(U_{\omega\mapsto 0})|$, i.e., *maxcons* is the total number of tokens consumed along the arcs of the tree by non-ω occurrences of transitions. Moreover, $k' \stackrel{\text{df}}{=} maxcons \cdot |P| \cdot (k+1)$.

In the base case, $v = v_0$ is the root of the tree and so $\mu(v) = M_0$. Suppose that the result holds for a node w, and that $w \xrightarrow{U} v$ with $\mu(w)[U\rangle M'$.

Let $Y \stackrel{\text{df}}{=} U_{\omega\mapsto k+1+k'}$ and $k'' \stackrel{\text{df}}{=} |^\bullet Y| + k' + k + 1$.

By the induction hypothesis, there exists a marking $M_1 \in [M_0\rangle$ such that $M_1 \Subset_{k''} \mu(w)$. Clearly, Y is enabled at M_1, and we denote by M_2 the marking satisfying $M_1[Y\rangle M_2$. We now observe that the following hold:

- $\mu(w)(p) < \omega$ and $p \notin U_\omega{}^\bullet$ implies $M_2(p) = M'(p)$.
- $\mu(w)(p) = \omega$ implies $M_2(p) > k$.
- $p \in U_\omega{}^\bullet$ implies $M_2(p) > k$.

From the construction of $\mu(v)$ it follows that $\mu(v)(p) < \omega$ implies $\mu(v)(p) = M'(p)$. Thus M_2 satisfies the required condition for all p such that $\mu(v)(p) < \omega$ or $\mu(w)(p) = \omega$ or $p \in U_\omega{}^\bullet$ (note that $p \in U_\omega{}^\bullet$ implies $\mu(v)(p) = \omega$). Therefore, the required condition may not be satisfied only if the set $R \stackrel{\text{df}}{=} \{r \in P \mid \mu(v)(r) = \omega \wedge \mu(w)(r) < \omega \wedge r \notin U_\omega{}^\bullet\}$ is non-empty.

If $R \neq \emptyset$ one needs to increase the numbers of tokens in the places of R on the basis of paths leading to the node v in the constructed tree. To explain the idea, let us assume that $r \in R$. In such a case, by the construction, we have that there is a node u and a path $u = w_1 \xrightarrow{U_1} w_2 \ldots w_n \xrightarrow{U_n} w_{n+1} = v$ (i.e., $w_n = w$ and $U_n = U$) in the tree such that $\mu(u) \sqsubset M'$ and $\mu(u)(r) < M'(r)$.

Let $W_i = (U_i)_{\omega\mapsto 0}$ for $i = 1, \ldots, n$. Following the same line of reasoning as in [2], one can show that the step sequence consisting of $k+1$ repetitions of $W_1 \ldots W_n$ is enabled at the marking M_2 and in the resulting marking M'', place r contains more than k tokens (and the required condition has not been lost for any other place). This procedure is repeated starting from M'' for another place still violating the required condition (if any), until all the places satisfy it. $\dashv$

Probabilistic Computational Trust

Karl Krukow[1], Mogens Nielsen[2], Vladimiro Sassone[3]
[1] *Trifork*
Denmark
kkr@trifork.com
[2] *University of Aarhus*
Denmark
mn@science.au.dk
[3] *University of Southampton*
UK
vs@ecs.soton.ac.uk

Abstract

We argue briefly for the role of computational trust in ubiquitous computing, and in particular for the need of a formal foundation for computational trust. We provide two examples towards such a foundation: a formal foundation for a some probabilistic approaches from the literature, and a formal framework for comparing probabilistic trust models.

Keywords: Computational trust, ubiquitous computing

1 Introduction

This paper surveys the notion of computational trust and illustrates some initial progress towards developing its theoretical underpinning. Computational trust refers to *decision-making* in computing applications where *uncertainty* is a dominant aspect. This may manifest itself in several different ways, including unpredictability in the execution environment, as typical of e.g. mobile and sensor networks; in the emergent behaviour of dynamic agglomerates of computational entities, as e.g. in self-configuring, highly-distributed systems; and in applications where the optimal strategy is itself uncertain, as typical of adaptation and situational awareness in autonomous agent systems. The tract common to these examples is the move away from the 'Eden' of certain and abundant information to a place where the only 'truths' available are those that can be experienced directly. This is analogous to higher living organisms, like we humans, which continuously have to assess the evidence around them and decide whether or not to rely on the information they determine from it. To provide a purposely naive example in realm of computing, on the Internet it is ultimately a matter of trust whether or not I expect that following a given URL will actually take me to my bank's website. Such is in fact the origin of the term 'trust' in our context: in the absence of complete and completely reliable information, computational entities must weigh the relative importance of different factors —that is, determine their level of trust in them— in their decision-making. Indeed, 'decision-making' remains the central keyword here, and one can think of computational trust as a computer abstraction underpinning it.

To develop such an abstraction is not as immediate as it might appear at first. It is common experience to enter a room filled with strangers, sit around a table and soon establish a level of trust sufficient to function cooperatively (well, at least in most cases). This does not come as naturally to machines, and it is in fact from our point of view a superbly sophisticated collective behaviour: to achieve a similar level of adaptability, awareness and responsiveness from computing agents requires computational structures, algorithms, middleware infrastructures, models and theoretical foundations, and is exactly what the work on computational trust aspires to.

An application field central to our interests is ubiquitous computing. That is a comprehensive term which indicates information processing through a computational infrastructure embedded seamlessly and pervasively in the surroundings. It encompasses very many issues of a quite different nature, ranging from the design and deployment of low-power, self-sustaining electronic devices in pervasive wireless networks, to the investigation of innovative user interfaces, from the development of semantic models of agent mobility and distribution, to the architecture of communication networks and the corresponding programming language primitives. Computational trust is intrinsic to ubiquitous computing in that entities on the 'global' network have intrinsically unverifiable identities, origins, and past histories.

The area of computational trust has produced several applications within ubiquitous computing with truly impressive experimental performance. However, we are not yet in a position where we understand why, when, and how a particular approach is applicable, as expressed e.g. in [20]. Such questions are typically formulated in terms of underlying models. In this paper, we focus on just one type of models considered within computational trust based on well known concepts from probability theory.

Probabilistic approaches have proved useful in science to formulate and test hypotheses over quantities not exactly known, as illustrated by e.g. Thomas Bayes, who developed his eponymous method by solving the so-called problem of 'inverse probability:' given that an outcome has been observed (e.g., a red ball has been extracted from the urn), what can be inferred about the model (e.g., number of red and yellow balls in the urn)? This adapts well to our 'decision-making' problem; e.g.: given that the URL led to my bank's website, what can be said about the rest of the information on this page to inform my future decisions? We shall illustrate in this paper how such an observation led to a powerful computational idea, which has been exploited in several computational trust algorithms. Indeed, as the field grows under the thrust of suggestive analogies like the one above, and experimental successes pile up, the need arises to understand at a deeper level and distil the essence of the methods.

Our aim in this paper is primarily to illustrate that it is possible to ask and to answer formally questions on the behaviour of systems expressed in underlying formal probabilistic trust models. We focus here on behavioural notions like *correctness*. Traditionally in software development, the notion of correctness is formulated in terms of a yes/no question: does a piece of software satisfy its specification? It is our position that in the setting of computational trust, we need to develop new formal frameworks for a more general notion of correctness, which allows us e.g. (i) to express and to argue *how well* a particular system behaves under various assumptions about the environments (i.e. in which application scenarios does the system do well?), and (ii) to express and argue *how robust* a particular system is with respect to changes in the environment. And in this paper we present ideas towards new frameworks for addressing both these issues. However, these are just examples of new types of formal frameworks to be developed within computational trust. We hope this paper can serve to illustrate our results on 'formal' computational trust and as a 'call-to-arms' to tackle the open ones.

Contents of paper: We present a formal foundation for two of the well known probabilistic approaches from the literature. We then illustrate how a new approach to correctness and robustness can be applied to answer questions on the relative performance of the two approaches. We first summarise some of the arguments for trust playing a role in ubiquitous computing, giving a brief historical account on the development from trust management systems. We then focus on a few probabilistic approaches to computational trust, and we illustrate how they can be understood and explained formally in terms of standard concepts from probability theory. Finally, we sketch ideas towards a theoretically well-founded technique for comparing probabilistic systems in various different environments. The paper is mainly based on selected parts of the PhD dissertation of the first author [13]. Also, some of the results have appeared in [21, 14].

2 The Role of Trust in Ubiquitous Computing

Many researchers have argued convincingly for the relevance of trust management in distributed systems security (cf., Blaze, Feigenbaum et al. [2]). We recapture some of these arguments, and try to highlight a number of properties of ubiquitous computing which make the trust management approach even more appealing, even if the limitations of the traditional technology and models must first be overcome.

2.1 The Access Control List

The unique dynamic properties of the Internet and, more generally, those envisioned for ubiquitous computing, imply that traditional theories and mechanisms for security and resource access-control are often inappropriate as they are of too static a nature. For example, traditional access control consists of a policy specifying which subjects (user identities) may access which objects (resources), e.g., a user accessing a file in a UNIX file system. Apart from inflexibility and lack of expressive power, this approach assumes that resources are only accessed by a static set of known subjects (and that resources themselves are fixed); an assumption incompatible with open dynamic systems.

Many modern distributed systems use a combination of access control lists and user authentication, usually implemented via some public key authentication protocol, i.e., deciding a request to perform an action is done by authenticating the public key, effectively linking the key to a user identity, and then looking up in the access control list to see whether that identity is authorised to perform the action [2]. Furthermore, the security of current systems is often not verified, i.e., proofs of soundness of the security mechanism are lacking (e.g., statements of the form "if the system authorises an action requested by a public key, then the key is controlled by user U, and U is authorised to perform that action according to the access control list").

In Internet applications there is an extremely large set of entities making requests, and this set is in constant change as entities join and leave networks. Furthermore, even if we could reliably decide *who* signed a given request, the problem of deciding whether or not access should be granted is not obvious: Should all requests from unknown entities be denied?

Blaze et al. [2] present a number of reasons why the traditional approach to authorisation is inadequate:

- *Authorisation = Authentication + Access Control List.* Authentication deals with establishing identity. In traditional static environments, e.g., operating systems, the identity of an entity is well-known. In ubiquitous computing applications this is often not the case. This means that if an access control list (ACL) is to be used, some form of authentication must first be performed. In distributed systems, often public-key based authentication protocol are used, which usually relies on centralised and global certification authorities.
- *Delegation.* Since the global scale of ubiquitous computing implies that each entity's security policy must encompass billions of entities, delegation is necessary to obtain scalable solutions. Delegation implies that entities may rely on other (semi) trusted entities for deciding how to respond to requests. In traditional approaches either delegation is not supported, or it is supported in an inflexible manner where security policy is only specified at the last step of a delegation chain.
- *Expressive power and Flexibility.* The traditional ACL-based approach to authorisation has proved not to be sufficiently expressive (with respect to desired security policies) or extensible. The result has been that security policies are often hard-coded into the application code, i.e., using the general purpose programming language that the application is written in. This has a number of implications: changes in security policy often means rewriting and recompilation, and security reasoning becomes hard as security code is intertwined with application code (i.e., violating the principle of 'separation of concerns').
- *Locality.* The autonomy of ubiquitous computing entities means that different entities have different trust requirements and relationships with other entities. Hence, ubiquitous computing entities should be able to specify local security and trust policies, and security mechanisms should not enforce uniform and implicit policies or trusting relations.

2.2 The Traditional Trust Management Approach

In contrast to the 'access control list' approach to authorisation, trust management is naturally distributed, and consists of a unified and general approach to specifying security policies, credentials and trusting relationships, backed up by general (application-independent) algorithms to implement these policies. The trust management approach is based on programmable security policies that specify access-control restrictions and requirements in a domain-specific programming language, leading to increased flexibility and expressive power.

Given a request r signed by a key k, the question we really want to answer is the following: "Is the knowledge about key k such that the request r should be granted?" In principle, we do not care about *who* signed r, only whether or not sufficient information can be inferred to grant the request. The trust management approach does not need to resolve the actual identity (e.g., the human-being believed to be performing the request) but, instead, deals with the following question, known as the compliance-checking problem: "Does the set C of credentials prove that the request r complies with the local security policy σ?" [2, 3]. Let us elaborate: a request r can now be accompanied by a set C of *credentials*. Credentials are signed policy statements, e.g., of the form "public key k is authorised to perform action a." Each entity that receives requests has its own security policy σ that specifies the requirements for granting and denying requests. Policies can directly authorise requests, or they can delegate to credential issuers that the entity expects have more detailed information about the requester.

Policy is separated from mechanism: a trust management *engine* takes as input a request r, a set of credentials C and a local policy σ; it outputs an authorisation decision (this could be 'yes'/'no', but also more general statements about, say, *why* a request is denied, or *what would be further needed* to grant the request). This separation of concerns supports security reasoning needed in distributed applications, which we believe to be an important feature for the ubiquitous computing challenge.

2.3 Security in Ubiquitous Computing

The above reasoning argues that traditional authorisation mechanisms are inadequate in modern distributed systems. When one considers ubiquitous computing applications, we can add further to this list. Most of the dynamic properties of ubiquitous computing entities (e.g., mobility, autonomy, ubiquity, global connectivity, ...) affect their security requirements. For example, mobility implies that an entity might find itself in a hostile environment, disconnected from its preferred security infrastructure, e.g., certification authorities. Further, the autonomy requirement means that even in this scenario, it must be able to assign privileges to other entities, privileges that are meaningful based on the usually incomplete information that the assigning entity has about the assigned entity.

- **Active decisions**
  Trust management systems focus on deciding how to respond to *requests*. However, ubiquitous computing entities do not only need to respond meaningfully to requests, i.e., taking *passive* security decisions, but often need to *actively* and autonomously select among equivalent services provided by a number of apparently similar providers. Such decisions may also affect security: interaction often entails exposing personal data, as well as requiring resources like time, computation, battery and storage. When taking active decisions, there are usually no credentials available; hence, other information, e.g., reputation information, must be taken into account to make meaningful decisions.
- **Information vs. credentials**
  Traditional trust management systems focus on *credentials* as the main source of proving compliance of a request with a policy. However, even when no delegation chain may establish sufficient information about a requesting entity, sometimes, collaboration may still be the most beneficial action. Notions of *risk* of an interaction, and *cost/benefit* of the outcome of an interaction are relevant concepts that are not considered in traditional trust management policies. For example, histories, that is, memory of past interactions with an entity, may contain enough information to risk interaction. This entails that trusting relationships change dynamically, based on information about the history of an entity.
- **Probability**
  Incomplete information leads naturally to probabilistic decision-making. Trust management systems that focus on information could consider probabilities explicitly (as an alternative to, or, to complement the establishing of credentials in the traditional sense). Ideally, security policies would be amenable to probabilistic yet rigorous reasoning which leads to more generality and flexibility. Additionally, as factors such as cost and benefit of interactions enters the equation, a notion of *risk* emerges as a product of cost/benefit and probability.

2.4 Computational Trust

In the arguments above, we have considered only a single notion of trust in ubiquitous computing, namely the concept of 'trust management' coined by Blaze, Feigenbaum and Lacy [3]. In fact, there are many different strands of research on trust addressing the challenges of ubiquitous computing. A whole range of trust based alternatives to existing technologies have appeared, collectively referred to as *computational trust*. For comprehensive surveys on computational trust, the reader is referred to e.g. [8, 12, 20, 19, 14].

In this paper we deal with just one particular approach within computational trust based on probability theory. Before focusing on that, we would like to comment briefly on some of the other approaches aiming specifically for a computational formalisation of the *human notion of trust*, i.e., trust as a sociological, psychological and philosophical concept (for a good survey of these, see [1]). However, the human concept of trust is elusive and its many facets make it hard to define formally [17, 5]. We believe that to live up to the ubiquitous computing challenge, it is necessary that the two concepts be merged in a 'unified' theory of trust which combines the strengths of both notions. To be more precise, our ideal would be to combine the *rigour* of traditional trust management with the *dynamics* and *flexibility* of the human notion. Let us elaborate: traditional trust management deals with credentials, policies, requests and the compliance-checking problem. Rigourous security *reasoning* is possible: the intended meaning of a trust management engine is formally specified, correctness proofs are feasible, and many security questions are effectively decidable [16]. In contrast, we have yet to see a system based on the human notion of trust which, with realistic assumptions, guarantees any sort of rigourous security property. On the other hand, such systems are capable of making *intuitively* reasonable decisions based on information such as evidence of past behaviour, reputation, recommendations and probabilistic models. A combination of these two approaches would lead to powerful frameworks for decision-making which incorporates more general information than credentials, yet which remains tractable to rigorous reasoning.

3 Probabilistic Computational Trust

In the survey [14], computational trust is characterised as being either *credential-based* (following the ideas of Blaze et al. above) or *experience-based*. The latter term refers to approaches to trust, where an entity's trust in another is based on past behaviour, covering many so-called *reputation-based* trust management systems, which are often used in peer-to-peer (P2P) and eCommerce applications.

3.1 Experience-based Trust

Consider a set $\mathcal{P}$ of principal identities. From time to time, principals will interact in a pair-wise manner, and such interactions result in each principal observing a set of time-stamped events. In the following we make a number of simplifications, but stay general enough to capture most of the principles of existing experience-based systems: we assume that each time p interacts with another principal, say $q \in \mathcal{P}$, the interaction generates only a single event e, drawn from some set E of events (left unspecified here).

Let $(T, \leq)$ be a totally ordered set of time-stamps, e.g., $T = \{0, 1, \ldots\}$ for discrete time. Principal p records its interactions with other principals so that at each point in time, $t_0 \in T$, there is a set $Hist^p(t_0)$ consisting of triples (q, t, e) where $q \in \mathcal{P}$, $e \in E$; and $t \in T$ satisfies $t \leq t_0$. To be clear, a triple $(q, t, e) \in Hist^p(t_0)$ represents that: "In an interaction between p and q, principal p has observed event e at time t." We write $Hist^p_q(t)$ for the q-projection, i.e., the set of pairs (t', e) such that $(q, t', e) \in Hist^p(t)$.

At any point in time, $t \in T$, the sets $Hist^p(t)$, for $p \in \mathcal{P}$, constitute the *basic* or *direct* data of an experience-based system at time t. When p needs to make a decision at time t, e.g., about a principal q, it does so based on information from the direct data of the system at time t. Usually, such information is incomplete: while p typically knows $Hist^p(t)$, the sets $Hist^r(t)$ for $r \neq p$ may not be known exactly. This may be due to several reasons, e.g.: p may only have $Hist^r(t')$ for some $t' < t$; when asked about $Hist^r(t')$, r may lie; principal p may not be able to obtain any information about $Hist^r(t')$; principal p may only see some abstracted version of r's direct data; and any combinations of the above.

Most experience-based systems work on some abstracted version of the direct data, denoted $AbsHist^p(t)$. Some systems are centralised, so that (abstract versions of) the direct data are stored on a global server, whereas other system are distributed. In the following we focus on models, not architectures (e.g., centralised vs distributed). Given our general model, an experience-based system is designed by *(i)* choosing if and how to abstract (or aggregate) the sets $Hist^p(t)$ to obtain the 'abstract' sets $AbsHist^p(t)$; *(ii)* choosing if and how each principal p will obtain information about $Hist^q(t)$ for $q \neq p$; *(iii)* optionally choosing how principals combine personal data with the data of others; and *(iv)* designing an architecture and algorithms (possibly distributed) to implement the system. The optional step *(iii)* often works in the following way: principal p computes for each other principal, say q, a 'score' or 'rating', $T_{pq} \in D$, for some set D of possible scores. The score T_{pq} is usually computed from some of the the abstracted versions of the direct data, i.e. $(AbsHist^r_q(t) \mid r \in I)$ for some $I \subseteq \mathscr{P}$, and represents q's trustworthiness (or reputation), seen from the point-of-view of p. Some systems have a uniform mechanism where $T_{pq} = T_{p'q}$ for all $p, p' \in \mathscr{P}$, i.e., q has a unique 'global' score.

A common example of an abstraction is the following. At time t_0, principal p is interested in information about principal q. Each record (q, t, e) is evaluated as either 'positive' or 'negative', and time is ignored; hence, $Hist^p_q(t_0)$ is abstracted to a pair consisting of the number of 'positive' interactions and the number of 'negative' interactions. Principals then obtain information about $AbsHist^r_q(t)$ by asking a central repository. Sets of records $(AbsHist^r_q(t) \mid r \in I \subseteq \mathscr{P})$ are combined into a single pair by adding-up the total number of 'negative' interactions, and similarly adding-up total number of 'positive' interactions. This example system is much like the eBay system.

Probabilistic computational trust refers to a particular kind of an experience-based approach, which assumes a probabilistic model, say λ, for the behaviour of principals. The goal is then to predict the behaviour of principals in future interactions, given the model λ and their behaviour in past interactions. The abstractions, i.e., $AbsHist(t)$, are then chosen to be as efficient as possible while preserving as much information as is relevant with respect to the model. For example, λ may specify that each principal (intrinsically) is either 'good' or 'bad', and that interaction with 'good' principals always results in event e, whereas interaction with 'bad' principals always results in event f. In this model, one only needs to interact with a principal once to know if he is a 'good' type or 'bad' type. Hence, the sets $AbsHist^p_q(t)$ need only have three values to preserve sufficient information: 'good', 'bad' or 'unknown.'

3.2 Two Probabilistic Models

Despotovic et al. [6, 7] propose a probabilistic system and an estimation algorithm based on *maximum likelihood*. It is assumed that peers interact with each other in a binary way: in each interaction they can either be 'honest' or 'cheat.' Furthermore, peers can report to other peers on past behaviour (and they are allowed to lie in their reports).

The probabilistic model of Despotovic et al., λ_{D}, assumes that each principal $j \in \mathcal{P}$ is 'probabilistic' in the sense that there is a fixed probability $\theta_j \in [0,1]$ of peer j acting honestly in any interaction. Note, this assumes that j is always honest with probability θ_j, independently of any other information we might have (e.g., the time, the past, etc.). The parameters, θ_j, are unknown and the goal is to estimate them. Furthermore, each principal $k \in \mathcal{P}$ can report on its past interactions with j; it is assumed that k's report is also probabilistic so that the probability of observing a report $y_k \in \{0,1\}$ ('0' means cheated, '1' means honest) from principal k is given by

$$P(Y_k = y_k \mid \theta_j, l_k) = \begin{cases} l_k(1-\theta_j)+(1-l_k)\theta_j & \text{if } y_k = 1; \\ l_k\theta_j + (1-l_k)(1-\theta_j) & \text{if } y_k = 0 \end{cases}$$

where l_k (like θ_j) are fixed parameters specifying the probability of k submitting a false report. Hence for each principal j, there are two parameters that probabilistically decides its behaviour: θ_j and l_j.

Let us write $AbsHist_j^{\times}(t)$ for the collection of information that a particular principal has about j. In the system, this collection consists of a number of reports $((y_1, p_1), (y_2, p_2), \ldots, (y_m, p_m))$ where $y_i \in \{0,1\}$, $p_i \in \mathcal{P}$, and (y_i, p_i) means that principal p_i has filed report y_i (we do not consider here how reports are obtained). Hence, time is abstracted away and events are 'rated' in a binary fashion.

Now given $AbsHist_j^{\times}(t) = \mathbf{Y} = ((y_1, p_1), (y_2, p_2), \ldots, (y_m, p_m))$ of independent reports, the so-called likelihood function is:

$$L(\theta_j, l) = P(\mathbf{Y} \mid \theta_j, l, \lambda_{\mathrm{D}}) = \prod_{i=1}^{m} P(Y_i = y_i \mid \theta_j, l_{p_i})$$

(note this expression depends also on l, which is not clear from the authors' presentation [6]). Given current estimates for l and the data $\mathbf{Y}$ the goal is to estimate the behaviour of principal j, i.e., to estimate θ_j.

The system uses a maximum likelihood procedure which seeks to find a θ_j which maximises the likelihood expression. In the computation of likelihood function, estimates for the l_k are based on past interactions, but it is unspecified exactly how these are computed. The authors also present an approach based on normal distributions instead of the fixed θ_j's. Similarly, the maximum likelihood techniques are used to estimate the parameters of the normal distribution.

Jøsang et al. [11] and Mui et al. [18] were among to first to (independently) develop reputation systems based on a Bayesian probabilistic approach with *beta priors*. In the following we recall the beta distribution and explain the underlying theoretical model for the beta-based reputation systems.

The beta family $Beta(\cdot, \times)$ is a parameterised collection of continuous probability density functions (pdfs) defined on the interval $[0, 1]$. There are two parameters $\alpha > 0$ and $\beta > 0$ that select a specific beta distribution from the family. The pdf $Beta(\alpha, \beta)$ is given by

$$f(\theta \mid \alpha, \beta) = \frac{1}{\mathbf{B}(\alpha, \beta)} \theta^{\alpha-1}(1-\theta)^{\beta-1} = \frac{\theta^{\alpha-1}(1-\theta)^{\beta-1}}{\int_0^1 \mathrm{d}t\, t^{\alpha-1}(1-t)^{\beta-1}}$$

where $\mathbf{B}$ is the beta function, and $\mathbf{B}(\alpha, \beta)^{-1}$ is a normalising constant. The expected value and variance are given by

$$\mathbf{E}_{f(\theta|\alpha,\beta)}(\theta) = \frac{\alpha}{\alpha+\beta}, \qquad \sigma^2_{f(\theta|\alpha,\beta)}(\theta) = \frac{\alpha\beta}{(\alpha+\beta)^2(\alpha+\beta+1)}$$

The beta distributions provide a so-called *family of conjugate prior distributions* for the family of distributions for Bernoulli trials. To explain the notion of conjugate priors, consider the general problem of estimating a parameter θ given some data x and background information I. Let H_θ be some hypothesis about parameter θ. The Bayesian approach (see the excellent book of Jaynes [9]), is to compute the posterior $P(H_\theta \mid xI)$ (i.e., the probability after seeing the data) from the prior $P(H_\theta \mid I)$ (the *a priori* probability, given only information I) and the likelihood function $H_\theta \mapsto P(x \mid H_\theta I)$, using Bayes' Theorem:

$$P(H_\theta \mid xI) = P(H_\theta \mid I)\frac{P(x \mid H_\theta I)}{P(x \mid I)}$$

Different priors $P(H_\theta \mid I)$ may make this probability more or less difficult to calculate, but certain choices of the prior lead to the the posterior $P(H_\theta \mid xI)$ having the same algebraic form as the prior. Now, a family of conjugate prior distributions for the family of distributions $H_\theta \mapsto P(x \mid H_\theta I)$ is a collection of distributions such that when the prior $P(H_\theta \mid I)$ belongs to the family, the posterior $P(H_\theta \mid xI)$ is also in that family (one might say that the family is Bayes-closed, i.e., is closed under the application of Bayes' Theorem).

3.3 Probabilistic Models and Trust

The reader may wonder what this has to do with trust and reputation. In the following we give our personal explanation of the Bayesian beta-based approach in reputation systems. The explanation is not explicitly presented in such detail in the papers describing beta-systems [11, 18, 4, 22], and, hence, the authors may have different perspectives.

Consider again sequences of independent experiments with binary outcomes, each yielding one of the outcomes with some fixed probability (i.e., Bernoulli trials). In systems where principal-interactions consists of binary outcomes (or where interactions are rated on a binary scale, e.g., 'cooperate' or 'defect'; 'success' or 'failure'), one can model repeated interaction (or repeated ratings) as Bernoulli trials. Let us be more precise: let $p, q \in \mathcal{P}$ be principals, and assume that p and q have interacted n times; that in each interaction q takes an action; and that the whole interaction is given a binary rating by p (which depends only on q's action). Let $X_i^{pq} \in \{0,1\}$, for $i = 1,2,\ldots,n$, be p's rating (i.e., subjective evaluation) of the ith interaction with q. Let us *assume* that principal q's behaviour is so that there is a fixed parameter such that at each interaction we have, *independently of anything we know about other interactions*, the probability θ for a 'success' and therefore probability $1-\theta$ for 'failure.' This gives us a probabilistic model, and let us call it the *beta model*. Note, this is like the model λ_{D} of Despotovic et al., except for the parameters l_k for $k \in \mathcal{P}$. Let $\lambda_{\mathbf{B}}$ denote a formal proposition representing the beta model, i.e., the assumptions about the behaviour of q; also, let $\theta \in [0,1]$ be the parameter determining success in the ith trial. Finally, let $\mathbf{X}$ be the conjunction of statements Z_i of the form

$$Z_i \equiv (X_i^{pq} = 0) \text{ or } Z_i \equiv (X_i^{pq} = 1),$$

so $\mathbf{X} = \wedge_{i=1}^{n} Z_i$, and let there be f statements of the first form and s statements of the second form (there is one statement for each i, so $s + f = n$). Then, by definition of our model $\lambda_{\mathbf{B}}$, we have the following likelihood.

$$P(\mathbf{X} \mid \theta\lambda_{\mathbf{B}}) = \prod_{i=1}^{n} P(Z_i \mid \theta\lambda_{\mathbf{B}}) = \theta^s (1-\theta)^f$$

Hence, we can obtain the posterior pdf as

$$\begin{aligned} g(\theta \mid \mathbf{X}\lambda_{\mathbf{B}}) &= g(\theta \mid \lambda_{\mathbf{B}}) \frac{P(\mathbf{X} \mid \theta\lambda_{\mathbf{B}})}{P(\mathbf{X} \mid \lambda_{\mathbf{B}})} \\ &= g(\theta \mid \lambda_{\mathbf{B}}) \frac{\theta^s(1-\theta)^f}{\int_0^1 \mathrm{d}\theta\, P(\mathbf{X} \mid \lambda_{\mathbf{B}}\theta) g(\theta \mid \lambda_{\mathbf{B}})} \\ &= g(\theta \mid \lambda_{\mathbf{B}}) \frac{\theta^s(1-\theta)^f}{\int_0^1 \mathrm{d}\theta\, \theta^s(1-\theta)^f g(\theta \mid \lambda_{\mathbf{B}})} \end{aligned}$$

(where $g(\theta \mid \lambda_{\mathbf{B}})$ is the prior pdf for θ —cf. Jaynes [9]). If we postulate the prior pdf $g(\theta \mid \lambda_{\mathbf{B}})$ to be $Beta(\theta \mid \alpha_0, \beta_0)$ (which in particular is the uniform distribution when $\alpha_0 = \beta_0 = 1$), then we can compute the posterior:

$$g(\theta \mid \mathbf{X}\lambda_{\mathbf{B}}) = g(\theta \mid \lambda_{\mathbf{B}}) \frac{\theta^s(1-\theta)^f}{\int_0^1 \mathrm{d}\theta\, \theta^s(1-\theta)^f g(\theta \mid \lambda_{\mathbf{B}})}$$

Since the normalizing constant in $g(\theta \mid \lambda_{\mathbf{B}})$ cancels out, we obtain

$$\begin{aligned} g(\theta \mid \mathbf{X}\lambda_{\mathbf{B}}) &= \frac{\theta^{\alpha_0+s-1}(1-\theta)^{\beta_0+f-1}}{\int_0^1 \mathrm{d}\theta\, \theta^{s+\alpha_0-1}(1-\theta)^{f+\beta_0-1}} \\ &= \frac{1}{\mathbf{B}(\alpha_0+s, \beta_0+f)} \theta^{\alpha_0+s-1}(1-\theta)^{\beta_0+f-1} \end{aligned}$$

which means that $g(\theta \mid \mathbf{X}\lambda_{\mathbf{B}})$ is $Beta(\theta \mid \alpha_0+s, \beta_0+f)$.

Now let $Z_{n+1} \equiv (X_{n+1}^{pq} = 1)$, i.e., the statement that the $(n+1)$st interaction is rated as a 'success', then $P(Z_{n+1} \mid \mathbf{X}\lambda_{\mathbf{B}})$ is a predictive probability: given no direct knowledge of θ, but only past evidence ($\mathbf{X}$) and the model ($\lambda_{\mathbf{B}}$), then $P(Z_{n+1} \mid \mathbf{X}\lambda_{\mathbf{B}})$ is the probability that the next interaction will be a 'good' one. We can compute it as follows.

$$\begin{aligned} P(Z_{n+1} \mid \mathbf{X}\lambda_{\mathbf{B}}) &= \int_0^1 \mathrm{d}\theta\, P(Z_{n+1} \mid \mathbf{X}\lambda_{\mathbf{B}}\theta) g(\theta \mid \mathbf{X}\lambda_{\mathbf{B}}) \\ &= \int_0^1 \mathrm{d}\theta\, \theta g(\theta \mid \mathbf{X}\lambda_{\mathbf{B}}) \\ &= \mathbf{E}_{g(\theta \mid \mathbf{X}\lambda_{\mathbf{B}})}(\theta) \end{aligned}$$

Now recall the expectation of beta distributions; then

$$P(Z_{n+1} \mid \mathbf{X}\lambda_{\mathbf{B}}) = \mathbf{E}_{g(\theta|\mathbf{X}\lambda_{\mathbf{B}})}(\theta) = \frac{\alpha_0 + s}{\alpha_0 + s + \beta_0 + f} \tag{1}$$

since $g(\theta \mid \mathbf{X}\lambda_{\mathbf{B}})$ is $Beta(\theta \mid \alpha_0 + s, \beta_0 + f)$.

To summarise, given the assumptions of the beta model, one can compute the probability of a success in the next interaction as the expectation of the beta pdf $g(\theta \mid \mathbf{X}\lambda_{\mathbf{B}})$ which results via Bayesian updating given the past history $\mathbf{X}$. Hence, the beta based systems (that deploy the technique we have described here) are mathematically well-founded on probability theory.

Jøsang et al. [11], Mui et al. [18], Buchegger et al. [4], Jennings et al. [22] all present systems based on the beta model. Buchegger et al. and Jennings et al. also propose mechanisms for dealing with lying reputation sources. Technically, all the systems work by maintaining the two parameters (α, β) of the current pdf $g(\theta \mid \mathbf{X}\lambda_{\mathbf{B}})$. However, the systems (except for [18] and [22]) deviate from the model $\lambda_{\mathbf{B}}$ in the following sense: the parameters (α, β) are adjusted as time passes; for instance, Jøsang uses exponential decay where α and β are multiplied by a constant $0 < u < 1$ each time parameters are updated (or a fixed time limit is exceeded). The intuition is that somehow information about more recent interactions should we considered more important than information about older interactions.

Several models are based on a notion of 'belief theory' which is related to probability theory: Yu et al. developed a distributed reputation system [23], and Jøsang developed the subjective logic of opinions [10]. Indeed, the subjective logic is closely linked to the probabilistic beta model [10].

4 Towards Formal Computational Trust

Sabater and Sierra argue in [20] that the field of computational trust is lacking a more formal foundation, including a way of comparing the qualities of the many proposed trust-based systems. Sabater and Sierra propose that our field develop "(...) test-beds and frameworks to evaluate and compare the models under a set of representative and common conditions" [20].

We fully agree with these views. Also, as mentioned earlier, we believe that computational trust needs new notions of the correctness of systems, as well as frameworks for talking about the robustness of systems relative to their environments.

As an example, consider the issue of correctness of the maximum likelihood algorithm of Despotovic et al. introduced above. The traditional notion of correctness of algorithms would require a proof that the algorithm is correct with respect to its specification, i.e., that it indeed computes the maximum likelihood as specified above. However, in the setting of ubiquitous computing, we are more interested in the question of how well the algorithm approximates θ in the chosen probabilistic model. In the following, we introduce a particular framework for formalising such questions: a generic measure of how well an algorithm approximates the behaviour of entities in a given probabilistic model. And this framework is then applied in order to compare the performance of the two probabilistic algorithms from above: the maximum likelihood algorithm of Despotovic et al. and the the beta-based algorithm of Mui et al.

Our generic measure is intended to 'score' specific probabilistic trust-based systems in a particular environment (i.e., "a set of representative and common conditions"). The score, which is based on the so-called Kullback-Leibler divergence, is a measure of how well an algorithm approximates the 'true' probabilistic behavior of principals.

Consider a probabilistic model of principal behavior, say λ. We consider only the behavior of a single fixed principal p, and we consider only algorithms that attempt to solve the following problem: suppose we are given an interaction history $\mathbf{X} = [(x_1, t_1), (x_2, t_2), \ldots, (x_n, t_n)]$ obtained by interacting n times with principal p, observing outcome x_i at time t_i. Suppose also that there are m possible outcomes $(y_1, \ldots, y_m)$ for each interaction. The goal of a probabilistic trust-based algorithm, say $\mathscr{A}$, is to approximate a distribution on the outcomes $(y_1, \ldots, y_m)$ given this history $\mathbf{X}$. That is, $\mathscr{A}$ satisfies:

$$\mathscr{A}(y_i \mid \mathbf{X}) \in [0,1] \text{ (for all } i\text{)}, \qquad \sum_{i=1}^{m} \mathscr{A}(y_i \mid \mathbf{X}) = 1.$$

We assume that the probabilistic model, λ, defines the following probabilities: $P(y_i \mid \mathbf{X}\lambda)$, i.e., the probability of "y_i in the next interaction given a past history of $\mathbf{X}$" and $P(\mathbf{X} \mid \lambda)$, i.e., the "*a priori* probability of observing sequence $\mathbf{X}$ in the model."

Therefore $(P(y_i \mid \mathbf{X}\lambda) \mid i = 1, 2, \ldots, m)$ defines the 'true' distribution on outcomes for the next interaction (according to the model); in contrast, $(\mathscr{A}(y_i \mid \mathbf{X}) \mid i = 1, 2, \ldots, m)$ attempts to approximate this distribution. The Kullback-Leibler divergence [15], which is closely related to Shannon entropy, is a measure of the distance from a true distribution to an approximation of that distribution. The Kullback-Leibler divergence from distribution $\hat{p} = (p_1, p_2, \ldots, p_m)$ to distribution $\hat{q} = (q_1, q_2, \ldots, q_m)$ on a finite set of m outcomes, is given by

$$\mathrm{D_{KL}}(\hat{p} \parallel \hat{q}) = \sum_{i=1}^{m} p_i \log_2 \left(\frac{p_i}{q_i} \right)$$

(any log-base could be used). The Kullback-Leibler divergence is almost a distance (in the mathematical sense), but the symmetry property fails. That is $\mathrm{D_{KL}}$ satisfies $\mathrm{D_{KL}}(\hat{p} \parallel \hat{q}) \geq 0$ and $\mathrm{D_{KL}}(\hat{p} \parallel \hat{q}) = 0$ only if $\hat{p} = \hat{q}$. The asymmetry comes from considering one distribution as 'true' and the other as approximating.

For each n let $\mathbf{O}^n$ denote the set of interaction histories of length n. Let us define, for each n, the nth *expected Kullback-Leibler divergence from* λ *to* $\mathscr{A}$:

$$\mathrm{D}^n_{\mathrm{KL}}(\lambda \parallel \mathscr{A}) \overset{(\mathrm{def})}{=} \sum_{\mathbf{X} \in \mathbf{O}^n} P(\mathbf{X} \mid \lambda) \mathrm{D_{KL}}(P(\cdot \mid \mathbf{X}\lambda) \parallel \mathscr{A}(\cdot \mid \mathbf{X})),$$

that is,

$$\mathrm{D}^n_{\mathrm{KL}}(\lambda \parallel \mathscr{A}) = \sum_{\mathbf{X} \in \mathbf{O}^n} P(\mathbf{X} \mid \lambda) \left(\sum_{i=1}^{m} P(y_i \mid \mathbf{X}\lambda) \log_2 \left(\frac{P(y_i \mid \mathbf{X}\lambda)}{\mathscr{A}(y_i \mid \mathbf{X})} \right) \right).$$

Note that, for each input sequence $\mathbf{X} \in \mathbf{O}^n$ to the algorithm, we evaluate its performance as $\mathrm{D_{KL}}(P(\cdot \mid \mathbf{X}\lambda) \parallel \mathscr{A}(\cdot \mid \mathbf{X}))$; however, we accept that some algorithms may perform poorly on very unlikely training sequences, $\mathbf{X}$. Hence, we weigh the penalty on input $\mathbf{X}$, i.e., $\mathrm{D_{KL}}(P(\cdot \mid \mathbf{X}\lambda) \parallel \mathscr{A}(\cdot \mid \mathbf{X}))$, with the intrinsic probability of sequence $\mathbf{X}$; that is, we compute the *expected* Kullback-Leibler divergence.

Due to the relation to Shannon's Information Theory, one can interpret $\mathrm{D}^n_{\mathrm{KL}}(\lambda \parallel \mathscr{A})$ quantitatively as the expected number of bits of information one would gain if one would know the true distribution instead of $\mathscr{A}$'s approximation on n-length training sequences.

4.1 An example

As an example of the applicability of our measure, we compare the beta-based algorithm of Mui et al. [18] with the maximum-likelihood algorithm of Despotovic et al. [6] introduced above. We can compare these because they both deploy the same fundamental assumptions:

> *Assume* that the behavior of each principal is so that there is a fixed parameter such that at each interaction we have,

> *independently of anything we know about other interactions*, the probability θ for a 'success' and therefore probability $1-\theta$ for 'failure.'

This gives us the *beta model*, $\lambda_{\mathbf{B}}$. Let s stand for 'success' and f stand for 'failure,' and let $\mathbf{X} \in \{s, f\}^n$ for some $n > 0$.

We have the following likelihood for any $\mathbf{X} \in \{s, f\}^n$:

$$P(\mathbf{X} \mid \lambda_{\mathbf{B}}\theta) = \theta^{\mathrm{N}_s(\mathbf{X})}(1-\theta)^{\mathrm{N}_f(\mathbf{X})},$$

where $\mathrm{N}_x(\mathbf{X})$ denotes the number of x occurrences in $\mathbf{X}$.

Let $\mathcal{M}$ denote the algorithm of Mui et al., and let $\mathcal{D}$ denote the algorithm of Despotovic et al. Then,

$$\mathcal{M}(s \mid \mathbf{X}) = \frac{\mathrm{N}_s(\mathbf{X})+1}{n+2} \quad \text{and} \quad \mathcal{M}(f \mid \mathbf{X}) = \frac{\mathrm{N}_f(\mathbf{X})+1}{n+2},$$

and it is easy to show that:

$$\mathcal{D}(s \mid \mathbf{X}) = \frac{\mathrm{N}_s(\mathbf{X})}{n} \quad \text{and} \quad \mathcal{D}(f \mid \mathbf{X}) = \frac{\mathrm{N}_f(\mathbf{X})}{n}.$$

For each choice of $\theta \in [0,1]$, and each choice of training-sequence length, we can compare the two algorithms by computing and comparing $\mathrm{D}^n_{\mathrm{KL}}(\lambda_{\mathbf{B}}\theta \parallel \mathcal{M})$ and $\mathrm{D}^n_{\mathrm{KL}}(\lambda_{\mathbf{B}}\theta \parallel \mathcal{D})$. For example:

Theorem 1 *If* $\theta = 0$ *or* $\theta = 1$ *then for all* n

$$\mathrm{D}^n_{\mathrm{KL}}(\lambda_{\mathbf{B}}\theta \parallel \mathcal{D}) \quad = \quad 0 \quad < \quad \mathrm{D}^n_{\mathrm{KL}}(\lambda_{\mathbf{B}}\theta \parallel \mathcal{M}),$$

and if $0 < \theta < 1$ *then for all* n

$$\mathrm{D}^n_{\mathrm{KL}}(\lambda_{\mathbf{B}}\theta \parallel \mathcal{M}) < \mathrm{D}^n_{\mathrm{KL}}(\lambda_{\mathbf{B}}\theta \parallel \mathcal{D}) = \infty.$$

Proof Assume that $\theta = 0$, and let $n > 0$. The only sequence of length n with non-zero probability is f^n, and we have $\mathcal{D}(f \mid f^n) = 1$; in contrast, $\mathcal{M}(f \mid f^n) = (n+1)/(n+2)$, and $\mathcal{M}(s \mid f^n) = 1/(n+2)$. Since $P(s \mid f^n \lambda_{\mathbf{B}}\theta) = \theta = 0 = \mathcal{D}(s \mid f^n)$ and $P(f \mid f^n \lambda_{\mathbf{B}}\theta) = 1-\theta = 1 = \mathcal{D}(f \mid f^n)$, we have

$$\mathrm{D}^n_{\mathrm{KL}}(\lambda_{\mathbf{B}}\theta \parallel \mathcal{D}) = 0.$$

Since $\mathrm{D}^n_{\mathrm{KL}}(\lambda_{\mathbf{B}}\theta \parallel \mathcal{M}) > 0$ we are done. The argument for $\theta = 1$ is similar. For the case $0 < \theta < 1$ the result follows from similar arguments, and the fact that $\mathcal{D}$ assigns probability 0 to s on input f^k (for all $k \geq 1$), which results in $\mathrm{D}^n_{\mathrm{KL}}(\lambda \parallel \mathcal{D}) = \infty$. ⊣

The degenerate property of algorithm $\mathscr{D}$ of Despotovic et al. stated in the Theorem above when $0 < \theta < 1$ follows from a general property of the Kullback-Leibler measure: given two distribution $\hat{p} = (p_1, \ldots, p_n)$ and $\hat{q} = (q_1, \ldots, q_n)$, if one of the 'real' probabilities, p_i is non-zero and the corresponding 'approximating' probability q_i is zero, then we have $\mathrm{D}_{\mathrm{KL}}(\hat{p} \parallel \hat{q}) = \infty$. To obtain stronger and more informative results, we shall consider a continuum of algorithms, denoted $\mathscr{A}_\epsilon$ for a real number $\epsilon > 0$, defined as

$$\mathscr{A}_\epsilon(s \mid \mathbf{X}) = \frac{\mathrm{N}_s(\mathbf{X}) + \epsilon}{n + 2\epsilon} \quad \text{and} \quad \mathscr{A}_\epsilon(f \mid \mathbf{X}) = \frac{\mathrm{N}_f(\mathbf{X}) + \epsilon}{n + 2\epsilon}.$$

Note that one can think of $\mathscr{A}_\epsilon$ as approximating $\mathscr{D}$ (which is $\mathscr{A}_0$) for small values of ϵ.

As an illustration of the type of comparison results one may obtain, we are going to show just two results formally addressing the question of how best to choose between this continuum of algorithms. The first theorem states that unless behaviour is completely random, for any fixed θ there exists one particular algorithm, which out-performes all other $\mathscr{A}_\epsilon$ algorithms.

Theorem 2 *For all* $\theta \in [0,1], \theta \neq \frac{1}{2}$, $\mathrm{D}^n_{\mathrm{KL}}(\lambda_{\mathbf{B}}\theta \parallel \mathscr{A}_\epsilon)$ *assumes a minimal value compared to all* $\mathscr{A}_\epsilon$ *algorithms* ($\epsilon > 0$) *for*

$$\epsilon = \frac{2\theta(1-\theta)}{(2\theta-1)^2}$$

Proof Follows form standard analysis of $\mathrm{D}^n_{\mathrm{KL}}(\lambda_{\mathbf{B}}\theta \parallel \mathscr{A}_\epsilon)$ as a function of ϵ. ⊣

Correspondingly, the following theorem states that for a fixed $\epsilon_0 > 0$, algorithm $\mathscr{A}_{\epsilon_0}$ is the optimal choice amongst all $\mathscr{A}_\epsilon$ algorithms for precisely two values of θ.

Theorem 3 *For any fixed real number* ϵ_0, *for all* $\epsilon > 0$,

$$\mathrm{D}^n_{\mathrm{KL}}(\lambda_{\mathbf{B}}\theta \parallel \mathscr{A}_{\epsilon_0}) \leq \mathrm{D}^n_{\mathrm{KL}}(\lambda_{\mathbf{B}}\theta \parallel \mathscr{A}_\epsilon) \quad \textit{iff} \quad \theta = \frac{1}{2} \pm \frac{1}{2\sqrt{2\epsilon_0 + 1}}$$

Proof Follows from the analysis in the proof of Theorem 2. ⊣

As an illustrative corollary of this theorem, we see that the algorithm $\mathscr{M}$ of Mui et al. is the optimal choice for precisely $\theta = \frac{1}{2} \pm \frac{1}{\sqrt{12}}$.

In fact, it is not so much the concrete comparison of algorithms $\mathcal{M}$ and $\mathcal{D}$ that interests us; rather, our message is that using probabilistic models *enables* such theoretical comparisons. We have focused here purely on the application of our framework in comparing two specific algorithms from the literature of computational trust, but it is clear to us that the framework in general opens up the *possibility* of formalising many other relevant questions. As an example, the question of robustness of say algorithm $\mathcal{M}$ can be asked in terms of how $D^n_{KL}(\lambda_B\theta \parallel \mathcal{M})$ varies with θ, and investigate by a standard analysis techniques on $D^n_{KL}(\lambda_B\theta \parallel \mathcal{M})$ as a function of θ.

5 Concluding Remarks

In this paper we retraced some of the fundamental steps in the development of the concept of computational trust. We illustrated how the analogy with the common notion of trust leads to a new powerful computational paradigm for decision-making in the absence of complete and reliable information, and explained why this has a potentially major impact in several important application fields, and in particular on the emerging ubiquitous computing infrastructure. We focussed on the most prominent probabilistic approaches to computational trust, and showed how they can be expressed formally in terms of basic concepts from probability theory. Indeed, the experimental evidence in favour of such techniques is compelling; we argued however that as a community we do not yet possess a sufficient understanding of how, when and why a particular approach is superior, or at least fit for purpose, and we presented a formal framework in which such questions can be asked. We exercised our framework on comparing probabilistic trust algorithms 'quality-wise,' and obtained some new insights; yet, our main message is that a formal framework empowers us to formulate and investigate all sort of new questions, in particular those concerned with quantitative and relative measures of correctness and robustness for probabilistic computational trust systems.

References

[1] A. Abdul-Rahman. *A Framework for Decentralised Trust Reasoning*. PhD thesis, University of London, Department of Computer Science, University College London, England, 2005.

[2] M. Blaze, J. Feigenbaum, J. Ioannidis, and A. D. Keromytis. The role of trust management in distributed systems security. In *Secure Internet Programming: Security Issues for Mobile and Distributed Objects*, volume 1603 of *Lecture Notes in Computer Science*, pages 185–210. Springer, 1999.

[3] M. Blaze, J. Feigenbaum, and J. Lacy. Decentralized trust management. In *Proceedings from the 17th Symposium on Security and Privacy*, pages 164–173. IEEE Computer Society Press, 1996.

[4] S. Buchegger and J.-Y. Le Boudec. A Robust Reputation System for Peer-to-Peer and Mobile Ad-hoc Networks. In *P2PEcon 2004*, 2004.

[5] V. Cahill, E. Gray, J.-M. Seigneur, C. D. Jensen, Y. Chen, B. Shand, N. Dimmock, A. Twigg, J. Bacon, C., English, W. Wagealla, S. Terzis, P. Nixon, G. M. Serugendo, C. Bryce, M. Carbone, K., Krukow, and M. Nielsen, Using trust for secure collaboration in uncertain environments. *IEEE Pervasive Computing*, 2(3):52–61, 2003.

[6] Z. Despotovic and K. Aberer. A probabilistic approach to predict peers' performance in P2P networks. In *Proceedings from the Eighth International Workshop on Cooperative Information Agents (CIA 2004)*, volume 3191 of *Lecture Notes in Computer Science*, pages 62–76. Springer, 2004.

[7] Z. Despotovic and K. Aberer. P2P reputation management: Probabilistic estimation vs. social networks. *Computer Networks*, 60(4):485–500, 2006.

[8] T. Grandison and M. Sloman. A survey of trust in internet applications. *IEEE Communications Surveys & Tutorials*, 3(4), 2000.

[9] E. T. Jaynes. *Probability Theory: The Logic of Science.* Cambridge University Press, 2003.

[10] A. Jøsang. A logic for uncertain probabilities. *International Journal of Uncertainty, Fuzziness and Knowledge-Based Systems*, 9(3):279–311, 2001.

[11] A. Jøsang and R. Ismail. The beta reputation system. In *Proceedings from the 15th Bled Conference on Electronic Commerce*, 2002.

[12] A. Jøsang, R. Ismail, and C. Boyd. A survey of trust and reputation systems for online service provision. In *Decision Support Systems* **43**(2), 618–644. Elsevier Science, 2006.

[13] K. Krukow. *Towards a Theory of Trust for the Global Ubiquitous Computer.* PhD thesis, University of Aarhus, Denmark, 2006. Available online: `http://www.brics.dk/~krukow`.

[14] K. Krukow, M. Nielsen, and V. Sassone. Trust models in ubiquitous computing. Philosophical Transactions of the Royal Society A, 366(1881), 3781-93, 2008.

[15] S. Kullback and R. A. Leibler. On information and sufficiency. *Annals of Mathematical Statistics*, 22(1):79–86, March 1951.

[16] N. Li, J. C. Mitchell, and W. H. Winsborough. Beyond proof-of-compliance: Security analysis in trust management. *Journal of the ACM*, 52(3):474–514, 2005.

[17] S. P. Marsh. *Formalising Trust as a Computational Concept.* PhD thesis, Department of Computer Science and Mathematics, University of Stirling, 1994.

[18] L. Mui, M. Mohtashemi, and A. Halberstadt. A computational model of trust and reputation (for ebusinesses). In *Proceedings from 5th Annual Hawaii International Conference on System Sciences (HICSS'02)*, 188. IEEE, 2002.

[19] S. D. Ramchurn, D. Huynh, and N. R. Jennings. Trust in multi-agent systems. *The Knowledge Engineering Review*, 19(1):1–25, 2004.

[20] J. Sabater and C. Sierra. Review on computational trust and reputation models. *Artificial Intelligence Review*, 24(1):33–60, 2005.

[21] V. Sassone, K. Krukow, and M. Nielsen. Towards a formal framework for computational trust. In *Proceedings from 5th International Symposium on Formal Methods for Components and Objects*, volume 2562 of *Lecture Notes in Computer Science*, pages 175–184. Springer, 2007.

[22] W. T. L. Teacy, J. Patel, N. R. Jennings, and M. Luck. Coping with inaccurate reputation sources: experimental analysis of a probabilistic trust model. In *AAMAS '05: Proceedings of the fourth international joint conference on Autonomous agents and multiagent systems*, pages 997–1004, ACM Press, 2005.

[23] B. Yu and M. P. Singh. An evidential model of distributed reputation management. In *Proceedings of the first international joint conference on Autonomous agents and multiagent systems*, pages 294–301, ACM Press, 2002.

Folding Systems of Communicating Agents

Kamal Lodaya and Soumya Paul
The Institute of Mathematical Sciences,
C.I.T. Campus, Chennai 600 113, India
{kamal,soumya}@imsc.res.in

Abstract

Thiagarajan showed that the event structures obtained by unfolding finite 1-safe nets can be labelled in a deterministic manner preserving the natural independence relation of the net, and called them regular trace event structures. He also showed that for this class $\mathscr{T}$ of trace event structures, the converse holds: that is, from a regular event structure in the class $\mathscr{T}$, a finite 1-safe Petri net (which we call a *folding*) can be constructed. In earlier work, we extended the converse to the slightly larger class $\mathscr{D}$ of deterministically labelled event structures.

Thiagarajan's conjecture is that the converse holds for all regular event structures. We first prove Thiagarajan's conjecture for the class $\mathscr{S}\mathscr{C}$ of event structures which can be seen as n sequential agents communicating by synchronization. That is, regular n-SCSA have a 1-safe folding. Using this result, we prove Thiagarajan's conjecture for the full class $\mathscr{E}\mathscr{S}$ of event structures. We also give stronger results for the classes $\mathscr{C}\mathscr{F}$, $\mathscr{S}\mathscr{F}$ and $\mathscr{A}\mathscr{F}$ of event structures which are confusion-free, free of symmetric and free of asymmetric confusion, respectively. In the latter case, the constructed net satisfies a natural property which we call *symmetric choice*.

Keywords: Petri nets, regular event structures, confusion-freeness

1 Introduction

It is well known that finite state transition systems can be unfolded into regular trees where the branching is bounded, and conversely boundedly branching regular trees can be folded into finite state transition systems. Milner's axiomatization [14] of bisimulation [19] over synchronization trees was based on this idea.

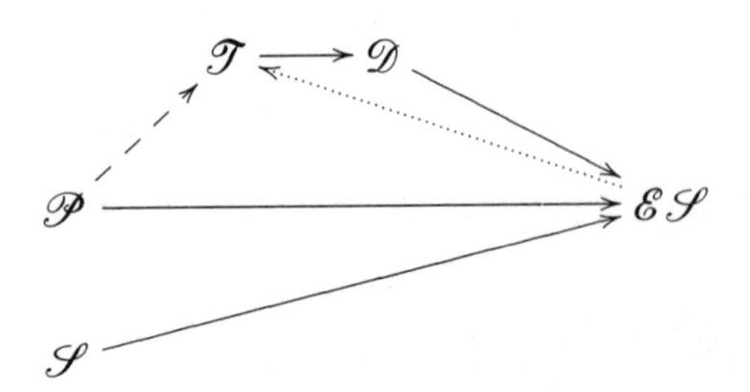

Figure 1: *Relationship between classes of event structures*

Similarly, (labelled) finite 1-safe Petri nets [20] unfold to (labelled) **regular** event structures [16], as defined by Thiagarajan [23]. Thiagarajan showed that the event structures obtained by unfolding nets can be labelled in a deterministic manner preserving the natural independence relation of the net (a **trace** labelling) [13], and called them **trace** event structures [22]. Conversely, consider a class of event structures, restricted to be regular. Thiagarajan showed that for the class $\mathscr{T}$ of trace event structures, the converse holds: that is, regular trace event structures can be folded into finite 1-safe Petri nets [22]. **Thiagarajan's conjecture** [23] is that the converse holds for *all* regular event structures. Equivalently it can be stated as the assertion that every regular event structure can be provided a trace labelling. The dotted line in Figure 1 represents the conjecture.

The conjecture was also proved for the class $\mathscr{P}$ of **conflict-free** event structures (or labelled posets), where the conflict relation of the event structure is empty [18]; for the class $\mathscr{S}$ of *sequential* (or concurrency-free) event structures, which have no concurrency [18]; and for the class $\mathscr{D}$ of *deterministic* event structures [9], where the labelling is deterministic (by definition of $\mathscr{T}$, we have $\mathscr{T} \subseteq \mathscr{D}$). The solid arrows in the figure show the inclusions among these classes. The dashed arrow from $\mathscr{P}$ to $\mathscr{T}$ refers to the proof in [18] which "embeds" conflict-free event structures by providing them a trace labelling.

The general case of Thiagarajan's conjecture is open. For a quick look at the setting of the problem, see our survey [8].

Thiagarajan suggested tackling the class $\mathscr{C}\mathscr{F}$ of **confusion-free** event structures as the next step in proving his conjecture. The class $\mathscr{C}\mathscr{F}$ allows conflict as well as concurrency, extending the conflict-free class $\mathscr{P}$ as well as the concurrency-free class $\mathscr{S}$.

A **confusion** is present in an event structure when concurrency is in close proximity to conflict. Traditional net theory [17] distinguishes between "symmetric" confusions, where events e_1 and e_2, as well as e_2 and e_3 are in conflict, but e_1 and e_3 are concurrent; and "asymmetric"

Figure 2: *Symmetric and asymmetric confusion*

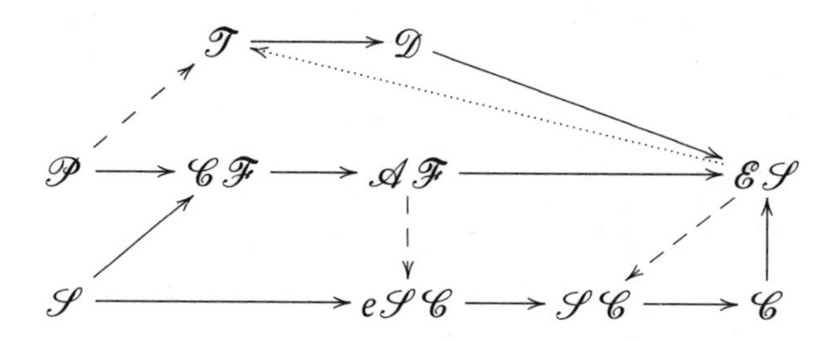

Figure 3: *Classes of event structures—the full picture*

confusions, where events e_1 precedes e_2, e_2 and e_3 are in conflict, and e_1 and e_3 are concurrent. These are illustrated by the two pictures in Figure 2.

Results. For any finite 1-safe Petri net, there is a bound n such that the event structure obtained can be partitioned into n **agents**, each of which is sequential (that is, has no concurrency). Such **systems of communicating sequential agents** (CSA) were defined in [11] and are shown to form a subclass of event structures, which let us call $\mathcal{C}$. (An earlier paper defined a smaller class called n-agent event structures [12] which acquired the name of systems of n **asynchrously** communicating sequential agents or n-ACSA in [11].)

The paper [11] also defined a subclass (SCSA) which communicate only by synchronization (they were called **synchronization structures** by Ramanujam [21]). SCSA can exhibit nondeterminism as well as concurrency, In fact, labelled finite 1-safe Petri nets unfold into the smaller class of regular SCSA. Our first result is the converse: a proof of Thiagarajan's conjecture for the class $\mathcal{SC}$ of SCSA, that is, from a regular system of n synchronously communicating agents we construct a finite 1-safe net whose unfolding has isomorphic behaviour.

Our second result is a proof of Thiagarajan's conjecture for the full class $\mathcal{ES}$ of all event structures by providing them with an agent mapping under which they are SCSA (see the dashed arrow in Figure 3). In case the event structures are confusion-free —respectively, free of symmetric confusion ($\mathcal{SF}$), free of asymmetric confusion ($\mathcal{AF}$)—the 1-safe net we construct is free choice (asymmetric choice, "symmetric choice", respectively).

2 Event structures, Petri nets and regularity

Event structures are a generalization of traces and labelled posets to include branching behaviour. Events can be related by causality ($\leq$ or $\geq$), conflict (#), or by concurrency: the **co** relation is defined to be the complement of causality and conflict.

We recall some terminology for posets (of events, in our case). We will use $<$ for the strict partial order (that is, $\leq \setminus =$). The **past** of an event e is the set of events which are below it ($<$) in the partial order. By also adding the event e, we get its past-closure $\downarrow e$. A poset is **finitary** if each event has a finite past.

Definition 1 *[Nielsen, Plotkin and Winskel] A Σ-**labelled event structure** $ES = (E, \ell, \leq, \#)$ consists of a Σ-labelled finitary poset $(E, \ell, \leq)$ with a binary irreflexive symmetric conflict relation # which is* **inherited**; *that is, if two events $e_1, e_2 \in E$ are in conflict, so are all events $e_1' \geq e_1$ and $e_2' \geq e_2$ above them.*

A **configuration** of an event structure is a past-closed and conflict-free set of events. Configurations are a notion of "state" in an event structure. In particular, the past-closure $\downarrow e$ of an event e is called a **local** configuration. We use C_{ES} for the set of configurations of an event structure ES.

An event e is **enabled** at a configuration c if $e \notin c$ and $c \cup \{e\}$ is a configuration. In this case, we write $c \overset{e}{\Rightarrow} c \cup \{e\}$. Let $en(c)$ be the set of events enabled at c and $\#(c)$ the set of events which are in conflict with events in c. We will restrict ourselves to event structures with **bounded enabling**, that is, there is a bound k such that for all configurations c of the event structure, $|en(c)| \leq k$. In particular, this restriction means that all configurations of interest are finite sets of events, and the conflict relation is inherited from an immediate conflict relation $\#_\mu$. We henceforth assume our event structures satisfy these properties.

We now define some subclasses of event structures, specializing a characterization of confusion-freedom to the absence of "symmetric" and "asymmetric" confusion.

Definition 2 *[Varacca, Völzer and Winskel] An event structure is* ***free of symmetric confusion*** *if immediate conflict is transitive and* ***free of asymmetric confusion*** *(or in the class $\mathcal{AF}$) if all events in immediate*

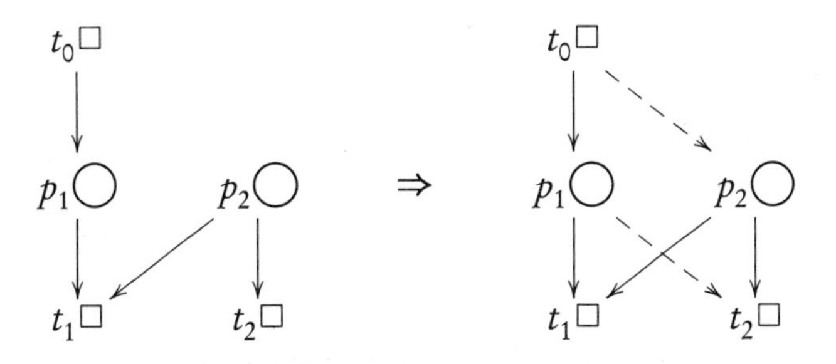

Figure 4: *The symmetric choice property*

conflict have the same past. It is **confusion-free** *(or in the class* $\mathcal{CF}$*) if it is free of both kinds of confusion.*

2.1 Petri nets

We use $\wp_{nf}(X)$ to stand for the finite nonempty subsets of X.

Definition 3 *[Petri] A* Σ*-**labelled net** consists of a tuple* $N = (P, T, \ell, pre, post, m_0)$ *of disjoint finite sets* P *of* **places** *and* T *of* **transitions**, *the latter labelled by* $\ell : T \to \Sigma$*, with two functions* $pre, post : T \to \wp_{nf}(P)$ *specifying the* **pre-** *and* **postconditions** *of a transition, and an* ***initial marking*** $m_0 \subseteq P$.

A marking is in general a multiset of places, and "firing" of transitions is described by a so-called token game [20]. In this article our attention will be restricted to nets which are **1-safe**, that is, where all reachable markings are sets.

We next define some subclasses of Petri nets.

Definition 4 *[Hack, Commoner] Two transitions* t_1 *and* t_2 *are said to be in* **conflict** *if* $pre(t_1)$ *and* $pre(t_2)$ *share a place. This conflict is said to be* **asymmetric choice** *if one of* $pre(t_1)$ *and* $pre(t_2)$ *is included in the other, and* **free choice** *if they are equal. A net is said to be asymmetric choice (resp. free choice) if all pairs of conflicting transitions are asymmetric choice (resp. free choice).*

We refer to the article by Best [2] and the book by Desel and Esparza [4] for details regarding these subclasses. The next definition appears to be new.

Definition 5 *A transition* t_0 *is said to be a* **cause** *for a transition* t_1 *if* $post(t_0)$ *and* $pre(t_1)$ *share a place. A conflict between two transitions* t_1 *and* t_2 *is said to be* **symmetric choice** *if, when the conflict is not free*

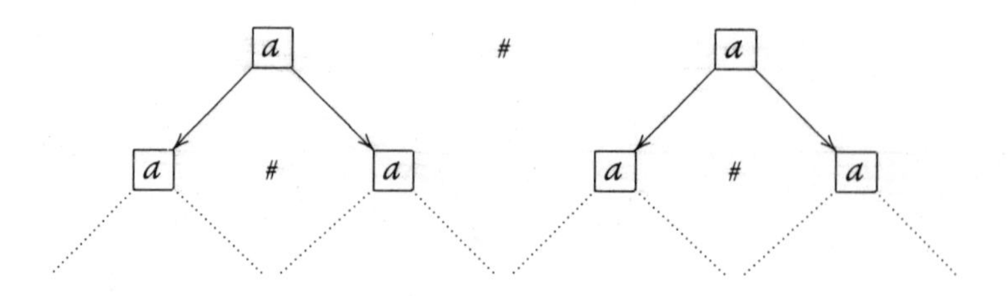

Figure 5: *The infinite binary tree*

choice, if a transition t_0 is a cause for t_1, then it is also a cause for t_2. A net is symmetric choice if all its conflicts are.

The dashed lines in Figure 4 illustrate two ways in which the symmetric choice property can be satisfied, one of them is not free choice.

2.2 Regularity in behaviour

In order to give a finite-state notion of concurrent branching-time behaviour, Thiagarajan extended some definitions from infinite trees (see, for example, the paper of Muller and Schupp [15]) to event structures [22, 23].

Definition 6 *[Thiagarajan] The* **residue** *of a configuration c in an event structure with events E is $E \setminus (c \cup \#(c))$. Two configurations are said to be* **right invariant** *(we denote the relation by R_{ES}) if their residues are isomorphic as labelled event structures. Given two residues in an R_{ES} class, let I_{ES} be the restriction of this isomorphism to their minimal events. An event structure ES is* **recognizable** *if the R_{ES} relation on its configurations is of finite index. A recognizable event structure with bounded enabling is said to be* **regular**.

As an example, consider the event structure in Figure 5 forming an infinite binary tree: that is, each event has a left and a right successor event in conflict with each other. Assume all events are labelled by the same letter, say a. This event structure has only one residue (itself) upto isomorphism. There are two I_{ES} classes corresponding to the left and right successors.

The behaviour of a net as an event structure [16] can be given by an **unfolding** construction [25]. It can be verified that if a net is free choice (resp. asymmetric choice, symmetric choice), its event structure unfolding is confusion-free (resp. free of symmetric confusion, free of asymmetric confusion).

Thiagarajan showed that 1-safe net unfoldings are regular trace event structures [22, 23]. Call a net N a **folding** of event structure ES if the unfolding of N is isomorphic to ES.

Conjecture 1 (Thiagarajan) *Every regular event structure has a 1-safe folding.*

3 Agent systems

Fix a finite alphabet Σ. Let Loc be a disjoint set of **locations** where events are executed. Unlike in traces [13], the locations are not connected to the letters of the alphabet; they can also be infinite in number.

Definition 7 *[with Ramanujam and Thiagarajan] A* **CSA** *(system of communicating sequential agents) over* (Σ, Loc) *is a tuple* $ES = (E, \ell, \leq, loc)$, *with* $(E, \ell, \leq)$ *a* Σ*-labelled finitary poset and* $loc : E \to \wp_{nf}(Loc)$ *mapping each event to the finitely many agents executing it (and we write* E_i *for* $loc^{-1}(i)$*), such that:*

- $\forall e \in E : \forall i \in Loc : \downarrow e \cap E_i$ *is totally ordered by* $\leq$ *(each agent is a tree).*

An **SCSA** *(synchronization structure) is a CSA which additionally satisfies the condition that:*

- $\leq = \bigcup_{i \in Loc} (\leq \cap (E_i \times E_i))$ *(communication takes place through synchronizations).*

An ***n*-SCSA** *(****n*-CSA***) is an SCSA (CSA) where* $|Loc| = n$. *The class* $\mathcal{SC}$ *consists of all the* n*-SCSA, for every* n*, but* not *the SCSA with* Loc *infinite. The class* $\mathcal{C}$ *is defined similarly.*

A CSA is easily seen to be an event structure by specifying that all unordered events within an agent are in conflict, and then applying inheritance. Thus Definition 6 on residues, regularity and other behavioural matters, applies to CSA. In the case of an SCSA, the minimal events in a residue of $\downarrow e$ are restricted to having an intersection with the agents in $loc(e)$, whereas in a CSA, they can range over all of Loc, but still restricted to a finite set of agents by the finitariness of the poset. For example, the synchronized products of Wöhrle and Thomas [26] unfold into regular SCSA, but maybe not into n-SCSA, for any n.

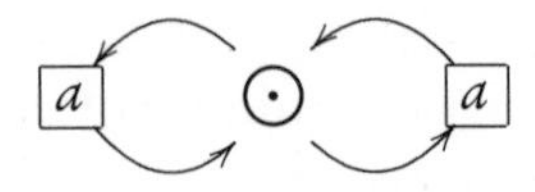

Figure 6: *Product net that generates the infinite binary tree*

On the other hand, isomorphism of CSA is a stronger notion than isomorphism of event structures: for example, the binary tree with a function loc_1 such that all events are in a single agent, and the same event structure with the function loc_2 which locates every event in a different agent, are not isomorphic as CSA.

Definition 8 *Given an event structure or a CSA ES, a function loc is called an* **agent mapping** *if (ES, loc) is a CSA, and a* **product mapping** *if (ES, loc) is an SCSA and the range of loc is finite—hence an n-SCSA, for some n. A product mapping loc, and more generally an n-SCSA (ES, loc), is said to be* **economical** *if whenever two events $e_1 \#_\mu e_2$ in immediate conflict belong to distinct agents, then there is an event $e_3 \#_\mu e_2$ such that e_2, e_3 share an agent.*

In case ES is a CSA, an agent mapping is really a "remapping". The function loc_1 above is an economical product mapping for (ES, loc_2), conversely loc_2 is an agent mapping for (ES, loc_1).

Below we will prove Thiagarajan's conjecture for n-SCSA. We fold an n-SCSA into a "synchronized product" net [1, 27, 10, 6], whose definition mildly generalizes earlier definitions by allowing multiple transitions with the same label, source and target nodes. The definition ensures the net is 1-safe.

Definition 9 *Let P be a finite set of* places *distributed by the function $dist : P \to Loc$ (and we write P_i for $dist^{-1}(i)$). For $L \subseteq Loc$, let $\Pi_L P$ be the functions $f : L \to P$ such that $dist(f(i)) = i$. Let T be a disjoint finite set of* transitions *labelled by $\ell : T \to \Sigma$ and distributed by $loc : T \to \wp_{nf}(Loc)$. A* **product net** *over (Σ, Loc) is given by $(P, T, \ell, pre, post, dist, loc, m_0)$, where $m_0 \in \Pi_{Loc} P$ is a global initial state, and $pre, post : T \to \bigcup_{L \subseteq Loc} \Pi_L P$ such that for a transition t, $pre(t), post(t) \in \Pi_{loc(t)} P$.*

The binary tree SCSA example we saw earlier is the unfolding of a product net (see the book by Esparza and Heljanko [6]) with a single place and two self-loops labelled by a, as shown in Figure 6. If we naïvely model this as a Zielonka automaton [27], which would allow only a single a-labelled self-loop, its unfolding would be an infinite line, which is not isomorphic to the binary tree.

3.1 A net construction

Given a configuration c of a CSA and an agent i, let $\downarrow_i c$ be $\downarrow e_i$ if e_i is the last event in c of agent i, and the empty set if there is no such event. A **global** configuration c of a CSA is of the form $\downarrow_1 c \cup \ldots \cup \downarrow_n c$.

We can now state our first result.

Theorem 10 *[Thiagarajan's conjecture for $\mathcal{SC}$] A regular n-SCSA has a product folding.*

Proof Given an SCSA $ES = (E, \ell, \leq, loc)$ over (Σ, Loc) (say $Loc = \{1, \ldots, n\}$), the product net is constructed as follows.

The local places $P_i = dist^{-1}(i)$ of the net are given by the set $\{([\downarrow e], i) \mid i \in loc(e)\} \cup \{([\emptyset], i)\}$, where $[c]$ is the equivalence class of configuration c under R_{ES}. Global states are products of these local places. Since no event has been executed initially, the global initial state $(([\emptyset], 1), \ldots, ([\emptyset], n))$ consists of residues of the empty configuration.

The transitions T are the I_{ES}-classes $[e]$ of the events of E. By label isomorphism, a function $loc : T \rightarrow \wp_{nf}(Loc)$ can be defined using the loc function of ES. For every $c \overset{e}{\Rightarrow} c'$ in ES, with $loc(e) = \{i_1, \ldots, i_k\}$, we set $pre([e]) = (p_{i_1}, i_1), \ldots, (p_{i_k}, i_k)$ and $post([e]) = (p'_{i_1}, i_1), \ldots, (p'_{i_k}, i_k)$, where p_j, p'_j are the R_{ES} classes of $\downarrow_j c$ and $\downarrow_j c'$ respectively. This also defines a transition relation on global states

$$(p_1, 1), \ldots, (p_n, n) \overset{[e]}{\rightarrow} (p'_1, 1), \ldots, (p'_n, n),$$

where $p'_j = p_j$ for $j \notin \{i_1, \ldots, i_k\}$. We also define a function $h : C_{ES} \rightarrow \Pi_{Loc} P$ by $h(c) = (([\downarrow_1 c], 1), \ldots, ([\downarrow_n c], n))$.

By the definition of the transition relation, if $c \overset{e}{\Rightarrow} c'$ then $h(c) \overset{[e]}{\rightarrow} h(c')$. For the other direction, if $m \overset{t}{\rightarrow} m'$, then we know that for some c, c', e such that $h(c) R_{ES} m$ and $t = [e]$, $c \overset{e}{\Rightarrow} c'$. By right invariance, we know that $h(c)$ and m have label-isomorphic residues. By the isomorphism I_{ES}, we know that there is an event e' with the same label corresponding to e enabled at m, and after performing it, $h(c')$ and m' will have isomorphic residues.

Thus every step of the product net corresponds to one from ES. Suppose the events $e_1, \ldots, e_k$ were concurrently enabled at c. By the definition of an SCSA, these events must operate on different agents. Hence the pre- and post-sets of the transitions corresponding to these events are disjoint and the transitions are independent in the constructed net. On the other hand, if e_1, e_2 are enabled but in immediate conflict at the configuration c, by definition of an SCSA, they must

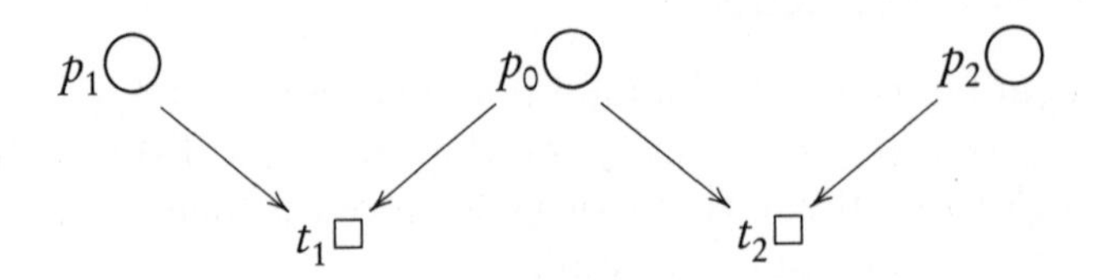

Figure 7: *Symmetric confusion*

share an agent and the pre-sets of the two corresponding transitions will not be disjoint. Hence the unfolding of the constructed net will be isomorphic to ES.

Confirming Thiagarajan's intuition, we find that it is the product mapping loc which is crucial to the construction.

3.2 The pursuit of confusion-freedom

The construction in Theorem 10 yields stronger results for SCSA which are free of different kinds of confusion.

Lemma 11 *A regular n-SCSA which is free of symmetric confusion has an asymmetric choice product folding.*

Proof Consider a regular n-SCSA ES which is free of symmetric confusion. By Theorem 10, such an SCSA folds into a product net.

Suppose this is not asymmetric choice. Then it has two transitions t_1 and t_2 in conflict, hence a place $p_0 \in pre(t_1) \cap pre(t_2)$, and by supposition a place $p_1 \in pre(t_1) \setminus pre(t_2)$ and a place $p_2 \in pre(t_2) \setminus pre(t_1)$. From the net construction in Theorem 10, this will be the case when $t_1 = [e_1]$, $t_2 = [e_2]$, $\{1,2\} \subseteq loc(e_1)$, $\{2,3\} \subseteq loc(e_2)$, $e_1 \in en(\downarrow_1 c) \setminus en(\downarrow_2 c)$, $e_2 \in en(\downarrow_3 c) \setminus en(\downarrow_2 c)$, $p_1 = [\downarrow_1 c]$, $p_2 = [\downarrow_3 c]$ and $p_0 = [\downarrow_2 c]$.

Now if there is an event e_1' with $1 \in loc(e_1')$ and $p_1 \in pre(e_1')$ we can assume $2,3 \notin loc(e_1')$ (otherwise the argument can be repeated), we would have $e_1' \#_\mu e_1$, $e_1 \#_\mu e_2$ and e_1', e_2 concurrent, which is a symmetric confusion. Symmetric arguments hold if we assume an event e_0' with $2 \in loc(e_0')$ and $p_0 \in pre(e_0')$ or an event e_2' with $3 \in loc(e_2')$ and $p_2 \in pre(e_2')$. Hence either $p_1 \in pre(t_2)$ or $p_2 \in pre(t_1)$, contradicting our supposition.

Lemma 12 *A regular n-SCSA which is free of asymmetric confusion has a symmetric choice product folding.*

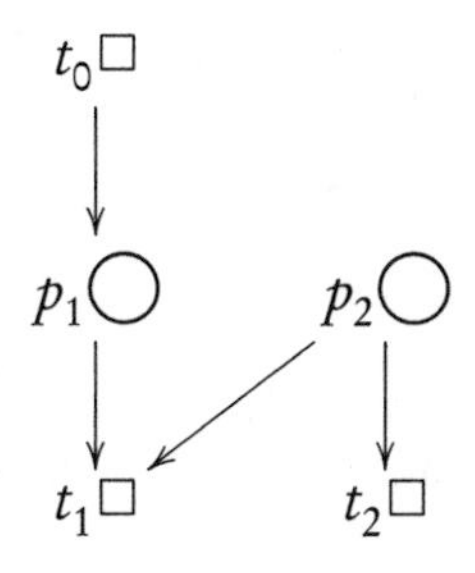

Figure 8: *Asymmetric confusion*

Proof Consider a regular n-SCSA ES which is free of asymmetric confusion. By Theorem 10, such an SCSA folds into a product net.

Consider two transitions t_1 and t_2 which are not free choice, that is, there is a place $p_1 \in pre(t_1) \setminus pre(t_2)$ and a place $p_2 \in pre(t_1) \cap pre(t_2)$. From the net construction of Theorem 10, this will be the case when $t_1 = [e_1]$, $t_2 = [e_2]$, $\{1,2\} \subseteq loc(e_1)$ and $2 \in loc(e_2)$. Without loss of generality, $e_1 \#_\mu e_2$.

Consider a transition t_0 which is a cause for t_1. We can assume $p_1 \in post(t_0)$, otherwise we can switch the argument to an appropriate set of places. Since ES was free of asymmetric confusion, e_1 and e_2 must have the same past and t_0 must also be a cause for t_2. Hence the net satisfies the symmetric choice condition.

Lemma 13 *An economical regular n-SCSA which is confusion-free has a free choice product folding.*

Proof Consider an economical regular n-SCSA ES which is confusion-free. By the previous lemma, such an SCSA folds into a symmetric choice product net.

Suppose this is not free choice. Then it has two transitions t_1 and t_2, a place $p_1 \in pre(t_1) \cap pre(t_2)$ and a place $p_2 \in pre(t_2) \setminus pre(t_1)$. From the net construction in Theorem 10, this will be the case when $t_1 = [e_1]$, $t_2 = [e_2]$, $1 \in loc(e_1)$, $\{1,2\} \subseteq loc(e_2)$, $e_1 \in en(\downarrow_1 c) \setminus en(\downarrow_2 c)$, $e_2 \in en(c)$, $p_1 = [\downarrow_1 c]$ and $p_2 = [\downarrow_2 c]$.

Since loc is economical, there is an event $e_3 \#_\mu e_2$ with $2 \in loc(e_3)$ and we can assume $1 \notin loc(e_3)$ (otherwise the argument can be repeated). If e_3 and e_1 were in conflict, by absence of asymmetric confusion we can without loss of generality take them to be in minimal conflict which contradicts the definition of $\#_\mu$ in an SCSA. They also cannot be causally related. Hence they must be concurrent and there is a symmetric confusion in ES, a contradiction.

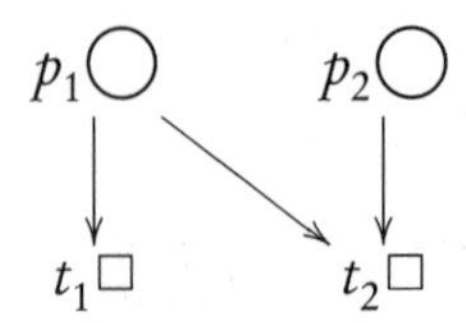

Figure 9: *Proof of Lemma 13*

4 The Pursuit of Mappings

Events in immediate conflict must belong to the same agent. We enlarge this idea to encompass other events which can be mapped to the same agent, using an idea similar to the **synchronization skeleton** of Emerson and Clarke [5].

Recall that a **clique** of a binary relation –for example, the edge relation in a graph– is a set of vertices, every pair of which is connected by an edge. In particular, a single disconnected vertex is also a clique. Let μ_{ES} be the set of $\#_\mu$-cliques of ES.

Definition 14 *A* ***concurrency-free layering*** *of an event structure ES is a forest structure on μ_{ES} (each tree of the forest being called a* **layer***), with the parent-child relationship defined as follows: If an event e is in a clique C, all its immediate successors form a clique D and have only e as an immediate predecessor, then the clique D is a child of C. A layering is* **maximal** *if the layers are maximal under inclusion.*

All the children D of an event e may have more than one event as predecessor (implying that the parents are in **co**). The definition above rules out this case. Since $\#_\mu$ is not necessarily transitive in a symmetric confusion, we can have $e_1 \#_\mu e_2 \#_\mu e_3$ and $e_1 \textbf{co} e_3$. In this case the event e_2 will appear in cliques in two layers. Hence a concurrency-free layering is a set of partial equivalence relations which cover ES (and an equivalence relation if ES is also free of symmetric confusion).

This is the basis for a procedure for providing a regular event structure with an agent mapping loc. Recall that a set of **lines** (maximal linear orders) **covers** a poset if every element is in some line and for every edge in the Hasse diagram of the poset, its source and target nodes are in one of the lines.

Figure 10 shows an event structure –actually a poset– which has an enabling bound of two, but every event is a layer and infinitely many lines are required to cover it.

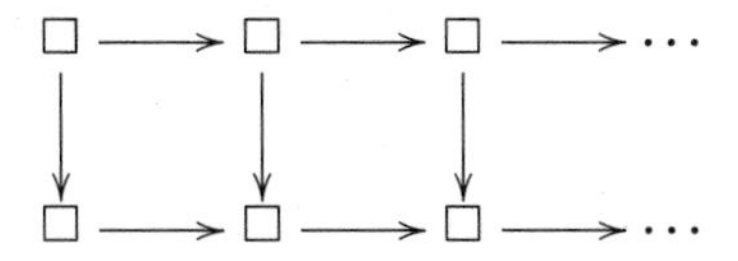

Figure 10: *A poset that requires infinitely many lines to cover it*

We now bring regularity into play. Extend R_{ES} by calling two cliques R_{ES}-equivalent if their enabling configurations are, and two layers R_{ES}-equivalent if their roots are.

Lemma 15 *A regular event structure has a product mapping.*
Proof Let $ES = (E, \ell, \leq, \#)$ be a regular event structure. Consider its (past-closed) prefix where if two events are ordered, they have different R_{ES}-classes. Because of bounded enabling, this prefix has bounded width (elements which are pairwise concurrent or in immediate conflict). By recognizability it is finite. Let $Lines$ be a finite set of lines covering the prefix, and $Layers$ a finite set of layers which cover the prefix. Define $Loc = Lines \cup Layers$.

We claim that the lines in Loc cover the entire event structure, by extending them so that they respect I_{ES}-classes, that is, if two events in the same right invariance class share a line, their successors which are in the same I_{ES} class fall on the same line. (This handles cases like the binary tree example we saw earlier.)

Define the function loc on ES by

$$\begin{aligned} loc(e) = \; & \{l \mid \exists C \in \mu_{ES} : e \in C \in l \in Layers\} \\ & \cup \{i \mid e \in i \in Lines\} \end{aligned}$$

mapping an event e to the layers and lines containing it. Since they cover the poset, loc will be total.

$(E, \ell, \leq, loc)$ is an SCSA, since two concurrent events in an event's past will not share lines, and any event must share a line with an immediate predecessor and with an immediate successor (if it has any). It is an n-SCSA where $n = |Loc|$.

Hence loc is a product mapping.

We can get a tighter construction when the event structure we start with is free of asymmetric confusion.

Definition 16 *Given a maximal layering of an event structure ES, its **skeleton** is a poset on the layers, ordering them whenever some events in the layers are ordered in ES.*

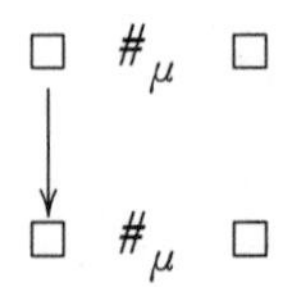

Figure 11: *Asymmetric confusion and layers*

In an asymmetric confusion (see Figure 11), the clique consisting of the top events is not a parent of the clique consisting of the lower events, but the above definition orders the layers containing them. It can be seen that in an event structure free of asymmetric confusion, if two layers are ordered, they do not contain concurrent events. Hence for ES free of asymmetric confusion, lines covering its skeleton are an adequate representation of lines covering ES and separating concurrency.

Lemma 17 *A regular event structure free of asymmetric confusion has an economical product mapping.*
Proof Let $ES = (E, \ell, \leq, \#)$ be a regular event structure free of asymmetric confusion. Fix a skeleton of a maximal layering of ES. Consider the past-closed prefix (of the skeleton) where if two layers are ordered, they have different R_{ES}-classes. Again, because of bounded enabling, the prefix has bounded width and by recognizability it is finite. Let $Lines$ be a finite set of lines covering this prefix, and this time we let $Loc = Lines$.

Define the function loc on the original event structure by

$$loc(e) = \{i \mid \exists C \in \mu_{ES}, l \in Layers : e \in C \in l \in i \in Lines\}$$

mapping an event e to the set of lines which pass through the layers in which the cliques containing e lie. Since the lines cover the poset, each event will get mapped to at least one agent. The mapping is economical. As in the proof of the previous lemma, the result is an n-SCSA.

This yields our final set of results, including Thiagarajan's conjecture for the full class $\mathcal{ES}$ of all event structures.

Theorem 18 *[Thiagarajan's conjecture] A regular event structure has a product folding. If it is free of asymmetric confusion, it has a symmetric choice product folding. If it is free of symmetric confusion, it has an asymmetric choice product folding. If it is confusion-free, it has a free choice product folding.*

Proof Let ES be a regular event structure. By Lemma 15, we get a product mapping loc under which ES is an SCSA, in particular, a regular SCSA.

If ES is free of symmetric confusion, Lemma 11 shows the net is asymmetric choice.

If ES is free of asymmetric confusion, applying the lemma above, we get an economical product mapping loc. Now, by Theorem 10, ES folds into a product net. If loc was economical, by Lemma 12, the net is symmetric choice. If ES is in addition confusion-free, Lemma 13 shows the net is free choice.

5 Conclusion

We have proved Thiagarajan's conjecture for the full class $\mathscr{E}\mathscr{S}$ of all event structures. We obtained a strengthened result for the subclasses $\mathscr{C}\mathscr{F}$, $\mathscr{S}\mathscr{F}$ and $\mathscr{A}\mathscr{F}$ by connecting them to the corresponding subclasses of free choice, asymmetric choice and "symmetric choice" nets.

Compared to Thiagarajan's use of a trace labelling which exposes the independence structure in an event structure, we have an agent mapping which exposes the communication structure. The result for sequential event structures is a subcase of that presented here (since $\mathscr{S} \subseteq \mathscr{S}\mathscr{C}$) where $|Loc|$ is one.

Our proofs are considerably simpler than earlier ones. In particular, we avoid an appeal to Zielonka's theorem [27], which was used in Thiagarajan's proof for the class $\mathscr{T}$. The proofs for deterministic event structures $\mathscr{D}$ and conflict-free event structures $\mathscr{P}$ we gave earlier [9] did not use Zielonka's theorem. They used algebraic techniques which collapse nondeterminism to determinism and hence cannot be applied here. The present proofs avoid algebra and are completely elementary.

Further, the nets we construct are "product" systems [1, 10, 6] and can be seen as a small generalization of Zielonka's asynchronous automata [27], necessitated by the fact that Thiagarajan's conjecture preserves event structure isomorphism, a stronger condition than acceptance of a (prefix-closed) trace language. Hence our proofs might be useful in factoring Zielonka's theorem into separate steps.

Acknowledgements

We discussed these ideas extensively with Antoine Meyer, during visits to Paris and Chennai in 2006 and 2008, under the Indo-French project Timed-DISCOVERI. He found a bug in an early purported proof of the full conjecture.

References

[1] A. Arnold. *Finite transition systems*, Prentice-Hall, 1994.

[2] E. Best. Structure theory of Petri nets: the free choice hiatus, *Proc. Adv. Petri nets '86*, Bad Honnef, Part 1 (W. Brauer, W. Reisig and G. Rozenberg, eds.), *LNCS* 254, pp. 168–205, 1987.

[3] F. Commoner. Deadlocks in Petri nets, Report CA-7206-2311, Applied Data Research, 1972.

[4] J. Desel and J. Esparza. *Free choice Petri nets*, Cambridge, 1995.

[5] E.A. Emerson and E.M. Clarke. Using branching time temporal logic to synthesize synchronization skeletons, *Sci. Comp. Program.* 2, pp. 241–266, 1982.

[6] J. Esparza and K. Heljanko. *Unfoldings: a partial order approach to model checking*, Springer, 2008.

[7] M.H.T. Hack. Analysis of production schemata by Petri nets, M.S. thesis, EE dept, MIT, 1972. Corrections in: Computation Structures Note 17, Project MAC, 1974.

[8] K. Lodaya. A regular viewpoint on processes and algebra, *Acta Cybernetica* 17(4), pp. 751–763, 2006.

[9] K. Lodaya. Petri nets, event structures and algebra, in *Formal models, languages and applications* (K.G. Subramanian, K. Rangarajan and M. Mukund, eds.), World Scientific, pp. 246–259, 2006.

[10] K. Lodaya. Product automata and process algebra, *Proc. 4th SEFM*, Pune (P.K. Pandya and D.v. Hung, eds.), IEEE, pp. 128–136, 2006.

[11] K. Lodaya, R. Ramanujam and P.S. Thiagarajan. Temporal logics for communicating sequential agents: I, *Int. J. Found. Comp. Sci.* 3(2), pp. 117–159, 1992.

[12] K. Lodaya and P.S. Thiagarajan. A modal logic for a subclass of event structures, *Proc. 14th ICALP*, Karlsruhe (T. Ottmann, ed.), *LNCS* 267, pp. 290–303, 1987. A correction in: Report DAIMI PB-275, University of Aarhus, 1989.

[13] A. Mazurkiewicz. Concurrent program schemes and their interpretations, Report DAIMI PB-78, Aarhus University, 1977.

[14] R. Milner. A complete inference system for a class of regular behaviours, *J. Comp. Syst. Sci.* 28(3), pp. 439–466, 1984.

[15] D.E. Muller and P. Schupp. The theory of ends, pushdown automata, and second-order logic, *Theoret. Comp. Sci.* 37, pp. 51–75, 1985.

[16] M. Nielsen, G. Plotkin and G. Winskel. Petri nets, event structures and domains, part 1, *Theoret. Comp. Sci.* 13, pp. 86–108, 1980.

[17] M. Nielsen and P.S. Thiagarajan. Degrees of nondeterminism and concurrency: a Petri net view, *Proc. 4th FSTTCS*, Bangalore (M. Joseph and R.K. Shyamasundar, eds.), *LNCS* 181, pp. 89–117, 1984.

[18] M. Nielsen and P.S. Thiagarajan. Regular event structures and finite Petri nets: the conflict-free case, *Proc. 23rd ATPN*, Adelaide (J. Esparza and C. Lakos, eds.), *LNCS* 2360, pp. 335–351, 2002.

[19] D. Park. Concurrency and automata on infinite sequences, *Proc. 5th GI conference*, Karlsruhe (P. Deussen, ed.), *LNCS* 104, pp. 167–183, 1981.

[20] C.-A. Petri. Fundamentals of a theory of asynchronous information flow, *Proc. 2nd IFIP congress*, Munich (C.M. Popplewell, ed.), North-Holland, pp. 386–390, 1962.

[21] R. Ramanujam. A local presentation of synchronizing systems, in *Structures in Concurrency Theory* (J. Desel, ed.), Springer, pp. 264–278, 1995.

[22] P.S. Thiagarajan. Regular trace event structures, BRICS Research Abstracts RS-96-32, 1996.

[23] P.S. Thiagarajan. Regular event structures and finite Petri nets: a conjecture, in *Formal and natural computing – essays dedicated to Grzegorz Rozenberg* (W. Brauer, H. Ehrig, J. Karhumäki and A. Salomaa, eds.), *LNCS* 2300, pp. 244–256, 2002.

[24] D. Varacca, H. Völzer and G. Winskel. Probabilistic event structures and domains, *Proc. 15th Concur*, London (P. Gardner and N. Yoshida, eds.), *LNCS* 3170, pp. 481–496, 2004.

[25] G. Winskel and M. Nielsen. Models for concurrency, in *Handbook of logic in computer science* IV (S. Abramsky, D. Gabbay and T.S.E. Maibaum, eds.), Oxford, pp. 1–148, 1995.
[26] S. Wöhrle, W. Thomas. Model checking synchronized products of infinite transition systems, *Log. Meth. Comp. Sci.* 3(4), 2007.
[27] W. Zielonka. Notes on finite asynchronous automata, *RAIRO Inf. Th. Appl.* 21(2), pp. 99–135, 1987.

Decidable Logics for Event Structures

P. Madhusudan

University of Illinois at Urbana-Champaign
madhu@cs.illinois.edu

Abstract

Event structures are a canonical representation of the behaviors of a concurrent system that depict causality, conflict (branching) and concurrency. Natural concurrency models such as finite 1-safe Petri nets give rise to regular trace event-structures, where regularity arises due to finiteness of the net and the trace-alphabet labeling stems from the events and their independence in the net.

It is known that model-checking monadic second-order logic (MSOL) on regular trace event structures is undecidable, in general. In 2003, I showed that first-order logic (FOL) and a new logic called monadic trace logic (MTL) (a fragment of MSOL where the quantification of sets is restricted to conflict-free events) are decidable [16].

In this article, I show a new and simpler proof of the above results, by interpreting FOL and MTL on event structures into MSOL on words.

Keywords: Event structures, logic, decidability

1 Introduction

Linear sequences of actions has become a widely accepted and popular depiction of the behaviors of a system, with tremendous impact in verification where system behaviors are specified and verified using linear sequences and automata over them [5, 26]. In order to define all behaviors of a system, and in particular to describe *non-determinism* and *choice*, the paradigm of describing the system in terms of a single tree depicting all possible linear behaviors along with the points of *choice* or *conflict* has become standard, and *branching-time* logics that specify properties of these trees have been useful in verification [5].

In the world of concurrent systems, linear sequences do not provide adequate depictions of behavior. For example, if a concurrent system executes two actions a and b, but the processes executing them are entirely different and independent of each other, then there is little sense in saying that b happened after a. A much more satisfactory description is one that says that a and b both happened, without ascribing any order between them. More precisely, we would like events to be related only by *causal* orderings, and not ordered by a global clock.

A popular notion that satisfies the above desire is that of (Mazurkiewicz) traces [17], a class of restricted labeled partial orders that extends words over a finite alphabet. One starts with a concurrent alphabet, which is a pair (Σ, I), where Σ is a finite alphabet and I is a binary independence relation over Σ that identifies which actions in Σ are independent of each other. A trace over (Σ, I) is a partial order that is labeled with letters in Σ and where the labeling respects the independence relation I (two events can be immediately causally related only if they are not independent and two events that are not independent must be causally related). Such a trace describes a single execution of a concurrent system and plays the role strings play in the world of sequential systems. This class of partial orders has very robust properties like that of words, and analogs of regularity, recognizability, and monadic second-order definability have been well studied [7, 8, 25, 9, 28].

In order to capture all behaviors of a concurrent system, including concurrency and choice, the class of structures known as *event structures* [27] is simple and pleasing. Event structures are the analog of trees for concurrent systems— an event structure is a single object that captures the concurrency, the causality, and the conflict (branching) information in a system. Event structures are partial orders equipped with a *conflict* relation that relates conflicting events of the system (events that cannot belong to the same execution of the system). A tree can be seen as an event structure where any two events (nodes) that do not belong to the same branch are in conflict. Event structures have been well studied and associations between the event-structure semantics of various concurrency models like Petri nets and distributed transition systems are well understood [27].

For sequential systems, trees are the natural analog, and the monadic-second order logic is decidable on (infinite) trees and constitutes a very robust and important decidability result [21]. Viewing an event structure of a finite-state concurrent system as a logical structure, a natural question is to ask which logics are decidable on it.

Several modal logics have been introduced for event structures, starting from an early paper by Lodaya and Thiagarajan [14], followed by several papers by Lodaya, Mukund, Ramanujam and Thiagarajan [13, 15]. These papers introduced modal logics for a variety of systems, with modal operators that could refer to conflicting, concurrent and causally related events of an event structure. These logics were shown to be decidable (more precisely, satisfiability was shown to be decidable) using sound and complete proof systems. Later work explicitly focused on model-checking, finding restricted modal logics on restricted classes of event structures that have a decidable model-checking problem. For instance, Penczek [20] identifies a modal logic on the class of *free-choice event structures* for which the model-checking problem is decidable.

In this paper, I consider the problem of model checking an event structure against classical (non-temporal) logics such as first-order logic (FOL) and monadic second-order logic (MSOL) specifications. The class of event structures I consider are *regular trace event structures*. This class of event structures can be presented by a regular prefix-closed language $\mathscr{L}$ of traces over a distributed alphabet. Such a language can be finitely presented (say as a finite automaton accepting the language) and we can associate with it a canonical labeled event structure such that the "executions" of the event structure is precisely $\mathscr{L}$. This event structure is regular where the notion of regularity is a natural one that intuitively says that there are only a finite number of non-isomorphic "suffixes" in the event structure. It is also a *trace* event structure in the sense that the labels on the events respect the dependency constraints in the concurrent alphabet.

The problem with event structures is that even simple finite-state concurrent systems can generate *grids*: structures where an infinite two-dimensional grid is formed by events, where the horizontal and vertical successors of an event on the grid can be expressed using the modalities of the event structure. Since *monadic second-order logic* is undecidable on any structure that has a definable grid, it follows that it is undecidable on event structures resulting from finite-state concurrent systems as well (this is pointed out by Thiagarajan in [22] and attributed to Walukiewicz).

In 2003, I showed that the model-checking problem for regular trace event-structures against first-order logic (FOL) and monadic trace logic (MTL) are decidable [16]. These logics are sub-logics of MSOL, and have atomic formulas that capture the causality and conflict relations. While FOL is the standard first-order logic on event structures, MTL has precisely the same syntax as MSOL, but its semantics is restricted so that set quantifiers can only range over *conflict-free* sets of events.

The proof of the above result was shown by defining a new notion of regular relations on traces, where the automaton works on a *braiding* of the traces in order to decide the relation. It was then shown that for any FOL formula with free variables, the relation defined by the formula is a regular relation on traces. Monadic trace logic was then shown to be decidable in a similar fashion, but where a singe trace with extra labels depicts a set of conflict-free events [16].

In this article, I show a much simpler and direct proof of the above results by interpreting FOL and MTL on event structures using MSOL on *words*. The decidability result proceeds by interpreting these logics using MSOL on the linear order $(\mathbb{N}, succ)$, the structure of natural numbers with the successor relation.

First-order logic on event structures turns out to be quite expressive in itself. (Note that, in the logics we consider, we have a relation in the signature that captures the causality relation, not the *immediate* causality relation; otherwise, FOL would be very weak as it would not be able to causally relate events unboundedly far away from each other.) The only work I know that explicitly deals with model checking event structures is that reported in [20]. We can show that all modal logics over event structures present in the literature (including the modal logic of [20, 15] and the tense logics in [18]) can easily be translated to FOL. While in [20] a modal logic is shown only to be decidable over the restricted class of free-choice event structures, our result generalizes this to a larger class.

The logic MTL is more powerful and can express properties such as "for every finite trace of the system, there is an extension of it that satisfies the MSOL property φ on traces" which is the analog of the CTL* property $\mathsf{AG}\ \mathsf{EF}\ \varphi$.

When the independence relation is empty (i.e., when there is no concurrency), traces degenerate to words and event-structures degenerate to trees. The logic MTL degenerates to monadic path logic (MPL) over trees, which has been studied in the literature. In [11], it was shown that MPL can capture CTL* over trees and hence is quite a powerful logic. I therefore believe that MTL is a very expressive logic and that any reasonable temporal logic over event structures should be embeddable into it.

The key idea in the new proof presented in this paper is the representation of traces using a variant of the Foata normal form. The Foata normal form has many nice properties that make it a powerful representation of traces. As I show, when traces are represented using this form and interpreted over an infinite word, the prefix and conflict relations between traces becomes MSOL definable, leading to an interpretation.

The fact that logics on event structures can be interpreted on a structure that has no "branching" may seem surprising. However, Elgot and Rabin have shown, already in the 60s, that the first-order theory of trees is interpretable using MSOL on words (in fact with the equal-level predicate thrown in) [10]. In a recent paper, Colcombet and Löding [6] study transformations of structures using set-interpretations in order to derive decidability of FOL of a structure using the decidability of MSOL on another. In fact, *automatic structures*, which are structures made up of objects represented by words and relations represented as *synchronized rational relations* on words, define a decidable FOL-theory (but not MSOL-theory), and this can be proved by using set interpretations into MSOL on $(\mathbb{N}, succ)$ (see [1, 6] and references therein). Our interpretability result can, in fact, be alternatively derived using two steps: we can first show that using Foata normal forms, event-structures become automatic, and then use the fact that FOL on automatic structures can be interpreted on a single infinite word. However, I prefer to give the direct interpretation on words in this paper, as I believe it is more accessible to a reader unfamiliar with the above topics.

2 Preliminaries

Let us fix some basic notation. For a finite alphabet Σ, let Σ^* denote the set of finite words of Σ, Σ^ω denote the set of ω length words over Σ and let $\Sigma^\infty = \Sigma^* \cup \Sigma^\omega$. For any $k \in \mathbb{N}$, let $[k]$ denote the set $\{i \mid 1 \leq i \leq k\}$.

Let $(S, \leq)$ be a partially ordered set (i.e. $\leq$ is a partial order on S). Then $<$ denotes the corresponding relation that is not reflexive, i.e. $< = \leq \setminus \{(s,s) \mid s \in S\}$. For any $s \in S$, $\downarrow s$ is the set of elements in S that are below s, i.e. the set $\{s' \in S \mid s' \leq s\}$. Also, let $\lessdot$ denote the immediate partial-order relation: $s \lessdot s'$ iff $s < s'$ and there is no s'' such that $s < s'' < s'$. In all partial orders $(S, \leq)$ that we consider in this paper, we will assume that for every $s \in S$, $\downarrow s$ is finite. Hence the reflexive and transitive closure of $\lessdot$ will be $\leq$.

Traces

A *trace alphabet* is a pair (Σ, I) where Σ is an alphabet containing a finite set of actions and $I \subseteq \Sigma \times \Sigma$ is an irreflexive and symmetric relation over Σ called the *independence* relation. The induced relation $D = (\Sigma \times \Sigma) \setminus I$ is called the *dependence* relation.

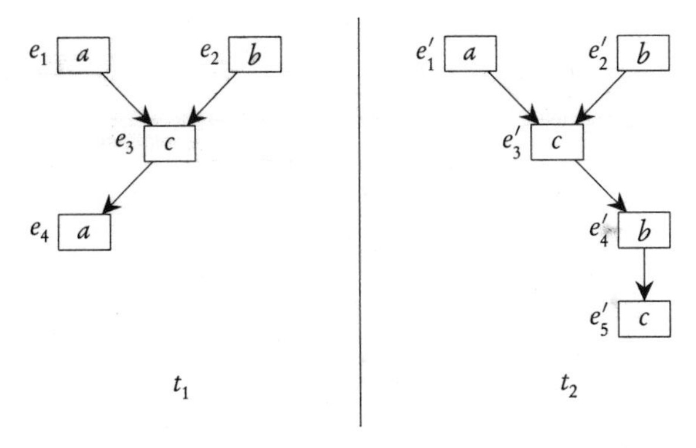

Figure 1: *Two traces t_1 and t_2 over $(\{a,b,c\},\{(a,b),(b,a)\})$*

A *(Mazurkiewicz) trace* over the trace alphabet (Σ, I) is a Σ-labeled poset $t = (E, \leq, \lambda)$ where E is a finite or countable set of events, $\leq$ is a partial order on E, called the causal order, and $\lambda : E \rightarrow \Sigma$ is a labeling function such that the following hold:

(T1) $\forall e \in E$, $\downarrow e$ is finite
(T2) $\forall e, e' \in E$, $e \lessdot e' \Rightarrow \lambda(e) D \lambda(e')$
(T3) $\forall e, e' \in E$, $\lambda(e) D \lambda(e') \Rightarrow e \leq e'$ or $e' \leq e$

(T2) says that events that are immediately causally related must correspond to dependent actions and (T3) says that any two events with dependent labels must be causally related. When $I = \emptyset$, all events are causally related and a trace degenerates to a word over Σ. See [7, 8] for further details on traces.

For example, Figure 2 depicts two traces t_1 and t_2. In the diagrams, I show only the immediate causal relation using arrows; the labels of events are shown inside the box while the event name is shown outside.

Let $\mathcal{T}r(\Sigma, I)$ denote the set of all traces over (Σ, I), and let $\mathcal{T}r^f(\Sigma, I)$ and $\mathcal{T}r^\omega(\Sigma, I)$ denote the set of finite and infinite traces over (Σ, I) (i.e. where the event set E is finite or infinite, respectively). A trace $t = (E, \leq, \lambda)$ is isomorphic to a trace $t' = (E', \leq', \lambda')$, denoted $t \approx t'$, if they are structurally isomorphic, i.e. there is a one-to-one correspondence between E and E' that preserves the partial order relations and the labeling functions. We do not distinguish between isomorphic traces.

If $t = (E, \leq, \lambda)$ is a trace, and $e \in E$, let $\Downarrow e$ be the sub-trace defined by $\downarrow e$, i.e. $\Downarrow e = (\downarrow e, \leq', \lambda')$ where $\leq'$ and λ' are $\leq$ and λ, respectively, restricted to $\downarrow e$. It is easy to see that $\Downarrow e$ is a trace and that if $e, e' \in E$ with $\Downarrow e \approx \Downarrow e'$, then $e = e'$.

Let us define concatenation over traces, which is basically obtained by taking the union of the two traces and causally relating events of the first trace to the events in the second trace that have dependent labels. Let $t_1 \in \mathcal{T}r^f(\Sigma, I)$, $t_2 \in \mathcal{T}r(\Sigma, I)$, with $t_1 = (E_1, \leq_1, \lambda_1)$ and $t_2 = (E_2, \leq_2, \lambda_2)$. Assume without loss of generality that E_1 and E_2 are disjoint. Then $t_1 \cdot t_2 = (E, \leq, \lambda)$ where $E = E_1 \cup E_2$, $\leq = (\leq')^*$, the reflexive and transitive closure of $\leq'$, where $\leq' = \leq_1 \cup \leq_2 \cup \{(e_1, e_2) \mid e_1 \in E_1, e_2 \in E_2, \lambda(e_1) D \lambda(e_2)\}$, and the labeling function is given by $\lambda(e) = \lambda_i(e)$ if $e \in E_i$, $i \in \{1, 2\}$. It is easy to verify that $t_1 \cdot t_2$ is a trace in $\mathcal{T}r(\Sigma, I)$.

We also define a prefix relation on traces: let $\sqsubseteq$ be the relation over traces over (Σ, I) such that $t_1 \sqsubseteq t_2$ iff there is some trace $t \in \mathcal{T}r(\Sigma, I)$ such that $t_1 \cdot t = t_2$. Equivalently, $t_1 \sqsubseteq t_2$ iff for every event e_1 in t_1, there is an event e_2 in t_2 such that $\Downarrow e_1 \approx \Downarrow e_2$. Note that $\sqsubseteq$ is a partial order on $\mathcal{T}r(\Sigma, I)$.

Two finite traces are said to be in conflict if they have no possible common extension. That is, t_1 is in conflict with t_2, denoted $t_1 \not\uparrow t_2$, if there is no trace $t \in \mathcal{T}r(\Sigma, I)$ such that $t_1 \sqsubseteq t$ and $t_2 \sqsubseteq t$. The conflict relation is clearly symmetric and irreflexive. Intuitively, conflicting traces describe two finite behaviors that are truly branching. For example, in Figure 2, t_1 and t_2 are in conflict.

For a trace $t = (E, \leq, \lambda)$, a *linearization* of t is a linearization (at most an ω-linearization) of its events that respects the partial order, i.e. a structure $(E, \leq', \lambda')$ where $\leq'$ is an (at most ω) linear order with $\leq \ \subseteq \ \leq'$. But a linearly ordered Σ-labeled structure is really a word over Σ. Hence the set of linearizations of t, denoted $lin(t)$, can be seen as a subset of Σ^*, if t is finite, or a subset of Σ^ω if t is infinite. Note that any infinite trace t has at least one ω linearization.

The conditions (T2) and (T3) make traces a very restricted kind of labeled partial order. In fact, linearizations of traces have as much information as the trace itself. One can verify, for example, that every word $\sigma \in \Sigma^\infty$ is a linearization of a unique trace over (Σ, I). This hence induces a natural equivalence relation over Σ^∞: $\sigma \equiv_I \sigma'$ iff σ and σ' are linearizations of the same trace. We sometimes refer to a trace using the notation $[\sigma]$, i.e. using the equivalence class of linearizations of the trace. For example, in Figure 2, $t_1 = [abca]$ since $abca$ is a linearization of t_1.

A trace language over (Σ, I) is a set of traces over (Σ, I). There are several ways to define regular languages of finite traces. The following definition is simple and convenient: a language of finite traces $\mathcal{L}$ over (Σ, I) is *regular* if the set of its linearizations, $lin(\mathcal{L})$ is a regular subset of Σ^*. (The class of regular languages of words is the standard notion and I assume the reader's familiarity with this.)

Event Structures

A Σ-*labeled (prime) event structure*, introduced first in [19], is a structure $\mathscr{ES} = (Ev, \preceq, \#, \eta)$ where Ev is a finite or countable set of *events*, $\preceq$ is a partial order on Ev called the *causality relation*, $\# \subseteq Ev \times Ev$ is an irreflexive and symmetric relation called the *conflict relation*, and $\eta : Ev \rightarrow \Sigma$ is a labeling function such that the following hold:

(ES1) If $e_1, e_2, e \in Ev$ and $e_1 \# e_2 \preceq e$, then $e_1 \# e$
(ES2) For every $e \in Ev$, $\downarrow e$ is finite.

(ES1) says that if two events are in conflict, then their future events are also in conflict. This also ensures that if $e \leq e'$, then $\neg(e\#e')$, i.e. causally related events are not in conflict. Hence $\downarrow e$, for any $e \in E$, is a conflict-free set. Note that we call elements of both traces as well as event-structures "events"; this will cause no confusion, however, as it will be clear from the context what me mean. The reader is referred to [27] for the literature on event structures.

Now consider a set $\mathscr{L}$ of traces over (Σ, I) that is prefix closed (i.e. $\forall t \in \mathscr{L}, \forall t' \sqsubseteq t, t' \in \mathscr{L}$). Then $\mathscr{L}$ can be seen to represent an event structure. In order to define this, we first need to define prime traces. A trace t is *prime* if it has exactly one maximal element. Let $pr(\mathscr{L})$ denote the set of prime traces in $\mathscr{L}$. Also, for a prime trace t, let $max(t)$ denote its maximal element.

If $\mathscr{L} \subseteq \mathscr{T}r(\Sigma, I)$ is a prefix-closed language of traces, let $\mathscr{ES}_{\mathscr{L}} = (Ev, \preceq, \#, \eta)$ where $Ev = pr(\mathscr{L})$, $\preceq = (\sqsubseteq \cap (Ev \times Ev))$, $\# = (\not\uparrow \cap (Ev \times Ev))$, and $\eta(t) = \lambda(max(t))$. In other words, we take the set of prime traces to be the events of the event structure, take the prefix and conflict relation on traces as the causality and conflict relations, respectively, and label each event of the event structure, which is a prime trace, using the label of the maximal element of the prime trace. The event structure $\mathscr{ES}_{\mathscr{L}}$ is strongly related to $\mathscr{L}$: for example, one can verify that $\downarrow t$ in the event structure is isomorphic to t and hence $pr(\mathscr{L})$ is recoverable from $\mathscr{ES}_{\mathscr{L}}$. See [22, 27] for a formal correspondence between $\mathscr{ES}_{\mathscr{L}}$ and $\mathscr{L}$.

Event structures thus obtained from languages of traces are special in the sense that the labels respect the causal and conflict relations. Let us identify this formally. First, let us define the *minimal* conflict relation as $e\#_{\mu}e'$ if and only if $e\#e'$ and there exists no $f \leq e$ with $f\#e'$ and there exists no $f' \leq e'$ with $e\#f'$.

A *trace event structure* over a trace alphabet (Σ, I) is a Σ-labeled event structure $\mathscr{ES} = (Ev, \preceq, \#, \eta)$ which satisfies the following: for every $e, e' \in Ev$,

(TES1) If $e \#_\mu e'$, then $\eta(e) \neq \eta(e')$
(TES2) If $e \prec\!\!\cdot\; e'$ or $e \#_\mu e'$, then $\eta(e) D \eta(e')$
(TES3) If $\eta(e) D \eta(e')$, then $e \preceq e'$ or $e' \preceq e$ or $e \# e'$.

Condition (TES1) is a determinacy condition that intuitively says that each prime trace is represented by at most one event (if $e \#_\mu e'$ and $\eta(e) = \eta(e')$, then $\downarrow e$ and $\downarrow e'$ define the same prime trace). (TES2) and (TES3) reflect the obvious connection between the labels, the causality and conflict relation to the independence relation.

We then have:

Proposition 1 *([22]) For any prefix-closed language of traces $\mathscr{L}$, $\mathscr{ES}_{\mathscr{L}}$ is a trace event structure.*

Finally, a *regular trace event structure* is an event structure $\mathscr{ES}$ such that $\mathscr{ES} = \mathscr{ES}_{\mathscr{L}}$ for some *regular* trace language $\mathscr{L}$. Regular trace event structures can thus be presented by a finite automaton over Σ^* that accepts $lin(\mathscr{L})$. We assume this presentation in the formulation of decidability results in this article.

For example, consider the alphabet $\Sigma = \{a, b, c\}$ and where $I = \{(a, b), (b, a)\}$. Consider the language of traces $\mathscr{L}_1 = ([abc] + [aabbc])^*$ (where regular expressions over traces have the obvious interpretation). Let $\mathscr{L}$ be the prefix-closure of $\mathscr{L}_1$. Note that traces such as $a.b$ are not prime traces in $\mathscr{L}$. Part of the event structure corresponding to $\mathscr{L}$ is shown in Figure 2. In the diagram, we denote the immediate causal relation using arrows and denote only the minimal conflict relation using a dashed edge.

Though the above definition of trace event structures and regularity is all that we need here, $\mathscr{L}$ and $\mathscr{ES}_{\mathscr{L}}$ are more strongly related. (see [22, 27] for details). There is in fact a very interesting open conjecture by Thiagarajan [24] that all event structures are regular in the sense that they have only a finite number of non-isomorphic "configurations" and satisfy a natural branching condition (which demands that there are only a bounded number of events enabled from any configuration) ought to be a regular *trace* event structure as well. This conjecture implies that every regular event structure can be defined by a finite 1-safe Petri net.

Logics over event structures

For any logical relational structure $(U, R_1, \ldots, R_n)$, where U is the universe and each R_i is a relation over U, the first order logic over the structure is the standard one that allows quantification over U,

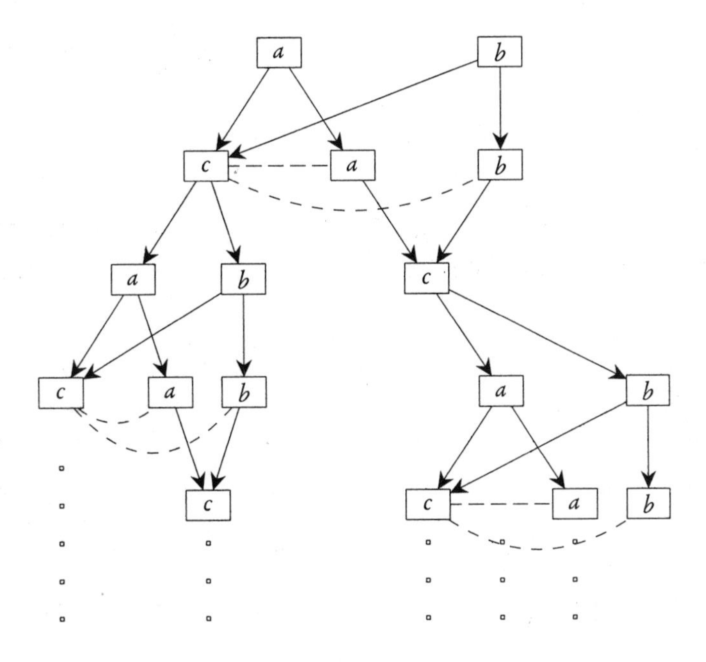

Figure 2: *Event structure corresponding to $\mathscr{L}$*

checking equality, checking the relations R_i ($i \in [n]$) between quantified elements and closed under boolean operations.

A labeled event structure $(Ev, \preceq, \#, \eta)$ can be viewed as a structure $(Ev, \preceq, \#, \eta_{a_1}, \ldots, \eta_{a_n})$ where $\Sigma = \{a_1, \ldots, a_n\}$ and each η_{a_i} is a unary predicate that identifies the elements labeled a_i.

FOL over event structures is then the logic:

$$\varphi ::= x = x' \mid x \preceq x' \mid x \# x' \mid \eta_a(x) \mid \neg\varphi \mid \varphi \wedge \varphi \mid \exists x.\varphi(x)$$

where x, x' are variables and $a \in \Sigma$.

Similarly, monadic second-order logic (MSOL) extends FOL in that it also allows quantification over *sets* of events:

$$\begin{aligned} \text{MSOL} : \varphi \quad ::= \quad & x = x' \mid x \preceq x' \mid x \# x' \mid \eta_a(x) \mid x \in X \mid \\ & \neg\varphi \mid \varphi \wedge \varphi \mid \exists x.\varphi(x) \mid \exists X.\varphi(X) \end{aligned}$$

The semantics of FOL and MSOL are the standard semantics on structures.

On event structures, we consider a new logic called *monadic trace logic* (MTL) which is an extension of FOL and a restriction of MSOL. The syntax of MTL is the same as that of MSOL; however, semantically, set-quantifiers are not interpreted over *all* possible sets of events but only over *conflict-free* sets of events. In other words, a variable X can be interpreted as a set of events $F \subseteq Ev$ only if F has no two events that are in conflict. Hence one can find a single trace such that all the events in F belong to the trace (and hence the name monadic trace logic).

For any of these logics LOG $\in$ {FOL, MSOL, MTL}, the model-checking problem for regular trace event structures against specifications in LOG is the following: Given a finite presentation of a regular trace event structure $\mathscr{ES}_{\mathscr{L}}$ and a formula $\varphi \in$ LOG, does $\mathscr{ES}_{\mathscr{L}} \models \varphi$?

It has been observed that MSOL is too expressive a logic and leads to undecidability (from [22], attributed to Walukiewicz):

Proposition 2 *[22] The model-checking problem for regular trace event structures against* MSOL *specifications is undecidable.*

The idea in the proof of the above proposition is to consider $\Sigma = \{a, b, c\}$, $I = \{(a, b), (b, a)\}$ and the prefix-closure of the language $\{[w] \mid w \in a^* b^* c\}$. The c-events then "stick out" at the grid points defined by a and b, and an MSOL formula can encode an undecidable tiling problem.

We can now state our main result:

Theorem 3 *The model-checking problem for regular trace event structures against* FOL *and* MTL *specifications is decidable.*

Before we move on, let me recall MSOL on the structure $(\mathbb{N}, succ)$, the monadic second-order logic on natural numbers with successor relation. Note that the $<$ relation on numbers can be defined using $succ$ in MSOL, and we can write a predicate $zero(x)$ that states that $x = 0$. It is well known that MSOL over $(\mathbb{N}, succ)$ is decidable ([3]). We will reduce the model-checking problem of FOL and MTL on regular event structures to the satisfiability problem for MSOL over $(\mathbb{N}, succ)$.

3 The Foata Normal Form and its Properties

The key insight in interpreting logics over event structures to logics on linear orders is to use the Foata normal form [4, 8]. Intuitively, every event in a trace event-structure is represented by a trace, and we can represent this trace using any of its linearizations. However, the Foata normal form has certain nice properties that help us in defining relations of causality and conflict of events using *synchronized regular relations* on the words representing two events. Synchronized regular relations are, loosely speaking, relations between words for which there is a regular finite automaton that can read a pair of words, reading alternatingly one letter from each word, and accepting the pair of words iff they belong to the relation. As we shall see, synchronized regular relations are a simple mechanism to interpret relations on a linear order.

For example, suppose aIb, aDc, and bDc. If we choose to represent the prime traces of the form $[a^n.b^m.c]$ (for all values of n and m) uniformly using the linearization $a^n.b^m.c$, then the causality relation that relates events of the form b^i ($i \leq m$) to traces of the form $a^j.b^{i+k}.c$ (for arbitrary j and k) would not be synchronized-regular as it would require counting the number of b's in the traces. Intuitively, it seems that alternating a's and b's would be a better representation, and in fact, the Foata normal form does this.

In this paper we will use a variant of the Foata normal form which encodes *multiple independent letters* simultaneously. The Foata normal form encoding, which will encode the trace $a^j.b^m$ ($j < m$) as $\{a,b\}^j \cdot \{b\}^{m-j}$, where both $\{a,b\}$ and $\{b\}$ are treated as single letters), and the trace b^i as $\{b\}^i$, makes the causal relation between these traces synchronized regular. This idea carries over to the general setting, making both the causal and conflict relations on Foata normal form traces a synchronized regular relation.

Let us now define the Foata normal form formally. Fix a trace alphabet (Σ, I). For a trace $t = (E, \leq, \lambda) \in \mathscr{T}r(\Sigma, I)$, let $Rank : E \rightarrow \mathbb{N}$ denote the (unique) function that satisfies the following:

- For any minimal event $e \in E$, $Rank(e) = 1$
- For any non-minimal event e, $Rank(e) = max\{Rank(f) \mid f \lessdot e\} + 1$

Intuitively, the rank of an event captures the length of the longest chain of causal predecessors of the event.

Let Π denote the set of all sets $S \subseteq \Sigma$ such that the elements of S are pairwise independent (i.e. $\forall a, b \in S.(a \neq b) \Rightarrow aIb$). The Foata normal form of a (finite or infinite) trace $t = (E, \leq, \lambda)$, denoted $FNF(t)$, is the infinite word $\pi_1\pi_2\pi_3 \ldots \in \Pi^\omega$, where $\pi_i = \{\lambda(e) \mid e \in E \text{ and } Rank(e) = i\}$.

Intuitively, the Foata normal form of a trace is obtained by collecting all the minimal elements of a trace to form the first letter, then removing these events from the trace, taking the resulting set of minimal events to form the second letter, removing them from the trace, and so on. Note that the events of the same rank have independent labels. Also, notice that for a finite trace t, its Foata normal form is a word such that if k is the maximal rank (and there will be a maximal rank), letters from $k+1$ onwards are the empty set. (Note that though we can represent finite traces using finite words, it will be convenient to represent it as an infinite word as above.)

For example, the Foata normal forms of the two traces in Figure 1 are $\{a, b\}\{c\}\{a\}$ and $\{a, b\}\{c\}\{b\}\{c\}$, respectively.

Not all elements of Π^ω correspond to the Foata normal form of some trace. However, it is easy to see the following:

Proposition 4 *The word $\pi_1\pi_2 \ldots \in \Pi^\omega$ is a Foata normal form of a trace iff it satisfies the following condition:*

- *For any $i > 1$, if $a \in \pi_i$, then there exists some $b \in \pi_{i-1}$ such that aDb.*

Moreover, $\pi_1\pi_2 \ldots \in \Pi^\omega$ is a Foata normal form of a finite *trace iff the above condition holds and for some $i \in \mathbb{N}$, $\pi_i = \emptyset$.*
From the above, the following is clear:

Proposition 5 *The set of all Foata normal forms of finite traces (i.e. FNF($\mathcal{T}r(\Sigma, I)$)), is regular.*
In fact, the following is not hard to see:

Proposition 6 *For any regular language $\mathcal{L}$ of finite traces over (Σ, I), the set of all Foata normal forms of the traces in the language (i.e. FNF($\mathcal{L}$)), is a regular ω-language.*

The above is easy to show: for example given a finite automaton A accepting all the linearizations of $\mathcal{L}$, we can build an automaton that reads a word over Π^ω and processes each set of letters read by simulating A on its contents in some order. We can then intersect this with the automaton accepting the set of words corresponding to Foata normal forms of finite traces.

The following lemma captures the usefulness of the Foata representation in capturing the prefix relation between traces.

Lemma 7 *Let t and t' be two traces, and let FNF(t) $= \pi_1\pi_2 \ldots$ and FNF(t') $= \pi'_1\pi'_2 \ldots$. Then $t \sqsubseteq t'$ iff*

(a) *$\forall i.(\pi_i \subseteq \pi'_i)$ and*
(b) *If $a \in \pi'_i \setminus \pi_i$ for any $i \in \mathbb{N}$, then for every $b \in \pi_j$ $(j > i)$, it must be that aIb.*

The above requires a proof, but is not hard to show. The forward direction follows from the following observation: intuitively, if $t \sqsubseteq t'$, when new events are added to t to obtain t', the rank of events in t do not change (hence condition (a) is met). Also, an event e in t' that is not in t cannot have a successor event in t; hence if its rank is i, then all events in t whose rank is greater than i must be independent of e (which makes (b) true). Conversely, if the latter conditions (a) and (b) of the lemma hold, then it is easy to see that the set of causal predecessors of any event in t is the same as the set of causal predecessors of the corresponding event in t'. Hence t must be a prefix of t'.

We can now similarly characterize the conflict relation on the Foata normal form representation:

Lemma 8 *Let t and t' be two traces, and let $FNF(t) = \pi_1\pi_2\ldots$ and $FNF(t') = \pi'_1\pi'_2\ldots$. Then $t \not\!\smile t'$ iff there exists $a, b \in \Sigma$, aDb, and $i, j \in \mathbb{N}$ with $i < j$ such that one of the following hold:*

(a) $a \in \pi'_i \setminus \pi_i$ *and* $b \in \pi_j$;
(b) $a \in \pi_i \setminus \pi'_i$ *and* $b \in \pi'_j$;

The two conditions are similar to the negation of the second condition in the characterization of the prefix relation of traces. Consider the backward implication. Intuitively, if t and t' are not in conflict, then it is easy to see that the Foata normal form of the minimal trace t'' which both t and t' are prefixes of is $\alpha = (\pi_1 \cup \pi'_1)(\pi_2 \cup \pi'_2)\ldots$. Since t_1 and t_2 are prefixes of t'', it follows from Lemma 7 that neither condition (a) nor condition (b) of the above lemma are met. Conversely, if the latter conditions (a) and (b) of the above lemma both does not hold, it is easy to see that the Foata normal form α above represents a trace t'', and by Lemma 7, $t \sqsubseteq t''$ and $t' \sqsubseteq t''$, proving t and t' are not in conflict.

The crucial aspect of the above two lemmas characterizing the prefix and causal relations on Foata-normal forms of traces is that they can be expressed in monadic second-order logic. We will exploit this in order to interpret an event of an event structure using a fixed tuple of sets over natural numbers (capturing the Foata normal form of the prime trace representing the event), and use MSOLover $(\mathbb{N}, succ)$ to capture the prefix and conflict relations.

4 Model-checking FOL Properties

We now consider model-checking FOL sentences on trace event structures. FOL is actually powerful enough to express most modal logics on event structures studied in the literature [18, 20]. Modal logics typically have modalities that refer to either future events, past events, conflicting events, minimally conflicting events or concurrent events. All such modalities can easily be expressed in FOL. Also, the "gossip" modality is sometimes used especially when the event structure explicitly comprises n communicating sequential agents [13, 12]; this modality can also be expressed in FOL. In fact, it seems to be hard to come up with any natural modality that is not expressible in FOL.

Recall that every regular trace event structure E is represented by a regular prefix-closed trace language $\mathscr{L}$. Also, any regular trace language $\mathscr{L}$ can be represented using a finite automaton that accepts the set of all linearizations of traces $Lin(\mathscr{L})$. We will hence assume that we are presented with a regular trace event structure through a finite automaton accepting a trace-closed prefix-closed language L.

The FOL model-checking problem is then to find, given a trace-closed prefix-closed language $L \subseteq \Sigma^*$ and a FOL-sentence φ, whether the regular trace event structure defined by L satisfies φ.

The new proof of model-checking FOL on regular event structures is achieved by interpreting FOL formulas using MSOL over the linear order $(\mathbb{N}, succ)$. Decidability of the latter implies decidability of the former. More formally, for any regular event structure R and a FOL sentence φ, we show how to constructively build an MSOL sentence φ' over $(\mathbb{N}, succ)$ such that φ holds in R iff φ' holds in $(\mathbb{N}, succ)$.

Representing events using a tuple of sets

An event e of a trace event structure is uniquely represented by the prime trace $\downarrow(e)$. In order to represent this trace on the linear order $(\mathbb{N}, succ)$, we represent e using $|\Sigma|$ sets $X_1, \ldots, X_n$ (we henceforth assume $|\Sigma| = n$). Let $\overline{X}$ denote the n-tuple of sets $X_1, \ldots X_n$.

More formally, let us fix an ordering over the letters in Σ, and let $\Sigma = \{a_1, \ldots a_n\}$ denote this ordered set. Let t be a trace over (Σ, I) and let $FNF(t) = \pi_1 \pi_2 \ldots$. Then we say that $(X_1, \ldots, X_n)$ is the interpretation of t on $(\mathbb{N}, succ)$ if $X_1, \ldots, X_n \subseteq \mathbb{N}$ and for every $m \in [1, n]$ and every $i \in \mathbb{N}$, $i \in X_m$ iff $a_m \in \pi_i$. For any trace t, let us denote its interpretation on $(\mathbb{N}, succ)$ as $Int(t)$.

It is obvious that the interpretation of a trace is a unique representation of it (i.e. Int is a one-to-one function).

First, note that the set of all interpretations of traces over (Σ, I) is regular, i.e. we can write an MSOL formula over $(\mathbb{N}, succ)$ $\psi_{FNF}(X_1, \ldots, X_n)$ such that for any tuple of sets $S_1, \ldots, S_n \subseteq \mathbb{N}$, $\psi_{FNF}(S_1, \ldots, S_n)$ holds iff there is some trace t such that $Int(t) = (S_1, \ldots, S_n)$. This formula is, in fact, using Proposition 4:

$$\psi_{FNF}(\overline{X}) = \psi_{Valid}(X_1, \ldots, X_n) \wedge \bigwedge_{m \in [1,n]} (\forall i.((i \in X_m) \Rightarrow (\forall j.(succ(j,i) \Rightarrow \bigvee_{m' | a_m D a_{m'}} (j \in X_{m'})))))$$

where

$$\psi_{Valid}(\overline{X}) = \bigwedge_{\substack{m, m' \in [1,n] \\ m \neq m', a_m D a_{m'}}} \forall i.(i \in X_m \Rightarrow \neg(i \in X_{m'}))$$

The formula ψ_{FNF} expresses the property in Proposition 4, while ψ_{Valid} checks that the elements of rank i, for any i, are independent.
Foata normal forms of finite traces can be captured by the following formula, again using Proposition 4:

$$\psi^{fin}_{FNF}(\overline{X}) = \psi_{FNF}(\overline{X}) \wedge (\exists i.(\bigwedge_{m \in [1,n]} \neg(i \in X_m)))$$

Let $\overline{X} = Int(t_1)$ and $\overline{Y} = Int(t_2)$ denote the interpretation of two traces t_1 and t_2. Then we can express various properties of these traces using MSOL on their interpretations:

Prefix We can express $t_1 \sqsubseteq t_2$ by expressing the two properties in Lemma 7. Let's denote this as the formula $\psi_{Prefix}(\overline{X}, \overline{Y})$:

$$\psi_{Prefix}(\overline{X}, \overline{Y}) = \bigwedge_{m \in [1,n]} (\forall i.(i \in X_m) \Rightarrow (i \in Y_m)) \wedge \bigwedge_{m, m' \in [1,n] | a_m D a_{m'}} \neg(\exists i \exists j\ (i < j \wedge i \in Y_m \wedge i \notin X_m \wedge j \in Y_{m'}))$$

Conflict Using properties in Lemma 8, we can express in MSOL that $t_1 \# t_2$; let's denote this using the formula $\psi_{Conflict}(\overline{X}, \overline{Y})$:

$$\psi_{Conflict}(\overline{X}, \overline{Y}) = \bigvee_{\substack{m, m' \in [1,n] \\ a_m D a_{m'}}} (\exists i\ \exists j\ (i < j \wedge ((i \in Y_m \wedge i \notin X_m \wedge j \in X_{m'}) \vee (i \in X_m \wedge i \notin Y_m \wedge j \in Y_{m'}))\ ))$$

Prime label For any trace t, we can write a formula that says that t is a prime trace and the final event in t is labeled a_m (where $a_m \in \Sigma$). We do this by demanding that there is a maximal $i \in \mathbb{N}$ that belongs to one of the sets in $\overline{X}$, and in fact this i belongs to X_m only.

$$\psi_{Label\text{-}m}(\overline{X}) = \psi_{FNF}(\overline{X}) \wedge \exists i.(i \in X_m \wedge \bigwedge_{m' \neq m} \neg(i \in X_{m'}) \wedge (\forall j.(j > i \Rightarrow \bigwedge_{m' \in [1,n]} j \notin X_{m'})))$$

Valid events in $\mathcal{L}$: Given any regular language of traces $\mathcal{L}$, recall that the set of words over Π^{ω} representing the Foata normal form sequences of traces in $\mathcal{L}$ is regular (Proposition 6), and hence expressible using MSOL over words. Using this formula, we can easily express whether an interpretation of an event belongs to the event structure, i.e. whether the trace corresponding to the event is a prime trace in $\mathcal{L}$. Let's denote this formula as $\psi_{\mathcal{L}}(\overline{X})$ (for any regular language of traces $\mathcal{L}$).

Interpretation of FOL on regular trace event structures

We are now ready to describe the main construction that maps any FOL sentence $\widehat{\varphi}$ over a regular trace event structure (presented as a regular trace language $\mathcal{L}$) to an MSOL sentence $\widehat{\psi}_{\mathcal{L},\varphi}$ over $(\mathbb{N}, succ)$ such that the regular trace event structure represented by $\mathcal{L}$ satisfies φ iff ψ holds in $(\mathbb{N}, succ)$. We will call this mapping h and it will be obviously constructible.

For every first-order variable x, let us fix a tuple of set variables $X_1, \ldots, X_n$ (i.e. we fix for each first-order variable, in lower-case, a tuple of set variables in the corresponding upper-case with indices running from 1 through n). We define h inductively such that any FOL formula φ with free-variables $x^1, \ldots x^k$ over an event structure is mapped by h to a formula with the corresponding free set variables $X_1^1, \ldots, X_n^1, X_1^2, \ldots, X_n^2, X_1^k, \ldots, X_n^k$. The function h will satisfy property that for any FOL formula $\varphi(x^1, \ldots, x^k)$ over the regular event structure represented by $\mathcal{L}$ with each x_i interpreted as e^i (with t^i being the prime trace representing e_i) satisfies φ iff $h(\varphi)(\overline{X^1}, \ldots \overline{X^k})$ holds in $(\mathbb{N}, succ)$ where each $\overline{X^i}$ is interpreted as $Int(t^i)$.

The function h is defined as follows:

- $h(x = y) = \bigwedge_{m\in[1,n]} X_m = Y_m$;
- $h(x \sqsubseteq y) = \psi_{Prefix}(\overline{X}, \overline{Y})$;
- $h(x \# y) = \psi_{Conflict}(\overline{X}, \overline{Y})$;
- $h(\eta_{a_m}(x)) = \psi_{Label\text{-}m}(\overline{X})$;
- $h(\neg\varphi) = \neg h(\varphi)$;
- $h(\varphi_1 \wedge \varphi_2) = h(\varphi_1) \wedge h(\varphi_2)$;
- $h(\exists x.\varphi) = \exists X_1, \ldots, X_n.(\psi_{\mathcal{L}}(\overline{X}) \wedge h(\varphi))$.

Given the correctness of the individual formulas, it is not hard to establish the following main lemma:

Lemma 9 *For any regular prefix-closed trace language $\mathcal{L}$ over (Σ, I) and a* FOL *sentence $\widehat{\varphi}$ over event structures over (Σ, I), the regular event structure corresponding to $\mathcal{L}$ satisfies $\widehat{\varphi}$ iff $(\mathbb{N}, succ)$ satisfies $h(\widehat{\varphi})$.*

Given that MSOL over $(\mathbb{N}, succ)$ is decidable, we have:

Theorem 10 *The model-checking problem for regular trace event structures against* FOL *sentences is decidable.*

Notice that we need only the decidability of the *weak* monadic second-order theory of $(\mathbb{N}, succ)$ [2] to derive the above result (as all set quantifications corresponding to prime traces are necessarily finite).

5 Model-checking MTL Properties

In this section, I show how to model check regular trace event structures against MTL specifications. Recall that in MTL one can quantify over sets of events, but any set of events that is quantified must be conflict-free. It is easy to see that if F is a conflict-free set of events of an event structure, then $\bigcup_{f\in F} \downarrow f$ is a subset of events that corresponds to a trace t_F, perhaps an infinite one, that contains all the events in F as subtraces.

In the decidability proof for FOL, recall that each first-order variable on the event structure was interpreted using an n-tuple of sets of natural numbers. The crux of the decidability proof for MTL is to interpret each set of conflict-free events using a $2n$-tuple of sets of natural numbers— n of these will capture a (possibly infinite non-prime) trace that contains all these events and the other n would be used to capture which prime subtraces of this trace belong to the set.

Formally, let T be a pairwise confict-free set of prime traces over (Σ, I). Then there is a unique minimal trace t such that all traces $t' \in T$ are prefixes of t. Let the Foata normal form of t be $FNF(t) = \pi_1 \pi_2 \ldots$. We say the $(X_1, \ldots, X_n, \widehat{X}_1, \ldots \widehat{X}_n)$ is the interpretation of the set of prime traces T on $(\mathbb{N}, succ)$ if $X_1, \ldots, X_n, \widehat{X}_1, \ldots \widehat{X}_n \subseteq \mathbb{N}$ and (a) $X_1, \ldots, X_n$ is the interpretation of the trace t (as defined in the last section), and (b) for every $i \in \mathbb{N}$ and $m \in [1, n]$, $i \in \widehat{X}_m$ iff there is a prime trace $t' \in T$ whose maximal element e has label a_m and rank i.

Intuitively, in order to represent T, we take a trace t that subsumes all events in T (this is possible only because T is conflict-free), take $X_1, \ldots, X_n$ to be the interpretation of t, and use $\widehat{X}_1, \ldots, \widehat{X}_n$ to pick the events in t whose downclosure trace belongs to T. It should be clear that the interpretation of any set T of conflict-free prime traces is unique.

Notice that the above scheme fails for interpreting an arbitrary set of events in an event structure. Intuitively, if $\overline{X}$ and $\overline{Y}$ represent two events t_1 and t_2, then using $\overline{Z}$ to represent $\{t_1, t_2\}$ where each $Z_i = X_i \cup Y_i$ does not work, as it represents traces which are neither t_1 nor t_2. Hence, a faithful set interpretation of an unbounded number of conflicting traces is hard. Of course, we expect no such interpretation, as the monadic theory of event structures is undecidable.

We are now ready to interpret MTL over trace event structures on $(\mathbb{N}, succ)$ using MSOL. For every first-order variable x, let us fix, as before, an n-tuple of set variables $X_1, \ldots, X_n$, and for every set-variable Z, let us fix a $2n$-tuple of set variables $Z_1, \ldots, Z_n, \widehat{Z}_1, \ldots, \widehat{Z}_n$. We define a function h that maps any formula MTL φ over an event structure with free variables $x^1, \ldots, x^k, Z^1, \ldots Z^l$ to an MSOL formula over $(\mathbb{N}, succ)$ with free variables $\{X_i^j \mid i \in [1, m], j \in [1, k]\} \cup \{Z_i^j, \widehat{Z}_i^j \mid i \in [1, m], j \in [1, l]\}$. Furthermore, $\varphi(x^1, \ldots, x^k, Z^1, \ldots, Z^l)$ holds when each x^i is interpreted as the trace t^i and each Z_j is interpreted as the conflict-free set T^j iff $h(\varphi)$ holds over $(\mathbb{N}, succ)$ when each set of variables $X_1^i, \ldots, X_m^i$ is the interpretation of t^i and each set of variables $Z_1^j, \ldots, Z_m^j, \widehat{Z}_1^j, \ldots, \widehat{Z}_m^j$ is the interpretation of T^j.

The function h is the same as the function h defined in the previous section, but extended in the following way:

- $h(x \in Z) = \psi_{Prefix}(X_1, \ldots, X_n, Z_1, \ldots, Z_n) \wedge$
  $\bigvee_{m \in [1,n]} (\exists i. (i \in X_m \wedge (\bigwedge_{m' \neq m} \neg (j \in X_{m'})) \wedge$

$$(\forall j.(succ(i,j) \Rightarrow \bigwedge_{m' \in [1,n]} \neg(j \in X_{m'}))) \wedge (i \in \widehat{Z}_m)))$$

- $h(\exists Z.\varphi) = \exists Z_1, \ldots, Z_n, \widehat{Z}_1, \ldots, \widehat{Z}_n.$
  $(\psi_{FNF}(Z_1, \ldots, Z_n) \wedge \bigwedge_{m \in [1,n]} (\widehat{Z}_m \subseteq Z_m) \wedge h(\varphi))$

Intuitively, when we want to quantify over a conflict-free set Z, we instead quantify over a $2n$-tuple of sets $\bar{Z}, \bar{\widehat{Z}}$, and demand that $\bar{Z}$ represents the Foata normal form of some (possibly infinite) trace, and that $\widehat{Z}$ chooses events only from this trace. In order to check whether $x \in Z$, we check whether trace representing x is a prefix of that trace which includes all events of Z and that $\widehat{Z}$ includes the event represented by x.

We can now show:

Lemma 11 *For any regular prefix-closed trace language $\mathcal{L}$ over (Σ, I) and a* MTL *sentence $\widehat{\varphi}$ over event structures over (Σ, I), the regular event structure corresponding to $\mathcal{L}$ satisfies $\widehat{\varphi}$ iff $(\mathbb{N}, succ)$ satisfies $h(\widehat{\varphi})$.*
Again, by decidability of $(\mathbb{N}, succ)$, we have:

Theorem 12 *The model-checking problem for regular trace event structures against* MTL *specifications is decidable.*

6 Conclusions

The main result of this paper is that FOL and MTL can be proved to be decidable logics for model-checking regular trace event structures by interpreting them to monadic second-order logic over the natural numbers with the successor relation. I believe this proof to be significantly simpler than the one presented in [16].

In the decidability proof of FOL, note that for the proof to go through, one just needed that the set of prime traces of the event structure represented in Foata normal form is regular. We can hence generalize the result to the entire class of structures that has the above property. In other words, for any set of prime traces $\mathcal{L}$ such that $FNF(\mathcal{L})$ is regular, FOL and MTL are decidable on the event-structure defined by $\mathcal{L}$. For instance, the event structure defined by the *non-regular* prefix-closure of the set of traces $a^n b^n c$, where aIb, aDc, bDc, does admit a decidable FOL model-checking problem, since the Foata normal representation of these events is $[\{a\}]^* + [\{b\}]^* \cup [\{a,b\}]^* \cdot [\{c\}]$.

Note that I have not considered the *satisfiability* problem for any of these logics. There are modal logics over event structures studied in the literature for which satisfiability is decidable [13, 15, 20].

However, we can show, using a variant of the MSOL model-checking undecidability result, that the satisfiability problem for trace event structures for FOL specifications is already undecidable.

In this work, I have reduced the expressive power of the logic to get decidability. Another approach is to restrict the class of event structures to get MSOL-decidability. An open problem in this vein is to characterize the precise class of event structures that have a decidable monadic theory. Thiagarajan has another conjecture on this [23]: he has identified a class of "grid-free" event structures (which can even be defined on the finite 1-safe Petri nets generating these event structures), and has conjectured that they constitute the *precise* class of trace event structures on which MSOL is decidable. I believe this conjecture ought to hold, and further, that a variant of the Foata normal form may be useful in proving it (we would of course expect an interpretation on MSOL over trees to prove this result as the class of event structures above includes trees).

References

[1] A. Blumensath and E. Grädel. Automatic structures. In *15th Annual IEEE Symposium on Logic in Computer Science*. IEEE Computer Society, 2000.

[2] J. R. Büchi and C. C. Elgot. Decision problems of weak second order arithmetics and finite automata, part 1. *Notices of the American Mathematical Society*, 5(834), 1958.

[3] J.R. Büchi. On a decision method in restricted second order arithmetic. In *Proc. Internat. Congr. Logic, Method and Philos. Sci. 1960*, pages 1–12, Stanford, 1962. Stanford University Press.

[4] Pierre Cartier and Dominique Foata. *Problèmes Combinatoires de Commutation et Réarrangements*, volume 85 of *Lecture Notes in Mathematics*. Springer, 1969.

[5] E.M. Clarke, O. Grumberg, and D. Peled. *Model Checking*. MIT Press, 1999.

[6] Thomas Colcombet and Christof Löding. Transforming structures by set interpretations. *Logical Methods in Computer Science*, 3(2), 2007.

[7] V. Diekert and G. Rozenberg, editors. *Book of Traces*. World Scientific, Singapore, 1995.

[8] Volker Diekert and Yves Métivier. Partial commutation and traces. *Handbook of Formal Languages, Vol. 3: Beyond Words*, pages 457–533, 1997.

[9] Werner Ebinger and Anca Muscholl. Logical definability on infinite traces. *Theoretical Computer Science*, 154:335–346, 1996.

[10] Calvin C. Elgot and Michael O. Rabin. Decidability and undecidability of extensions of second (first) order theory of (generalized) successor. *The Journal of Symbolic Logic*, 31(2):169–181, 1966.

[11] T. Hafer and W. Thomas. Computation tree logic CTL* and path quantifiers in the monadic theory of the binary tree. In *Proc. 14th International Coll. on Automata, Languages, and Programming*, volume 267 of *Lecture Notes in Computer Science*, pages 269–279. Springer-Verlag, 1987.

[12] M. Huhn, P. Niebert, and F. Wallner. Verification based on local states. In *Proc. of Tools and Algorithms for Construction and Analysis of Systems, (TACAS)*, volume 1384 of *LNCS*. Springer, 1998.

[13] K. Lodaya, R. Ramanujam, and P.S. Thiagarajan. Decidability of a partial order based temporal logic. In *Proc. of Int'l coll. on Automata, Languages and Programming, (ICALP)*, volume 700 of *LNCS*. Springer, 1993.

[14] K. Lodaya and P.S. Thiagarajan. A modal logic for a subclass of event structures. In *Proc. of Int'l coll. on Automata, Languages and Programming, (ICALP)*, volume 267 of *LNCS*. Springer, 1987.

[15] Kamal Lodaya, Madhavan Mukund, R Ramanujam, and P S Thiagarajan. Models and logics for true concurrency. *Sadhana*, 17(1):131–165, 1992.

[16] P. Madhusudan. Model-checking trace event structures. In *Proc. 18th IEEE Symposium on Logic in Computer Science (LICS 2003)*, pages 371–380. IEEE Computer Society, 2003.

[17] A. Mazurkiewicz. Concurrent program schemes and their interpretations. Technical Report DAIMI Rep. PB 78, Aarhus University, Denmark, 1977.

[18] M. Mukund and P.S. Thiagarajan. A logical characterization of well branching event structures. *Theoretical Computer Science*, 96(1):35–72, 1992.

[19] M. Nielsen, G. Plotkin, and G. Winskel. Petri nets, event structures and domains, part 1. *Theoretical Computer Science*, 13(1):85–108, 1981.

[20] W. Penczek. Model-checking for a subclass of event structures. In *Proc. of Tools and Algorithms for Construction and Analysis of Systems, (TACAS)*, volume 1217 of *LNCS*. Springer, 1997.

[21] M.O. Rabin. Decidability of second order theories and automata on infinite trees. *Transaction of the AMS*, 141:1–35, 1969.

[22] P. S. Thiagarajan. Regular trace event structures. Technical Report RS-96-32, BRICS Tech report, Denmark, September 1996. 34 pp.

[23] P. S. Thiagarajan. Personal communication. 2007.

[24] P.S. Thiagarajan. Regular event structures and finite Petri nets: A conjecture. In *Formal and Natural Computing*, volume 2300. Springer, 2002.

[25] W. Thomas. On logical definability of trace languages. In *Proc. ESPRIT Basic Research Action No. 3116: Workshop on Algebraic and Syntactic Methods in Computer Science (ASMICS), Technical Report TUM-I9002, Technical University of Munich*, pages 172–182, 1990.

[26] M.Y. Vardi and P. Wolper. Reasoning about infinite computations. *Information and Computation*, 115(1):1–37, November 1994.

[27] Glynn Winskel and Mogens Nielsen. Models for concurrency. *Handbook of Logic in Computer Science (Vol. 4): Semantic Modelling*, pages 1–148, 1995.

[28] W. Zielonka. Notes on finite asynchronous automata. *R.A.I.R.O. Inform. Théor. Appl.*, pages 99–135, 1987.

A Look at the Control of Asynchronous Automata

Anca Muscholl, Igor Walukiewicz and Marc Zeitoun*
LaBRI, Bordeaux University, France
{anca,igw,mz}@labri.fr

Abstract

In the controller synthesis problem, one is given a system, referred to as a plant, and is asked to find a controller such that the controlled system satisfies a given specification. In this paper, we investigate the case when both the plant and the controller are distributed systems modelled by asynchronous automata. Two forms of distributed control have been studied in the literature: *process-based control*, where each process declares which actions it permits to execute and *action-based control*, where each action is declared executable or not by considering all processes in which it occurs. We show how to reduce process-based control to action-based control and examine to what extent results about distributed control can be transferred form one setting to the other.

Keywords: Controller synthesis, asynchronous automata, distributed control

1 Introduction

In the simplest case, the controller synthesis problem asks to find a model for a given specification. So it is just the satisfiability problem. In a more refined version one is given a system, referred to as a plant, and is asked to find a controller such that the controlled system satisfies a given specification. Here, we are interested in the case when both the plant and the controller are distributed systems. More precisely, when they are modelled by asynchronous automata.

There are numerous versions of the distributed synthesis problem [14, 10, 16, 8, 11, 1, 13, 6, 12]. We focus on two variants that are

*Work supported by project DOTS (ANR-06-SETI-003)

most rooted in the theory of Mazurkiewicz traces. They correspond to two different, and quite intuitive, ways of controlling a distributed system. The interest in these two control problems is also motivated by the fact that their decidability is still open. This contrasts with most of the other settings where one very quickly hits the undecidability barrier [14, 11, 2].

In the two problems we consider, the goal is to control the behaviour of an asynchronous automaton. This is an automaton with some fixed number of components, or processes, executing in parallel. Each input letter has a preassigned set of processes it acts on. In this way, if two letters have disjoint sets of assigned processes we can consider that they can occur in parallel. The objective is to control an asynchronous automaton so that every possible run satisfies a given specification. The control consists in forbidding some actions of the automaton, but not every action can be forbidden, and the control has also to ensure that the system does not block completely.

In the context of asynchronous automata, it is not reasonable to consider a controller that at every moment of the execution has a complete knowledge of the state of all the processes. This would amount to eliminating all concurrency in the automaton and controlling the resulting finite automaton. Control of finite automata is well studied and much simpler than distributed control [7, 3]. It is much more interesting to try to control an asynchronous automaton without forcing an ad hoc sequentialisation of its behaviours. A natural assumption is that the controller has to use the same communication architecture as the plant.

Here, we consider the following two ways of controlling an asynchronous automaton. In the *process-based* version, each process declares which actions it permits to execute. Then, an action can execute if all the processes it involves permit it. In the *action-based* version, each action is declared executable or not by regarding all its processes. This gives more power to the controller as it can look at several processes at the same time.

The process-based version of control was introduced by Madhusudan, Thiagarajan and Yang [12]. The action-based variant was proposed by Gastin, Lerman and Zeitoun [6]. Madhusudan et al. show the decidability of the control problem for asynchronous automata communicating connectedly (roughly speaking, this requires that if two processes do not synchronise during a long amount of time, then they won't synchronise ever again). The proof proceeds by coding the problem into monadic second-order theory of event structures and showing that this theory is decidable when the criterion holds. Gastin et al. show the decidability of action-based control for asynchronous automata, with a restriction on the form of dependencies between letters of the input alphabet.

Although these two forms of control have been studied for some time, they have never been put side by side. In this paper we have reworded their definitions in order to underline their similarities. Based on this reformulation, we give a reduction from process-based to action-based control. The resemblance of the two definitions allows us also to examine to which extent the above mentioned results can be transferred form one setting to the other.

2 Preliminaries

2.1 Traces and event structures

A *trace alphabet* is a pair (Σ, D), where Σ is a finite set of *actions* and $D \subseteq \Sigma \times \Sigma$ is a *dependence relation* that is reflexive and symmetric. We will use $I = \Sigma \times \Sigma \setminus D$ to denote the *independence relation.*

This alphabet induces a congruence relation $\sim_I$ over the set of words Σ^* over Σ. This is the smallest congruence such that $ab \sim_I ba$ for all $a, b \in I$. An equivalence class of $\sim_I$ is called a (Mazurkiewicz) *trace.* We will use t for denoting traces and write $[u]_I$ for the trace containing $u \in \Sigma^*$ (or simply $[u]$ if I is clear from the context).

A trace t is *prime* if all words it contains end with the same letter. A *prefix* of a trace t' is any trace $t = [v]$ where v is a prefix of some $u \in t'$. Write $t \leq t'$ when t is a prefix of t'; in this case we also say that t' is an *extension* of t. For any two traces t, t' that have some common extension, we write $t \sqcup t'$ for their least common extension. For a detailed introduction to the theory of traces, see the book [4].

Let L be a prefix closed set of traces. In the same way as a prefix closed set of words forms a tree, a prefix closed set of traces forms an *event structure* that we denote by $ES(L)$. The latter is a tuple $\langle E, \leq, \#, \lambda \rangle$, where:

- $E = \{e \in L : e \text{ prime trace}\}$,
- $e \leq e'$ if e is a prefix of e',
- $e \# e'$ if e and e' do not have a common extension in L,
- $\lambda(e)$ is the last letter of a word in e.

Observe that if $\lambda(e)$ and $\lambda(e')$ are dependent then either $e \# e'$, or the two events are comparable with respect to the $\leq$ relation.

For an event $e \in E$ we use $e\!\downarrow = \{e' : e' \leq e\}$ for the set of events below e. More generally, for a set $C \subseteq E$ we write $C\!\downarrow$ for the set

$\bigcup_{e \in C} e\downarrow$. A *configuration* C of an event structure is a conflict-free, downward-closed subset $C \subseteq E$. That is, $C\downarrow = C$ and no events $e, e' \in C$ satisfy $e \# e'$. Notice that every configuration C corresponds to a trace. We can even have a bijection between configurations of $ES(L)$ and traces in L in the case when L is both prefix and extension closed; the latter means that for every two traces $t, t' \in L$ having some common extension, also $t \sqcup t' \in L$.

2.2 Asynchronous automata

Let $\mathbb{P}$ be a finite set of *processes*. Consider an alphabet Σ and a function $loc : \Sigma \to (2^{\mathbb{P}} \setminus \emptyset)$. A (deterministic) *asynchronous automaton* is a tuple

$$\mathcal{A} = \langle \{S_p\}_{p \in \mathbb{P}}, s_{in}, \{\delta_a\}_{a \in \Sigma} \rangle,$$

where

- S_p is a finite set of (local) states of process p,
- $s_{in} \in \prod_{p \in \mathbb{P}} S_p$ is a (global) initial state,
- $\delta_a : \prod_{p \in loc(a)} S_p \dot{\to} \prod_{p \in loc(a)} S_p$ is a transition relation; so on a letter $a \in \Sigma$ it is a partial function on tuples of states of processes in $loc(a)$.

The location mapping loc defines in a natural way an independence relation I: two actions $a, b \in \Sigma$ are independent if they use different processes, i.e., if $loc(a) \cap loc(b) = \emptyset$. An asynchronous automaton can be seen as a sequential automaton with the state set $S = \prod_{p \in \mathbb{P}} S_p$ and transitions $s \xrightarrow{a} s'$ if $((s_p)_{p \in loc(a)}, (s'_p)_{p \in loc(a)}) \in \delta_a$, and $s_q = s'_q$ for all $q \notin loc(a)$. By $L(\mathcal{A})$ we denote the set of words labelling runs of this sequential automaton that start from the initial state. This definition has an important consequence. If $(a, b) \in I$ then the same state is reached on the words ab and ba. More generally, whenever $u \sim_I v$ and $u \in L(\mathcal{A})$ then $v \in L(\mathcal{A})$, too. This means that $L(\mathcal{A})$ is trace closed. By definition, $L(\mathcal{A})$ is also prefix closed, thus it defines an event structure $ES(L(\mathcal{A}))$, which we also denote as $ES(\mathcal{A})$.

3 Two variants of distributed control problems

As in the standard setting of Ramadge and Wonham [15] for finite sequential automata, we can divide the input alphabet of an asynchronous automaton into controllable and uncontrollable actions. We can then ask if there is a way of choosing controllable actions so that a given specification is satisfied. The important parameter here is the mechanism of controlling actions. A most general solution is to ask for a global controller that makes decisions by looking at the global state of the system. This approach completely ignores the distributed aspect of the system. A more interesting approach is to ask for a controller that respects the concurrency present in the automaton. There are several ways of formalising this intuition. Here we present two. The first one was proposed by Madhusudan et al. [12], the second one by Gastin et al. [6].

3.1 Process-based control

A *plant* is a deterministic asynchronous automaton together with a partition of the input alphabet into actions controlled by the *system* and actions controlled by the *environment*: $\Sigma = \Sigma^{sys} \cup \Sigma^{env}$.

A plant defines a game arena, with plays corresponding to initial runs of $\mathcal{A}$. Since $\mathcal{A}$ is deterministic, we can view a play as a word from $L(\mathcal{A})$. Let *Plays*$(\mathcal{A})$ denote the set of traces associated with words from $L(\mathcal{A})$.

The p-view of a play u, denoted $view_p(u)$, is the smallest trace prefix of u containing all the p-events of u (a p-event is an occurrence of some $a \in \Sigma$ with $p \in loc(a)$). We write $Plays_p(\mathcal{A})$ for the set of plays that are p-views:

$$Plays_p(\mathcal{A}) = \{view_p(u) : u \in Plays(\mathcal{A})\}.$$

A *strategy* for a process p is a function $f_p : Plays_p(\mathcal{A}) \to 2^{\Sigma_p}$; where $\Sigma_p = \{a \in \Sigma : p \in loc(a)\}$ is the set of actions of process p. A *process-based strategy* is a family of strategies $\{f_p\}_{p \in \mathbb{P}}$, one for each process.

The set of plays respecting a strategy $\sigma = \{f_p\}_{p \in \mathbb{P}}$, denoted $Plays(\mathcal{A}, \sigma)$, is the smallest set containing the empty play ε, and such that for every $u \in Plays(\mathcal{A}, \sigma)$:

- if $a \in \Sigma^{env}$ and $ua \in Plays(\mathcal{A})$ then ua is in $Plays(\mathcal{A}, \sigma)$.
- if $a \in \Sigma^{sys}$ and $ua \in Plays(\mathcal{A})$ then $ua \in Plays(\mathcal{A}, \sigma)$ provided that $a \in f_p(view_p(u))$ for all $p \in loc(a)$.

Intuitively, the definition says that actions of the environment are always possible, whereas actions of the system are possible only if they are allowed by the strategies of all the involved processes.

Before defining specifications, we need to make it precise what are infinite plays that are consistent with a given strategy σ. Let X be an infinite set of traces from $Plays(\mathcal{A}, \sigma)$ such that for any two traces t, t' in X, there is some trace t'' in X with $t \leq t''$ and $t' \leq t''$ (such a set is called directed). The limit of an infinite, directed set X, denoted $\sqcup X$, is the least ω-trace with $t \leq \sqcup X$ for every $t \in X$ (for the formal definition of ω-traces see [4]). We write $Plays^{\omega}(\mathcal{A}, \sigma)$ for the set of ω-traces of the form $\sqcup X$, where X is a maximal, directed subset of $Plays(\mathcal{A}, \sigma)$.

Specifications for controllers are given by ω-regular languages of traces. Such languages can be defined as ω-regular word languages L that are closed under the equivalence $\sim_I$: if $u_i \sim_I v_i$ for all i, then $u_0 u_1 \cdots \in L$ iff $v_0 v_1 \cdots \in L$. A process-based *controller* satisfying a specification *Spec* is a process-based strategy σ, such that $Plays^{\omega}(\mathcal{A}, \sigma) \subseteq Spec$.

In order to avoid trivial solutions we need also to impose another restriction. A strategy is *non-blocking* if every play in $Plays(\mathcal{A}, \sigma)$ having an extension in $Plays(\mathcal{A})$ also has an extension in $Plays(\mathcal{A}, \sigma)$. Intuitively, the system is not allowed to stop the execution of the plant by forbidding all possible actions.

The *process-based control problem* is to determine for a given plant and specification if there is a process-based controller for the plant that is non-blocking and satisfies the specification.

Madhusudan et. al. introduced the notion of connectedly communicating plants, meaning that in every prime trace of $L(\mathcal{A})$, each of its events has at most $|\mathcal{A}|$ many concurrent events. They showed:

Theorem 1 *[Madhusudan & Thiagarajan & Yang] The process-based control problem is decidable for connectedly communicating plants.*

3.2 Action-based control

As in the process-based version, a plant is a deterministic[†] asynchronous automaton $\mathcal{A}$ with a partition of the input alphabet. The difference is that with an action-based control we need to define the views of actions from Σ. The a-view of a play u, denoted $view_a(u)$, is the smallest trace prefix of u that contains all p-events of u, for every $p \in loc(a)$. Notice that a trace u in $Plays_a(\mathcal{A})$ need not be prime, in

[†]In [6], non-deterministic plants were admitted. Moreover, non-blocking controlled behaviour was not required (but the specification could enforce it). These minor differences with our presentation are motivated in order to present a unified framework.

general, but ua is. We write $Plays_a(\mathcal{A})$ for the set of plays which are a-views:

$$Plays_a(\mathcal{A}) = \{view_a(u) : u \in Plays(\mathcal{A})\}.$$

A *strategy* for an action $a \in \Sigma^{sys}$ is a function $g_a : Plays_a(\mathcal{A}) \to \{tt, ff\}$. An *action-based strategy* is a family of strategies $\{g_a\}_{a\in\Sigma^{sys}}$, one for each action in Σ^{sys}.

The set of plays respecting a distributed strategy $\rho = \{g_a\}_{a\in\Sigma^{sys}}$ is the smallest set containing ε, and such that if $u \in Plays(\mathcal{A}, \rho)$ we also have:

- if $a \in \Sigma^{env}$ and $ua \in Plays(\mathcal{A})$ then ua is in $Plays(\mathcal{A}, \rho)$.
- if $a \in \Sigma^{sys}$ and $ua \in Plays(\mathcal{A})$ then $ua \in Plays(\mathcal{A}, \rho)$ provided that $g_a(view_a(u)) = tt$.

The definitions of $Plays^{\omega}(\mathcal{A}, \rho)$ and of non-blocking controllers are exactly as before. The notion of a controller satisfying a specification is also the same.

The *action-based control problem* is to determine for a given plant and specification if there is a non-blocking action-based controller for the plant satisfying the specification.

Decidability of the action-based control problem for plants over trace alphabets of a special form has been shown, first for reachability conditions [6], later for all ω-regular specifications [9]. A trace alphabet (Σ, D) can be seen as a graph with letters from Σ as nodes and the edges given by D. Since D is symmetric this graph is undirected; D is also reflexive but we will not put self loops in the graph. A *co-graph* is a graph that can constructed from singletons using parallel and sequential products. Another characterisation is that it is a graph without an induced P_4 subgraph, i.e. without a graph of the form $x_1 - x_2 - x_3 - x_4$ as an induced subgraph.

Theorem 2 *[Gastin & Lerman & Zeitoun] The action-based control problem for plants over trace alphabets that are co-graphs and ω-regular specifications is decidable.*

3.3 An example

We give a small example showing the difference between process-based and action-based control. Consider a system of two processes $\mathbb{P} = \{1, 2\}$ with actions $\Sigma = \{a_1, b_1, a_2, b_2, c, d\}$ distributed as follows:

$$loc(c) = loc(d) = \{1, 2\} \quad loc(a_p) = loc(b_p) = \{p\} \quad \text{for } p = 1, 2$$

Intuitively, it means that actions c and d are common to two processes, while a_p and b_p are local to process p. The plant will be very simple. It will allow each process to do either a_p or b_p followed by one common action which can be either c or d. Common actions are controllable by the system, while a_p and b_p are environment actions. Formally we have

$$\mathscr{A} = \langle \{S_p\}_{p \in \mathbb{P}}, s_{in}, \{\delta_a\}_{a \in \Sigma} \rangle$$

with $S_p = \{s_p, t_p, r_p\}$, $s_{in} = (s_1, s_2)$ and the transition relation defined by:

$$\delta_{a_p}(s_p) = \delta_{b_p}(s_p) = t_p \quad \delta_c(t_1, t_2) = \delta_d(t_1, t_2) = (r_1, r_2)$$

Consider a specification saying that if the two processes do action a then the system should do an action c, otherwise it should do d. Formally the specification language L is a prefix and trace closure of $\{a_1 a_2 c, a_1 b_2 d, a_2 b_1 d, b_1 b_2 d\}$. It means that, for example a_1 and $b_2 a_1 d$ also belong to L. We will show that there is no process-based strategy for the system while there is an action-based one.

A process-based strategy for system $\mathscr{A}$ is a pair of functions f_p : $Plays_p(\mathscr{A}) \to 2^{\Sigma_p}$; for $p = 1, 2$. In our case $Plays_p(\mathscr{A}) = (a_p + b_p)(\varepsilon + c + d)$. Among this four strings it is only important to define f_p on a_p and b_p. If the environment chooses to do a_1 and b_2 then the system should reply with d. For this we should have $d \in f_1(a_1)$ and $d \in f_2(b_2)$. By similar reasoning we should have $d \in f_1(b_1)$ and $d \in f_2(a_2)$. But then the play $a_1 a_2 d$ would also be consistent with the strategy. As this is a string not in L, no process-based strategy can satisfy the specification.

An action-based strategy is a pair of functions $g_x : Plays_x(\mathscr{A}) \to \{tt, ff\}$ for $x \in \{c, d\}$; one for each action of the system. In this case we have that $Plays_c(\mathscr{A}) = Plays_d(\mathscr{A})$ is the trace closure of $(a_1 + b_1)(a_2 + b_2)$. We can define

$$g_c(a_1 a_2) = g_c(a_2 a_1) = tt \quad \text{and } ff \text{ otherwise}$$
$$g_d(a_1 a_2) = g_d(a_2 a_1) = ff \quad \text{and } tt \text{ otherwise}$$

It is easy to verify that all the plays respecting this strategy satisfy the specification.

We can see with this example that the difference between process and action-based control lies in the information available when the decision is taken. The intuitive reason is that an action-based strategy has more information available, since it can read the state of every process involved in an action in order to decide whether to allow it or not. In contrast, process-based strategies look at each process separately. The example shows that this difference is crucial.

4 Reduction from process-based to action-based control

Action-based strategies are more powerful than process-based ones. It is easy to see that a process-based controller for a given $(\mathcal{A}, Spec)$ can be converted into an action-based one. The converse is not true, as shown by the example above. Below, we show that the process-based control problem can be reduced to the action-based problem.

Fix a plant $\mathcal{A}$ and a specification *Spec*. We will construct $\overline{\mathcal{A}}$ and $\overline{Spec}$ such that: there is a process-based controller for $(\mathcal{A}, Spec)$ iff there is an action-based controller for $(\overline{\mathcal{A}}, \overline{Spec})$. Informally, the reduction works as follows: for each process we use additional local system actions, one for every subset of Σ^{sys}. The execution of such an action corresponds to a declaration of the value of the process-based strategy at this process. Then we will have another set of actions, this time for the environment, that will allow to choose one of the declared actions.

We set:

$$\begin{aligned}\overline{S}_p &= S_p \cup (S_p \times 2^{\Sigma^{sys}}) \cup (S_p \times \Sigma^{sys}),\\ \overline{\Sigma}^{sys} &= \Sigma^{sys} \cup (\mathbb{P} \times 2^{\Sigma^{sys}}) \cup \{\top\},\\ \overline{\Sigma}^{env} &= \Sigma^{env} \cup (\mathbb{P} \times \Sigma^{sys}) \cup \{\bot\}.\end{aligned}$$

We also need to specify the locations of the new letters. For all $p \in \mathbb{P}$, $a \in \Sigma^{sys}$ and $\theta \in 2^{\Sigma^{sys}}$ we set:

$$\begin{aligned}&loc((p,\theta)) = loc((p,a)) = \{p\},\\ &loc(\top) = loc(\bot) = \mathbb{P}.\end{aligned}$$

The idea is that by playing its local action (p, θ_p), process p declares that it is willing to enable any action of θ_p. Later on, the environment will have to pick some action a present in all sets θ_p with $p \in loc(a)$. Thus, a move on action $a \in \Sigma^{sys}$ in $\mathcal{A}$ is simulated in $\overline{\mathcal{A}}$, up to commutation of independent actions, by a sequence of the form $(p_1, \theta_1), \ldots, (p_i, \theta_i), (p_1, a), \ldots, (p_i, a), a$, where $loc(a) = \{p_1, \ldots, p_i\}$ and $a \in \theta_j$ for all j. Actions $\bot$ and $\top$ will be used to "punish" the system or the environment for an "unfair" behaviour captured by the following definitions:

- We say that a global state $\prod_{p\in\mathbb{P}}(s_p,\theta_p)$ of $\overline{\mathcal{A}}$ is *s-blocking* if there is an action of $\mathcal{A}$ possible from $\prod_{p\in\mathbb{P}} s_p$ but for every such action a we have $a \notin \theta_p$ for some $p \in loc(a)$.
- We say that a global state $\prod_{p\in\mathbb{P}}(s_p,a_p)$ of $\overline{\mathcal{A}}$ is *e-blocking* if there is an action of $\mathcal{A}$ possible from $\prod_{p\in\mathbb{P}} s_p$ but for every such action a we have $a \neq a_p$ for some $p \in loc(a)$.

An s-blocking state represents a situation when some actions are possible in the plant, but all of them are blocked due to the choice made by the system. The e-blocking state is similar, but here the reason is that the environment has not made a consistent choice of a next action.

Now we define the transition function of $\overline{\mathcal{A}}$:

$$\overline{\delta}_{(p,\theta)}(s_p) = \{(s_p,\theta)\} \qquad \text{if } s_p \in S_p,\ \theta \in 2^{\Sigma^{sys}}$$

$$\overline{\delta}_{(p,a)}((s_p,\theta)) = \{(s_p,a)\} \qquad \text{if } s_p \in S_p,\ a \in \theta$$

$$\overline{\delta}_a\Big(\prod_{p\in loc(a)}(s_p,a)\Big) = \delta_a\Big(\prod_{p\in loc(a)} s_p\Big) \qquad \text{if } a \in \Sigma^{sys}$$

$$\overline{\delta}_a(s) = \delta_a(s) \qquad \text{if } a \in \Sigma^{env} \text{ and } s \in \prod_{p\in loc(a)} S_p$$

$$\overline{\delta}_\top(s) = s \qquad \text{if } s \in \prod_{p\in\mathbb{P}}(S_p \times \Sigma^{sys}) \text{ is e-blocking}$$

$$\overline{\delta}_\bot(s) = s \qquad \text{if } s \in \prod_{p\in\mathbb{P}}(S_p \times 2^{\Sigma^{sys}}) \text{ is s-blocking}$$

The specification $\overline{Spec}$ contains an ω-trace t if: (i) it does not have an occurrence of $\bot$, and (ii) either it has an occurrence of $\top$, or its projection $t_{|\Sigma}$ on Σ is in *Spec*. In general, a projection of a trace may not be a trace. For example, the words $a\bot b$ and $b\bot a$ are not trace equivalent, but their projections are if a and b are independent. Fortunately, in our case $u_{|\Sigma}$ is a trace when u has neither $\top$ nor $\bot$.

Lemma 3 *If* $(\mathcal{A}, Spec)$ *has a process-based controller then* $(\overline{\mathcal{A}}, \overline{Spec})$ *has an action-based controller.*

Proof Given a process-based controller $\sigma = \{f_p\}_{p\in\mathbb{P}}$ for $\mathcal{A}$, we construct an action-based controller $\rho = \{g_a\}_{a\in\overline{\Sigma}^{sys}}$ for $\overline{\mathcal{A}}$ as follows:

$$g_{(p,\theta)}(v) = tt \text{ iff } \theta = f_p(v_{|\Sigma}), \text{ for all } v \in Plays_{(p,\theta)}(\overline{\mathcal{A}}),$$
$$g_a(v) = tt \text{ for every } v \text{ in } Plays_a(\overline{\mathcal{A}}),$$
$$g_{\top}(v) = tt \text{ for every } v \text{ in } Plays_{\top}(\overline{\mathcal{A}}).$$

We want to show that if $v \in Plays(\overline{\mathcal{A}},\rho)$ and contains neither $\top$ nor $\bot$, then $v_{|\Sigma} \in Plays(\mathcal{A},\sigma)$. This will show the same statement for infinite executions: any $v \in Plays^{\omega}(\overline{\mathcal{A}},\rho)$ either contains $\top$ or $\bot$, or $v_{|\Sigma} \in Plays^{\omega}(\mathcal{A},\sigma)$. We will then get that ρ satisfies $\overline{Spec}$ if σ satisfies *Spec*. We also need to show that ρ is non-blocking and that it avoids $\bot$.

Let $state(\overline{\mathcal{A}},v)$ denote the global state reached by $\overline{\mathcal{A}}$ after reading v. Similarly for $state(\mathcal{A},v)$. We write $\prod_{p\in\mathbb{P}} s_p \sim \prod_{p\in\mathbb{P}} \overline{s}_p$ if for every $p \in \mathbb{P}$, either $s_p = \overline{s}_p$, or s_p is the first component of $\overline{s}_p$. By induction on the length of $v \in Plays(\overline{\mathcal{A}},\rho)$ we show that if v contains neither $\top$ nor $\bot$ then $v_{|\Sigma} \in Plays(\mathcal{A},\sigma)$ and $state(\mathcal{A},v_{|\Sigma}) \sim state(\overline{\mathcal{A}},v)$.

If v ends with a letter (p,a) or (p,θ) then the statement is immediate from the induction assumption. For the case of $v = ua$ with $a \in \Sigma^{env}$ we use the induction hypothesis $state(\overline{\mathcal{A}},u) \sim state(\mathcal{A},u_{|\Sigma})$. From $u, ua \in Plays(\overline{\mathcal{A}},\sigma)$ we deduce that the transition of $\overline{\mathcal{A}}$ from $state(\overline{\mathcal{A}},u)$ is possible. The definition of $\overline{\mathcal{A}}$ tells us that on environment actions the transitions of $\overline{\mathcal{A}}$ are the same as that of $\mathcal{A}$. Hence the transition of $\mathcal{A}$ on a is possible from $state(\mathcal{A},u_{|\Sigma})$. Thus $ua \in Plays(\mathcal{A},\sigma)$, and the constraint on states holds, since the same transition is taken by the two automata. Next, we consider $v = ua$ with $a \in \Sigma^{sys}$. As $Plays(\overline{\mathcal{A}},\rho)$ is trace closed, we can assume that u ends with the sequence $(p_{i_1},\theta_{i_1}),\ldots,(p_{i_k},\theta_{i_k}),(p_{i_1},a),\ldots,(p_{i_k},a)$, where $loc(a) = \{p_{i_1},\ldots,p_{i_k}\}$. From the definition of the strategy ρ it follows that $f_{p_{i_j}}(view_{p_{i_j}}(u_{|\Sigma})) = \theta_{i_j}$ and $a \in \theta_{i_j}$ for all $j = 1,\ldots,k$. This means that $u_{|\Sigma}a \in Plays(\mathcal{A},\sigma)$ and $state(\overline{\mathcal{A}},v) \sim state(\mathcal{A},v_{|\Sigma})$.

We need also to ensure that if σ is non-blocking then ρ is also non-blocking and that $\bot$ is not reachable. Take $v \in Plays(\overline{\mathcal{A}},\rho)$ and suppose that $v_{|\Sigma}$ has a prolongation in $Plays(\mathcal{A},\sigma)$. From the above we know that $state(\overline{\mathcal{A}},v) \sim state(\mathcal{A},v_{|\Sigma})$. The first case is when $state(\overline{\mathcal{A}},v)$ has at least one component in a state from $\mathcal{A}$, say

$s_p \in S_p$ on the p-th component. In this case it is possible to extend v either by an environment action or by the system action (p, θ_p), where $\theta_p = f_p(view_p(v_{|\Sigma}))$. The next case is when all components are of the form (s_p, θ_p), observe that we have to have $\theta_p = f_p(view_p(v_{|\Sigma}))$ as any execution needs to respect the strategy ρ. As σ is non-blocking, $\overline{\mathcal{A}}$'s transition on $\perp$ is not applicable (the state $(s_p, \theta_p)_{p \in \mathbb{P}}$ is not s-blocking). Next, suppose that there is at least one component in a state of the form (s_p, θ_p) with $\theta_p \neq \emptyset$. Then (s_p, a) with $a \in \theta_p$ is a possible next action. Finally, assume that all components are in states of the form (s_p, a_p) or $(s_p, \emptyset)$. Then either a letter from Σ^{sys} or $\top$ is possible.

It remains to show the converse.

Lemma 4 *If there is an action-based controller for* $(\overline{\mathcal{A}}, \overline{Spec})$ *then there is a process-based controller for* $(\mathcal{A}, Spec)$.

Proof We first need an observation about action-based controllers. We call a strategy $\rho = \{g_a\}_{a \in \overline{\Sigma}^{sys}}$ *deterministic* if for all $p \in \mathbb{P}$ and $v \in Plays_{(p,\theta)}(\overline{\mathcal{A}})$, there is at most one $\theta \in 2^{\Sigma^{sys}}$ such that $g_{(p,\theta)}(v)$ is true. We claim that if there is a controller then there is a deterministic one. To see this suppose that we have a controller ρ such that $g_{(p,\theta_1)}(v)$ and $g_{(p,\theta_2)}(v)$ hold for some v, p, and two distinct θ_1, θ_2. Then we modify the strategy into ρ' where we set $g_{(p,\theta_2)}(v)$ to false. Clearly $Plays(\overline{\mathcal{A}}, \rho')$ is not bigger than $Plays(\overline{\mathcal{A}}, \rho)$. So all the sequences admitted by ρ' are in $\overline{Spec}$. The new strategy is also non-blocking if ρ was non-blocking.

Let us now suppose that $\rho = \{g_a\}_{a \in \overline{\Sigma}^{sys}}$ is a deterministic strategy. Thanks to determinism, for every $p \in \mathbb{P}$ we can introduce an auxiliary function F_p defined by $F_p(v) = \theta$ iff $g_{(p,\theta)}(v)$ holds. For a trace u in $Plays(\mathcal{A})$ we define a trace $\overline{u}$ in the following way:

$$\overline{ua} = \overline{u}a \qquad \text{for } a \in \Sigma^{env},$$

$$\overline{ua} = \overline{u}(p_1, F_{p_1}(view_{p_1}(\overline{u}))) \cdots (p_k, F_{p_k}(view_{p_k}(\overline{u})))(p_1, a) \cdots (p_k, a)a$$
$$\text{for } a \in \Sigma^{sys} \text{ with, for easier notation, } loc(a) = \{p_1, \ldots, p_k\}.$$

Naturally the definition is the same also for actions that have the sets of locations of a different form than $\{p_1, \ldots, p_k\}$.

We assume that $\overline{ua}$ is undefined if so is one of the expressions of the right hand side. We construct the process-based strategy $\sigma = \{f_p\}_{p \in \mathbb{P}}$:

$$f_p(u) = F_p(view_p(\overline{u})) \quad \text{if } \overline{u} \text{ defined and in } Plays(\overline{\mathcal{A}}).$$

By a simple induction on the size of u, one can show that if $u \in Plays(\mathcal{A}, \sigma)$ then $\overline{u} \in Plays(\overline{\mathcal{A}}, \rho)$ and $state(\mathcal{A}, u) = state(\overline{\mathcal{A}}, \overline{u})$. This implies that if all infinite plays consistent with ρ satisfy $\overline{Spec}$, then all infinite plays consistent with σ satisfy *Spec*. It remains to check that σ is non-blocking if ρ is. We take $u \in Plays(\mathcal{A}, \sigma)$ and the corresponding word $\overline{u} \in Plays(\overline{\mathcal{A}}, \rho)$. Suppose that $\mathcal{A}$ can do an action from $state(\mathcal{A}, u)$. As ρ is non-blocking and winning (in particular, ρ can avoid $\perp$), $\overline{u}$ has an extension $\overline{u} v b \in Plays(\overline{\mathcal{A}}, \rho)$ with $b \in \Sigma$ and $v \in (\overline{\Sigma} \setminus \Sigma)^*$. Then $u b \in Plays(\mathcal{A}, \sigma)$.

With these two lemmas we get:

Theorem 5 *For every* $(\mathcal{A}, Spec)$ *one can construct* $(\overline{\mathcal{A}}, \overline{Spec})$ *with the property that: there is a process-based controller for* $(\mathcal{A}, Spec)$ *iff there is an action-based controller for* $(\overline{\mathcal{A}}, \overline{Spec})$.

This theorem says that in general it is easier to solve process control problems. Unfortunately, it is not universally applicable. For instance, we cannot use Theorem 2 to get decidability of the process-based control for trace alphabets that are co-graphs. Our reduction extends the alphabet, and this extension does not preserve the property of being a co-graph. For example, in the original alphabet we can have just two actions with locations $\{p_1, p_2\}$ and $\{p_2, p_3\}$. In the extended alphabet we add actions on each process so we will have an action with location $\{p_1\}$ and one with location $\{p_3\}$. This results in a P_4 induced subgraph.

5 Control problems and MSOL over event structures

In this section we encode the two control problems into the satisfiability problem of monadic second order logic (MSOL) over event structures. This encoding gives a decidability of the control problem for plants $\mathcal{A}$ such that $ES(\mathcal{A})$ has decidable MSOL theory. This is for example the case for connectedly communicating processes. As we will see, while sufficient, the decidability of the MSOL theory of $ES(\mathcal{A})$ is not a necessary condition for the decidability of control problems.

Monadic second order logic over event structures has two binary relations $<$ and #, and unary relations R_a, one for each letter in the alphabet. The interpretation of a relation R_a is the set of events labelled with a. The interpretation of $<$ and # is, as expected: the partial order, and the conflict relation between events.

5.1 Encoding for the process-based control

We describe in this section the reduction of the process-based control problem to the satisfiability problem of MSOL over event structures, as provided in [12]. For a given specification *Spec* we want to write a formula φ_{Spec} such that for every plant $\mathscr{A}$: the process-based control problem for $(\mathscr{A}, Spec)$ has a solution iff $ES(\mathscr{A}) \models \varphi_{Spec}$.

As a preparation, let us see how one can talk about plays from $\mathscr{A}$ inside $ES(\mathscr{A})$. A play is a trace and it corresponds to a configuration inside the event structure. Such a configuration can be described by its maximal events. We can also talk about $view_p(u)$ inside the event structure, as it is the smallest configuration containing all events from u that are labelled with letters having p in their domain.

A process-based strategy $\sigma = \{f_p\}_{p \in \mathbb{P}}$ can be encoded into an event structure with the help of second-order variables Z^a_p, for every $p \in \mathbb{P}$ and $a \in \Sigma^{sys}$ such that $p \in loc(a)$. Observe that elements of $ES(\mathscr{A})$ are prime traces, and that $Plays_p(\mathscr{A})$ are exactly the prime traces that end with an action from p. We define Z^a_p as the set of all $u \in Plays_p(\mathscr{A})$ such that $a \in f_p(u)$. An event (prime trace) $u = u'a$ may of course belong to more than one Z^a_p, ranging over processes $p \in loc(a)$. Clearly there is a bijection between strategies and such assignments to variables Z^a_p.

Now we need to define $Plays(\mathscr{A}, \sigma)$ inside the event structure. The first observation is that this set is determined by the prime traces in it. Indeed, a configuration is in the set iff all its prime subconfigurations are in the set. Then we write a constraint that an event (prime trace) ua with $u \in Plays(\mathscr{A}, \sigma)$ and $a \in \Sigma^{sys}$ is in the set iff for all $p \in loc(a)$ we have $view_p(u) \in Z^p_a$. Similarly for $a \in \Sigma^{env}$.

Finally, let us see how to talk about infinite traces in $Plays^{\omega}(\mathscr{A}, \sigma)$ and about satisfying the specification *Spec*. As for finite plays, an infinite configuration X in $Plays^{\omega}(\mathscr{A}, \sigma)$ is determined by its prime subconfigurations. Moreover, we need to state that such a set X is maximal (no event $u \in Plays(\mathscr{A}, \sigma)$ can be added to X such that a larger configuration is obtained). Using [5], we can assume that the specification *Spec* is given by a monadic second-order formula ψ_{Spec} over ω-traces. Then it suffices to ensure that any set X of events that describes an infinite trace in $Plays^{\omega}(\mathscr{A}, \sigma)$ satisfies ψ_{Spec}. We also need to say that a strategy is non-blocking. For this it suffices to say that every configuration inside $Plays(\mathscr{A}, \sigma)$ that is not maximal in the event structure has a proper extension in $Plays(\mathscr{A}, \sigma)$.

5.2 Encoding for the action-based control

We show now how to adapt the encoding presented in the previous section to action-based control.

Proposition 6 *Given a ω-regular trace specification Spec, one can write an MSOL formula φ_{Spec} such that for every plant $\mathscr{A}$, the action-based control problem for $(\mathscr{A}, Spec)$ has a solution iff $ES(\mathscr{A}) \models \varphi_{Spec}$.*

Notice first that the a-views of configurations (or of traces in $Plays(\mathscr{A})$) can be described in a similar way as p-views. Namely, $view_a(u)$ is the smallest configuration containing every event of every process $p \in loc(a)$.

The encoding of an action-based strategy $\rho = \{g_a\}_{a\in\Sigma^{sys}}$ is even simpler than in the process-based case. We use second-order variables Z^a, $a \in \Sigma^{sys}$, with the following interpretation: an event e with label a belongs to Z^a iff the prime trace ua associated with e satisfies $g_a(u) = tt$. Now we describe the set of finite plays $Plays(\mathscr{A}, \rho)$. As before, a configuration is in $Plays(\mathscr{A}, \rho)$ iff all its prime subconfigurations are in this set. We write that a prime configuration ua belongs to $Plays(\mathscr{A}, \rho)$ iff $u \in Plays(\mathscr{A}, \rho)$ and moreover $ua \in Z_a$ if $a \in \Sigma^{sys}$.

For the encodings of $Plays^{\omega}(\mathscr{A}, \rho)$ and $Spec$ we proceed exactly as in the process-based case.

Madhusudan et al. [12] showed that if $\mathscr{A}$ is a connectedly communicating asynchronous automaton, then $ES(\mathscr{A})$ has a decidable MSOL theory. Together with the encoding above this gives decidability of the action-based control for such plants.

Proposition 7 *The action-based control problem for connectedly communicating plants is decidable.*

We can now observe that the decidability of the MSOL theory of $ES(\mathscr{A})$ is not a necessary condition for the decidability of controller problems, neither for the process nor for the action variant. Indeed it suffices to take a three letter alphabet $\{a, b, c\}$ with dependencies induced by the pairs (a, c) and (b, c). This alphabet is a co-graph so by Theorem 2 the action-based control problem is decidable. It is not difficult to check that the extended alphabet used in the translation from Section 4 is still a co-graph. Hence, by Theorem 2, the process-based control problem for this three letter alphabet is also decidable. Now, consider $\mathscr{A}$ that generates all the traces over this alphabet. The MSOL theory of $ES(\mathscr{A})$ is undecidable as it contains a grid as a subgraph.

References

[1] A. Arnold, A. Vincent, and I. Walukiewicz. Games for synthesis of controllers with partial observation. *Theoretical Computer Science*, 303(1):7–34, 2003.

[2] A. Arnold and I. Walukiewicz. Nondeterministic controllers of nondeterministic processes. In J. Flum, Erich Grädel, and Thomas Wilke, editors, *Logic and Automata*, volume 2 of *Texts in Logic and Games*, pages 29–52. Amsterdam University Press, 2007.

[3] C. G. Cassandras and S. Lafortune. *Introduction to Discrete Event Systems*. Kluwer Academic Publishers, 1999.

[4] V. Diekert and G. Rozenberg, editors. *The Book of Traces*. World Scientific, 1995.

[5] W. Ebinger and A. Muscholl. Logical definability on infinite traces. *Theoretical Computer Science*, 154(3):67–84, 1996.

[6] P. Gastin, B. Lerman, and M. Zeitoun. Distributed games with causal memory are decidable for series-parallel systems. In *FSTTCS*, volume 3328 of *Lecture Notes in Computer Science*, pages 275–286, 2004.

[7] R. Kumar and V. K. Garg. *Modeling and control of logical discrete event systems*. Kluwer Academic Pub., 1995.

[8] O. Kupferman and M.Y. Vardi. Synthesizing distributed systems. In *Proc. 16th IEEE Symp. on Logic in Computer Science*, 2001.

[9] B. Lerman. *Vérification et Spécification des Systèmes Distribués*. PhD thesis, Université Denis Diderot - Paris VII, 2005. `http://tel.archives-ouvertes.fr/tel-00322322/fr/`.

[10] F. Lin and M. Wonham. Decentralized control and coordination of discrete-event systems with partial observation. *IEEE Transactions on automatic control*, 33(12):1330–1337, 1990.

[11] P. Madhusudan and P.S. Thiagarajan. Distributed control and synthesis for local specifications. In *ICALP'01*, volume 2076 of *Lecture Notes in Computer Science*, pages 396–407, 2001.

[12] P. Madhusudan, P. S. Thiagarajan, and S. Yang. The MSO theory of connectedly communicating processes. In *FSTTCS*, volume 3821 of *lncs*, pages 201–212, 2005.

[13] S. Mohalik and I. Walukiewicz. Distributed games. In *FSTTCS'03*, volume 2914 of *Lecture Notes in Computer Science*, pages 338–351, 2003.

[14] A. Pnueli and R. Rosner. Distributed reactive systems are hard to synthesize. In *31th IEEE Symposium Foundations of Computer Science (FOCS 1990)*, pages 746–757, 1990.

[15] P. J. G. Ramadge and W. M. Wonham. The control of discrete event systems. *Proceedings of the IEEE*, 77(2):81–98, 1989.

[16] K. Rudie and W. Wonham. Think globally, act locally: Decentralized supervisory control. *IEEE Trans. on Automat. Control*, 37(11):1692–1708, 1992.

A Sampling Approach to the Analysis of Metric Temporal Logic

Paritosh K. Pandya*
Tata Institute of Fundamental Research
Colaba, Mumbai, India 400005
pandya@tifr.res.in

Abstract

Metric Temporal Logic is a well studied real-time logic. It can be defined with various notions of time such as continuous, sampled (pointwise) or discrete. Moreover time can be strictly monotonic or weakly monotonic. We give a uniform presentation of these metric logics. We propose sampling and digitization techniques for abstracting the undecidable continuous timed MTL into decidable discrete time MTL. The sampling abstraction relies upon a Nyquist-like oversampled representation of continuous timed behaviours.

Keywords: Real-time logics, Sampling, Digitization, MTL, Decidability.

1 Introduction

Timed behaviours capture how the system state evolves with time. Temporal logics specify properties of such behaviours. Real-time logics specify quantitative timing properties of timed behaviours. Metric Temporal Logic (MTL) is one such logic [10]. It was originally formulated for behaviours with continuous (real-valued) time. Variants of this logic were defined for sampled timed behaviours (i.e. timed words) as well as discrete timed behaviours [2, 3, 6]. Moreover, time can be chosen to be strictly monotonic or weakly monotonic. With weakly monotonic time a finite sequence of state changes can take place at the same time point.

We present a uniform formulation of MTL, called $MTL[M]$, where the behaviours are always over dense and weakly monotonic time but

*This work was partially supported by the project *Laboratory for Construction, Analysis and Verification of Embedded Systems.*

the set of observable time points within each behaviour is restricted to a subset M. By suitably restricting the class of behaviours and choosing M we can obtain different flavours of MTL. Moreover, we can also define morphisms which map one form of timed behaviour to another. In the paper, we investigate invariance and transformation of MTL formulae under such morphisms.

Specifically, we propose a *well sampling morphism* which records the value of the continuous timed behaviour at each of its change points, each integer valued point and it "oversamples the behaviour by 1" by recording system state also at the midpoint between two consecutive above type (integer or change) points. Such a notion of sampling leads to a point on the real line being represented by a nearby sampled point which is a short but bounded distance away from the original point. Hence, there is small sampling error in measurement of time distances. By suitably relaxing the formulae to adjust for the sampling error, we define two transformations of MTL formulae which respectively preserve the examples and counter examples of MTL formulae under the sampling morphism. This approximates the validity of continuous time MTL by sampled time MTL.

As a subsequent move, we consider the digitization morphism from sampled time behaviour to integer timed behaviour. This morphism was originally suggested by Henzinger, Manna and Pnueli [7]. Digitization shifts the change points within a behaviour to a nearby integral point while (weakly) preserving the relative ordering of their fractional parts. This results into quantization error in the time stamps of the change points. By suitably relaxing the formulae to adjust for the quantization error, we give two transformations of MTL formulae which respectively preserve the examples and counter-examples under digitization.

Thus, we give techniques of approximating continuous time formulae of the undecidable MTL by discrete time formulae of the decidable MTL. This provides a partial method for checking validity and model checking of MTL formulae.

Related work Several authors have dealt with the problem of reasoning about continuous time behaviour by its sampled or discrete timed approximation. Manna, Henzinger and Pnueli first proposed the notion of digitization and its use in deciding language containment of timed regular languages. Ouaknine and Worrell [11] gave a rigorous proof that timed automata with "closed" guards are closed under digitization and time automata with "open" guards are closed under inverse digitization. The notion of sampling and well sampling were proposed by the current author and his colleagues [13, 15].

Agrawal and Thiagarajan [4] have investigated a discretization of hybrid systems which work under an implicit sampling scheme due to digital nature of the controller. They have called the resulting model as lazy linear hybrid automata. Krcál, Mokrushin, Thiagarajan, and Wang Yi [9] have formulated timed automata incorporating synchronisation of events in time-triggered architectures using a notion akin to digitization. The current author and his collaborators have embarked on a program of investigating and applying sampling and digitization approximations to a highly expressive real-time logic called Duration Calculus [5]. Their experiments show that the method provides a partial but often practical technique to deciding real-time logics [13, 15].

An important question is the choice of the form of time in timed behaviours which is selected for modelling and property specification. The well established timed and hybrid automata work from a sampled view of time. However, a notion such a well sampling is missing in these models. Real-time logics have been defined with different forms of time [10, 17, 2, 3, 6]. Weakly monotonic time is useful in modelling behaviours under concurrency and interleaving [2], synchronous circuits and synchronous programs [12], and when dealing with finite precision digital clocks [7].

Trakhtenbrot as well as Hirshfeld and Rabinovich have argued that continuous time is preferable for modelling and analysis of real-time systems [16], [8]. In our own experience, sampled time logics without any notion of sufficient sampling (such as our well sampling) seem counter-intuitive and error-prone for specifying properties. On the other hand, sampled time behaviours seem to admit a wide variety of analysis techniques such as timed/hybrid automata based algorithms, bounded model checking and digitization; which are not currently available for continuous timed behaviours In this paper, we have attempted to build bridges between behavioural descriptions with different forms of time.

2 Behaviours with Weakly Monotonic Time

Metric Temporal Logic (MTL) was originally defined for continuous time behaviours by Koymans [10]. The time was strictly monotonic, i.e. the system could be only in one state at each real numbered time point. Since then, it has been adapted to other views of time. In this Section, we give a unified treatment of MTL under various forms of time.

We first define a notion of observed behaviour and we interpret MTL over such behaviours. The definitions are adapted from similar such definitions originally formulated for an interval temporal logic[15]. The behaviours incorporate weakly monotonic time. (See [12] for motivations for using this form of time.)

Let $(\Re^0, <)$ be the set of non-negative real-numbers with usual order. We will use $t, t_1, \ldots$ to range over reals. Let $Pvar$ be the set of observable propositions. Consider $F \subseteq \Re^0$ which has either the form $[t]$ denoting a singleton set, or the form $[t_1, t_2)$ with $t_1 < t_2$ denoting a non-singular convex set of reals which is half open. Here t_2 can also be ∞. We call such a subset F of reals a *phase*. Let $\nwarrow F$ and $\nearrow F$ denote the left and right end-points (limits) of F. Let a *frame* $\mathscr{F}$ be a finite or countably infinite sequence of adjacent phases which partition $\Re^0$. Formally, $\mathscr{F} = (F_1, F_2, \ldots)$ such that $\nearrow F_i = \nwarrow F_{i+1}$ and $(\cup_i F_i) = \Re^0$. Also, if F_k is the last element of $\mathscr{F}$ then $\nearrow F_k = \infty$. Let $dom(\mathscr{F})$ denote the set of indices (positions) of $\mathscr{F}$. For example $\mathscr{F} = [0, 1.2)[1.2, 3.4)[3.4][3.4, 6)[6, \infty)$ is a frame which partitions the reals into 5 phases. Hence, $dom(\mathscr{F}) = \{1, \ldots, 5\}$. Notice that in $\mathscr{F}$ two phase changes take place at the time point 3.4 which belongs to multiple phases. Let $FRAM$ denote the set of all frames.

Let $Points(\mathscr{F}) \stackrel{\text{def}}{=} \{(t,i) \mid t \in F_i\}$ give the set of *points* in a frame $\mathscr{F}$. Each point (t,i) consists of the time stamp t and the phase number i. It is easy to see that under the point-wise ordering $(t_1, i_1) \leq (t_2, i_2) \iff (t_1 \leq t_2) \wedge (i_1 \leq i_2)$, this set is linearly ordered. We shall use b, e, z to range over points. Also, we define *distance* $d(b,e)$ between two points as absolute difference between their timestamps, i.e. $d((t,i),(t',i')) = |t - t'|$.

A *behaviour* θ over a frame $\mathscr{F}$ has the form $\theta \in dom(\mathscr{F}) \rightarrow 2^{Pvar}$. Thus, a behaviour assigns to each phase F_i of $\mathscr{F}$ a state giving the truth values of the propositions. Such a behaviour encodes a finitely variable evolution of system state with time, where only finitely many state changes take place in a finite time interval. However, we do allow multiple state changes to take place at the same time [12]. This is analogous to the timed state sequences model with superdense time [6]. We shall denote the set of all behaviours over a frame $\mathscr{F}$ by $BEH(\mathscr{F})$.

Example 1 A behaviour $(\mathscr{F}, \theta)$ is given below.

$$\begin{array}{lllllllll}
\mathscr{F} & = & [0,1.5) & [1.5,2.4) & [2.4] & [2.4] & [2.4,3) & [3,4.3) & [4.3,\infty) \\
\theta & = & \neg P, & P, & \neg P, & P, & \neg P, & P, & \neg P
\end{array}$$

In order to accommodate and integrate various notions of time such as sampled (pointwise) or continuous, we associate with each behaviour a set of observable points $S \subseteq Points(\mathscr{F})$. Let $\bar{0}$ denote the initial time point $(0,1)$ with time stamp 0 in the initial phase 1. We assume that this point is always in S, i.e. point $(0,1) \in S$. The tuple $(S, \mathscr{F}, \theta)$ is called an observed behaviour or *o-behaviour*. We will use $\rho, \rho_1 \ldots$ to range over o-behaviours. A collection (class) of o-behaviours is denoted by M. We shall define a generic version of logic MTL which can be interpreted over given M.

Syntax of MTL Let p range over $Pvar$, i,j,k over natural numbers $\aleph$ and let ϕ, ψ denote MTL formulae. Let $I = \langle i,j \rangle$ denote an interval with natural numbered end points i,j with the possible exception that j can also be ∞. The interval can be open, closed or half open and these are denoted using [,) etc. in the usual manner†. Each such interval denotes a convex subset of reals. Let $k + \langle i,j \rangle$ denote the interval $\langle k+i, k+j \rangle$. Let $\top$ denote the formula "true". Logic MTL is given by the abstract syntax:

$$\top \mid p \mid \phi \wedge \psi \mid \neg \phi \mid \phi \mathscr{S}_I \psi \mid \phi \mathscr{U}_I \Psi$$

Semantics For a given o-behaviour $(S, \mathscr{F}, \theta)$, and a point $b \in S$, let $S, \mathscr{F}, \theta, b \models \phi$ denote that formula ϕ evaluates to true in o-model $S, \mathscr{F}, \theta, b$. Omitting the usual boolean cases, this is inductively defined below.

$S, \mathscr{F}, \theta, b \models p$ iff $p \in \theta(i)$ where $b = (t, i)$
$S, \mathscr{F}, \theta, b \models \phi \mathscr{U}_I \psi$ iff for some $e \in S : b \leq e$.
  $d(b,e) \in I$ and $S, \mathscr{F}, \theta, e \models \psi$ and
  for all $z \in S : b \leq z < e$. $S, \mathscr{F}, \theta, z \models \phi$
$S, \mathscr{F}, \theta, b \models \phi \mathscr{S}_I \psi$ iff for some $e \in S : e \leq b$.
  $d(e,b) \in I$ and $S, \mathscr{F}, \theta, e \models \psi$ and
  for all $z \in S : e < z \leq b$. $S, \mathscr{F}, \theta, z \models \phi$

Note that the definitions of $\mathscr{U}$ and $\mathscr{S}$ are relativized to the observable points S. Recall that $\bar{0}$ denotes the initial time point $(0,1)$ with time stamp 0 in the initial phase 1. Define $S, \mathscr{F}, \theta \models \phi \stackrel{\text{def}}{=} S, \mathscr{F}, \theta, \bar{0} \models \phi$. Also, given a set of o-behaviours M, we have $M \models \phi$ iff $\rho \models \phi$ for all $\rho \in M$. When formulae of MTL are interpreted over a set of o-behaviours M, we denote this by $MTL[M]$.

---

†Thus, in a generic interval $\langle i,j \rangle$, symbol $\langle$ ranges over [or (and symbol $\rangle$ ranges over] or).

3 A Variety of Metric Temporal Logics

Metric Temporal Logics available in the literature can be defined as special cases of $MTL[M]$ by appropriately choosing the set of o-behaviors M. We derive many such logics below.

3.1 Continuous Time Metric Temporal Logic (MTL_{ct})

This logic, MTL_{ct}, was investigated in [10] for strictly monotonic time. Here we generalise it to weakly monotonic time.

Let $M_{ct} \stackrel{\text{def}}{=} \{(Points(\mathscr{F}), \mathscr{F}, \theta) \mid \mathscr{F} \in FRAM,\ \theta \in BEH(\mathscr{F})\}$, i.e. models where the set of observable points is fixed as $Points(\mathscr{F})$, the set of all points. Logic $MTL_{ct} = MTL[M_{ct}]$. We shall abbreviate $Points(\mathscr{F}), \mathscr{F}, \theta, b \models \phi$ by $\mathscr{F}, \theta, b \models_{ct} \phi$.

Theorem 2 *Validity of MTL_{ct} formulae is undecidable [2].*

3.2 Pointwise MTL (MTL_{pt})

In this logic, the continuous behaviour is only observed at a countable set of sampling points. This logic was investigated by Alur and Henzinger [2].

Given a frame $\mathscr{F}$, let $Beg(\mathscr{F}) \subseteq Points(\mathscr{F})$ be the set of beginning points of all phases in $\mathscr{F}$. Let $S_{\mathscr{F}} \subseteq Points(\mathscr{F})$ be such that $Beg(\mathscr{F}) \subseteq S_{\mathscr{F}}$ and $S_{\mathscr{F}}$ is countably infinite and time divergent. Thus $S_{\mathscr{F}}$ represents a countably infinite set of *sampling points* which includes all change points between phases. Such an $S_{\mathscr{F}}$ is called *adequate*.

Example 3 Consider the behaviour $(\mathscr{F}, \theta)$ in Example 1. An adequate set of sampling points for this behaviour is as follows.

$$
\begin{aligned}
Beg(\mathscr{F}) &= \{(0,1),(1.5,2),(2.4,3),(2.4,4),(2.4,5),(3,6),(4.3,7)\} \\
S_{\mathscr{F}} &= \{(0,1),(1.1,1),(1.5,2),(2.2,2),(2.4,3),(2.4,4), \\
&\qquad (2.4,5),(3,6),(3.3,6),(4.3,7)\} \\
&\quad \cup \{(4.4,7),(5.5,7),(6.6,7),\ldots\}
\end{aligned}
$$

Let M_{pt} be the collection of all adequately sampled behaviours, i.e. $M_{pt} = \{(S_{\mathscr{F}}, \mathscr{F}, \theta) \mid \mathscr{F} \in FRAM,\ \theta \in BEH(\mathscr{F})$ and $S_{\mathscr{F}}$ is adequate$\}$. Then, logic MTL_{pt} can be defined as $MTL[M_{pt}]$[‡].

[‡]It should be noted that the original MTL_{pt} [2] was formulated using timed words as models. Here, we reformulate this as continuous behaviour together with a set of

Theorem 4 *The validity of MTL_{pt} is undecidable [2].*

3.3 Well Sampled MTL (MTL_{ws})

This is a special case of MTL_{pt} where the continuous time behaviour is sampled at the beginning of every phase and at every integer valued point. Moreover the behaviour is also 1-oversampled by including the midpoint between every consecutive pair of above sampling points. This provides a reasonably faithful method of sampling continuous behaviours.

Formally, given a behaviour $(\mathcal{F}, \theta)$ let $Beg(\mathcal{F})$ be the set of beginning points of phases in $\mathcal{F}$ as in case of M_{pt}. Let $\aleph$ be the set of natural numbers. Let $Int(\mathcal{F}) = \{ (t,i) \in Points(\mathcal{F}) \mid t \in \aleph \}$ be the set of integer valued points. Let $BI(\mathcal{F}) = Beg(\mathcal{F}) \cup Int(\mathcal{F})$. Let $Mid(\mathcal{F}) = \{ ((t_1+t_2)/2, i) \mid (t_1,i),(t_2,j)$ are consecutive points in $BI(\mathcal{F})\}$. Define $WS(\mathcal{F}) = BI(\mathcal{F}) \cup Mid(\mathcal{F})$. The set $WS(\mathcal{F})$ is called the set of *well-sampling points with 1-oversampling* [15]. Here, 1-oversampling refers to the fact that we add one additional point between every pair of consecutive elements of $BI(\mathcal{F})$. Note that $WS(\mathcal{F})$ is uniquely determined by $\mathcal{F}$.

Example 5 For the behaviour $(\mathcal{F}, \theta)$ of Examples 1 and 3, we have

$$\begin{array}{rcl} Beg(\mathcal{F}) & = & \{(0,1),(1.5,2),(2.4,3),(2.4,4),(2.4,5),(3,6),(4.3,7)\} \\ Int(\mathcal{F}) & = & \{(0,1),(1,1),(2,2),(3,6),(4,6),(5,7),(6,7),\ldots\} \\ Mid(\mathcal{F}) & = & \{(0.5,1),(1.25,1),(1.75,2),(2.2,2),(2.7,5),(3.5,6), \\ & & (4.15,6),(4.65,7),(5.5,7),(6.5,7),(7.5,7),\ldots\} \\ WS(\mathcal{F}) & = & Beg(\mathcal{F}) \cup Int(\mathcal{F}) \cup Mid(\mathcal{F}) \end{array}$$

Let $M_{ws} = \{(WS(\mathcal{F}), \mathcal{F}, \theta) \mid \mathcal{F} \in FRAM,\ \theta \in BEH(\mathcal{F})\}$ consisting of the set of o-behaviours where observable points are exactly the set of well sampled points. Such models provide one way of canonically representing the continuous behaviour by sampling. Then logic MTL_{ws} is obtained as $MTL[M_{ws}]$. We shall abbreviate $WS(\mathcal{F}), \mathcal{F}, \theta, b \models \phi$ by $\mathcal{F}, \theta, b \models_{ws} \phi$.

The following theorem is a variant of Theorem 4 of Alur *et al* and the proof is also similar.

Theorem 6 *The validity of MTL_{ws} is undecidable.*

---

observable points. It is easy to see that the two formulations are equivalent.

3.4 Integer Time Metric Temporal Logic (MTL_Z)

This is the discrete time variant of logic MTL_{ct}. In MTL_Z behaviours each phase change happens at an integer valued time point. Formally, a frame $\mathscr{F}$ is called discrete if $Beg(\mathscr{F}) \subseteq Int(\mathscr{F})$. Moreover, the set of sampling points is also integer valued. Thus, $M_Z = \{(S_{\mathscr{F}}, \mathscr{F}, \theta) \mid \mathscr{F}$ is discrete and $S_{\mathscr{F}} \subseteq Int(\mathscr{F})\}$. Logic MTL_Z can be obtained as $MTL[M_Z]$. Further, a o-behaviour $(S_{\mathscr{F}}, \mathscr{F}, \theta)$ is called full if $S_{\mathscr{F}} = Int(\mathscr{F})$. Let M_{ZF} be the class of all full o-behaviours from M_Z. Logic MTL_{ZF} is obtained as $MTL[M_{ZF}]$.

Example 7 Let $\mathscr{F} = [0,1)[1][1][1,\infty)$. Then, an adequate set of sampling points is $S_{\mathscr{F}} = \{(0,1),(1,2),(1,3),(1,4),(2,4),(3,4),\ldots\}$.

Theorem 8 *Satisfiability of MTL_Z (and also MTL_{ZF}) is decidable and EXPSPACE-complete (assuming binary encoding of integer constants) [2, 6].*

4 Sampling Abstraction

Sampling morphism transforms a continuous time behaviour to a well sampled view of the same behaviour. We investigate the effect of such a sampling on the truth of a MTL formula ϕ.

We first define a function f mapping points of a continuous time behaviour to the well sampled points within the behaviour. Given a frame $\mathscr{F}$ recall the definition of well sampled points $WS(\mathscr{F})$ which includes all the change points, integer valued points (these together constitute $BI(\mathscr{F})$) as well as the midpoints between every two consecutive $BI(\mathscr{F})$ points. We now define a sampling function $f : Points(\mathscr{F}) \rightarrow WS(\mathscr{F})$ which represents every point in $\mathscr{F}$ by a "nearby" sampling point in $WS(\mathscr{F})$. Let $f(b) = b$ if $b \in BI(\mathscr{F})$. Also let $b = (t_1, i)$ and $e = (t_2, j)$ with $b < e$ be two consecutive points in $WS(\mathscr{F})$. Then define $f(z) = ((t_1 + t_2)/2, i)$ for all $b < z < e$. Thus, each point in $BI(\mathscr{F})$ maps to itself whereas the open interval between two successive $BI(\mathscr{F})$ points maps to its midpoint. Note that f is a total function.

Example 9 Let $\mathscr{F} = [0,1.5)[1.5,2.4)[2.4][2.4][2.4,3)[3,4.3)[4.3,\infty)$. Consider the behaviour $(\mathscr{F}, \theta)$ from of Examples 1 and 3 for which the set of well sampled points $WS(\mathscr{F})$ was given in Example 5. Recall that any point has form (t, i) with time stamp t and phase number i. Then, $f((0,1)) = (0,1)$ and $f((1,1)) = (1,1)$ as they are integer valued points and $f(b) = (0.5,1)$ for $(0,1) < b < (1,1)$. Similarly,

$f((2,2)) = (2,2)$ and $f((2.4,3)) = (2.4,3)$ as these are beginning points of a phase. Also, $f(b) = (2.2,2)$ for $(2,2) < b < (2.4,3)$, and so on. ⊣

Lemma 10 *The sampling function f has the following properties.*

- *f is total and onto.*
- *f is weakly order preserving, i.e. $b \leq e \Rightarrow f(b) \leq f(e)$. However, f is not strictly order preserving, i.e. it is possible that $b < e$ but $f(b) = f(e)$.*
- *$f(t,i) = (t',i)$ i.e. f maps a point of a phase F_i to a point within the same phase. Hence, the state does not change over the closed interval $[b, f(b)]$ (or $[f(b), b]$).*
- *$d(b, f(b)) < 0.5$, i.e. a point does not shift very far by sampling.*

We promote the sampling function f from points to frames and models. Let f map a continuous time model $(Points(\mathscr{F}), \mathscr{F}, \theta, b)$ to a sampled time model $(WS(\mathscr{F}), \mathscr{F}, \theta, f(b))$. We will study how the formula truth is affected by this f transformation. For this, we will define two maps α^+ and α^- giving weak and strong approximations of the original formula which respectively preserve the examples and counter-examples under the f map.

Definition 11 *Given a (open, closed or half-open) interval $I = \langle i, j \rangle$ let I^+ be $(i-1, j+1)$ if $i - 1 \geq 0$ and be $[0, j+1)$ otherwise. Also, let I^- be $[i+1, j-1]$ if $i + 1 \geq j - 1$, and be undefined otherwise.*

Definition 12 *Let $\alpha^+, \alpha^- : MTL \rightarrow MTL$ be inductively defined as follows.*

- $\alpha^+(p) = \alpha^-(p) = p$.
- $\alpha^+(\phi \wedge \psi) = \alpha^+(\phi) \wedge \alpha^+(\psi)$ *and* $\alpha^-(\phi \wedge \psi) = \alpha^-(\phi) \wedge \alpha^-(\psi)$.
- $\alpha^+(\neg\phi) = \neg\alpha^-(\phi)$ *and* $\alpha^-(\neg\phi) = \neg\alpha^+(\phi)$.
- $\alpha^+(\phi\, \mathscr{U}_I \psi) \;=\; \alpha^+(\phi)\; \mathscr{U}_{I^+}\; \alpha^+(\psi)$.
- $\alpha^-(\phi\, \mathscr{U}_I \psi) \;=\; \alpha^-(\phi)\; \mathscr{U}_{I^-}\; \alpha^-(\psi)$.
- $\alpha^+(\phi\, \mathscr{S}_I \psi) \;=\; \alpha^+(\phi)\; \mathscr{S}_{I^+}\; \alpha^+(\psi)$.
- $\alpha^-(\phi\, \mathscr{S}_I \psi) \;=\; \alpha^-(\phi)\; \mathscr{S}_{I^-}\; \alpha^-(\psi)$.
- $\alpha^-(\phi\, \mathscr{U}_I \psi)$ *and* $\alpha^-(\phi\, \mathscr{S}_I \psi)$ *are $false$ when I^- is undefined.*

Theorem 13 *[Sampling Abstraction]*

1. $\mathscr{F}, \theta, b \models_{ct} \phi \;\Rightarrow\; \mathscr{F}, \theta, f(b) \models_{ws} \alpha^+(\phi)$.
2. $\mathscr{F}, \theta, b \models_{ct} \phi \;\Leftarrow\; \mathscr{F}, \theta, f(b) \models_{ws} \alpha^-(\phi)$

The proof of the theorem is by induction on the structure of the formula. While we do not give here the full proof of the above theorem, the following reasoning explains the case of $\mathcal{U}_I$ operator. Let $I = \langle i, j \rangle$.

- For part (1) of the theorem, let x be in interval I relative to b, i.e. let $x \in b + I$, that is $x \in \langle b+i, b+j \rangle$. Then, by Lemma 10, $f(b) \in (b - 0.5, b + 0.5)$ and $f(x) \in (b + i - 0.5, b + j + 0.5)$. Hence, by interval arithmetic, $f(x) - f(b) \in (b + i - 0.5 - (b + 0.5), b + j + 0.5 - (b - 0.5)) = (i - 1, j + 1) = I^+$.
- For part (2) of the theorem, let $f(x) \in f(b) + [i + 1, j - 1]$, i.e. $f(x) \in [f(b) + i + 1, f(b) + j - 1]$. Then, by lemma 10, $b \in (f(b) - 0.5, f(b) + 0.5)$ and $x \in (f(x) - 0.5, f(x) + 0.5)$. Substituting for $f(x)$ its interval, we get $x \in (f(b) + i + 1 - 0.5, f(b) + j - 1 + 0.5)$, that is $x \in (f(b) + i + 0.5, f(b) + j - 0.5)$, Hence, by interval arithmetic, $x - b \in (f(b) + i + 0.5 - (f(b) + 0.5), f(b) + j - 0.5 - (f(b) - 0.5)) = (i, j)$.

Corollary 14

1. $\models_{ws} \alpha^-(\phi) \;\Rightarrow\; \models_{ct} \phi$,
2. $\mathcal{F}, \theta, f(b) \not\models_{ws} \alpha^+(\phi) \;\Rightarrow\; \mathcal{F}, \theta, b \not\models_{ct} \phi$.

5 Digitization Abstraction

A digitization morphism transforms a sampled time behaviour into a set of integer timed behaviours. This morphism was originally proposed by Henzinger *et al* [7]. Let $frac(a) \stackrel{\text{def}}{=} a - \lfloor a \rfloor$ denote the fractional part of a real number a.

Definition 15 *[digitization]*

- *Let $x \in \Re^0$ and let $0 \leq \epsilon < 1$. Then,* $x \downarrow \epsilon \stackrel{\text{def}}{=} \left(\begin{array}{c} \lfloor x \rfloor \;\; if \;\; frac(x) \leq \epsilon \\ else \;\; \lceil x \rceil \end{array} \right.$.
- *Let $(S, \mathcal{F}, \theta, b)$ be a sampled time model in M_{pt}. Its ϵ-digitization, denoted $(S, \mathcal{F}, \theta, b) \downarrow \epsilon$, is given by $(S \downarrow \epsilon, \mathcal{F} \downarrow \epsilon, \theta, b \downarrow \epsilon)$ where $S \downarrow \epsilon = \{(t \downarrow \epsilon, i) \mid (t, i) \in S\}$. Also, $\mathcal{F} \downarrow \epsilon = (F_1 \downarrow \epsilon, F_2 \downarrow \epsilon, \ldots)$ where $\mathcal{F} = (F_1, F_2, \ldots)$. For each phase, we have $[t] \downarrow \epsilon = [t \downarrow \epsilon]$, and $[t_1, t_2) \downarrow \epsilon = [t_1 \downarrow \epsilon, t_2 \downarrow \epsilon)$ if $t_1 \downarrow \epsilon < t_2 \downarrow \epsilon$ and $[t_1 \downarrow \epsilon]$ otherwise. It is easy to check that $\mathcal{F} \downarrow \epsilon$ satisfies all the requirements of being a frame.*
- *The set of digitizations $Z((S, \mathcal{F}, \theta, b)) = \{(S, \mathcal{F}, \theta, b) \downarrow \epsilon \mid 0 \leq \epsilon < 1\}$.*

Note that in digitization of $\mathcal{F}$ with ϵ, the number and ordering of phases remains unchanged. Only the end-points of each phase are shifted to nearby integral points. Hence, the ordering between points is also weakly preserved by digitization. Moreover, the assigment of state to each phase (index) is also not changed.

Example 16 Let $\mathcal{F} = [0,1.5)[1.5,2.4)[2.4][2.4][2.4,3)[3,4.3)[4.3,\infty)$ from Example 1. Then, $\mathcal{F} \downarrow 0.45 = [0,2)[2][2][2][2,3)[3,4)[4,\infty)$. Note the phase collapse $[1.5,2.4) \downarrow 0.45 = [2]$.

We define weak and strong approximations β^+ and β^- of MTL formulae which respectively preserve examples and counter examples under digitization.

Definition 17 *Given an interval I let $I \uparrow$ be the smallest closed interval containing I and let $I \downarrow$ be the largest open interval contained in I. If I singleton closed interval $[i,i]$ then $I \downarrow$ is undefined. For example, $[2,4) \uparrow = [2,4]$ and $[2,4) \downarrow = (2,4)$.*

Definition 18 *Let $\beta^+, \beta^- : MTL \rightarrow MTL$ be inductively defined as follows.*

- $\beta^+(p) = \beta^-(p) = p$.
- $\beta^+(\phi \wedge \psi) = \beta^+(\phi) \wedge \beta^+(\psi)$ *and* $\beta^-(\phi \wedge \psi) = \beta^-(\phi) \wedge \beta^-(\psi)$.
- $\beta^+(\neg\phi) = \neg\beta^-(\phi)$ *and* $\beta^-(\neg\phi) = \neg\beta^+(\phi)$.
- $\beta^+(\phi\, \mathcal{U}_I \psi) = \beta^+(\phi)\ \mathcal{U}_{I\uparrow}\ \beta^+(\psi)$.
- $\beta^-(\phi\, \mathcal{U}_I \psi) = \beta^-(\phi)\ \mathcal{U}_{I\downarrow}\ \beta^-(\psi)$.
- $\beta^+(\phi\, \mathcal{S}_I \psi) = \beta^+(\phi)\ \mathcal{S}_{I\uparrow}\ \beta^+(\psi)$.
- $\beta^-(\phi\, \mathcal{S}_I \psi) = \beta^-(\phi)\ \mathcal{S}_{I\downarrow}\ \beta^-(\psi)$.
- $\beta^-(\phi\, \mathcal{U}_I \psi)$ *and* $\beta^-(\phi\, \mathcal{S}_I \psi)$ *are $false$ when $I \downarrow$ is undefined.*

In the following theorem we use the notiation $\models_Z$ and $\models_{pt}$ to denote $\models$ within logics MTL_Z and MTL_{pt} respectively. We omit the proof of the theorem.

Theorem 19 *[Digitization Abstraction]*

1. $S, \mathcal{F}, \theta, b \models_{pt} \phi \ \Rightarrow\ Z(S, \mathcal{F}, \theta, b) \models_Z \beta^+(\phi)$,
2. $S, \mathcal{F}, \theta, b \models_{pt} \phi \ \Leftarrow\ Z(S, \mathcal{F}, \theta, b)) \models_Z \beta^-(\phi)$.

Corollary 20

1. $\models_Z \beta^-(\phi) \ \Rightarrow\ \models_{pt} \phi$,
2. $S, \mathcal{F}, \theta, b \not\models_Z \beta^+(\phi) \ \Rightarrow\ S, \mathcal{F}, \theta, b \not\models_{pt} \phi$.

Applying successively the sampling abstraction and the digitization abstraction, we can approximate the validity of MTL over continuous time by its validity over discrete time.

6 Discussion

This paper proposes a sampling and digitization technique to approximate continuous time MTL validity by discrete time MTL validity. The technique reminds us of the Nyquist-like sampling analysis in signal processing. The sampling abstraction relies upon representation of continuous time behaviour by a sufficiently oversampled behaviour and it is subject to small sampling error whereas digitization leads to quantization error in the time stamps. We have given the weak and strong approximations of a MTL formula which relax the constants in the formula in the correct direction to respectively preserve the examples and counter examples under sampling and digitization. The Sampling and digitization abstractions together provide a partial technique for reasoning about continuous time MTL. Based on experience with other real-time logics, we believe that the technique should be quite applicable in practice [15].

While we have proposed the well sampling scheme as a theoretical device for representing continuous time behaviour by sufficiently oversampled behaviour, this scheme is not easy to build into sampling hardware. Other *periodic* sampling schemes need to be studied which may work together with assumptions of stability, such as any state remains stable for at least δ time, to give desirable properties similar to well sampling. This is a topic of our ongoing work.

Acknowledgements The author thanks Kamal Lodaya for his helpful comments.

References

[1] R. ALUR AND D.L. DILL. Automata for modeling real-time systems. *Proc. 17th ICALP*, LNCS 443, Springer-Verlag, (1990) 332-335.

[2] R. ALUR AND T.A. HENZINGER. Real-time logics: complexity and expressiveness. *Information and Computation*, **104**, (1993) 35–77.

[3] R. ALUR AND T.A. HENZINGER. A Really Temporal Logics. *Journal of the ACM*, **41**, (1994) 181–204.

[4] MANINDRA AGRAWAL AND P.S. THIAGARAJAN. The Discrete Time Behavior of Lazy Linear Hybrid Automata. *Proc. HSCC 2005*, LNCS 3414, Springer-Verlag, (2005) 55–69.

[5] G. CHAKRAVORTY AND P.K. PANDYA. Digitizing Interval Duration Logic. *Proc. of 15th CAV*, LNCS 2725, Springer-Verlag, (2003) 167–179.

[6] T.A. HENZINGER. Its about time: Real-time Logics Reviewed. *Proc. 9th CONCUR*, LNCS 1466, Springer-Verlag, (1998).

[7] T.A. HENZINGER, Z. MANNA, AND A. PNUELI. What good are digital clocks?. *Proc. 19th ICALP*, LNCS 623, Springer-Verlag, (1992) 545–558.

[8] Y. HIRSHFELD, AND A. RABINOVICH. A framework for decidable metrical logics. *Proc. ICALP'99*, LNCS 1644, Springer-Verlag, (1999) 422–432.

[9] PAVEL KRCÁL, LEONID MOKRUSHIN, P.S. THIAGARAJAN, AND WANG YI. Timed vs. time-triggered automata. *Proc. CONCUR 2004*, LNCS 3170, Springer-Verlag, (2004) 340–354.

[10] R. KOYMANS. Specifying real-time properties with metric temporal logic. *Real-time Systems*, **2**(4), (1990) 255–299.

[11] J. OUAKNINE AND J. WORRELL. Revisiting Digitization, Robustness and Decidability for Timed Automata. *Proc. 18th IEEE Symposium on LICS*, IEEE Computer Society, (2003) 198–207

[12] P. PANDYA AND D.V. HUNG. Duration calculus of weakly monotonic time. *Proc. FTRTFT 1998*, LNCS 1486, Springer-Verlag, (1998).

[13] P.K. PANDYA, S.N. KRISHNA AND K. LOYA. On Sampling Abstraction of Continuous Time Logic with Durations. *Proc. TACAS 2007*, LNCS 4424, Springer-Verlag, (2007) 246–260.

[14] B. SHARMA, P.K. PANDYA AND S. CHAKRABORTY. Bounded Validity Checking of Interval Duration Logic. *Proc. 11th TACAS*, LNCS 3440, Springer-Verlag, (2005) 301–316.

[15] P.K. PANDYA. All Those Duration Calculi: An Integrated Approach. *Proc. GM R&D Workshop: Next Generation Design and Verification Methodologies for Distributed Embedded Control Systems*, Bangalore, January 2007, Springer, (2007) 67–81.

[16] B.A. TRAKHTENBROT. Understanding Basic Automata Theory in Continuous Time. *Fundament a Informatic*, **62**(1), (2004) 69–121,.

[17] ZHOU CHAOCHEN, C.A.R. HOARE AND A.P. RAVN. A Calculus of Durations. *Info. Proc. Letters*, **40**(5), (1991).

Abstract Switches: A Distributed Model of Communication and Computation

Sanjiva Prasad
Indian Institute of Technology Delhi
sanjiva@cse.iitd.ac.in

Abstract

I describe a model of computation and communication consisting of a network of abstract switching elements, which communicate messages between one another. Message are sequences of identifiers, and can be read, rewritten and communicated by the abstract switches as specified according to the entries in their control tables. Each abstract switch executes a "universal forwarder loop, in which it receives a message on some input port and places rewritten messages on some of its output ports. The control tables specify how *prefixes* of messages on particular input ports should be rewritten and onto what output ports the modified messages should be copied. Four typical patterns of message prefix rewriting are push, pop, nop and swap, which are sufficient to express the variety of communication protocols. The framework covers both point-to-point and broadcast communication models.
We have in earlier work illustrated how this model seems adequate for expressing the basic message deliverability properties of wide variety of communication protocols. Various simple models of computation (finite state automata, push-down automata) also seem to be expressible as restricted versions of this model. The model thus provides a framework in which both computation and communication are expressible.
We show that using a small set of simple axioms of reachability, we can define fundamental notions of communication such as names, addresses, peers, name spaces etc. These notions provide us modular constructions for composing networks of switches (and their corresponding name spaces) via two kinds of constructions — *layering* and *gateway*. The systematic composition of name spaces is arguably the basis for sound and robust communication protocol design.

Keywords: Formal models of computing, communicating systems, distributed systems

1 Introduction

Computer communication networks, and the Internet in particular, are quintessential distributed systems, displaying concurrency in operation, spatial distribution of computing entities, partial failures and independent modes of failure. The Internet has come to occupy a central role in computing frameworks, particularly with the dizzying spread of telecommunication as well as web-based applications. The Internet is therefore a challenging object of study for those interested in modelling and reasoning about concurrent and distributed systems.

Traditionally, the Internet is modelled as a graph, where each node implements a set of protocol layers and each edge corresponds to a physical communication link. The basic purpose of the Internet is to support the ability of a node to send a message to another node, by *naming or addressing* the destination node. In the traditional model, nodes are addressed by one or more static IP addresses. For reasons elucidated by Clark [4], the most appropriate architecture for the Internet, given the goals of robust and scalable interconnection of heterogeneous networks, came be a *packet-switching network* with a *connectionless* or *datagram* mode of service. A defining feature of the classical internet architecture is that state information is consolidated at the endpoints of the network whereas the intermediate packet-switching nodes simply store and forward messages based on a simple logic. End systems implement a simple five-layer stack, with applications using a transport layer to access IP, which is layered on the data link and physical layers. Packet forwarding decisions are made purely on the basis of IP "routing" tables. Moreover, a protocol layer at any node only inspects packet headers associated with that layer, obeying strict rules in how it interfaces with other layers.

Over the years, however, communication networks have supported an increasing variety of services, and over a plethora of media. Consequently, the fundamental assumptions of the internet architecture have often been violated. For example:

- DHCP, anycast, multicast, NAT, mobile IP and others break the static association between a node and its IP address.
- Nodes implement more layers, including IP or VLAN tunnels, overlays, and shims, such as MPLS.
- Forwarding decisions are made not only by IP routers, but also by VLAN switches, MPLS routers, NAT boxes, firewalls, and wireless mesh routing nodes.
- Middleboxes and cross-layered nodes such as NATs, firewalls, and load balancers violate layering.

The complications induced by such changes render even elementary concepts such as naming and addressing difficult to define, let alone understand. And yet, the collective system surprisingly works! In [8, 9], Martin Karsten, Srinivasan Keshav and I tried to understand the basic principles and essential invariants of the *de facto* architecture of the Internet, which allow it to support connectivity in a "reasonable" manner. Our approach was to propose an axiomatic framework using which one could define fundamental concepts of reachability in such dynamic scenarios, and then *derive* definitions of concepts such as names, addresses, name spaces, etc. We then used this framework to describe and explain, using a purely formal integrated model, packet progress across multiple layers of communication protocols.

In this article, I revisit the model presented in [9] (abstracting it from the particular context of computer networks) as I believe it constitutes a fairly general model of computation (apart from being a model of communication), and one which generates a number of interesting problems for researchers interested in applying formal methods, particularly algebraic and logical techniques, to the analysis of large, dynamic networks. I am presenting work at a very early stage of development (more concepts and definitions than theorems) mainly to try to interest the readership of this volume, which has considerable expertise in the areas of automata, logic, algebra, verification techniques (and all their myriad interconnections) into examining an area of computer science which could clearly use their insights in developing more reliable and *reasonable* networks and system architectures.

The main results reported here concern *naming*, and are inspired by the seminal work of Saltzer and others on the crucial role of names as the essential "glue" used for building all manner of systems by connecting together the three fundamental abstractions of processors or interpreters, storage elements, and communication links.

Our model consists of a network of nodes which are *abstract switching elements* that communicate *messages* between one another over *communication links*. Messages can be read, rewritten and communicated by the abstract switches, based on the message structure, and according to the actions specified in the entries in the *control tables* located at the switches. Each abstract switch is equipped with a set of input and output ports, and with a control table, the entries of which can change during the course of execution. It executes a "universal forwarder loop", in which it receives a message on some input port and places rewritten messages on some of its output ports. The control tables specify how prefixes of messages on particular input ports should be rewritten and onto what output ports the modified messages should be copied.

The model is a truly distributed one and permits operations on different messages at different nodes at the same time. The framework is scalable, and is expressive enough and sufficiently generic to describe a variety of network protocols. In particular, it covers both point-to-point and broadcast communication models. Moreover, it encompasses as special cases some well-studied formal automata models. The model incorporates the usual slew of issues of interest to us: concurrency, asynchrony, non-determinism, etc. What is interesting about the model is that in it the notion of connectivity depends on a *distributed network state*; the deliverability of a message depends on whence it originates, on its structure (i.e., some prefix of its header), as well as the possibly changing state of network forwarding tables. Thus *reachability analysis*, a cornerstone of a lot of verification work, is not trivial in this model. I reiterate that such reachability analysis is fundamental for establishing the correct functioning of a communication network (which exists, after all, to deliver messages), but is also germane to reasoning about reference chains in e.g., storage systems.

In Section 1.1, I recap the major features of the model and provide an informal description of its operation. I mention the four message rewriting idioms, and how they are used in network protocols.I then briefly survey some of the related work in Section 1.2. Section 2 provides the crucial definition used in the paper, the "leads-to" relation, which is inductively defined using a small number of axioms. The "leads-to" relation is an abstract way of expressing the connectivity in a network, "abstract" in that it is oblivious of time and time-dependent behaviour. While such an abstraction is perhaps too drastic for accurately answering certain questions about message delivery as it may grossly over-approximate the actual behaviour of systems, it is nonetheless very useful in dealing with many simple questions, and provides a tractable foundation for purely formal reasoning about message deliverability. In fact, our earlier paper [9] contains a Hoare-style logic specialized to deal with this particular relation, which allows us to formally verify a generic implementation of the forwarding behaviour of a generic protocol element, using which we can then derive reliable protocol implementations, almost "for free".

Another significant contribution, reported in Section 3, arising from the axiomatic specification of "leads-to" is that one can recover more robust definitions of elementary referencing and communication concepts such as *names*, *addresses*, *peers* and *tunnels*. I then use these concepts in defining notions of equivalence of names, which is very similar to equivalence of strings recognized by states in traditional automata. More interestingly, one can define an equivalence relation over switching elements based on how they interpret names. This notion induces the fundamental concepts of *scope of a name* and *name spaces*. What we therefore have is a *nominal* framework, in the sense of the seminal work of Saltzer [16], which deals with issues of names and how they are interpreted, bindings and their contexts; moreover, our framework operates in the presence of the dynamics of networks and is therefore a pretender to being a generalization of Saltzer's single-system framework to networked, distributed systems.

The concept of a name space, which is a set of names that have identical scope, and are consistently interpreted at all nodes in the scope, is *the* fundamental notion that allows us to build in a disciplined, systematic fashion increasingly complex network architectures and protocols. In Section 4, I give a formal characterization of the two canonical constructions — *overlay* and *gateway* — of complex networks from simpler ones, in a way that preserves the kinds of invariant nominal properties on which protocols rely. The main new results of the paper are these characterizations of the constructions, together with their corresponding name-space theorems (Thm 2 and Thm 3). The theorems are given without their easy proofs (the more interesting part lay in cobbling up the right definitions).

Finally, in Section 5, I conclude by summarizing and mentioning what I believe are interesting problems that can be addressed in the future.

1.1 The model

The constituents of the model are:

- **Nodes = abstract switches** An *abstract switching element (ASE)* is an entity that participates in communication and relays messages. It generalizes a simple switching element, such as the table-based crossbar switch, to an abstract object that can, in addition to switching messages, carry out more complex actions, such as rewriting header labels of messages and encapsulating a message into another. We will use letters A, B, C and X, Y, Z to denote ASEs (known or unknown).

- **Ports** An ASE has *named* input and output communication *ports*. In the case of computer networks, these ports may be physical network interfaces as well as logical ports in protocol interfaces. For simplicity, we use the identifier of a "connected"

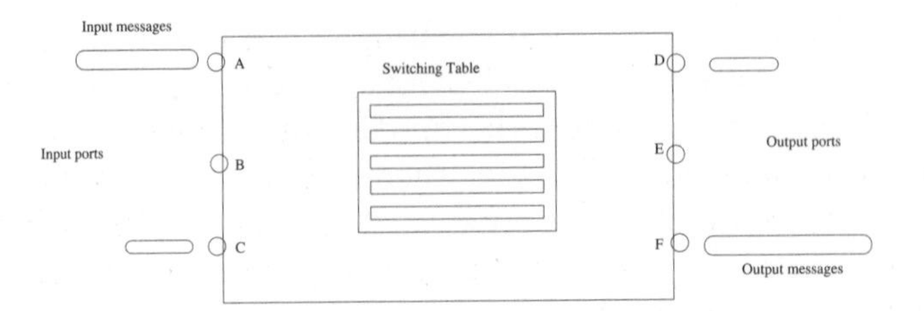

Figure 1: *Schematic view of an abstract switch*

ASE to identify a port*. At ASE B, the input port from a *predecessor* ASE A is denoted as ${}^{A}B$ and the output port to a *successor* ASE C is B^{C}. We use a lower case superscript x such as ${}^{x}B$ or B^{x} when talking of an unspecified port. Different logical ports may actually refer to the same physical entity.

- **Edges = Direct connections between ports** The output port A^{B} is connected by a directed edge to the input port ${}^{A}B$. Simplex *direct communication* from the output port of an ASE to the input port of an adjacent ASE is accomplished by any of a variety of mechanisms (such as shared memory or physical media such as wires), the details of which are left unspecified in our model. In particular, we do not assume that communication is synchronous between the vertices of an edge, nor do we necessarily assume reliability in the generalization of the model presented here (see remarks below). With reference to Figure 2, where each rectangle represents an ASE, examples of direct communication are between the TCP and IP ASEs on the same machine and two Ethernet ASEs on the same shared medium.

- **Messages** The unit of communication is a *message*, which in our abstract model is merely a *string of identifiers* drawn from an arbitrary alphabet Σ. The identifiers used in messages may not be the same as those used for denoting ASEs. A message m that exists at a port x is denoted as $m@x$. Typical messages are denoted $m, m', \ldots$ and prefixes of messages are usually denotes as $p, p', \ldots$

*The set of identifiers used for referring to ASEs is some meta-level global set — which is not part of the model — whereas the identifiers used for ports are locally employed by an ASE. There is no reason that the latter names should be drawn from the former.

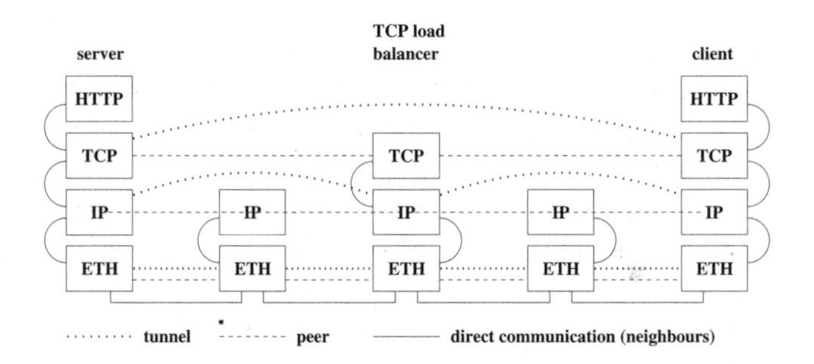

Figure 2: *Protocol layers expressed using a network of ASEs*

- **Creating/consuming messages** We denote with ${}^{0}B$ the logical port that is equivalent to *creating* a message at ASE B and B^{0} the logical port that *consumes* a message. Creation and consumption refer to a transformation of the message in or out of the realm of this theoretical model. In Figure 2, the HTTP ASEs create and consume application-level messages, and the TCP ASEs create and consume acknowledgement messages.

- **Switching tables** An ASE maintains a private set of mappings, called its local *switching table.* The switching table at ASE B is denoted as S_B and contains mappings $\langle A, p\rangle \mapsto \{\langle C_i, p_i'\rangle\}$ from a ASE-and-string pair $\langle A, p\rangle$ to a **set** of ASE-and-string pairs $\langle C_i, p_i'\rangle$. The switching table can be queried through an exact lookup operation $\mathsf{S}_B[A, p]$. If no exact mapping exists in the switching table, the message is discarded. Examples of a switching table in computer networks are Ethernet or IP forwarding tables with $p_i' = p$. The switching table specifies a local prefix rewriting system at that ASE. Note that the switching tables at different ASEs may be quite different, and the model does not mandate strong constraints on the contents of the tables. However, in the context of particular systems, e,g., computer network protocols, one has to ensure that the distributed state of these tables collectively satisfies certain invariant properties, so as to imply correctness properties of the protocols such as absence of forwarding loops.

- **Updates and control messages** System or network dynamics, such as mobility, necessitate dynamic updates to switching tables. The extended version of the model allows for updates to the entries of the switching tables. New entries can be added to a switching table, and existing entries modified or deleted, and these changes can be made in response to the arrival of messages of particular "shape" at the ASE. Such message are usually called *control messages*, and contain specific "*opcodes*", though in some protocols, data messages too can trigger table updates. In fact, the primary challenge in reasoning about protocols involving dynamism such as mobility lies in proving that despite such changes, crucial invariant properties about the distributed state of the switching tables are preserved.

The operation of model is then informally describable as follows. When a message pm appears on the input port ${}^{A}B$ such that S_B has an entry $\langle A, p\rangle \mapsto \{\langle C_i, p_i'\rangle\}$, then for each i, a message $p_i'm$ is placed on the output port B_i^C. The operational semantics can easily be specified in the style of an abstract machine. In this paper, however, we will specify this operational intuition in a more abstract fashion as an inductively defined predicate using simple axiom schemata. The axiomatic formulation should be viewed as an abstraction in which the precise time-dependency of the execution is forgotten.

Forwarding idioms

In theory, the transformation from p to p_i' is unrestricted, but in practice it is either a *push*, *pop*, *swap*, or *nop* operation, as described next. In case of *push*, a new prefix q is prepended and $p_i' = qp$. In case of *pop*, p is removed and $p_i' = \epsilon$. In case of *swap*, p is replaced by p_i', which usually is of equal length. With *nop*, p remains unchanged and $p_i' = p$. Given these transformations, it is possible to identify corresponding forwarding operations that cover a wide range of forwarding techniques used in communication protocols.

Nop - Forwarding If no modification of the current name takes place, the message is just forwarded. In the context of communication networks, Ethernet bridging or IP forwarding are prominent examples, but circuit switching also trivially falls in this category.

Push - Encapsulation The push operation stacks a new name "on top" of the existing "stack". This is used at as a general mechanism for protocol layering.

Pop - Decapsulation The pop operation removes the "topmost" name from the stack. It is used at the receiving side of layered protocols, as well as when forwarding a source-routed message.

Swap - Label Switching The swap operation replaces the current name. It can, of course, be simulated by a pop and push operation at the cost of additional ASEs. Swap is used when translating addresses when traversing different administrative zones.

Remarks The prefix rewriting that occurs at an ASE may be realized by using standard pattern-matching techniques derived from automata theory. Indeed, one possible way of classifying and studying classes of protocols implemented in a class of ASEs within a network may be by considering the precise class of transducers that can be employed to implement the rewriting.

At another level, the model of abstract switches collectively operates like an automaton in that at an ASE a message prefix is examined and according to the switching table, the message is transferred to another ASE. This is analogous to the change of state in an automaton based on the input letter – each ASE behaving as a state, and the message as the input string. However, in our model the switching decision may depend on the port at which the message is present, which is related to the predecessor ASE. Further, the message may be relayed to multiple ASEs. Finally, the prefix of a message (on the basis of which a switching decision is made) need not consist of a single identifier, and may get rewritten into another string.

Thus, in the special cases where the switching table entries are *oblivious* of the predecessor ASE, and when the prefixes on which switching decisions are made consist of singleton identifiers, we obtain transition relations of what are essentially simple automata. In the special case where prefix p is a singleton a, and message am is rewritten to m (pop) [or is not rewritten at all (nop)], we have the transition relations of familiar (finite) state automata [possibly with ϵ moves] for regular languages. It is also seems possible to simulate the working of a pushdown automaton by a network of ASEs by employing a more complex alphabet, and allowing additionally the ability to push (i.e., prepend) symbols onto the front of the message. Switching table entries that discriminate based on the predecessor ASE can capture a small degree of history-sensitivity. In this sense, we claim that familiar models of *computation* are encompassed by our communication model of abstract switches. In the future, we will be investigating these connections in greater detail, particularly algebraic characterizations of the behaviour of various kinds of switching networks.

1.2 Related work

We recall in brief various existing models of computing that deal with communication as computation, in which names or strings of identifiers are the primary data type, and which employ some form of switching as a primary computational operation.

Several computational frameworks are based on the idea of identifying computation and communication. Process calculi such as CCS [13, 14], CSP [6], and ACP [1] employ this idea in developing their model of *interacting* concurrent processes. This interaction is formulated as some form of *synchronization*, be it that of "hand-shaking" communication in CCS or co-action as in CSP. The formulation of the interacting processes is reminiscent of automata, in particular a variation on the idea of transducers (although they are presented in the "programming language style" of abstract grammars). In contrast, synchronization is not a central feature of our model, and the form of interaction between ASEs is loosely coupled, and mediated by messages.

Among such calculi are a class in which *names* are the primary data structure. Such "nominal" process calculi include the π-calculus [15] (which evolved from CCS) and Ambient calculi [3]. In contrast to our model, where the primary data are strings of identifiers, the primary data in the π-calculus are *atomic* names.

The idea of switching is fundamental to a wide variety of computational models. The primary differences lie in *what* is switched (control or data) and whether the switching is a synchronous or asynchronous with respect to the participating elements. We have already discussed how automata of various kinds in a very direct sense employ switching based on the current state and the letter at the head of the input string. Petri Nets also present a model where tokens are switched from a collection of places according to transitions on which the places are incident. The model of computing underlying actual physical devices in general-purpose computers is based on switching in logic circuits. Combinatory Logic [2] and its variants such as Director Strings [11] can be considered functional models of computing (akin to the λ-calculus) which also employ a form of switching of inputs according to the definition of the combinators.

Finally, Kahn Networks [7, 10] present an interesting model of asynchronous processes that communicate via *streams* of data. The model has some commonalities with our proposed model — distributed processing units that are connected by point to point links, and with sequences (or streams) being the salient data structure. However, these sequences are of data elements, which are the primary data for the processes. In contrast, our model operates on discrete datagrams, the structure of which is a sequence of identifiers.

Our work draws from and is related to a handful of other attempts to bring clarity to Internet architecture, and connections with logic and the algebraic structure of paths and routing. Clark's seminal paper [4] succinctly lays out the design principles of the classical Internet, but does not provide a basis for formal reasoning about its properties. Recently, Griffin and Sobrinho have used formal semantics, and concepts from universal algebra in particular, to model routing [5] and Loo et al. have used a declarative approach to describe routing protocols [12]. I believe that there is a rich lode of problems in the networks domain that can be subject to formal techniques based on logic, algebra and language theory.

2 The Axioms

We now present axiomatically an abstract relation called "leads-to" that characterizes the operational behaviour of message relaying in a store-and-forward network of abstract switches. The "leads-to" relation is denoted as $\rightarrow$ and defined by the following four axiom schemata:

LT1. (Direct Communication)
$\forall A, B, m : \exists A^B, {}^A B \Longrightarrow m@A^B \rightarrow m@{}^A B.$

LT2. (Local Switching)
$\forall A, B, C, m, p, p' : \exists {}^A B, B^C \wedge \langle C, p' \rangle \in \mathsf{S}_B[A, p]$
$\Longrightarrow pm@{}^A B \rightarrow p'm@B^C.$

LT3. (Transitivity)
$\forall x, y, z, m, m', m'' : (m@x \rightarrow m'@y) \wedge (m'@y \rightarrow m''@z)$
$\Longrightarrow m@x \rightarrow m''@z.$

LT0. (Reflexivity) $\forall m, x : m@x \rightarrow m@x$

Axiom LT1 describes direct communication between ASEs. A^B and ${}^B A$ exist if and only if A can directly communicate with B. Axiom LT1 states that a message on an output port of an ASE (eventually) appears on the input port of the connected ASE (at the other end of the "wire").

Axiom LT2 expresses the lookup and switching capability of an ASE. Note that a message pm is logically split into a header prefix p and the (opaque) rest of the message m during each local switching step. LT2 also covers any form of multi-recipient forwarding, such as multicast, since $\mathsf{S}_B[A, p]$ may have multiple elements.

Axiom LT3 splices individual forwarding steps together. These three axioms LT1-3 naturally capture the simplex forwarding process in a communication network, where, potentially, at each forwarding step, a forwarding label is altered.

For a variety of reasons, such as for simplifying formal proofs of certain reachability properties, we include the reflexivity axiom LT0, which says that a message at a port leads to (vacuously) that message being at that port.

Two examples from computer networks readily illustrate the applicability of the model. Consider the special case of LT2 where $p = p'$. In this case, the "leads-to" relation describes a single-layer forwarding system based on global destination addresses, such as IP forwarding. As another special case, consider the absence of prefixes altogether. Then, the axioms describe a forwarding model based on input and output ports, such as circuit switching. In general, as we claim and illustrate in [9], the "leads-to" relation can describe arbitrarily complex multi-layer forwarding systems.

Discussion Note that we have chosen our atomic predicate to be $m@x \rightarrow m'@y$, namely the *deliverability* of messages, rather than the time-dependent property of a message m being present at a particular port x. Indeed, our formalization abstracts from time-dependent behaviour, and instead concentrates on connectivity – or how messages *can* move through the network based on its connectivity. This is for the very simple reason that in designing forwarding schemes, network protocols and architectures, one is not particularly interested in proving properties of particular messages being at particular ports, but rather in more abstract invariant properties about network structure. Indeed, proving properties about message m being at a particular port x (at a particular time t) would be a tedious task, amounting to tracing a pilgrim message's progress through the network. Such an approach would neither be scalable nor would it be tractable in establishing the correctness of a protocol or architecture.

Since we are abstracting over time and time-dependent behaviour of the network, we should therefore read the predicate $m@x \rightarrow m'@y$ as "if message m ever appears at port x, then message m' *may eventually* appear at port y". Under such a reading, this abstraction is intuitively correct with respect to the operational behaviour, provided the ASEs do not make non-monotonic time-dependent switching decisions†. First, note that axiom schema LT1 does not imply that the transfer of the message from the output port A^B to input port B^A is synchronous. Neither is LT2 a synchronous switching decision. Moreover, an asynchronous interpretation admits transitivity and reflexivity in a natural manner.

†There are now networks such as delay-tolerant networks where mutually incompatible routing and forwarding decisions are made based on the time of day. A different, less abstract, notion of connectivity would be needed for studying deliverability in such networks.

An interesting observation is that the axioms do not imply that a message actually "moves" from one locus to another. They merely document that the appearance of a message at a particular port eventually leads to messages appearing at other ports. Once at a port, a messages can persist there for ever after — thus justifying the reflexivity axiom. This monotonicity is useful in proofs of connectivity. In particular, one can treat the removal of a message from the source port when moving it to a destination port as an implementation issue involving efficient resource usage, akin to "garbage collection".

A word about actual proofs involving the "leads-to" relation: The axiom schemata reflect that reachability is the reflexive (LT0), transitive (LT3) closure of the connectivity made either at the switches (LT1) or across edges (LT2). Proving that a message can be delivered from x to y will need considering whether a given ASE A may be involved or not — there is a path from x to y either if there is a direct path not going through A, or else there is a path from x to A, and a path from A to y. The formal reasoning involve using properties of an idempotent semiring (with union and concatenation as its operations).

3 Names and Name Spaces

3.1 Nominal concepts

Based on the "leads-to" relationship, we can succinctly define and explain a number of well-known communication concepts, beginning with a formal definition of a name:

Name Suppose A, B are ASEs, and $p \neq \epsilon$ a prefix (string). p is a name for B at A if $\forall m : \; pm@^x A \to p'm@^y B \to m@B^z$ for some x, y, z and $p' \neq \epsilon$.

The name of an ASE is the prefix that is removed when the message is transmitted to this ASE. Note that p can be a name at A for multiple ASEs. The condition $p' \neq \epsilon$ ensures that B is indeed the ASE where the prefix or any residual of it is removed. Any string that "leads to" a particular ASE B from origin ASE A is considered a name for B at A. By default, names are local and *relative* to the ASE from which they originate. For example, a message header could contain a sequence of labels, each of which identifies forwarding state at subsequent ASEs (also known as *source routing*) along a particular path from a source

ASE to a destination ASE. The complete sequence of labels would then be a name for the destination ASE relative to the source ASE.

Address Suppose A, B are ASEs, and $p \neq \epsilon$ a prefix (string). p is an address for B at A if $\forall m: \; pm@^x A \rightarrow pm@^y B \rightarrow m@B^z$ for some x, y, z.

We define an address as a special kind of name that does not change along the path. In other words, if an ASE writes a name into a message with the assumption that at least some other ASEs interpret the string to send it to the same destination, it becomes an address. Every address is also a name, but the reverse is not true. Note that p can be an address for multiple ASEs.

Peer ASEs A, B are *peers* if there exist prefixes p, p' such that $\forall m:$ $pm@^x A \rightarrow p'm@^y B$ and $p' \neq \epsilon$ and $\mathsf{S}_A[x, p] \neq \emptyset$ and $\mathsf{S}_B[y, p'] \neq \emptyset$.

Tunnel There is a tunnel from ASE A to B if there exist prefixes p, p' such that $\forall m: m@^w A \rightarrow pm@A^x \rightarrow p'm@^y B \rightarrow m@B^z$ and $p' \neq \epsilon$.

The difference between direct neighbours, peer, and tunnel is illustrated in Figure 2. Peer ASEs operate on an identical portion of the message prefix, which typically corresponds to the same protocol header field(s). Note that tunnels are between two peers, but may traverse additional peers. Taking examples from the domain of computer networks again, an IP sender, IP routers, and an IP destination are peers, but not direct neighbours because (a) they all operate on the same IP header, and maintain a local switching table indexed by the IP destination address found in the IP header and (b) IP ASEs never directly communicate with each other; they communicate through a link-layer ASE. The IP sender and IP destination form a tunnel, because the definition of tunnel is satisfied for $m =$ the IP payload, $p =$ the IP header. A pair of connected TCP endpoints also form a tunnel for similar reasons. A pair of connected Ethernet NICs can be considered as both direct neighbours and peers.

3.2 Name-induced equivalences and scopes

Let $[\![n]\!]_A = \{C | n$ is a name for C at $A\}$ denote the *set* of ASEs for which x is a name at A. Let $Dom(n)$ denote the set of ASEs A where $[\![n]\!]_A \neq \emptyset$. The notation lifts to sets of names N in the usual manner:

$[\![N]\!]_A = \bigcup\{[\![n]\!]_A \mid n \in N\}$, and $Dom(N) = \bigcup_{n \in N} Dom(n)$.
Note that if $pm@^x A \to p'm@^y B$, then $[\![p']\!]_B \subseteq [\![p]\!]_A$.

Names p and p' are *indistinguishable* at A if $[\![p]\!]_A = [\![p']\!]_A$, which we will denote as $p \asymp_A p'$. Two names are treated equivalent with respect to a set of ASEs Q, written $p \asymp_Q p'$ if for all ASEs $A \in Q$, $p \asymp_A p'$, and we write $p \asymp p'$ if for all ASEs A, $p \asymp_A p'$. This notion of indistinguishable strings suggests a development of automata-theoretic concepts for this model of computation. Observe that $\asymp_A$ is a right congruence.

Language of an ASE We can also propose a notion of language recognized by a network $\mathcal{N}$, starting from an ASE A:

$$L_A = \{p \mid \exists B, x : \forall m : pm@^x A \to m@B^0\}$$

That is, those prefixes that leave no residual while ensuring that a message is *consumed* at some ASE B. Note that the notion of language of a network is not static, since there may be updates to the network state, i.e., the entries in the switching tables at some or all of the ASEs can be modified during the course of operation of the network.

Note that if $x \asymp_A y$ and $x \in L_A$, then $y \in L_A$.

Dually, names induce equivalences between ASEs: Two ASEs, A, B can be considered equivalent with respect to a name n, denoted as $A \sim_n B$, if they interpret n in the same way, namely, if $[\![n]\!]_A = [\![n]\!]_B$.

One can also extensionally observe in a game-playing sense how a sequence of names are interpreted by ASEs. Two ASEs A, B may be extensionally indistinguishable based on observations over the same sequence of names, though for any name $[\![n]\!]_A$ and $[\![n]\!]_B$ may not be identical. This leads naturally to the notion of bisimulation.

Bisimulation A *bisimulation* is a symmetric binary relation R between ASEs, such that $A\ R\ B$ if for all n, whenever $C \in [\![n]\!]_A$ there exists ASE C' such that $C' \in [\![n]\!]_B$ and $C\ R\ C'$. Let $\approx$ denote the largest bisimulation relation, which is the union of all bisimulation relations, and is an equivalence.

$\approx_n$ We write $A \approx_n B$ if for a given n: whenever $C \in [\![n]\!]_A$ there exists ASE C' such that $C' \in [\![n]\!]_B$ and $C \approx C'$, and symmetrically for any $C' \in [\![n]\!]_B$. That is, there is a bisimulation relating all ASEs named n at A with ASEs named n at B. We write $A \approx_N B$ if for all names $n \in N$, $A \approx_n B$.

The notion of bisimulation can be used in the minimization of a network, by folding together bisimulation equivalence classes into representative ASEs.

Name scopes The *scope* of a name is the equivalence classes induced by it, that is, those ASEs in a network $\mathcal{N}$ which interpret a name in a consistent way. Sometimes the terminology "scope of a name n" is used to mean a particular equivalence class of $\sim_n$, with respect to a (sometimes implicit) "target set" β. If such a (target-related) name scope encompasses only one ASE, i.e., is a trivial equivalence class, we call such a scope *local*. The notion of scope can also be extended to messages, since messages are processed based on their prefix: the outermost portion of the name for the destination of the message determines a set of ASEs that interpret the message in a consistent manner.

Name scope The scope of a name n (for a set of ASEs β) in a network $\mathcal{N}$ is denoted as $\sigma_n^{(\beta)}$ and defined as follows:
$\exists\beta : \forall$ ASEs A_i : n is a name for β at A_i implies $A_i \in \sigma_n^{(\beta)}$.

Message scope The scope ρ_m of a message (at a port) $m@^y x$ is defined as the scope of the outermost destination name:
$m@^y x = pl@^y x \wedge \mathsf{S}_x[y, p] \neq \emptyset$ implies $\rho_m = \sigma_p$.

3.3 Name spaces and routing

A set of names or addresses with the same scope is called a *name space* in a network $\mathcal{N}$ or *address space* (in a network $\mathcal{N}$) respectively. Given a set of names N, we write $A \sim_N B$ if for all $n \in N$: $A \sim_n B$, and omit the subscript, writing $A \sim B$ if for all n: $A \sim_n B$. We write $n \frown p$ if the names n and p inhabit the same name space, i.e., if $\sim_n = \sim_p$.

We posit that name spaces are in fact fundamental in the architecture of distributed networks and indeed of architectures of most complex software systems. The protocol designer constructs a set of names with a particular syntactic format, all of which have the same scope. The switching tables at each ASE in this common scope are initialized so that each name in the name space is interpreted the same way at all ASEs, and this consistency of interpretation is maintained by the invariants of the protocol, even if entries at the ASEs can get updated. *Observe that belonging to a name space does not mean that all the names have identical interpretations, but rather that they have identical scopes.* Also note that the notion of a name space is with respect to a given network $\mathcal{N}$; if the network is altered, extended or combined with another, then assumptions of what set of names constitutes a name space may "break". Indeed, the challenge in developing dynamic protocols, e.g., those dealing with mobile devices, lies in repairing or restoring the name space assumptions within the protocol.

Routing is the *control* process that creates and maintains consistent forwarding *within* a name space in the presence of network dynamics. More formally, the goal of routing is to establish, at some set of ASEs γ, appropriate local switching table entries so that each member of γ has a name for each other member of γ with scope γ. Routing is thus defined based on a name scope. A change in network topology results in a breakage of the "leads-to" relation at some ASEs, which invalidates names, or changes their meaning. Dynamic routing re-establishes the "leads-to" properties throughout γ. By establishing consistent forwarding for a set of ASEs and names, routing effectively transforms all names into addresses and enables *datagram forwarding.* Note that the name space N of a *routing domain* such as IP satisfies the property that for each $n \in N$, each name in the set of names $N\{n\} = \{pn \mid p \in N\}$ is indistinguishable within $Dom(N)$ from the name n.

4 Combining Name Spaces

The heterogeneous nature of networked environments results in different network technologies with different assumptions, goals, and design strategies. This naturally results in a diversity of approaches for naming network elements. Different naming schemes are appropriate for different requirements. Due to the combination of network technologies, a message may travel through many name spaces before reaching the destination.

Let N_1 and N_2 be name spaces in a network $\mathcal{N}$. It is very easy to show that $N_1 \cap N_2$ is also a name space in network $\mathcal{N}$.

Note that name spaces are not closed under union: the equivalence classes with respect to names in N_1 may not coincide with the equivalence classes with respect to names in N_2.

However, if both N_1 and N_2 are name spaces in the same network $\mathcal{N}$ and $N_1 \cap N_2 \neq \emptyset$, then $N_1 \cup N_2$ is a name space in network $\mathcal{N}$.

Suppose N_1 and N_2 are sets of strings. Let $N_1 N_2$ denote $\{n_1 n_2 \mid n_1 \in N_1 \; \& \; n_2 \in N_2\}$. We say that $n_1 n_2$ interpreted at A is *prefix-conformant* if $Y \in [\![n_1 n_2]\!]_A$ implies $Y \in \bigcup\{[\![n_2]\!]_X \mid X \in [\![n_1]\!]_A\}$. Intuitively this means that there does not exist a switching table entry at A (or subsequently) that interprets $n_1 n_2$ in any way other than by "going through" the ASEs named n_1 at A, and then via n_2. Prefix-conformance is enforced in real networks by initializing all switching tables to be so, and maintaining this condition as an invariant.

Note that the equivalence classes of $\sim_p$ are not prefix-closed: $A \sim_{n_1 n_2} A'$ does not imply that $A \sim_{n_1} A'$, since $[\![n_1 n_2]\!]_A$ may be the same as $[\![n_1 n_2]\!]_{A'}$, but $[\![n_1]\!]_A$ may not coincide with $[\![n_1]\!]_{A'}$.

Theorem 1 Name composition
Let N_1 and N_2 be name spaces in a network $\mathcal{N}$, such that for every ASE $Z \in \mathcal{N}$, every name in $N_1 N_2$ is prefix-conformant at Z. If $A \sim_{N_1} B$, then $A \sim_{N_1\, N_2} B$.

Note that without the condition of prefix-conformance, $N_1\ N_2$ is not trivially a name space, since it is possible for $A \sim_{N_1} B$ but not $A \sim_{N_1\, N_2} B$ due to a particular name $n_1 n_2$ having a "divergent" entry in the table at A that leads to an ASE which cannot be reached from the ASEs in $[\![n_1]\!]_A$.

Note also, in general, that while for $n_1, n_1' \in N_1, A \sim_{n_1} B$ if and only if $A \sim_{n_1'} B$, and if $n_2, n_2' \in N_2$, $A \sim_{n_1} B$ implies both $A \sim_{n_1 n_2} B$ and $A \sim_{n_1 n_2'} B$ (and similarly for n_1'), it does not follow that if $A \sim_{n_1 n_2} B$ then $A \sim_{n_1' n_2'} B$. Thus, a result stating "if N_1 and N_2 are name spaces, then so is $N_1\ N_2$" is not achievable. However, if the interpretations of all names in N_2 are *absolute* at all ASEs named by N_1, i.e., if $\forall n_2 \in N_2, \forall A \in \mathcal{N}, \forall n_1 \in N_1, \forall B, B' \in [\![n_1]\!]_A : [\![n_2]\!]_B = [\![n_2]\!]_{B'}$, then $N_1\ N_2$ is a name space.

4.1 Overlay composition of Networks

A fundamental construction in network architectures and also those of most complex software systems is *overlay*, wherein new abstractions are built atop existing ones by "stacking" a protocol layer over a lower or "underlay" layer.

Overlay In the overlay model, islands of an overlay network are connected with each other via tunnels across an underlay network (see Figure 3). This corresponds to pushing underlay names on the header stack that enable the transmission of a message from the ingress tunnel endpoint to the egress. The underlay network is oblivious to the overlay naming scheme and thus provides a transparent service between tunnel endpoint. This

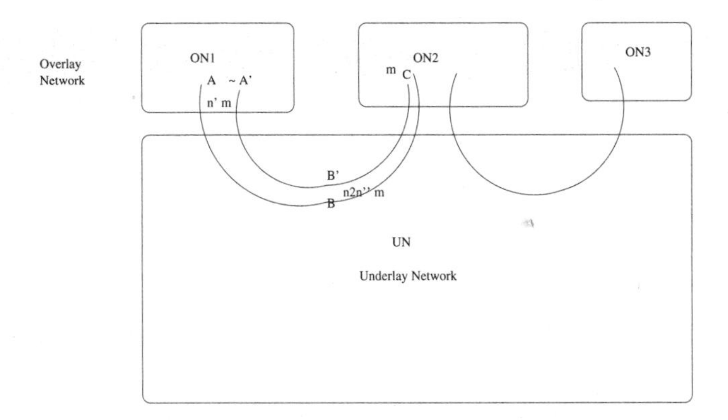

Figure 3: *The overlay composition of networks*

model uses the encapsulation and decapsulation mechanisms presented in Section 1.1. Examples are TCP over IP, I3 over IP, and MPLS over Ethernet.

Formally the overlay construction of networks can be described as follows: Suppose we are given an network $\mathcal{UN}$, the so-called "underlay network", with name space UN, the scope of which is all the ASEs in $\mathcal{UN}$. Also assume that for any set of names M, all names in $UN\ M$ are prefix-conformant at all ASEs in $\mathcal{UN}$.

Consider a network $\mathcal{N}$ which extends $\mathcal{UN}$ with a set of ASEs $\mathcal{ON}$, the "overlay network", connected to ASEs in $\mathcal{UN}$ such that certain properties described below hold.

1. For any message m, if $nm@^x A \to pm@^y B$ where $B \in \mathcal{UN}$ then $p = n_2 n''$ where $n_2 \in UN$, that is, whenever a message goes from the overlay to a node in the underlay, then a name in the underlying name space UN has been pushed onto the head in the message's residual.
2. Let ON be a set of names n' such that whenever n' is a name for C at $A \in \mathcal{ON}$, then $C \in \mathcal{ON}$. That is, ON is a set of names for ASEs in the overlay network used at other ASEs in the overlay network. (Clearly we are interested in the largest set satisfying this property).
3. $\forall m$, $\forall n' \in ON, \forall A, A' \in \mathcal{ON}, \forall n_2 \in UN, \forall B \in \mathcal{UN}$: *whenever* $n'm@^x A \to n_2 n'' m@^y B$ then $n'm@^z A' \to n_2 n'' m@^w B'$ and $B \sim_{UN} B'$ for some $B' \in \mathcal{UN}$. That is, different ASEs in the overlay identically transform an overlay name into an underlay name when handing a message to ASEs in the underlay, which in turn must interpret underlay name n_2 identically.

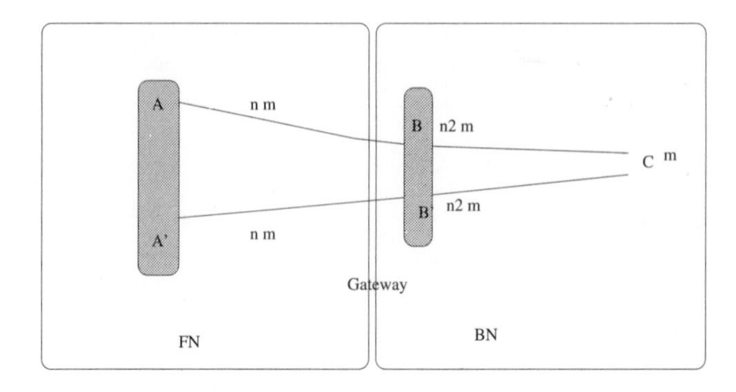

Figure 4: *The gateway composition of networks*

We can then prove the following theorem, which establishes that *ON* is a name space for the overlay network $\mathcal{ON}$.

Theorem 2 Overlay composition
Suppose we are given an overlay composition of networks satisfying the conditions given above. Then, for all $A, A' \in \mathcal{ON}$, $\forall n', n'' \in ON$*:* $[\![n']\!]_A = [\![n']\!]_{A'}$ *if and only if* $[\![n'']\!]_A = [\![n'']\!]_{A'}$.

A recent interesting development in computer networks is the concept of self overlay, especially of IP, yielding IP over IP tunnels. Note that such a construction is smoothly handled by the above formalization.

4.2 Gateway composition of networks

Another fundamental construction used in communication networks is *gateway composition* of one name space with another through an "address translation" (see Figure 4).

Gateway In the gateway model, name spaces are connected by replacing the name from the sender name space with a name that is valid in the receiver name space. Examples are NAT, MPLS, or DNS/IP translation.

Formally the gateway construction of networks can be described as follows: Suppose we are given an network $\mathcal{BN}$, the so-called "back network", with name space BN, the scope of which is the ASEs in $\mathcal{BN}$. Also assume that for any set of names M, all names in $BN\ M$ are prefix-conformant at all ASEs in $\mathcal{BN}$.

Consider a network $\mathcal{N}$ which extends $\mathcal{BN}$ with a set of ASEs $\mathcal{FN}$, the "front network", connected to one another and to ASEs in $\mathcal{BN}$ such that certain properties described below hold.

Consider a set of names FN such that $\forall n \in FN, \forall A \in \mathcal{FN}$:

1. If n is a name for B at A and $B \in \mathcal{FN}$, then n is a name for B at all other $A' \in \mathcal{FN}$;
2. If $nm@^x A \rightarrow pm@^y B$ where $B \in \mathcal{BN}$ then $p \in BN$; that is, a name in the front name space FN has been translated into a name in the back name space BN in the message's residual.

The gateway construction also needs to satisfy the following condition:

Let FN be a set of names n' satisfying the conditions stated above. $\forall m, \forall n \in FN, \forall A, A' \in \mathcal{FN}, \forall n_2 \in BN, \forall B \in \mathcal{BN}$: *whenever* $nm@^x A \rightarrow n_2 m@^y B$ then $nm@^z A' \rightarrow n_2 m@^w B'$ and $B \sim_{BN} B'$ for some $B' \in \mathcal{BN}$. That is, different ASEs in the front network identically transform an name into a target name when handing a message to ASEs in the back network, which then interpret name n_2 identically.

We can then prove the following theorem, which establishes that FN is a name space for the front network $\mathcal{FN}$.

Theorem 3 Gateway composition
Suppose we are given a gateway composition of networks satisfying the conditions given above. Then, for all $A, A' \in \mathcal{FN}$, $\forall n', n'' \in FN$: $A \sim_{n'} A'$ *if and only if* $A \sim_{n''} A'$.

In general, networks when composed through a gateway translate names from one network to the other in both directions. This requires a symmetric gateway construction.

5 Concluding Remarks

I have so far described a simple but reasonably dynamic communication model of switches communicating asynchronously with one another using messages. The model is expressive, and my collaborators and other network researchers believe that a large variety of protocols can be succinctly described, explained and reasoned about using this model (or minor variants). The communication model is asynchronous, and the model of abstract switches is truly distributed and scalable, and exhibits nondeterminism.

Moreover, the model seems to be a reasonable one to study: its seems to be a not-too-distant relative of the traditional automata, and for this reason, I hope a variety of language-theoretic and algebraic techniques can be brought to bear on it. The particular problem of interest is the most basic requirement of communication networks: connectivity or deliverability of messages, which involves a reachability analysis. The reachability analysis looks somewhat like rewriting a prefix to the empty string, except that the rewriting takes places at different locations, each equipped with a possibly different set of rules. While it may be possible to encode such networks into prefix rewriting systems, and thence finite state transducers, I believe such an encoding would essentially be a specification of the operational semantics of a particular network with all its minutiae.

From a small and simple set of axioms, I presented an inductively defined relation that allowed us to express deliverability of messages, but which abstracted from time-dependent behaviour (this allows one to be able to reason without having to resort to modal and temporal logics). Using this relation, I described how a variety of communication concepts can be reconstructed and characterized, and presented a nominal framework, on the basis of which large-scale distributed systems that use names as their "glue" can be built. The notions of name scopes and name spaces makes it possible to do so in a modular, systematic fashion. While the formal results are relatively easy to establish, I believe that this is the first formal account of the correctness of the canonical constructions used in building layered architectures and communication protocols, or at least one that accounts for the actual dynamics of present day networks.

Future work Some directions of future work are: first, to formally encode the simpler automata models in the presented model (this should be a routine exercise), and then go on to provide appropriate definitions for language acceptance, and to properly characterize and explore issues such as determinization, minimization and the like. After that, suitable connections with algebraic structures could be explored, in analogy with the developments for various kinds of finite state automata. Finally, a suitable logic, e.g., for expressing properties related to progress of a message through a network, could be explored, again on the lines of what has been done in formal language theory.

Another aspect that deserves greater study is the relationship of this model with the (automata) models used to study distributed computing issues. Indeed, a straightforward generalization of the model presented here will involve the notion of an *adversary* that has access to connection edges, and which can be invested with a variety of capabilities — the adversary's capabilities characterizing the operational environment under which different distributed computing problems are to be addressed. In the present paper (and in its predecessors) we assume a trivial adversary which faithfully relays the message from the output port of one ASE to the corresponding input port of the other. Adversaries with greater capabilities should be considered in the future.

References

[1] J.A. Bergstra and J.W. Klop. Algebra of communicating processes with abstraction. *Theoretical Computer Science*, 37(1):77–121, 1985.

[2] H. B. Curry and R. Feys. *Combinatory Logic*. North-Holland, 1958.

[3] Luca Cardelli and Andrew D. Gordon. Mobile ambients. In *Foundations of Software Science and Computation Structures: First International Conference, FOSSACS '98*. Springer-Verlag, Berlin Germany, 1998.

[4] D. Clark. The Design Philosophy of the DARPA Internet Protocols. In *Proceedings of SIGCOMM '88, Stanford, CA, USA*, pages 106–114, August 1988.

[5] T. G. Griffin and J. L. Sobrinho. Metarouting. In *Proceedings of SIGCOMM '05, Philadelphia, PA, USA*, pages 1-12, August 2005.

[6] C. A. R. Hoare. *Communicating sequential processes*. Prentice-Hall, Inc., Upper Saddle River, NJ, USA, 1985.

[7] G. Kahn. The semantics of a simple language for parallel programming. *Information Processing*, 74, 1974.

[8] M. Karsten, S. Keshav, and S. Prasad. An Axiomatic Basis for Communication. In *5th ACM SIGCOMM Workshop on Hot Topics in Networks (HotNets V), Irvine, CA, USA*, pages 19–24, November 2006.

[9] Martin Karsten, S. Keshav, Sanjiva Prasad, and Mirza Beg. An axiomatic basis for communication. *SIGCOMM Comput. Commun. Rev.*, 37(4):217–228, 2007.

[10] G. Kahn and D. B. MacQueen. Coroutines and networks of parallel processes. *Information Processing*, 77, 1977.

[11] Richard Kennaway and Ronan Sleep. Director strings as combinators. *ACM Trans. Program. Lang. Syst.*, 10(4):602–626, 1988.

[12] B. Loo, J. Hellerstein, I. Stoica, and R. Ramakrishnan. Declarative Routing: Extensible Routing with Declarative Queries. In *Proceedings of SIGCOMM '05, Philadelphia, PA, USA*, pages 289–300, August 2005.

[13] Robin Milner. *A Calculus of Communicating Systems*. Springer Verlag, Berlin, 1980.

[14] Robin Milner. *Communication and Concurrency*. Prentice Hall, International Series in Computer Science, 1989.

[15] Robin Milner. *Communicating and Mob:'e Systems: the Pi-Calculus*. Cambridge University Press, 1999.

[16] J. H. Saltzer. Naming and Binding of Objects. *Lecture Notes in Computer Science*, 60:99–208, 1978.

Dynamic Logic of Tree Composition

R. Ramanujam and Sunil Simon
The Institute of Mathematical Sciences
C.I.T. Campus, Chennai 600 113, India.
{jam,sunils}@imsc.res.in

Abstract

While computation tree logics (CTL) quantify globally over paths in Kripke structures, we suggest that it is useful to consider quantification over paths generated by finite state programs as in dynamic logic (PDL). We propose a dynamic logic whose programs are built over finite trees, and whose modalities quantify universally or existentially over frontiers. The logic is elementarily decidable, while proof of completeness (of an axiom system) combines techniques from CTL and PDL.

Keywords: Temporal logic, dynamic logic, path logics, decidability, completeness

1 Summary

In [HT99], Henriksen and Thiagarajan propose an extension of Linear Time Temporal Logic (LTL) called *Dynamic LTL* (DLTL), in which the eventuality operator is indexed by regular programs of propositional dynamic logic. The resulting logic is expressively complete with respect to the monadic second order theory of ω-sequences. The authors show that DLTL is decidable in exponential time and present a complete axiomatization. DLTL is a pleasant extension of LTL, and the ability to constrain computation paths to be those generated by finite state programs leads to a rich spcification language.

DLTL belongs in the tradition of *Process Logics* ([Pra79], [Par78], [Nis80], [HKP82]), which were motivated by the need to extend dynamic logic with the ability to reason about computation paths. An assertion like "The variable x takes the value 0 at some time in the computation" cannot be expressed in dynamic logic, whereas LTL is designed for just such specifications. In process logic, one reasons about computation paths whereby one can constrain initial and final segments of computations, much as in interval logic. Coupled with the ability to constrain paths by regular programs, this results in an expressively rich logic.

While process logic extends dynamic logic to (linear time) reasoning about computation paths, DLTL extends linear time logic by restricting paths to follow programs. In both cases, we shift from studying Kripke structures to paths over Kripke structures, which are further constrained by a syntax of programs.

In this context, there is a natural relation to another logic, which studies paths over Kripke structures: the Computation Tree Logic (CTL), proposed by Clarke and Emerson [EC82], in which formulas are interpreted on the tree unfolding of Kripke structures and hence we can reason about the branching patterns of the system. The foregoing discussion leads us naturally to the idea of constraining the tree unfolding by regular programs, much as paths in DLTL or process logic are constrained. This means that branching patterns in the Kripke structure are constrained to follow the tree structure dictated by regular programs, resulting in a combination of dynamic logic and CTL, which we call *Dynamic logic of tree composition* (DLTC) in this paper.

Why is program constrained tree unfolding interesting ? We suggest that such a logic is worthy of study from three viewpoints. Firstly, DLTC can be seen as a (kind of) process logic, and many questions on the relationships between program logics and process logics await answers ([Hon04]), especially when extended to concurrent programs. Secondly, CTL is interpreted over Kripke structures which are themselves typically generated by (finite state) programs and hence branching structure in the Kripke structure is determined by program behaviour. Incorporating (abstract) program structure into the specification itself can be of some importance, especially for synthesis: the branching patterns imposed by the specifications need to be achieved by finite state programs ([NV96]). Thirdly, CTL specifications are often interpreted on systems acting in the presence of an uncertain environment, and branching behaviour is intended to quantify over the possibilities introduced by such an environment. In such a situation, the **strategies** available to the system can be thought of as finite state programs, naturally leading us to trees generated by such strategies embedded in the Kripke structure.

We need to be more specific about DLTC now: consider a Kripke structure M and a node s in its tree unfolding T_M. We consider two modalities $\langle\pi\rangle^{\forall}\alpha$ and $\langle\pi\rangle^{\exists}\alpha$ asserted at s. The former asserts that the finite tree T_π defined by π is embedded in T_M, and the formula α

holds in all the frontier nodes of this finite tree. $\langle\pi\rangle^{\exists}\alpha$ is similar, but asserts α at some frontier node. With the dual modalities, we get an ability to constrain path properties considerably.

The DLTC modalities are weaker than CTL in the sense that they do not quantify over the entire set of paths in T_M; thus they refer to choices "local" to programs. On the other hand, indexing by programs allows us to refine the specifications considerably. Importantly, sequential composition allows to compose trees (much as paths are composed in process logics), and with iteration, we can express many interesting properties. Note that sequential composition acts very differently here from the way it does in dynamic logic, since we compose not paths but trees. In particular, $\pi_1; \pi_2$ does not denote the computation tree of the composite program, but to the tree obtained by attaching the tree of π_2 to the leaves of the tree of π_1; it is here that the two modalities mentioned above play significant roles. If π_1 and π_2 are "iteration free", i.e. they define finite trees then tree composition can be coded up using composition over paths. However this is no longer true in the presence of iteration. Consider the formula $\langle\pi^*\rangle^{\forall}\alpha$ where π is iteration free. This talks about a tree generated by some bounded unfolding of π in which α holds at all the leaf nodes. In contrast, the PDL modality $\langle\pi^*\rangle\alpha$ asserts that α holds along a computation path of π and $[\pi^*]\alpha$ says that α holds along all unfoldings.

In this paper, we present DLTC and prove that it is elementarily decidable. We also present a complete axiomatization; the completeness proof is an interesting combination of the techniques used for dynamic logic ([KP81]) and those used for CTL ([EH85]). While determining the expressiveness of the logic precisely is an interesting question, we have no answers yet.

2 Preliminaries

Finite trees

Let Σ be a finite set of actions and T be a Σ (edge) labelled tree denoted by $T = (S, \Longrightarrow, s_0)$ where S is the set of nodes, $s_0 \in S$ is the root of the tree and $\Longrightarrow$: $S \times \Sigma \rightarrow S$ is the edge relation. For a node $s \in S$, let $\vec{s} = \{a \in \Sigma \mid \exists s' \in S \text{ where } s \overset{a}{\Longrightarrow} s'\}$. A node s is called a *leaf node* if $\vec{s} = \emptyset$ and the set of all leaf nodes in the tree is denoted by *frontier*(T). Non-deterministic programs can be easily modelled as trees where the nodes correspond to program states and branching arises due to the non-determinism in the program. The formulas of the logic will refer to such non-deterministic programs in an explicit manner. One convenient way of representing the tree is to specify it using the following syntax:

Syntax for finite trees

Let *Nodes* be a countable set. The finite tree is specified using the syntax:

$$G := x \mid \Sigma_{a_m \in J}(x, a_m, t_{a_m})$$

where $J \subseteq \Sigma$, $x \in Nodes$ and $t_{a_m} \in G$.

Given $t \in G$ we define the tree T_t generated by t inductively as follows.

- $t \equiv (x)$: $T_t = (S_t, \Longrightarrow_t, s_x)$ where $S_t = \{s_x\}$.
- $t \equiv (x, a_1, t_{a_1}) + \cdots + (x, a_k, t_{a_k})$: Inductively we have trees $T_1, \ldots T_k$ where for $j : 1 \leq j \leq k$, $T_j = (S_j, \Longrightarrow_j, s_{j,0})$. Define $T_t = (S_t, \Longrightarrow_t, s_x)$ where
  - $S_t = \{s_x\} \cup S_{T_1} \cup \ldots \cup S_{T_k}$.
  - $\Longrightarrow_t = (\bigcup_{j=1,\ldots,k} \Longrightarrow_j) \cup \{s_x \overset{a_j}{\Longrightarrow}_j s_{j,0} \mid 1 \leq j \leq k\}$

3 The Logic

Syntax

Let P be a countable set of propositions. The syntax of the logic is given by:

$$\Gamma := t \in G \mid \pi_1; \pi_2 \mid \pi_1 \cup \pi_2 \mid \pi^* \mid \beta?$$

$$\Phi := p \in P \mid \neg\alpha \mid \alpha_1 \vee \alpha_2 \mid \langle\pi\rangle^{\forall}\alpha \mid \langle\pi\rangle^{\exists}\alpha$$

where $\pi \in \Gamma$ and $\beta \in \Phi$. As a convention we use $\top \overset{\text{def}}{=} p \vee \neg p$. We also define $[\pi]^{\exists}\alpha \overset{\text{def}}{=} \neg\langle\pi\rangle^{\forall}\neg\alpha$ and $[\pi]^{\forall}\alpha \overset{\text{def}}{=} \neg\langle\pi\rangle^{\exists}\neg\alpha$.

In the syntax of the logic, the program π represents regular expressions over finite trees. The intuitive meaning of the modalities are as follows:

- $\langle\pi\rangle^{\forall}\alpha$: There exists a π tree and in at least one of these trees, it holds that all the frontier nodes satisfy the formula α.
- $\langle\pi\rangle^{\exists}\alpha$: There exists a π tree and in at least one of these trees, it holds that there exists a frontier node which satisfies α.

The dual modalities get the following interpretation.

- $[\pi]^{\exists}\alpha$: For all π trees, there exist a frontier node where α holds.
- $[\pi]^{\forall}\alpha$: For all π trees, α holds at all frontier nodes.

We also make use of the following abbreviation. For $a \in \Sigma$, let e_a denote the single edge tree $e_a = (x, a, y)$.

- $\langle a\rangle\alpha \stackrel{\text{def}}{=} \langle e_a\rangle^{\exists}\alpha$.

From the semantics it will be clear that this results in the usual interpretation of $\langle a\rangle\alpha$. i.e. $\langle a\rangle\alpha$ holds at a state u iff there is a state w such that $u \xrightarrow{a} w$ and α holds at w.

Semantics

Model

A model $M = (W, \longrightarrow, V)$ where W is the set of states, the relation $\longrightarrow \subseteq W \times \Sigma \times W$ and $V : W \to 2^P$ is the valuation function.

The truth of a formula $\alpha \in \Phi$ in a model M and a position u (denoted $M, u \models \alpha$) is defined as follows:

- $M, u \models p$ iff $p \in V(u)$.
- $M, u \models \neg\alpha$ iff $M, u \not\models \alpha$.
- $M, u \models \alpha_1 \vee \alpha_2$ iff $M, u \models \alpha_1$ or $M, u \models \alpha_2$.
- $M, u \models \langle\pi\rangle^{\forall}\alpha$ iff $\exists(u, X) \in R_{\pi}^{\forall}$ such that $\forall w \in X$ we have $M, w \models \alpha$.
- $M, u \models \langle\pi\rangle^{\exists}\alpha$ iff $\exists(u, X) \in R_{\pi}^{\exists}$, $\exists w \in X$ such that $M, w \models \alpha$.

where $R_{\pi}^{\forall} \subseteq W \times 2^W$ and $R_{\pi}^{\exists} \subseteq W \times 2^W$.

The relations $R_{\pi}^{\forall}$ and $R_{\pi}^{\exists}$ specify the interpretation for programs, and once we generate the relations for compositie programs from those for atomic programs (namely, finite trees), the semantics of the logic is given.

One natural way to define this is to associate each composite program π with a set of Σ edge labelled trees (denoted $[[\pi]]^{\forall}$ and $[[\pi]]^{\exists}$), defined inductively. For the atomic case, $t \in G$ we have $[[t]]^{\forall} = [[t]]^{\exists} = \{T_t\}$. This is extended to composite programs in the obvious manner: $[[\pi_1;\pi_2]]^{\forall}$ would correspond to attaching π_2 trees at each of the leaf nodes of a π_1 tree whereas for $[[\pi_1;\pi_2]]^{\exists}$, a π_2 tree need be attached to one of the leaf nodes of a π_1 tree. Note that the definition of $[[\pi]]$ does not depend on the model.

Now given a model M, and a state u, we can then define $(u,X) \in R^{\forall}_{\pi}$ iff there exists a tree $T_\pi \in [[\pi]]^{\forall}$ and an embedding of T_π at u (denoted $M_u \upharpoonright T_\pi$) such that X is the set of frontier nodes of $M_u \upharpoonright T_\pi$.

However, this definition does not capture the situation where programs π_1 and π_2 are composed based on the outcome of executing π_1. To achieve this, rather than building a composite tree structure and then restricting the model with respect to this structure, one needs to consider the model restriction at the atomic level and perform composition over the trees generated as a result of this restriction.

Example

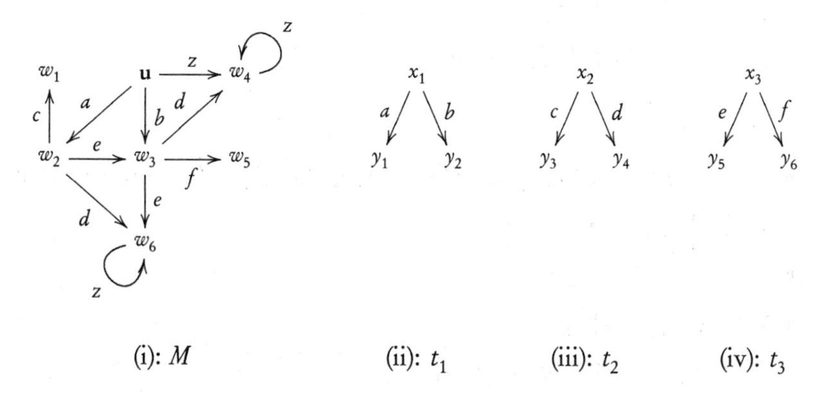

Figure 1: *Model restriction at atomic level*

Consider the Kripke structure given in Fig. 1(i) and the atomic programs t_1, t_2 and t_3 given in Fig. 1(ii), (iii) and (iv) respectively. Let $\pi = t_1;(p?;t_2 \cup \neg p?;t_3)$. The intended meaning of π is: after executing program t_1, if proposition p holds then execute t_2 else execute t_3. The program t essentially identifies a set of states in M from which depending on whether p holds, t_2 or t_3 is executed. Let $V(p) = \{w_2\}$ and $V(q) = \{w_1, w_5, w_6\}$. For $\beta \equiv \langle\pi\rangle^{\forall} q$ it will be clear from the semantics that $M, u \models \beta$. This illustrates that the ability to interleave programs and formulas is of interest in branching time.

Such an assertion cannot be made in CTL where the reasoning is done with respect to the branching structure of the entire model. In particular we have $M, u \not\models \forall \Diamond q$.

Assertions that utilise sequential tree composition of bounded length can be translated into PDL in terms of the computation paths. However in the presence of iteration this is not true, for instance in a formula of the form $\langle (t_1;(p?;t_2 \cup \neg p?;t_3))^* \rangle^{\forall} \alpha$.

Interpretation of composite programs

For $\pi \in \Gamma$, we have relations $R^{\forall}_{\pi} \subseteq W \times 2^W$ and $R^{\exists}_{\pi} \subseteq W \times 2^W$ defined as follows: Assume that for atomic programs $t \in$ G, a relation $R_t \subseteq W \times 2^W$ is defined.

- $R^{\forall}_t = R_t$.
- $R^{\forall}_{\pi_1;\pi_2} = \{(u,X) \mid \exists Y = \{v_1,\ldots,v_k\}$ such that $(u,Y) \in R^{\forall}_{\pi_1}$ and $\forall v_j \in Y$, there exists $X_j \subseteq X$ such that $(v_j,X_j) \in R^{\forall}_{\pi_2}$ and $\bigcup_{j=1,\ldots,k} X_j = X\}$.
- $R^{\forall}_{\pi_1 \cup \pi_2} = R^{\forall}_{\pi_1} \cup R^{\forall}_{\pi_2}$.
- $R^{\forall}_{\pi^*} = \bigcup_{n \geq 0} (R^{\forall}_{\pi})^n$.
- $R^{\forall}_{\beta?} = \{(u,\{u\}) \mid M,u \models \beta\}$

- $R^{\exists}_t = R_t$.
- $R^{\exists}_{\pi_1;\pi_2} = \{(u,X) \mid \exists Y \subseteq W$ such that $(u,Y) \in R^{\exists}_{\pi_1}$ and $\exists v_j \in Y$ such that $(v_j,X) \in R^{\exists}_{\pi_2}\}$.
- $R^{\exists}_{\pi_1 \cup \pi_2} = R^{\exists}_{\pi_1} \cup R^{\exists}_{\pi_2}$.
- $R^{\exists}_{\pi^*} = \bigcup_{n \geq 0} (R^{\exists}_{\pi})^n$.
- $R^{\exists}_{\beta?} = \{(u,\{u\}) \mid M,u \models \beta\}$

For atomic programs $t \in$ G, we want a pair (u,X) to be in R_t if the program structure can be embedded at state u. We give a formal definition below.

Restriction on trees

For $w \in W$, let $T_w = (S^w_M, \Longrightarrow_M, \lambda_M, s_w)$ denote the tree unfolding of M starting at w. $\lambda_M : S^w_M \rightarrow 2^P$ is the valuation function derived from V. For $t \in$ G, let $T_t = (S_t, \Longrightarrow_t, s_{t,0})$. The restriction of T_w with respect to t (denoted $T_w \upharpoonright t$) is the subtree of T_w which is generated by the structure specified by T_t. The restriction is defined inductively as follows: $T_w \restriction t = (S, \Longrightarrow, \lambda, s_0, f)$ where $f : S \rightarrow S_t$. Initially $S = \{s_w\}$, $s_0 = s_w$ and $f(s_w) = s_{t,0}$.

For any $s \in S$, let $f(s) = s_x \in S_t$. Let $\{a_1, \ldots, a_k\}$ be the outgoing edges of s_x. For each a_j, let $\{s_j^1, \ldots, s_j^m\}$ be the nodes in S_M^w such that $s \stackrel{a_j}{\Longrightarrow}_M s_j^l$ for all $l : 1 \leq l \leq m$. Add nodes $s_j^1, \ldots, s_j^m$ to S and the edges $s \stackrel{a_j}{\Longrightarrow} s_j^l$ for all $l : 1 \leq l \leq m$. Also set $f(s_j^l) = t_j$.

We say that a program t is enabled at w (denoted *enabled*(t, w)) if the tree $T_w \restriction t = (S, \Longrightarrow, \lambda, s_0, f)$ has the following property:

- $\forall s \in S, \vec{s} = \overrightarrow{f(s)}$.

Atomic program trees

For atomic programs $t \in G$ we define the relation $R_t \subseteq W \times 2^W$ as follows:

- $R_{(x)} = \{(u, \{u\})\}$.
- $R_t = \{(u, X) \mid enabled(t, u) \text{ and } frontier(T_u \restriction t) = X\}$.

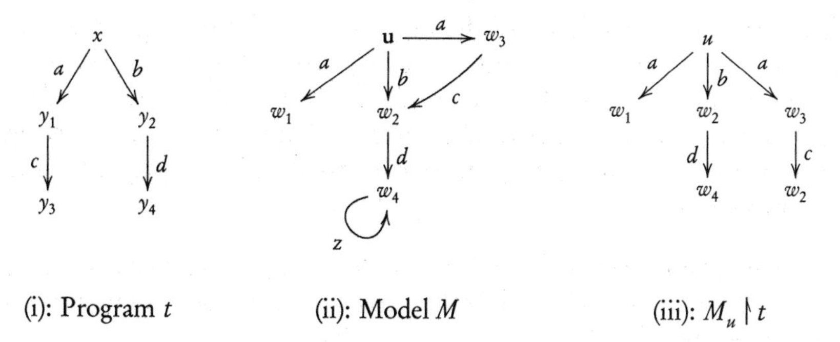

(i): Program t (ii): Model M (iii): $M_u \restriction t$

Figure 2: *t is not enabled at u*

To illustrate the restriction operation and the semantics, consider the program t given in Fig. 2 (i). Let the Kripke structure M be as shown in Fig. 2(ii). For a proposition p, let $V(p) = \{w_2, w_4\}$. The structure $M_u \restriction t$ is shown in Fig. 2(iii). According to the definition of the restriction operator, we have $f(w_1) = f(w_3) = y_1$. Therefore $\overrightarrow{f(w_1)} = \{c\}$ whereas $\vec{w_1} = \emptyset$ and thus we get that in M, program t is *not* enabled at state u. This implies that $M, u \not\models \langle t \rangle^{\forall} p$.

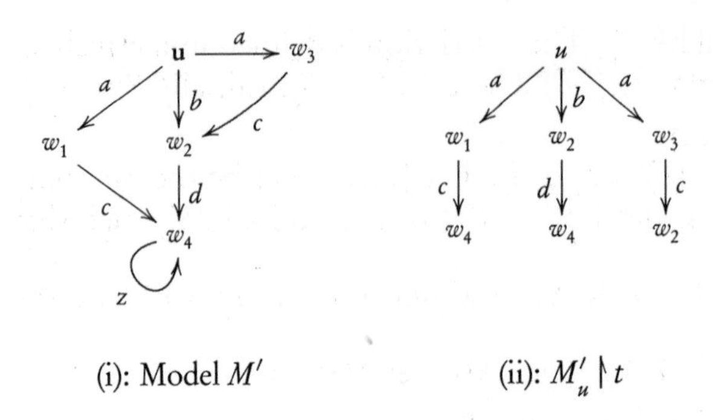

(i): Model M' (ii): $M'_u \restriction t$

Figure 3: *t is enabled at u*

Now consider the structure M' given in Fig. 3(i). This is the same as M except for the additional edge $w_1 \xrightarrow{c} w_4$. From the definition, it is easy to see that *enabled*(t, u) holds in M'. We have *frontier*$(M'_u \restriction t) = \{w_2, w_4\} \subseteq V(p)$ and therefore $M, u \models \langle t \rangle^{\forall} p$.

4 Examples

For a finite set Σ let $\pi^u = ((\bigcup_{a \in \Sigma} e_a)^*)$ be the universal program. Recall that for $a \in \Sigma$, $e_a = (x, a, y)$ is the single edge tree. The formula $[\pi^u]^{\forall}\alpha$ says that for all paths it is always the case that α holds. This corresponds to the CTL formula $\forall\Box\alpha$. The dual $\langle \pi^u \rangle^{\exists}\alpha$ says that there exists a path ρ, and a state u in ρ such that u satisfies α. This corresponds to the CTL formula $\exists\Diamond\alpha$. The above two modalities quantify over all states of the system. This is useful in asserting safety properties. Interesting properties that can be expressed using these modalities include:

- Partial correctness: $\phi \supset [\pi^u]^{\forall}(halt \supset \psi)$. This says that starting in a state where ϕ holds if the program terminates then ψ holds at termination.
- Global invariance: $[\pi^u]^{\forall}\psi$. This says that ψ holds at all reachable states.
- Deadlock freedom: $[\pi^u]^{\forall}\langle \pi^u \rangle^{\exists}\psi$. This states that from any state, we can always get to a state where ψ holds.

For a composite program π, the formula $\langle \pi \rangle^{\forall}\alpha$ on the other hand asserts that there is a π-tree of finite depth where α holds at all the frontier nodes.

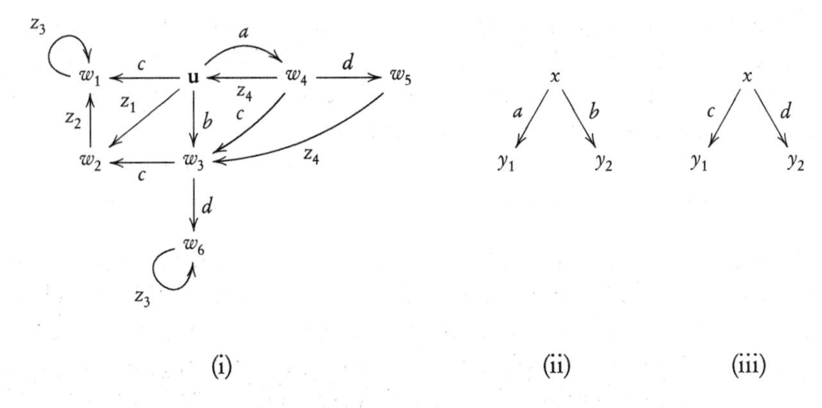

Figure 4: *The difference from CTL*

Consider the Kripke structure M given in Fig. 4(i). For a proposition p, let $V(p)=\{w_2, w_3, w_5, w_6\}$. Starting at state u, there is a path where p never holds ($uw_1w_1\cdots$), thus the CTL formula $\forall\Diamond p$ does not hold at u. Let programs π_1 and π_2 be as shown in Fig. 4(ii) and Fig. 4(iii) respectively. It is easy to check that $M, u \models \langle(\pi_1 \cup \pi_2)^*\rangle^{\forall} p$.

To see the difference between the CTL modality $\forall\Diamond$ and the DLTC modality $\langle\pi\rangle^{\forall}$, consider the formula:

- $\beta \equiv [\pi^u]^{\forall}(\mathit{request} \supset \langle(\pi_1 \cup \pi_2)^*\rangle^{\forall} p_{ack})$.

Here $\pi_{ack} = (\pi_1 \cup \pi_2)^*$ can be thought of as the program (or module) which is responsible for sending the acknowledgment. π_{ack} might have internal choices and therefore be nondeterministic. The formula β asserts that in all computation paths generated by π_{ack}, eventually p_{ack} holds. Unlike in CTL, here we are making use of the fact that the global transition system is synthesized from the local program structure. Assertions are made about the local branching possible in these individual programs and global assertions are then derived from the compositional structure.

System synthesis from behavioural specification presented in a hardware description language such as VHDL illustrates the necessity for incorporating program structure into the specifications. Verifying such systems against branching time temporal specifications has been looked at in [NV96]. Typically these are synchronous systems consisting of a data path and a controller. The controller itself consists of a set modules each implemented by a finite state machine (FSM). A special module called the *Main* FSM controls the execution cycle. It takes input from the environment through a set of flags and triggers calls to other modules as required.

When the *Start* signal is raised by the environment, the *Main* FSM issues control to read the input signals. Let π_R denote the program which implements the read and *Main.RI* denote the fact that the input signals has been read. Consider the following assertion:

$$\alpha \equiv Start \supset \langle \pi^*_{read} \rangle^{\forall} Main.RI$$

This says that iteration of π_{read} ensures that eventually all input signals are read. In contrast, the CTL assertion $\forall \Diamond Main.RI$ fails when there are other interleaved actions present in the system. However, α makes use of the program structure in the assertion explicitly.

After the input is read, the main process triggers the *Process* FSMs by executing the invoke process module π_{IP}. Let α_{exec} hold when the process FSMs start executing. The sequence of operations is captured by the assertion

$$\langle \pi^*_{read}; Main.RI?; \pi_{IP} \rangle^{\forall} \alpha_{exec}.$$

This first takes the pruning of the system tree by considering only those paths which conform to the iteration of π_{read}. At all leaf nodes thus reachable, the test is carried out to check whether *Main.RI* holds. From all nodes where *Main.RI* holds, the program π_{IP} is executed.

After all the *Process* FSMs suspend, the "All Process Suspend Signal" (*APS*) is raised. The main process then raises *US* to update all signals lines. Let π_{US} denote the program resetting the signal lines. The assertion

$$[\pi^u]^{\forall}(Main.APS \supset \langle \pi^*_{US} \rangle^{\forall} Main.US)$$

states that it is always the case that if *APS* holds then the iteration of π_{US} will ensure that all the signal lines are cleared.

5 Axiom System

We now present an axiomatization of the valid formulas of the logic. We will use the following notations:

For tree $t \in G$, we use the formula $t^{\surd}$ to denote that the tree structure t is enabled. This is defined inductively on the structure of t as:

- if $t = (x)$ then $t^{\surd} \equiv \top$.

- if $t=(x,a_1,t_{a_1})+\ldots+(x,a_k,t_{a_k})$ then
  - $t^{\checkmark}\equiv(\bigwedge_{j=1,\ldots,k}(\langle a_j\rangle\top\wedge[a_j]t^{\checkmark}_{a_j}))$.

The axiom schemes

(A1) Propositional axioms:

(a) All the substitutional instances of tautologies of PC.

(A2) (a) $\langle a\rangle^{\forall}(\alpha_1\vee\alpha_2)\equiv\langle a\rangle^{\forall}\alpha_1\vee\langle a\rangle^{\forall}\alpha_2$.
(b) $\langle\pi\rangle^{\exists}(\alpha_1\vee\alpha_2)\equiv\langle\pi\rangle^{\exists}\alpha_1\vee\langle\pi\rangle^{\exists}\alpha_2$.
(c) $\langle\pi\rangle^{\exists}(\alpha_1\wedge\alpha_2)\supset\langle\pi\rangle^{\exists}\alpha_1\wedge\langle\pi\rangle^{\exists}\alpha_2$.
(d) $\langle\pi\rangle^{\exists}(\bot)\equiv\bot$.

(A3) Dynamic logic axioms:

(a) $\langle\pi_1\cup\pi_2\rangle^{\forall}\alpha\equiv\langle\pi_1\rangle^{\forall}\alpha\vee\langle\pi_2\rangle^{\forall}\alpha$.
(b) $\langle\pi_1;\pi_2\rangle^{\forall}\alpha\equiv\langle\pi_1\rangle^{\forall}\langle\pi_2\rangle^{\forall}\alpha$.
(c) $\langle\pi^*\rangle^{\forall}\alpha\equiv\alpha\vee\langle\pi\rangle^{\forall}\langle\pi^*\rangle^{\forall}\alpha$.
(d) $\langle\beta?\rangle^{\forall}\alpha\equiv\beta\supset\alpha$.

(A4) (a) $\langle\pi_1\cup\pi_2\rangle^{\exists}\alpha\equiv\langle\pi_1\rangle^{\exists}\alpha\vee\langle\pi_2\rangle^{\exists}\alpha$.
(b) $\langle\pi_1;\pi_2\rangle^{\exists}\alpha\equiv\langle\pi_1\rangle^{\exists}\langle\pi_2\rangle^{\exists}\alpha$.
(c) $\langle\pi^*\rangle^{\exists}\alpha\equiv\alpha\vee\langle\pi\rangle^{\exists}\langle\pi^*\rangle^{\exists}\alpha$.
(d) $\langle\beta?\rangle^{\exists}\alpha\equiv\beta\supset\alpha$.

(A5) $\langle t\rangle^{\forall}\alpha\equiv t^{\checkmark}\wedge push_{\forall}(t,\alpha)$.
(A6) $\langle t\rangle^{\exists}\alpha\equiv t^{\checkmark}\wedge push_{\exists}(t,\alpha)$.

Inference rules

$$(MP)\ \frac{\alpha,\ \alpha\supset\beta}{\beta}\qquad (NG)\ \frac{\alpha}{[a]\alpha}$$

$$(IND_{\forall})\ \frac{\langle\pi\rangle^{\forall}\alpha\supset\alpha}{\langle\pi^*\rangle^{\forall}\alpha\supset\alpha}\qquad (IND_{\exists})\ \frac{\alpha\supset[\pi]^{\forall}\alpha}{\alpha\supset[\pi^*]^{\forall}\alpha}$$

Definition of push

For atomic trees $t=(x)$, we have

(C1) $push_{\forall}((x),\alpha) \equiv \alpha$.
(C2) $push_{\exists}((x),\alpha) \equiv \alpha$.

For $t = (x,a_1,t_{a_1}) + \ldots + (x,a_k,t_{a_k})$ we have

(C3) $push_{\forall}(t,\alpha) \equiv \bigwedge_{a_m \in A}[a_m]push_{\forall}(t_{a_m},\alpha)$.
(C4) $push_{\exists}(t,\alpha) \equiv \bigvee_{a_m \in A}\langle a_m\rangle push_{\exists}(t_{a_m},\alpha)$.

Axiom (A2a) does not hold for general trees, (i.e. $\pi \in \Gamma$). In particular $\langle\pi\rangle^{\forall}(\alpha_1 \vee \alpha_2) \supset \langle\pi\rangle^{\forall}\alpha_1 \vee \langle\pi\rangle^{\forall}\alpha_2$ is not valid. (A2a) is sound only because it asserts properties about a single edge.

Since the relation $R^{\forall}$ is synthesised over tree structures, the interpretation of sequential composition is quite different from the standard one. Consider the usual relation composition semantics for $R^{\forall}_{\pi_1;\pi_2}$, i.e. $R^{\forall}_{\pi_1;\pi_2} = \{(u,X) | \exists Y$ such that $(u,Y) \in R^{\forall}_{\pi_1}$ and for all $v \in Y$, $(v,X) \in R^{\forall}_{\pi_2}\}$. It is easy to see that under this interpretation the formula $\langle\pi_1\rangle^{\forall}\langle\pi_2\rangle^{\forall}\alpha \supset \langle\pi_1;\pi_2\rangle^{\forall}\alpha$ is not valid.

Soundness arguments for axioms (A3b) and (A4b) are given in the appendix.

6 Completeness

To show completeness, we prove that every consistent formula is satisfiable. Let α_0 be a consistent formula, and $CL(\alpha_0)$ denote the subformula closure of α. Let $\mathcal{AT}(\alpha_0)$ be the set of all maximal consistent subsets of $CL(\alpha_0)$, referred to as atoms. We use u, w to range over the set of atoms. Each $u \in \mathcal{AT}$ is a finite set of formulas, we denote the conjunction of all formulas in u by $\widehat{u}$. For a nonempty subset $X \subseteq \mathcal{AT}$, we denote by $\widetilde{X}$ the disjunction of all $\widehat{u}, u \in X$. Define a transition relation on $\mathcal{AT}(\alpha_0)$ as follows: $u \xrightarrow{a} w$ iff $\widehat{u} \wedge \langle a\rangle\widehat{w}$ is consistent. The valuation V is defined as $V(w) = \{p \in P \mid p \in w\}$. The model is $M = (W, \longrightarrow, V)$ where $W = \mathcal{AT}(\alpha_0)$. For the sake of clarity, we present the proof for test free programs.

Lemma 1 *For all $t \in G$, for all $u, w \in W$, if $\widehat{u} \wedge \langle t\rangle^{\exists}\widehat{w}$ is consistent then $\exists X \subseteq W$ such that $(u,X) \in R^{\exists}_t$ and $\vdash \widehat{w} \supset \widetilde{X}$.*

The proof is given in the appendix.

The following two lemmas can be shown by using techniques similar to those developed for dynamic logic. Detailed proofs can be found in the appendix.

Lemma 2 *For all $\pi \in \Gamma$, for all $u, w \in W$, if $\widehat{u} \wedge \langle \pi \rangle^{\exists} \widehat{w}$ is consistent then $\exists X \subseteq W$ such that $(u, X) \in R^{\exists}_{\pi}$ and $\vdash \widehat{w} \supset \widetilde{X}$.*

Lemma 3 *For all $\langle t \rangle^{\exists} \alpha \in CL(\alpha_0)$ and for all $u \in W$ if there exists $(u, X) \in R^{\exists}_{t}$ and $w \in X$ such that $\alpha \in w$ then $\widehat{u} \wedge \langle t \rangle^{\exists} \alpha$ is consistent.*

Lemma 4 *For all $\langle \pi \rangle^{\exists} \alpha \in CL(\alpha_0)$ and for all $u \in W$, $\widehat{u} \wedge \langle \pi \rangle^{\exists} \alpha$ is consistent iff $\exists (u, X) \in R^{\exists}_{\pi}$, $\exists w \in X$ such that $\alpha \in w$.*

Proof ($\Rightarrow$) Let $X_{\alpha} = \{w \mid \widehat{w} \wedge \alpha$ is consistent$\}$. Suppose $\widehat{u} \wedge \langle \pi \rangle^{\exists} \alpha$ is consistent. From axiom (A2b) we get $\exists w \in X_{\alpha}$ such that $\widehat{u} \wedge \langle \pi \rangle^{\exists} \widehat{w}$ is consistent. From lemma 2, there exists $X \subseteq W$ such that $(u, X) \in R^{\exists}_{\pi}$ and $\vdash \widehat{w} \supset \widetilde{X}$. Since $\vdash \widehat{w} \supset \alpha$, we have $\exists (u, X) \in R^{\exists}_{\pi}$, $\exists w \in X$ such that $\alpha \in w$.

($\Leftarrow$) Suppose $\exists (u, X) \in R^{\exists}_{\pi}$, $\exists w \in X$ such that $\alpha \in w$. We need to show that $\widehat{u} \wedge \langle \pi \rangle^{\exists} \alpha$ is consistent. This is done by induction on the structure of π.

- The case when $\pi \equiv t \in G$ follows from lemma 3. For $\pi \equiv \pi_1 \cup \pi_2$ the result follows from axiom (A4a).
- $\pi \equiv \pi_1; \pi_2$: Suppose $(u, X) \in R^{\exists}_{\pi_1;\pi_2}$ and $\exists w \in X$ such that $\alpha \in w$. From the definition of $R^{\exists}$ we get that there exists $Y \subseteq W$ such that $(u, Y) \in R^{\exists}_{\pi_1}$ and $\exists v \in Y$ such that $(v, X) \in R^{\exists}_{\pi_2}$. By induction hypothesis we have $\widehat{v} \wedge \langle \pi_2 \rangle^{\exists} \alpha$ is consistent. By definition of closure we have $\langle \pi_2 \rangle^{\exists} \alpha \in CL(\alpha_0)$. Therefore we get $\langle \pi_2 \rangle^{\exists} \alpha \in v$. Again applying induction hypothesis we get that $\widehat{u} \wedge \langle \pi_1 \rangle^{\exists} \langle \pi_2 \rangle^{\exists} \alpha$ is consistent. From (A4b) we get $\widehat{u} \wedge \langle \pi_1; \pi_2 \rangle^{\exists} \alpha$ is consistent.
- $\pi \equiv \pi_1^*$: From definition of $R^{\exists}$ there must be sets $Y_1, \ldots, Y_k$ such that $u \in Y_1, X = Y_k$ and for all $j : 1 < j < k, \exists v_j \in Y_j$ such that $(v_j, X_{j+1}) \in R^{\exists}_{\pi_1}$. From (A4c) we get $\widehat{w} \wedge \langle \pi_1^* \rangle^{\exists} \alpha$ is consistent. By definition of closure, we have $\langle \pi_1 \rangle^{\exists} \langle \pi_1^* \rangle^{\exists} \alpha \in CL(\alpha_0)$. By induction hypothesis, $\widehat{v}_{k-1} \wedge \langle \pi_1 \rangle^{\exists} \langle \pi_1^* \rangle^{\exists} \alpha$ is consistent and therefore from (A4c) $\widehat{v}_{k-1} \wedge \langle \pi_1^* \rangle^{\exists} \alpha$ is consistent. Continuing in this manner we get $u \wedge \langle \pi_1^* \rangle^{\exists} \alpha$ is consistent.

Lemma 5 *For all $t \in G$, for all $X \subseteq W$ and for all $u \in W$ the following holds:*

1. *if* $(u,X) \in R^{\forall}_{t}$ *then* $\widehat{u} \wedge \langle t \rangle^{\forall} \widetilde{X}$ *is consistent.*
2. *if* $\widehat{u} \wedge \langle t \rangle^{\forall} \widetilde{X}$ *is consistent then there exists* $X' \subseteq X$ *such that* $(u,X') \in R^{\forall}_{t}$.

A detailed proof can be found in the appendix. Item 1 follows from the axioms and the fact that closure of α_0 is rich enough that it has the tree structure coded into it as required by the axioms. For item 2, we show the following:

- The tree t can be embedded at state u.
- The frontier of $T_u \upharpoonright t$ is $X' \subseteq X$.

Lemma 6 *For all* $\pi \in \Gamma$, *for all* $X \subseteq W$ *and* $u \in W$, *if* $\widehat{u} \wedge \langle \pi \rangle^{\forall} \widetilde{X}$ *is consistent then there exists* $X' \subseteq X$ *such that* $(u,X') \in R^{\forall}_{\pi}$.

The proof is given in the appendix.

Lemma 7 *For all* $\langle \pi \rangle^{\forall} \alpha \in CL(\alpha_0)$, *for all* $u \in W$, $\widehat{u} \wedge \langle \pi \rangle^{\forall} \alpha$ *is consistent iff there exists* $(u,X) \in R^{\forall}_{\pi}$ *such that* $\forall w \in X$, $\alpha \in w$.

Proof ($\Rightarrow$) Follows from lemma 6 by considering the set $X_\alpha = \{w \in W \mid \alpha \in w\}$.

($\Leftarrow$) Suppose $\exists (u,X) \in R^{\forall}_{\pi}$ such that $\forall w \in X$, $\alpha \in w$. We need to show that $\widehat{u} \wedge \langle \pi \rangle^{\forall} \alpha$ is consistent, this is done by induction on the structure of π.

- The case when $\pi \equiv t \in G$ follows from lemma 5. For $\pi \equiv \pi_1 \cup \pi_2$ the result follows from axiom (A3a).
- $\pi \equiv \pi_1; \pi_2$: Since $(u,X) \in R^{\forall}_{\pi_1;\pi_2}$, there exists $Y = \{v_1, \ldots, v_k\}$, there exists sets $X_1, \ldots, X_k \subseteq X$ such that $\bigcup_{j=1,\ldots,k} X_j = X$, for all $j : 1 \leq j \leq k$, $(v_j, X_j) \in R_{\pi_2}$ and $(u,Y) \in R^{\forall}_{\pi_1}$. By induction hypothesis, for all j, $\widehat{v_j} \wedge \langle \pi_2 \rangle^{\forall} \alpha$ is consistent. Since v_j is an atom and $\langle \pi_2 \rangle^{\forall} \alpha \in CL(\alpha_0)$, we get $\langle \pi_2 \rangle^{\forall} \alpha \in v_j$. Again by induction hypothesis we have $\widehat{u} \wedge \langle \pi_1 \rangle^{\forall} \langle \pi_2 \rangle^{\forall} \alpha$ is consistent. Hence from (A3b) we have $\widehat{u} \wedge \langle \pi_1; \pi_2 \rangle^{\forall} \alpha$ is consistent.
- $\pi \equiv \pi_1^*$: If $u \in X$ then $\vdash \widehat{u} \supset \widetilde{X}$. We have $\vdash \widetilde{X} \supset \alpha$ and hence we get $\widehat{u} \wedge \alpha$ is consistent. From axiom (A3c) we have $\widehat{u} \wedge \langle \pi_1^* \rangle^{\forall} \alpha$ is consistent.
  Else we have $(u,X) \in R^{\forall}_{\pi_1;\pi_1^*}$. Let $Z_0 = X$ and $Z_{n+1} = Z_n \cup \{w \mid (w,Z') \in R^{\forall}_{\pi_1}, Z' \subseteq Z_n\}$. Take the least m such that $u \in Z_m$. We have for all $w \in Z_{m-1}$, $\vdash \widehat{w} \supset \langle \pi_1^* \rangle^{\forall} \widetilde{X}'$ for some $X' \subseteq X$. We

also have $(u, Z'_m) \in R^{\forall}_{\pi_1}$ for some $Z'_m = \{v_1, \dots, v_k\} \subseteq Z_m$. Let $X_1, \dots, X_k \subseteq X$ such that $\forall j : 1 \leq j \leq k$, we have $(v_j, X_j) \in R^{\forall}_{\pi_1^*}$ and $X' = \bigcup_{j=1,\dots,k} X_j$. By an argument similar to the previous case we can show that $\widehat{u} \wedge \langle \pi_1 \rangle^{\forall} \langle \pi_1^* \rangle^{\forall} \widetilde{X}'$ is consistent. Hence we get $\widehat{u} \wedge \langle \pi_1; \pi_1^* \rangle^{\forall} \alpha$ is consistent. Therefore from axiom (A3c) we have $\widehat{u} \wedge \langle \pi_1^* \rangle^{\forall} \alpha$ is consistent.

Theorem 8 *For all $\beta \in CL(\alpha_0)$, for all $u \in W$, $M, u \models \beta$ iff $\beta \in u$.*

The theorem follows from lemma 7 and lemma 4 by a routine inductive argument.

When the test operator is part of the program structure, the above theorem can be shown by performing simultaneous induction on the structure of programs and formulas of the logic.

Decidability

For a formula α_0, the size of $CL(\alpha_0)$ is linear in the size of α_0. Since atoms are maximal consistent subsets of $CL(\alpha_0)$, we have $|\mathcal{AT}(\alpha_0)|$ is exponential in the size of α_0. From the completeness theorem it follows that if α_0 is satisfiable, then it has a model of exponential size, i.e. $|W| = \mathcal{O}(2^{|\alpha_0|})$. Given a model M and a state u, checking whether $M, u \models \alpha_0$ can be done in time exponential in the size of the model (details given in the appendix). Therefore it follows that the logic is decidable in nondeterministic double exponential time.

References

[EC82] E. A. Emerson and E. M. Clarke. Using branching time logic to synthesize synchronization skeletons. *Science of Computer Programming*, 2:241–266, 1982.

[EH85] E. A. Emerson and J. Y. Halpern. Decision procedures and expressiveness in the temporal logic of branching time. *Journal of Computer and System Sciences*, 30:1–24, 1985.

[HKP82] D. Harel, D. Kozen, and R. Parikh. Process logic: Expressiveness, decidability, completeness. *Journal of Computer and System Sciences*, 25(2):144–170, October 1982.

[Hon04] K. Honda. From process logic to program logic. In *Proceedings of the ninth International Conference on Functional Programming*, pages 163–174. ACM, 2004.

[HT99] J. G. Henriksen and P. S. Thiagarajan. Dynamic linear time temporal logic. *Annals of Pure and Applied Logic*, 96(1-3):187–207, 1999.

[KP81] D. Kozen and R. Parikh. An elementary proof of the completeness of PDL. *Theoretical Computer Science*, 14:113–118, 1981.

[Nis80] H. Nishimura. Descriptively complete process logic. *Acta Informatica*, 14(4):359–369, 1980.

[NV96] N. Narasimhan and R. Vemuri. Specification of control flow properties for verification of synthesized VHDL designs. In *Proceedings of the First International Conference on Formal Methods in Computer-Aided Design*, volume 1166 of *Lecture Notes In Computer Science*, pages 327–345. Springer-Verlag, 1996.

[Par78] R. Parikh. A decidability result for second order process logic. In *Proceedings of the 19th Symposium on Foundations of Computer Science*, pages 177–183. IEEE, 1978.

[Pra79] V. R. Pratt. Process logic. In *Proceedings of the 6th ACM Symposium on Principles of Programming Languages*, pages 93–100. ACM, 1979.

A Appendix

Soundness

Axiom (A3b)

Suppose $\langle\pi_1;\pi_2\rangle^{\forall}\alpha \supset \langle\pi_1\rangle^{\forall}\langle\pi_2\rangle^{\forall}\alpha$ is not valid. Then there exists M and u such that $M,u \models \langle\pi_1;\pi_2\rangle^{\forall}\alpha$ and $M,u \not\models \langle\pi_1\rangle^{\forall}\langle\pi_2\rangle^{\forall}\alpha$. Since $M,u \models \langle\pi_1;\pi_2\rangle^{\forall}$, from semantics we have there exists $(u,X) \in R^{\forall}_{\pi_1;\pi_2}$ such that $\forall w \in X, M,w \models \alpha$. From definition of $R^{\forall}$, $\exists Y = \{v_1,\ldots,v_k\}$ such that $(u,Y) \in R^{\forall}_{\pi_1}$ and $\forall v_j \in Y$ there exists $X_j \subseteq X$ such that $(v_j,X_j) \in R^{\forall}_{\pi_2}$ and $\bigcup_{j=1,\ldots,k} X_j = X$. Therefore $\forall v_k \in Y$ we have $M,v_k \models \langle\pi_2\rangle^{\forall}\alpha$ and hence from semantics, $M,u \models \langle\pi_1\rangle^{\forall}\langle\pi_2\rangle^{\forall}\alpha$. This gives the required contradiction.

Suppose $\langle\pi_1\rangle^{\forall}\langle\pi_2\rangle^{\forall}\alpha \supset \langle\pi_1;\pi_2\rangle^{\forall}\alpha$ is not valid. Then there exists M and u such that $M,u \models \langle\pi_1\rangle^{\forall}\langle\pi_2\rangle^{\forall}\alpha$ and $M,u \not\models \langle\pi_1;\pi_2\rangle^{\forall}\alpha$. We have $M,u \models \langle\pi_1\rangle^{\forall}\langle\pi_2\rangle^{\forall}\alpha$ iff there exists $(u,Y) \in R^{\forall}_{\pi_1}$ such that $\forall v_k \in Y, M,v_k \models \langle\pi_2\rangle^{\forall}\alpha$. $M,v_k \models \langle\pi_2\rangle^{\forall}\alpha$ iff there exists $(v_k,X_k) \in R^{\forall}_{\pi_2}$ such that $\forall w_k \in X_k$, $M,w_k \models \alpha$. Let $X = \bigcup_k X_k$, from definition of $R^{\forall}$ we get $(u,X) \in R^{\forall}_{\pi_1;\pi_2}$. Hence from semantics $M,u \models \langle\pi_1;\pi_2\rangle^{\forall}\alpha$.

Axiom (A4b)

Suppose $\langle\pi_1;\pi_2\rangle^{\exists}\alpha \supset \langle\pi_1\rangle^{\exists}\langle\pi_2\rangle^{\exists}\alpha$ is not valid. Then there exists M and u such that $M,u \models \langle\pi_1;\pi_2\rangle^{\exists}\alpha$ and $M,u \not\models \langle\pi_1\rangle^{\exists}\langle\pi_2\rangle^{\exists}\alpha$. Since $M,u \models \langle\pi_1;\pi_2\rangle^{\exists}$, from semantics we have there exists $(u,X) \in R^{\exists}_{\pi_1;\pi_2}$, there exists $w \in X$ such that $M, w \models \alpha$. From definition of $R^{\forall}$, there exists $Y = \{v_1, \ldots, v_k\}$ such that $(u, Y) \in R^{\forall}_{\pi_1}$ and there exist $v_j \in Y$ such that $(v_j, X) \in R^{\forall}_{\pi_2}$. Therefore we get $M, v_j \models \langle \pi_2 \rangle^{\exists} \alpha$ and hence from semantics $M, u \models \langle \pi_1 \rangle^{\exists} \langle \pi_2 \rangle^{\exists} \alpha$.

Suppose $\langle \pi_1 \rangle^{\exists} \langle \pi_2 \rangle^{\exists} \alpha \supset \langle \pi_1; \pi_2 \rangle^{\exists} \alpha$ is not valid. Then there exists M and u such that $M, u \models \langle \pi_1 \rangle^{\exists} \langle \pi_2 \rangle^{\exists} \alpha$ and $M, u \not\models \langle \pi_1; \pi_2 \rangle^{\exists} \alpha$. We have $M, u \models \langle \pi_1 \rangle^{\exists} \langle \pi_2 \rangle^{\exists} \alpha$ iff there exists $(u, Y) \in R^{\exists}_{\pi_1}$, there exists $v_j \in Y$ such that $M, v_j \models \langle \pi_2 \rangle^{\exists} \alpha$. $M, v_j \models \langle \pi_2 \rangle^{\exists} \alpha$ iff there exists $(v_j, X) \in R^{\exists}_{\pi_2}$ and there exists $w \in X$ such that $M, w \models \alpha$. From the definition of $R^{\exists}$ we get $(u, X) \in R^{\exists}_{\pi_1; \pi_2}$. Hence from semantics $M, u \models \langle \pi_1; \pi_2 \rangle^{\exists} \alpha$.

Detailed Proofs

Lemma 1: For all $t \in \mathrm{G}$, for all $u, w \in W$, if $\widehat{u} \wedge \langle t \rangle^{\exists} \widehat{w}$ is consistent then $\exists X \subseteq W$ such that $(u, X) \in R^{\exists}_t$ and $\vdash \widehat{w} \supset \widetilde{X}$.

Proof By induction on the structure of t.

- $t \equiv (x)$: From axiom (A6) case (C2) we get $\langle (x) \rangle^{\forall} \alpha \equiv \alpha$. The lemma follows from this quite easily.
- $t = (x, a_1, t_{a_1}) + \ldots + (x, a_k, t_{a_k})$: Suppose $\widehat{u} \wedge \langle t \rangle^{\exists} \widehat{w}$ is consistent, from axiom (A6) we get $\widehat{u} \wedge t^{\checkmark}$ is consistent. Therefore there exists sets $Y_1, \ldots, Y_k$ such that $\forall j : 1 \leq j \leq l$, for all $v^l_j \in Y_j$ we have $u \xrightarrow{a_j} v^l_j$. From (A6) case (C4) we get $\widehat{u} \wedge (\bigvee_{a_j \in A} \langle a_m \rangle \langle t_{a_m} \rangle^{\exists} \widehat{w})$ is consistent. Therefore there exists v^r_m such that $u \xrightarrow{a_m} v^r_m$ and $v^r_m \wedge \langle t_{a_m} \rangle^{\exists} \widehat{w}$ is consistent. By induction hypothesis, for all j, l we get $\exists X^l_j$ such that $(v^l_j, X^l_j) \in R^{\exists}_{t_{a_j}}$ and there exists X^r_m such that $(v^r_m, X^r_m) \in R^{\exists}_{t_{a_m}}$, $\vdash \widehat{w} \supset \widetilde{X^r_m}$. Let $X = \bigcup_{j=1,\ldots,k} \bigcup_{l=1,\ldots,|Y_j|} X^l_j$, from semantics we get $(u, X) \in R^{\exists}_t$. We also have $\vdash \widetilde{X^r_m} \supset \widetilde{X}$ and $\vdash \widehat{w} \supset \widetilde{X^r_m}$ and thus $\vdash \widehat{w} \supset \widetilde{X}$ as required.

Lemma 2: For all $\pi \in \Gamma$, for all $u, w \in W$, if $\widehat{u} \wedge \langle \pi \rangle^{\exists} \widehat{w}$ is consistent then $\exists X \subseteq W$ such that $(u, X) \in R^{\exists}_{\pi}$ and $\vdash \widehat{w} \supset \widetilde{X}$.

Proof By induction on the structure of π.

- $\pi \equiv t \in G$: The result follows from lemma 1.
- $\pi \equiv \pi_1 \cup \pi_2$: The result follows from axiom (A4a).
- $\pi \equiv \pi_1;\pi_2$: Suppose $\widehat{u} \wedge \langle \pi_1;\pi_2\rangle^{\exists}\widehat{w}$ is consistent. From (A4b) we have $\widehat{u} \wedge \langle \pi_1\rangle^{\exists}\langle \pi_2\rangle^{\exists}\widehat{w}$ is consistent. This is equivalent to $\bigvee(\widehat{u}\wedge\langle\pi_1\rangle^{\exists}(\widehat{v}\wedge\langle\pi_2\rangle^{\exists}\widehat{w}))$ where the join is taken over all atoms. Therefore we have there exists a v such that $\widehat{u}\wedge\langle\pi_1\rangle^{\exists}(\widehat{v}\wedge\langle\pi_2\rangle^{\exists}\widehat{w})$ is consistent. From (A2c) and (A2d) we get $\widehat{u} \wedge \langle\pi_1\rangle^{\exists}\widehat{v}$ is consistent and $\widehat{v} \wedge \langle\pi_2\rangle^{\exists}\widehat{w}$ is consistent. By induction hypothesis, $\exists Y$ such that $(u,Y) \in R^{\exists}_{\pi_1}$, $\vdash \widehat{v} \supset \widetilde{Y}$ and $\exists X$ such that $(v,X) \in R^{\exists}_{\pi_2}$, $\vdash \widehat{w} \supset \widetilde{X}$. Therefore we get $\exists X$ such that $(u,X) \in R^{\exists}_{\pi_1;\pi_2}$ and $\vdash \widehat{w} \supset \widetilde{X}$.
- $\pi \equiv \pi_1^*$: We have by induction hypothesis, for all u, w, if $\widehat{u} \wedge \langle\pi_1\rangle^{\exists}\widehat{w}$ is consistent then $\exists X \subseteq W$ such that $(u,X) \in R^{\exists}_{\pi_1}$ and $\vdash \widehat{w} \supset \widetilde{X}$. Let Z be the least set containing u and closed under the condition: for all $u' \in Z$ if there exists w' such that $\widehat{u'} \wedge \langle\pi_1\rangle^{\exists}\widehat{w'}$ is consistent then $X' \subseteq Z$ where $(u',X') \in R^{\exists}_{\pi_1}$ and $\vdash \widehat{w'} \supset \widetilde{X'}$. It is easy to see from the closure condition on Z that for all $u' \in Z$, if there exists w' such that $\widehat{u'} \wedge \langle\pi_1\rangle^{\exists}\widehat{w'}$ is consistent then $w' \in Z$. If $w \in Z$ then we are done. Suppose this is not the case, then we have $\vdash \widetilde{Z} \supset \neg\widehat{w}$ and $[\pi_1^*]^{\forall}\widetilde{Z} \supset [\pi_1^*]^{\forall}\neg\widehat{w}$.

  Claim 9 $\vdash \widetilde{Z} \supset [\pi_1]^{\forall}\widetilde{Z}$.

  Proof Suppose the claim is not true, then $\widetilde{Z} \wedge \langle\pi_1\rangle^{\exists}\neg\widetilde{Z}$ is consistent. Let $Z' = \mathscr{AT}(\alpha_0)\setminus Z$, then we have $\widetilde{Z} \wedge \langle\pi_1\rangle^{\exists}\widetilde{Z'}$ is consistent. From (A2b), there exists $u' \in Z$ and $w' \in Z'$ such that $\widehat{u'}\wedge\langle\pi_1\rangle^{\exists}\widehat{w'}$ is consistent. By the closure condition on Z, we get that $w' \in Z$ which contradicts the fact that $w' \in \mathscr{AT}(\alpha_0)\setminus Z$. Applying the induction rule ($IND_{\exists}$), we get $\vdash \widetilde{Z} \supset [\pi_1^*]^{\forall}\widetilde{Z}$. Thus $\vdash \widetilde{Z} \supset [\pi_1^*]^{\forall}\neg\widehat{w}$ and $\vdash \widehat{u} \supset [\pi_1^*]^{\forall}\neg\widehat{w}$. Therefore we have $\widehat{u} \wedge \neg\langle\pi_1^*\rangle^{\exists}\widehat{w}$ is consistent which is a contradiction.

Lemma 3: For all $\langle t\rangle^{\exists}\alpha \in CL(\alpha_0)$ and for all $u \in W$ if there exists $(u,X) \in R^{\exists}_{t}$ and $w \in X$ such that $\alpha \in w$ then $\widehat{u} \wedge \langle t\rangle^{\exists}\alpha$ is consistent.

Proof By induction on the structure of t.

- $t \equiv (x)$: Suppose there exists u such that $(u, \{u\}) \in R^{\exists}_{(x)}$ and $\alpha \in u$. We have $\widehat{u} \wedge \alpha$ is consistent and from (A6) we get $\widehat{u} \wedge \langle (x) \rangle^{\exists} \alpha$ is consistent as required.
- $t = (x, a_1, t_{a_1}) + \ldots + (x, a_k, t_{a_k})$: Suppose $\exists (u, X) \in R^{\exists}_t$ and $\exists w \in X$ such that $\alpha \in w$. Since *enabled*(t, u) holds, we get $\exists Y_1, \ldots Y_k$ such that $\forall j : 1 \leq j \leq k$, for all $v^l_j \in Y_j$, we have $u \xrightarrow{a_j} v^l_j$. Since we have $(u, X) \in R^{\exists}_t$ we get that $\forall j : 1 \leq j \leq k$, $\forall l : 1 \leq l \leq |Y_j|$, there exists X^l_j such that $(v^l_j, X^l_j) \in R^{\exists}_{t_{a_j}}$ and $\bigcup_{j=1,\ldots,k} \bigcup_{l=1,\ldots,|Y_j|} X^l_j = X$. We also get that $w \in X^r_m$ for one of the sets. By induction hypothesis, for all j, l we have $t^{l\surd}_j$ holds and $\widehat{v^r_m} \wedge \langle t^r_m \rangle^{\exists} \alpha$ is consistent. Therefore from (A6), we get $\widehat{u} \wedge \langle t \rangle^{\exists} \alpha$ is consistent.

Lemma 5: For all $t \in$ G, for all $X \subseteq W$ and for all $u \in W$ the following holds:

1. if $(u, X) \in R^{\forall}_t$ then $\widehat{u} \wedge \langle t \rangle^{\forall} \widetilde{X}$ is consistent.
2. if $\widehat{u} \wedge \langle t \rangle^{\forall} \widetilde{X}$ is consistent then there exists $X' \subseteq X$ such that $(u, X') \in R^{\forall}_t$.

Proof For atomic tree $(x) \in$ G, from axiom (A5) case (C1) we get $\langle (x) \rangle^{\forall} \alpha \equiv \alpha$. The lemma follows from this quite easily.
Let $t = (x, a_1, t_{a_1}) + \ldots + (x, a_k, t_{a_k})$.

Suppose $(u, X) \in R^{\forall}_t$. Since *enabled*(u, t) holds, there exists sets $Y_1, \ldots, Y_k$ such that for all $j : 1 \leq j \leq k$, for all $w^l_j \in Y_j$, $u \xrightarrow{a_j} w^l_j$. We get $\exists X^l_j \subseteq X$ such that $(w^l_j, X^l_j) \in R^{\forall}_{t_{a_j}}$ and $X = \bigcup_{j=1,\ldots,k} \bigcup_{l=1,\ldots,|Y_j|} X^l_j$. Applying induction hypothesis and the fact that $\vdash \widetilde{X^l_j} \supset \widetilde{X}$ for all j, l we get that $\widehat{w^l_j} \wedge \langle t^l_j \rangle^{\forall} \widetilde{X}$ is consistent. Hence from axiom (A5) case (C3) we conclude $\widehat{u} \wedge \langle t \rangle^{\forall} \widetilde{X}$ is consistent.

Suppose $\widehat{u} \wedge \langle t \rangle^{\forall} \widetilde{X}$ is consistent. From axiom (A5), we get that $\widehat{u} \wedge t^{\surd}$ is consistent. This implies that there exists sets $Y_1, \ldots, Y_k$ such that for all $j : 1 \leq j \leq k$, for all $w^l_j \in Y_j$ we have $u \xrightarrow{a_j} w^l_j$. From (A5) case (C3) we have $\widehat{u} \wedge (\bigwedge_{a_j \in A} [a_j] \langle t_{a_j} \rangle^{\forall} \widetilde{X})$ is consistent. Therefore for all j

such that $u, \xrightarrow{a_j} w_j^l$, we have $w_j^l \wedge \langle t_{a_j} \rangle^{\forall} \widetilde{X}$ is consistent. By induction hypothesis, there exists $X_j^l \subseteq X$ such that $(w_j^l, X_j^l) \in R^{\forall}_{t_{a_j}}$. Let $X' = \bigcup_{j=1,\ldots,k} \bigcup_{k=1,\ldots,|Y_j|} X_j^l$, by definition of $R^{\forall}$ we have $(u, X') \in R^{\forall}_t$.

Lemma 6: For all $\pi \in \Gamma$, for all $X \subseteq W$ and $u \in W$, if $\widehat{u} \wedge \langle \pi \rangle^{\forall} \widetilde{X}$ is consistent then there exists $X' \subseteq X$ such that $(u, X') \in R^{\forall}_{\pi}$.

Proof By induction on the structure of π.

- $\pi \equiv t \in G$: Suppose $\widehat{u} \wedge \langle t \rangle^{\forall} \widetilde{X}$ is consistent. From lemma 5 item 2, it follows that there exists $X' \subseteq X$ such that $(u, X') \in R^{\forall}_{\pi}$.
- $\pi \equiv \pi_1 \cup \pi_2$: By axiom (A3a) we get $\widehat{u} \wedge \langle \pi_1 \rangle^{\forall} \widetilde{X}$ is consistent or $\widehat{u} \wedge \langle \pi_2 \rangle^{\forall} \widetilde{X}$ is consistent. By induction hypothesis there exists $X_1 \subseteq X$ such that $(u, X_1) \in R^{\forall}_{\pi_1}$ or there exists $X_2 \subseteq X$ such that $(u, X_2) \in R^{\forall}_{\pi_2}$. Hence we have $(u, X_1) \in R^{\forall}_{\pi_1 \cup \pi_2}$ or $(u, X_2) \in R^{\forall}_{\pi_1 \cup \pi_2}$.
- $\pi \equiv \pi_1; \pi_2$: By axiom (A3b), $\widehat{u} \wedge \langle \pi_1 \rangle^{\forall} \langle \pi_2 \rangle^{\forall} \widetilde{X}$ is consistent. Hence $\widehat{u} \wedge \langle \pi_1 \rangle^{\forall} (\bigvee (\widehat{w} \wedge \langle \pi_2 \rangle^{\forall} \widetilde{X}))$ is consistent, where the join is taken over all $w \in Y = \{w \mid w \wedge \langle \pi_2 \rangle^{\forall} \widetilde{X}$ is consistent $\}$. So $\widehat{u} \wedge \langle \pi_1 \rangle^{\forall} \widetilde{Y}$ is consistent. By induction hypothesis, there exists $Y' \subseteq Y$ such that $(u, Y') \in R^{\forall}_{\pi_1}$. We also have that for all $w \in Y$, $\widehat{w} \wedge \langle \pi_2 \rangle^{\forall} \widetilde{X}$ is consistent. Therefore we get for all $w_j \in Y' = \{w_1, \ldots, w_k\}$, $\widehat{w}_j \wedge \langle \pi_2 \rangle^{\forall} \widetilde{X}$ is consistent. By induction hypothesis, there exists $X_j \subseteq X$ such that $(w_j, X_j) \in R^{\forall}_{\pi_2}$. Let $X' = \bigcup_{j=1,\ldots,k} X_k \subseteq X$, we get $(u, X') \in R^{\forall}_{\pi_1;\pi_2}$.
- $\pi \equiv \pi_1^*$: Let Z be the least set containing X and closed under the condition: for all w, if $\widehat{w} \wedge \langle \pi_1 \rangle^{\forall} \widetilde{Z}$ is consistent, then $w \in Z$. By definition of Z and induction hypothesis, we get for all $w \in Z$, there exists $X_w \subseteq X$ such that $(w, X_w) \in R^{\forall}_{\pi_1^*}$. It is also easy to see that $\vdash \widetilde{X} \supset \widetilde{Z}$. Using standard techniques, it is also easy to show that $\vdash \langle \pi_1 \rangle^{\forall} \widetilde{Z} \supset \widetilde{Z}$.
  Applying the induction rule ($IND_{\forall}$), we have $\vdash \langle \pi_1^* \rangle^{\forall} \widetilde{Z} \supset \widetilde{Z}$. By assumption, $\widehat{u} \wedge \langle \pi_1^* \rangle^{\forall} \widetilde{X}$ is consistent. So $\widehat{u} \wedge \langle \pi_1^* \rangle^{\forall} \widetilde{Z}$ is consis-

tent. Hence $\widehat{u} \wedge \widetilde{Z}$ is consistent and therefore $u \in Z$. Thus we have $(u, X') \in R^{\forall}_{\pi_1^*}$ for some $X' \subseteq X$.

Truth checking

Given a model M, a state u and a formula α_0, the truth checking (or model checking) problem is to determine if $M, u \models \alpha_0$.

The non-trivial part in solving the truth checking problem is in constructing the relation $R^{\forall}_{\pi}$ and $R^{\exists}_{\pi}$. Once the relation is explicitly constructed, we can employ the usual labelling algorithm. Below we show how the relation $R^{\forall}_{\pi}$ can be constructed, $R^{\exists}_{\pi}$ can be constructed in a similar manner.

Let $PG(\alpha_0)$ denote the "sub-program" closure of α_0 which is the least set containing all the programs occuring in α_0 and closed under the following conditions:

- $(x, a_1, t_{a_1}) + \ldots + (x, a_k, t_{a_k}) \in PG(\alpha_0)$ implies $\{t_{a_1}, \ldots, t_{a_k}\} \subseteq PG(\alpha_0)$.
- $\pi_1 \cup \pi_2 \in PG(\alpha_0)$ implies $\{\pi_1, \pi_2\} \subseteq PG(\alpha_0)$.
- $\pi_1 ; \pi_2 \in PG(\alpha_0)$ implies $\{\pi_1, \pi_2\} \subseteq PG(\alpha_0)$.
- $\pi^* \in PG(\alpha_0)$ implies $\pi \in PG(\alpha_0)$.

The labelling function $l_1 : W \times PG(\alpha_0) \rightarrow 2^{2^W}$ is used to construct the relation $R^{\forall}_{\pi}$. We proceed as follows: Enumerate all the programs in $PG(\alpha_0)$ in the order of increasing complexity. For each program π in the enumeration, and a state u, the labelling function $l_1(u, \pi)$ is defined inductively on the structure of π.

- $\pi = (x)$ (i.e. a single node): $l_1(u, \pi) = \{\{u\}\}$.
- $\pi = (x, a_1, t_{a_1}) + \ldots + (x, a_k, t_{a_k})$: For $j : 0 \leq j \leq k$, let $Z_j = \{w \mid u \xrightarrow{a_j} w\}$ and $Z = \bigcup_{j=1,\ldots,k} Z_j$ be the set $\{w_1, \ldots, w_m\}$. $l_1(u, \pi) = \{X_1 \cup \ldots \cup X_m \mid X_i \in l_1(w_i, t_{a_j})$ where $w_i \in Z_j, 0 \leq i \leq m\}$.
- $\pi = \pi_1 \cup \pi_2$: $l_1(u, \pi) = l_1(u, \pi_1) \cup l_1(u, \pi_2)$.
- $\pi = \pi_1 ; \pi_2$: Let $l_1(u, \pi_1) = \{Y_1, \ldots, Y_k\}$ where $Y_i = \{w_i^1, \ldots, w_i^m\}$. $l_1(u, \pi) = \bigcup_{Y_i \in l_1(u, \pi_1)} \{X_i^1 \cup \ldots \cup X_i^m \mid X_i^j \in l_1(w_i^j, \pi_2), 0 \leq j \leq m\}$
- $\pi = \pi_1^*$: Let $Z_0 = l_1(u, \pi_1)$ and inductively assume that Z_i is defined for $i \geq 0$. Pick any $Y = \{w^1, \ldots, w^m\} \in Z_i$. Let $X =$

$\{X^1 \cup \ldots \cup X^m \mid X^j \in l_1(w^j, \pi_1), 0 \leq j \leq m\}$. We set $Z_{i+1} = Z_i \cup X$.
Let f be the least index such that $Z_f = Z_i$ for all $i > f$. We have $l_1(u, \pi_1^*) = Z_f$.

For each program $\pi \in PG(\alpha_0)$ and state u we have $|l_1(u, \pi)| = \mathcal{O}(2^{|W|})$. Therefore the relation $R_\pi^\forall$ can be explicitly constructed in time exponential in the size of the model.

Axiomatization of a Class of Parametrised Bisimilarities

Pranav Singh[1], **S. Arun-Kumar**[2]
[1] *Department of Applied Mathematics and Theoretical Physics*
University of Cambridge, Wilberforce Road
Cambridge, CB3 0WA, United Kingdom
`singh.pranav@gmail.com`
[2] *Department of Computer Science and Engineering*
Indian Institute of Technology Delhi
Hauz Khas, New Delhi 110 016,India
`sak@cse.iitd.ernet.in`

Abstract

The question of when two nondeterministic concurrent systems are behaviourally related has occupied a large part of the literature on process algebra and has yielded a variety of equivalences (and congruences) and preorders (and precongruences) all based on the notion of bisimulations.
Recently one of the authors has tried to unify a class of these bisimulation based relations by a parametrised notion of bisimulation and shown that the properties of the bisimilarity relations are often inherited from those of the underlying relationships between the observables. In addition to the usual strong and weak bisimilarity relations, it is possible to capture some other bisimilarity relations – those sensitive to costs, performance, distribution or locations etc – by parametrised bisimulations.
In this paper we present an equational axiomatization of all equivalence relations that fall in the class of parametrised bisimilarities without empty observables. Our axiomatization has been inspired by the axiomatization of *observational congruence* by Bergstra and Klop and attempts to extend it for parametrised bisimilarities. The axiomatization has been proven to be complete for finite process graphs relative to a complete axiomatization for the relations on observables. In the process, we also show that in the absence of empty observables, all preorders and equivalence relations are also precongruences and congruences, respectively.

Keywords: axiomatization, bisimulation, concurrency, process-graphs, proof-system

1 Introduction

It is a well known and accepted fact that the notion of trace equivalence defined in automata theory is too coarse for comparing processes and cannot capture the interactive or communicating behavior that characterize the notion of processes. In particular, it cannot distinguish between the processes $a.(0+b.0)$ and $a.0+a.b.0$. While after taking an a-step the second process can always take a b-step, the first process might become 0 and stop. The notion of strong bisimilarity [10], however, distinguishes between these processes. Clearly, this difference between trace equivalence and strong bisimilarity can be seen to stem from the difference in their handling of the meaning of the '+' operator and not from any particular structure of the alphabet of *actions* or *observables* $\{a, b, c \ldots\}$ that they might assume. The notion of observational equivalence [10] follows a definition similar to that of strong bisimilarity, with the additional assumption that the set of actions contains a unique invisible action τ. It differs from trace equivalence in terms of its handling of the semantics of '+' as well as in its special handling of the invisible action τ.

The observation that the structural nature of the definitions of such relations on processes (including their handling of the semantics of operators like '+','.','|' and recursion), and the structure on observables (for example the special handling of τ in observational equivalence) can be viewed separately in this manner, formed the motivation for a generalized notion called (ρ,σ)-bisimilarities [2]. In [2] and [5], many relations such as strong bisimilarity, observational equivalence, performance prebisimulation [7] and pomset prebisimulation [1] have been shown to be special cases of the (ρ,σ)-bisimilarities. While the strength of this generalization is obvious from the wide variety of relations that have been shown to fall under its umbrella, perhaps of greater importance is the manner in which it cleanly separates properties that are arrived at by virtue of the structural nature of its definition (including the semantical interpretation of the choice '+', prefix '.', parallel '|' and recursion) from the properties induced by the nature of relations on observables ρ and σ.

In the large class of (ρ,σ)-bisimilarities, the structure of the definition remains fixed and is similar to that of strong bisimilarity. A large number of properties of these relations can be deduced simply from the nature of this definition, making little or no assumptions about the observables or relations on the observables. Another class of properties are those inherited from the properties of the relations on observables ρ and σ. Of particular interest is the property that a (ρ,σ)-bisimilarity is an equivalence relation if and only if ρ is a preorder and $\sigma = \rho^{-1}$ [5].

The question of when two processes are equal in the context of a particular equivalence relation has occupied a large part of the literature in process algebra. A complete axiomatization of an equivalence relation yields both, a neat algebraic handle on processes and a proof system within which the equality of any two processes expressible in the system can be proven formally. While complete axiomatizations have been provided for some of the most common relations in process algebra, they have all had to be proven sound and complete individually, each involving a distinct method of proving completeness. Also, while the similarity in the axioms of these relations has often been noted, it has hardly been studied in detail. A schema of complete axiomatizations for (ρ, σ)-bisimilarities would automatically provide an axiomatization for any relation that can be proven to fit within the generalized framework of (ρ, σ)-bisimilarities. Apart from saving a great deal of effort in arriving at an axiomatization for each such relation and proving its soundness and completeness separately, such an axiomatization schema highlights the properties that are arrived at as a consequence of the structural nature of processes and relations.

Salomaa's axiomatization of regular expressions [11] has served as a motivation for the axiomatizations of many relations in process algebra. Milner's axiomatization [9] $\mathscr{A}$ for strong bisimilarity over CCS is one. A comparison of these two axiomatizations highlights the difference in the handling of the structural operator '+'. The left distributivity of '.' over '+' that is present in Salomaa's axiomatization cannot be proven in Milner's $\mathscr{A}$ and captures the fundamental difference between strong bisimilarity and trace equivalence. Another axiomatization of strong bisimilarity BPA_{LR} by Bergstra and Klop [6] uses the notion of linear-recursive (LR) expressions instead of the CCS notation and is closer to [11] in terms of its use of recursion equations. BPA_{LR} also closely corresponds to Milner's $\mathscr{A}$ as proven in [6], and was intended as a stepping stone towards formulating $BPA_{\tau LR}$, a complete axiomatization for observational congruence. At around the same time as [6], Milner gave a complete axiomatization [8] $\mathscr{A}_\tau$ for observational congruence on CCS processes.

In this paper we present the axiomatization (schema) $\mathscr{A}_{(\rho,\rho^{-1})}$ where ρ is a preorder. Our axiomatization is built on top of the axiomatization BPA_{LR} for strong bisimilarity [6] and instead of CCS uses LR expressions, which are recursion equations and can also be thought of as term rewrite systems. Our choice is motivated by the close and intuitive correspondence of LR expressions with process graphs, as well as the fact that our proof of completeness relies on a method of saturation similar to the method of Δ-saturation employed for the completeness proof of $BPA_{\tau LR}$ in [6]. However, we see no reason why a simple CCS counterpart of the axiomatization cannot be formulated. We also assume that the set of observables does not contain any empty action.

On lines similar to those of BPA_{LR} and $BPA_{\tau LR}$, we focus on finite or regular process graphs and do not handle the parallel composition construct or the notion of communication. The axiomatization $\mathscr{A}_{(\rho,\rho^{-1})}$ is proved to be complete for all finite process graphs and characterizes all equivalence relations that fall in the class of (ρ,σ)-bisimilarities. The completeness of the axiomatization is proved relative to a complete axiomatization of the relation ρ on observables. The completeness also depends on the completeness of the axiomatization BPA_{LR} for strong bisimilarity, which is presented and proven complete over finite process graphs in [6]. Similarity of the structural definitions of strong bisimilarity and (ρ,σ)-bisimilarities leads to an axiomatization which makes minimal assumptions about the nature of ρ and closely corresponds to BPA_{LR} .

The rest of the paper is organized as follows. In section 2, we establish the notations for labelled transition systems and process graphs [6] used in this paper. In section 3, we summarize the notion of (ρ,σ)-bisimilarities and some results of [2] and [5] that will be of particular relevance to us. Using a very weak assumption, we show that all (ρ,σ)-bisimilarities that are preorders are precongruences while those that are equivalences are congruences. We define the terms of BPA_{LR} and therefore of $\mathscr{A}_{(\rho,\rho^{-1})}$ in section 4, while also quoting some of the important properties of LR expressions from [6]. Having established the required notations, we present the axiomatization $\mathscr{A}_{(\rho,\rho^{-1})}$ in section 5. The proof of soundness of the axiomatization, which largely follows from the soundness of BPA_{LR}, is treated in section 6, while sections 7 and 8 deal solely with the proof of completeness, with the former introducing our notion of graph saturation. In section 8 we formally prove the properties of saturated graphs within the context of the axiomatization, which leads to a proof of completeness. Section 9 is the concluding section.

2 Process Graphs

The notion of process graphs here is based on [6] to which we refer the reader for a more detailed study than is possible to provide here. Process graphs, or graphs for short, are connected, rooted labelled multidigraphs. That is, a process graph g has a root g_r or $root(g)$ and a set of nodes $nodes(g)$ (ranged over by $p,q,\ldots$) with labelled and directed edges between the nodes such that there may be several edges between any two nodes and every node is accessible from the root g_r. The edges in a process graph are labelled by *labels* from the set of observables $\mathcal{O}$ (ranged over by $a, b, \ldots$). A *transition* is a tuple (p, a, q) consisting of a *starting node* p, an *ending node* q and a label a and is also written as $p \xrightarrow{a} q$ for convenience. A transition $p \xrightarrow{a} q$ is a *transition in the graph* g if $p, q \in nodes(g)$ and there is an edge labelled a starting at p and ending at q in the graph g. A node is a *terminal state* if it has no outgoing edges. Note that transitions and edges are considered different. $p \xrightarrow{a} q$ is a transition whereas $\xrightarrow{a}$ is an edge between p and q. $trans(g)$ (ranged over by $\pi, \chi, \ldots$) is the set of all transitions in the graph g.

We sometimes prefer the use of labelled transition system (LTS) notation over process graphs. In this notation a process graph g is written as $\langle nodes(g), \mathcal{O}, trans(g), root(g) \rangle$ with $nodes(g)$ forming the set of states of the LTS, $\mathcal{O}$ being the set of symbols, $trans(g)$ the transition relation and $root(g)$ the initial state.

If $p \in nodes(g)$ then $(g)_p$ is the subgraph of g with p as the root and all the nodes and edges accessible from p in the obvious way. We weaken this boundary by identifying the subgraph $(g)_p$ with the node p. We therefore consider both process graphs and nodes as representation of processes. We weaken the boundary further by identifying them with the processes they represent. Two graphs g and h may have the same node names, in which case we use the subscripts g and h for distinction. p_g is a node in g and p_h is a node in h. Note that these nodes represent different processes $(g)_p$ and $(h)_p$ respectively. Graphs differing only in their naming of the nodes are considered to be identical.

$\mathbb{G}$ (ranged over by $g, h, \ldots$) forms the set of all process graphs while $\mathbb{R}$ is the set of all *finite process graphs* [6] or *regular process graphs*. In this paper we restrict ourselves primarily to the study of finite process graphs.

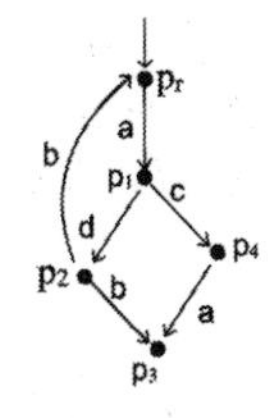

Figure 1: *A process graph*

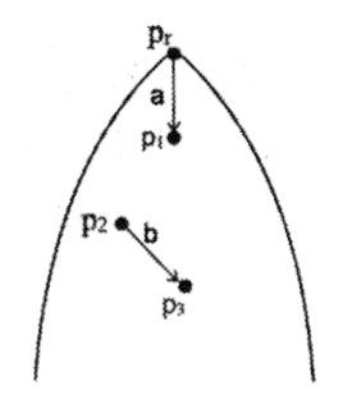

Figure 2: *Partial view of a process graph*

While drawing process graphs we may either depict all nodes and edges (as in figure 1), or choose to depict only the parts that are of immediate concern to us (as in figure 2).

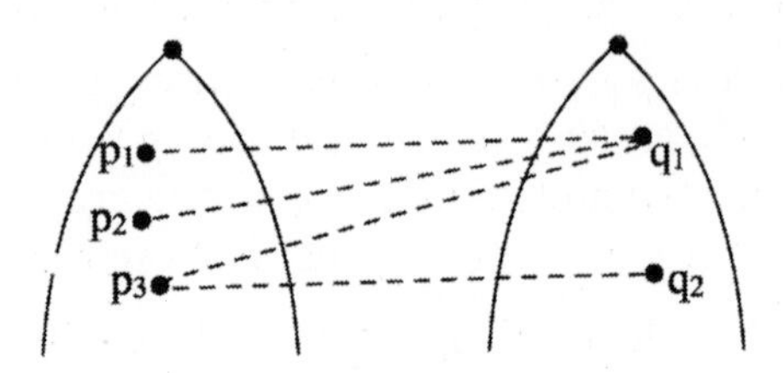

Figure 3: *Depicting relations*

We indicate nodes in two graphs to be related by drawing a dashed or dotted line or arc between them 3. Other notational conventions we use are the following.

- $\equiv$ for the identity relation on a set. It may be used in the context of observables, processes and also process graphs.
- id_g for the identity relation on the nodes of the process graph g. If two process graphs g and h have the same node names then id_g (or equivalently id_h) could also be used to mean the bijection between these nodes.
- $id_{trans(g)}$ for the identity relation on the transitions of the process graph g.
- $R : pSq$ to indicate $R \subseteq S$ and $(p,q) \in R$ where the relation R may be seen to act as a "proof" of the relation pSq.
- $\circ$ to denote relational composition.

3 (ρ, σ)-Bisimulations

Definition 1 *Let* **P** *be the set of processes and let ρ and σ be binary relations on $\mathcal{O}$. A binary relation $R \subseteq \mathbf{P} \times \mathbf{P}$ is a (ρ,σ)-***induced bisimulation** *or simply a (ρ,σ)-***bisimulation** *if pRq implies the following conditions.*

$$\forall a \in \mathcal{O}[p \xrightarrow{a} p' \Rightarrow \exists b, q'[a\rho b \wedge q \xrightarrow{b} q' \wedge p'Rq']] \tag{1}$$

and

$$\forall b \in \mathcal{O}[q \xrightarrow{b} q' \Rightarrow \exists a, p'[a\sigma b \wedge p \xrightarrow{a} p' \wedge p'Rq']] \tag{2}$$

The largest (ρ,σ)-bisimulation (under set containment) is called (ρ,σ)- **bisimilarity** *and denoted $\sqsubseteq_{(\rho,\sigma)}$. A $(\equiv,\equiv)$-induced bisimulation will sometimes be called a* **natural bisimulation***.

In proposition 2 we quote a few important properties of (ρ,σ)-bisimulations and refer the reader to [2] and [5] for their proofs.

Proposition 2 (Properties)

1. **Monotonicity** *If $(\rho,\sigma) \subseteq (\rho',\sigma')$ pointwise then every (ρ,σ)-bisimulation is also a (ρ',σ')-bisimulation and hence $\sqsubseteq_{(\rho,\sigma)} \subseteq \sqsubseteq_{(\rho',\sigma')}$.*
2. **Inversion** *Since $R : p\sqsubseteq_{(\rho,\sigma)}q$ implies $R^{-1} : q\sqsubseteq_{(\sigma^{-1},\rho^{-1})}p$, we have $\sqsubseteq_{(\rho,\sigma)} = \sqsubseteq_{(\sigma^{-1},\rho^{-1})}$.*
3. **Composition** $\sqsubseteq_{(\rho_1,\sigma_1)} \circ \sqsubseteq_{(\rho_2,\sigma_2)} \subseteq \sqsubseteq_{(\rho_1\circ\rho_2,\sigma_1\circ\sigma_2)}$ *since $R_1 : p\sqsubseteq_{(\rho_1,\sigma_1)}q$ and $R_2 : q\sqsubseteq_{(\rho_2,\sigma_2)}r$ implies $R_1 \circ R_2 : p\sqsubseteq_{(\rho_1\circ\rho_2,\sigma_1\circ\sigma_2)}r$.*
4. **Reflexivity** *If ρ and σ are both reflexive then the identity relation $\equiv$ on $\mathbf{P}$ is a (ρ,σ)-bisimulation and consequently $\sqsubseteq_{(\rho,\sigma)}$ is reflexive.*
5. **Symmetry** *ρ and σ are both symmetric implies the converse of each (ρ,σ)-bisimulation is a (σ,ρ)-bisimulation. In addition, if $\rho=\sigma$ then $\sqsubseteq_{(\rho,\sigma)}$ is a symmetric relation.*
6. **Transitivity** *If ρ and σ are both transitive then the relational composition of (ρ,σ)-bisimulations is a (ρ,σ)-bisimulation, and $\sqsubseteq_{(\rho,\sigma)}$ is also transitive.*
7. **Preorder characterisation** *$\sqsubseteq_{(\rho,\sigma)}$ is a preorder iff ρ and σ are both preorders.*
8. **Equivalence characterisation** *$\sqsubseteq_{(\rho,\sigma)}$ is an equivalence iff ρ and σ are both preorders and $\sigma=\rho^{-1}$.*

$\dashv$

A question that arises in the context of (ρ,σ)-bisimilarities is whether there is a characterization of the smallest congruence or pre-congruence that contains a particular (ρ,σ)-bisimilarity. We answer this question to a very large extent in lemma 5. Lemma 5 demands an accurate definition of what is considered an empty observable.

Definition 3 *An observable ϵ is considered* **empty** *if $\forall p,p' \in \mathbf{P}[p \equiv p' \iff p \xrightarrow{\epsilon} p']$ and corresponds to an extended operational*

*A strong bisimulation on CCS processes with $\mathcal{O} = Act$ is an example of a *natural bisimulation*.

$$\begin{array}{ll} \text{Act} & a.e \xrightarrow{a} e \\ \text{Sum1} & e \xrightarrow{a} e' \Rightarrow e+f \xrightarrow{a} e' \\ \text{Sum2} & e \xrightarrow{a} e' \Rightarrow f+e \xrightarrow{a} e' \\ \text{Rec} & e\{\mu\, x : e/x\} \xrightarrow{a} e' \Rightarrow \mu\, x : e \xrightarrow{a} e' \end{array}$$

Figure 4: *Operational Semantics of CCS expressions*

semantics of CCS with the additional rule Empty: $p \xrightarrow{\epsilon} p$ *along with a restriction of the prefix rule* Act *in figure 4 on* $a \in \mathcal{O} - \{\epsilon\}$.

Lemma 4 *Some consequences of having an empty observable.*

1. *A transition may take place without destroying choice. That is,* $\exists \epsilon \in \mathcal{O} : [p+q \xrightarrow{\epsilon} p+q]$.
2. *A transition of a child of a choice expression need not result in a transition of the choice expression. That is,* $\exists \epsilon \in \mathcal{O} : [p \xrightarrow{\epsilon} p' \wedge p+q \stackrel{\epsilon}{\not\longrightarrow} p']$.
3. *There is a special observable that cannot occur as a label of transitions between arbitrary nodes. That is,* $\exists \epsilon \in \mathcal{O} : \not\exists\, p, p' \in \mathbf{P} : [p \not\equiv p' \wedge p \xrightarrow{\epsilon} p']$.

⊣

Lemma 5 *If* $\mathcal{O}$ *does not contain an empty observable, then over the CCS operators '+', '.' and 'μ'*

1. **Precongruence** *Every* $\sqsubseteq_{(\rho,\sigma)}$ *that is a preorder is also a precongruence.*
2. **Congruence** *Every* $\sqsubseteq_{(\rho,\sigma)}$ *that is an equivalence is also a congruence.*

⊣

The proof of lemma 5 is along the lines of a similar congruence proof in [4]. In the presence of empty observables, the proof does not work out because of part 2 of lemma 4. It should be noted that the characterization of observational equivalence provided in [5] uses $\mathcal{O} = Act^*$ where the empty string can be easily seen to be an empty observable. This is the reason for observational equivalence not being a congruence. In this paper we will concern ourselves only with the cases where $\mathcal{O}$ has no empty observables. While the preorders and equivalences have been proven to be congruences over operators of CCS, it is not hard to see that the same holds for these over LR expressions. Our preference for CCS notation over LR expressions here, is for the sake of brevity.

4 Syntactical Representations

While process graphs are convenient graphical representations of processes and provide the semantics of a process, we also need a linear representation of processes that is syntactically convenient. Process algebras have traditionally evolved with a stronger focus on the syntactical representations because of the convenience of representation and preciseness of argument they allow. The notation of CCS introduced by Milner in [10] and LR expressions introduced by Bergstra and Klop in [6] are more convenient syntactically because of their compactness of representation that allows a more precise reasoning than graphically rich representations such as process graphs do. The semantics of processes are given here in terms of process graphs.

The axiomatizations of relations, even if such relations are defined over process graphs, are therefore given in one of the syntactical representations. We introduce Linear Recursive (LR) expressions, which have been used in [6] and will be the preferred syntactical representation of processes in this paper.

4.1 LR Expressions

LR expressions, where LR stands for Linear-Recursive, are the recursive terms of the system BPA_{LR} introduced in [6]. Since the axiomatization $\mathscr{A}_{(\rho,\rho^{-1})}$ is built on top of the system BPA_{LR}, LR expressions along with the non-recursive terms of BPA_{LR} will be the choice of syntactical representation of processes in this paper.

Let *VAR* be a denumerably infinite set of variables $\{X, Y, Z \ldots\}$. $\mathscr{O}$ is the set of observables. The set of *non-recursive terms* T over $\mathscr{O}$ is defined by the following BNF:

$$T ::= a \mid X \mid T + T \mid T.T$$

where $a \in \mathscr{O}$ and $X \in VAR$. We may omit the concatenation operator '.' when it is obvious. We do not have a 0 in BPA_{LR}, and instead of a process $a.0$ we simply have the process a. While in CCS we may only prefix a process expression by an observable a, here we have a full sequential composition instead, allowing us to sequentially compose any two terms in BPA_{LR}. The reader is referred to [6] for a deeper insight into the notion of full sequential composition.

The free variables $fv(T)$ of a term T are defined in the usual way. $fv(a) = \emptyset, fv(X) = \{X\}, fv(T_1.T_2) = fv(T_1+T_2) = fv(T_1) \cup fv(T_2)$. A term is *closed* if it has no free variables. We also write $T(\overline{X})$, where $\overline{X} = X_1X_2 \ldots X_n$ is the vector of free variables of T, to indicate that T may depend at most on $X_1, X_2, \ldots, X_n$. The substitution of free variables $T\{p_1/x_1\}\{p_2/x_2\}\ldots\{p_n/x_n\}$, where $\overline{p} = p_1p_2\ldots p_n$ is the vector of non-recursive substitution terms, is also written as $T(\overline{p})$.

The free variables of the non-recursive term $T = abX + cY(X + Z)Y$ are X, Y, Z and we write $T(X, Y, Z)$ instead of T to indicate this fact. $T(a, b + Z, YY) = aba + c(b + Z)(a + YY)(b + Z)$ and $T(a, b, c) = aba + cb(a + c)b$ are obtained by the substitutions of the free variables X, Y, Z by $a, b + Z, YY$ and a, b, c respectively. $T(a, b, c)$ is a closed term.

Definition 6 *A non-recursive term T is **guarded** if every occurrence of a variable in T is preceded by some $a \in \mathcal{O}$. Formally:*

1. *$a \in \mathcal{O}$ is guarded*
2. *if T is guarded and T' is an arbitrary non-recursive term, then $T.T'$ is guarded*
3. *if T, T' are both guarded then so is $T + T'$*

Definition 7 *A non-recursive term T is said to be **linear** if all occurrences of variables are "at the end". Formally:*

1. *$X \in$ VAR is linear*
2. *Closed non-recursive terms are linear*
3. *if T, T' are both linear then so is $T + T'$*
4. *if T is a closed non-recursive term and T' is linear, then $T.T'$ is linear*

Definition 8 *A non-recursive term is **strictly linear** if it is of the form $\sum a_i + \sum b_j.X_j$.*

The terms $abX + cY(X + Z)Y, a, b + c(X + e)$ are guarded while $X + cY(X + Z)Y, a + Y, b + c(X + e) + X(Y + a)$ are not. The terms $X, a, a+Y, b+c(X+e), abX$ are linear while $cY(X+Z)Y, abYY, b+c(X+e)+X(Y+a)$ are not. The terms $a, a+bY$ and $a+b+cY+dZ$ are strictly linear while the linear terms $X, a+Y$ and $c(X+e)$ are not strictly linear.

Definition 9 *LR-expressions are syntactical constructs of the form $\langle X_1 | E \rangle$ where $X_i \in$ VAR, $E = \{X_i = T_i(\overline{X}) \mid i = 1, \ldots, n\}$ is a set of recursion equations, $\overline{X}$ is the vector of variables $X_1, X_2 \ldots X_n$ and for every i,*

1. *$T_i(\overline{X})$ is guarded*
2. *$T_i(\overline{X})$ is linear*
3. *$T_i(\overline{X})$ may contain variables only from $\overline{X} = X_1, \ldots, X_n$.*

Definition 10 *An equation $X_i = T_i(\overline{X})$ is* ***superfluous*** *if X_i is not accessible from X_1 in the obvious sense. Superfluous equations in E may be omitted.*

Definition 11 ***Canonical LR-expressions*** *are LR-expressions where every $T_i(\overline{X})$ is strictly linear and E does not contain any superfluous equations.*

It is understood that LR expressions that differ only by a renaming of variables are identical. The semantics of the terms of BPA_{LR} are provided by a mapping $[\,] : Ter(BPA_{LR}) \longrightarrow \mathbb{R}$ where $Ter(BPA_{LR})$ stands for the set of all terms of BPA_{LR}. This mapping is termed *intermediate semantics* in [6] and has been quoted in definition 12. In lemma 15, we provide a procedural mapping $[[\,]]$ restricted to canonical LR expressions that highlights their intuitive correspondence with process graphs. This is termed the *direct intermediate semantics* in [6]. The direct intermediate semantics will prove useful for the purpose of providing graphical intuition. However, the true semantics of a term in BPA_{LR} is given by the intermediate semantics $[\,]$. We proceed to establish the intuitive correspondence of canonical LR expressions with process graphs and refer the reader to the construction in lemma 15 for the precise correspondence.

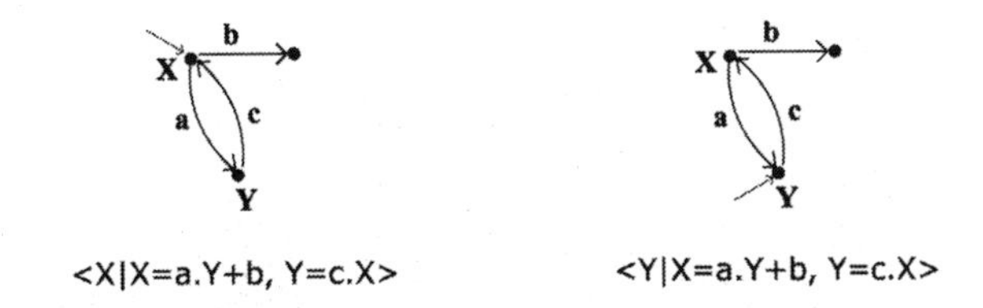

Figure 5: *The same set of equations with different root nodes*

The LR expression $\langle X_1 | E \rangle$ represents a process graph with X_1 as the root. The expression $\langle X_i | E \rangle$ represents the same graph but rooted at X_i instead. Note that in this case the rest of the graph might not be reachable anymore and some equations might become superfluous. In figure 5, we consider two canonical LR expressions $\langle X | X = a.Y + b, Y = c.X \rangle$ and $\langle Y | X = a.Y + b, Y = c.X \rangle$ which differ only in their root nodes. The nodes have been labelled X and Y for the purpose of naming and this convention should not be confused with the process chart notation [9], [6] where nodes decorated with variables have special semantics. $Y = c.X$ intuitively implies that there is an edge labelled c that starts at the node named Y and ends at the node named X. In a similar way $X = a.Y + b$ implies that there is an edge labelled a starting at X and ending at Y. In addition, the lone b in $X = a.Y + b$ implies that there is an outgoing edge labelled b that goes to a node that has not been named, or alternatively to a terminal node. In the LR expression $\langle X|X = a + bY, Y = cY + dX, Z = eZ\rangle$, $Z = eZ$ is superfluous because Z is not accessible from X. In the LR expression $\langle Z|X = a + bY, Y = cY + dX, Z = eZ\rangle$, $X = a + bY$ and $Y = cY + dX$ are superfluous.

We use the notation E_{-k} for the set of equations in E except the kth equation, $E_{-k} = \{X_i = T_i(\overline{X}) \mid 1 \leq i \neq k \leq n\}$. This is useful for focusing attention on a particular equation. For example, if we want to focus on the kth equation in $\langle X_1|E\rangle$, we may instead write this LR expression as $\langle X_1|X_k = T_k(\overline{X}), E_{-k}\rangle$. All the variables $X_1, X_2, \ldots X_n$ in $\langle X|E\rangle$, where $E = \{X_i = T_i(\overline{X}) \mid i = 1, \ldots, n\}$, are bound in $\langle X|E\rangle$. If a variable is bound in a term, it is not free and since $T_i(\overline{X})$ may contain variables only from $\overline{X} = X_1, \ldots, X_n$, every variable in a LR expression $\langle X|E\rangle$ is bound. Hence every LR expression $\langle X|E\rangle$ is a closed term.

Definition 12 *The intermediate semantics [6] is an inductively defined mapping* $[\,] : Ter(BPA_{LR}) \longrightarrow \mathbb{R}$ *of the terms of* BPA_{LR} *into the domain of finite process graphs*

1. $[a] = \rightarrow\circ\stackrel{a}{\longrightarrow}\circ$
2. $[S + T] = [S] \oplus [T]$
3. $[S.T] = [S] \odot [T]$
4. $[\langle X|E\rangle] = \omega(c(t(p(\langle X|E\rangle))))$

where $\oplus$ *is the addition of process graphs,* $\odot$ *is their full sequential multiplication,* $\omega : \mathbb{G} \longrightarrow \mathbb{G}$ *is the* ***root-unwinding*** *operator*[†], $c : \mathbb{G} \longrightarrow \mathbb{G}$ *is the* ***collaps*** *operator,* t *is the* ***tree*** *operator which defines the mapping of LR expressions into the domain of process trees* $\mathbb{T}$ *(note that* $\mathbb{T} \subseteq \mathbb{G}$*) and* p *is the* ***prefix*** *operator which makes a LR expression prefix free. The notation* $\rightarrow\circ_p\stackrel{a}{\longrightarrow}\circ_q$ *is used to denote the process graph that has a root* p *which has a single edge labelled 'a' leading to a terminal node* q.

[†]In [6] the letter ρ is used in place of ω, however this conflicts with our convention of using ρ for relations on observables.

We will not concern ourselves with a detailed study of the mapping [ ] and refer the reader to [6] for the definitions of root-unwinding, collaps, tree and prefix operators as well as the properties quoted in proposition 13. For the purpose of this paper, only the intermediate semantics of a and $a+b$ as given by $[a] = \rightarrow\circ\xrightarrow{a}\circ$ and $[a+b] = [a] \oplus [b]$ are important as they are required in the soundness proof. Without defining $\oplus$ in detail, we note that $\rightarrow\circ_{p_1}\xrightarrow{a}\circ_{q_1} \oplus \rightarrow\circ_{p_2}\xrightarrow{b}\circ_{q_2}$ is the process graph obtained by combining the roots p_1 and p_2 into a single node p which is the new root. In this process graph p will have two outgoing edges, one labelled a going to q_1 and another labelled b going to q_2.

Proposition 13 (Properties)

1. *LR expressions are always closed.*
2. $[\langle X|E_1\rangle.\langle Y|E_2\rangle] \sqsubseteq_{(\equiv,\equiv)} [\langle Z|E_3\rangle]$ *for some* Z *and* E_3.
3. $[\langle X|E_1\rangle + \langle Y|E_2\rangle] \sqsubseteq_{(\equiv,\equiv)} [\langle Z|E_3\rangle]$ *for some* Z *and* E_3.
4. $[T] \sqsubseteq_{(\equiv,\equiv)} [\langle Z|E\rangle]$ *for some* Z *and* E *if* T *is a closed non-recursive term.*
5. *For every LR expression there is a corresponding canonical LR expression, such that the intermediate semantical mappings of the two are strongly bisimilar.*

⊣

Proposition 13.4 states that a closed non-recursive term can also be represented by an appropriate LR expression. The proof of this is trivial, with $E = \{Z = T\}$. From proposition 13.2 and 13.3, we also know that expressions formed over LR expressions with the '+' and '.' operators also have an equivalent LR expression. As a consequence any process expressible in BPA_{LR} can be represented by an LR expression. Every LR expression, in turn, can be converted to an equivalent canonical form upto strong bisimilarity, as stated in proposition 13.5.

In order to highlight the correspondence between canonical LR expressions and process graphs we provide here two correspondences, [[ ]] from canonical LR expressions to process graphs and || || from process graphs to canonical LR expressions. These can be seen to have the interesting properties that each process graph g is strongly bisimilar to [[||g||]]. Also, as pointed out in [6] $[[\langle X|E\rangle]]$ is strongly bisimilar to $[\langle X|E\rangle]$.

Lemma 14 (LTS to canonical LR) *There is a mapping || || from the domain of finite rooted LTS into the domain of canonical LR Expressions such that, for each finite rooted LTS* $\mathcal{L} = \langle Q, \Sigma, \longrightarrow, p_k\rangle$ *there is a* *canonical LR Expression $||\mathscr{L}|| \equiv \langle X_k|E\rangle$ with $[\langle X_k|E\rangle]$ strongly bisimilar to $\mathscr{L}$.*

Proof If the non-terminal states of $\mathscr{L}$ are named $p_1, p_2, \ldots, p_n$ for some n, we create a corresponding LR expression in the following manner.

1. Create and identify a variable X_i with each non-terminal state p_i.
2. Start with each T_i being empty.
3. If $p_i \xrightarrow{a} q$ where q is some terminal state, convert $X_i = T_i$ to $X_i = T_i + a$.
4. If $p_i \xrightarrow{a} p_j$ where p_j is some non-terminal state, convert $X_i = T_i$ to $X_i = T_i + a.X_j$.

The required canonical LR expression is $\langle X_k|E\rangle$ where $E = \{X_i = T_i(\overline{X}) \mid i = 1, \ldots, n\}$. It is not hard to see that $[\langle X_k|E\rangle]$ is strongly bisimilar to $\mathscr{L}$. The mapping $||\ ||$ is defined through this construction, with $||\mathscr{L}|| \equiv \langle X_k|E\rangle$. ⊣

Lemma 15 (Canonical LR to LTS)

There is a mapping $[[\ ]]$ from the domain of canonical LR expressions into the domain of finite rooted LTSs such that, for each canonical LR Expression $\langle X_1|E\rangle$ there is a finite rooted LTS $\mathscr{L} \equiv [[\langle X_1|E\rangle]]$ with $[\langle X_1|E\rangle]$ strongly bisimilar to $\mathscr{L}$.

Proof Given a canonical LR expression $\langle X_1|E\rangle$ where $E = \{X_i = T_i(\overline{X}) \mid i = 1, \ldots, n\}$ and $T_i = \sum_j a_{ij} + \sum_k b_{ik}.X_{ik}$, construct a rooted LTS $\mathscr{L}$ such that

1. The set of states of $\mathscr{L}$ is $\{X_i | i = 1, \ldots n\} \cup \{X_T\}$ where we introduce a variable X_T which stands for a terminal state.
2. If $T_i = \sum_j a_{ij} + \sum_k b_{ik}.X_{ik}$, then add the transitions $X_i \xrightarrow{a_{ij}} X_T$ and $X_i \xrightarrow{b_{ik}} X_{ik}$ for every j and k.
3. The set of observables in $\mathscr{L}$ can be taken to be the set of all observables appearing in $\langle X_1|E\rangle$.
4. X_1 is the root of $\mathscr{L}$.

Clearly, $\mathscr{L}$ obtained here is finite. It is not hard to see that $[\langle X_1|E\rangle]$ is strongly bisimilar to $\mathscr{L}$. The mapping $[[\ ]]$ is defined through this construction, with $[[\langle X_1|E\rangle]] = \mathscr{L}$.

⊣

Lemma 16 **(Semantics)**

1. $[[\langle X|E\rangle]] \sqsupseteq_{(\equiv,\equiv)} [\langle X|E\rangle]$
2. $[[||g||]] \sqsupseteq_{(\equiv,\equiv)} g$

$\dashv$

5 The Axiomatization $\mathscr{A}_{(\rho,\rho^{-}1)}$

Using $\mathscr{A}_{(\rho,\sigma)}$ to denote a complete axiomatization for $\sqsupseteq_{(\rho,\sigma)}$, we present in this section the axiomatization $\mathscr{A}_{(\rho,\rho^{-1})}$ where ρ is assumed to be a preorder. The axiomatization is proven complete for $\sqsupseteq_{(\rho,\rho^{-1})}$ over finite process graphs $\mathbb{R}$ relative to a complete axiomatization of ρ. A necessary condition for soundness is that $\mathscr{O}$ does not contain empty observables. In this case, $\sqsupseteq_{(\rho,\rho^{-1})}$ also turns out to be a congruence, allowing us to give a more succinct formulation of the axiomatization. It should be noted that all equivalence relations in the class of (ρ,σ)-bisimilarities have $\sigma = \rho^{-1}$ and ρ as a preorder. Thus $\mathscr{A}_{(\rho,\rho^{-}1)}$ is in fact a complete axiomatization of all such equivalences defined over finite process graphs $\mathbb{R}$ with no empty observables.

Since ρ is a preorder, from proposition 2 we know that $\sqsupseteq_{(\rho,\rho^{-1})}$ contains $\sqsupseteq_{(\equiv,\equiv)}$. Taking $\mathscr{O} = Act$, strong bisimilarity has been shown to be equivalent to the natural bisimilarity relation $\sqsupseteq_{(\equiv,\equiv)}$ in [2]. We therefore refer to $\sqsupseteq_{(\equiv,\equiv)}$ as strong bisimilarity as well. The axiomatizations we consider are for relations no finer than strong bisimilarity. For this reason, we build our axiomatization upon the axiomatization BPA_{LR} shown in figure 6 which is a complete axiomatization for strong bisimilarity (see [6]).

It should be noted that the equality $=$ in BPA_{LR} is the relation on terms of BPA_{LR} that corresponds to strong bisimilarity $\sqsupseteq_{(\equiv,\equiv)}$ on process graphs. Similarly $=_{\rho}$ will stand for the relation on terms of $\mathscr{A}_{(\rho,\rho^{-}1)}$ that corresponds to $\sqsupseteq_{(\rho,\rho^{-1})}$ on process graphs.

We briefly present the outline of one of the ways of building $\mathscr{A}_{(\rho,\rho^{-1})}$ on top of BPA_{LR} before moving on to our preferred formulation. The purpose of presenting this formulation is only to provide an alternative structuring of the axiomatization which might prove easier to extend. In particular, we believe that axiomatizations of other

$x+y=y+x$	$A1$
$(x+y)+z=x+(y+z)$	$A2$
$x+x=x$	$A3$
$(x+y)z=xz+yz$	$A4$
$(xy)z=x(yz)$	$A5$

$$\frac{p_i=\langle X_i|E\rangle,\ i=1,\ldots,n}{p_1=T_1(\overline{p})} \qquad R1$$

$$\frac{p_i=T_i(\overline{p}),\ i=1,\ldots,n}{p_1=\langle X_1|E\rangle} \quad T_i(\overline{X}) \text{ is guarded} \qquad R2$$

Figure 6: BPA_{LR} *- a proof system for* $\underline{\square}_{(\equiv,\equiv)}$

relations that are not congruences might have to proceed in this manner as this formulation does not rely on the congruence of $\underline{\square}_{(\rho,\rho^{-1})}$.

The first formulation provides an axiom that cannot work in contexts, and is applicable only for canonical LR expressions. Every closed term of BPA_{LR}, whether recursive, non-recursive or an expression formed over these with the operators '.' and '+', can be proven to be strongly bisimilar to a canonical LR expression within the framework of BPA_{LR}. This boils down to proving parts 2,3,4 and 5 of proposition 13 within the scope of the BPA_{LR}, the proof of which is provided in [6]. Since strong bisimilarity will be the finest relation on processes considered here except for identity on processes, we may restrict our attention to the quotient structure $\mathbb{R}/\underline{\square}_{(\equiv,\equiv)}$. Instead of needing to deal with closed terms of arbitrary forms, this allows us to deal with equivalent canonical LR expressions which have a very convenient form. At the same time, it allows us to add duplicate edges between two nodes while staying in the same equivalence class. On the quotient structure $\mathbb{R}/\underline{\square}_{(\equiv,\equiv)}$, strong bisimilarity reduces to identity $\equiv$ on the equivalence classes. In this formulation, the full axiomatization would be composed of two "layers", BPA_{LR} forming the inner part and a second "layer" consisting of axioms that work on the quotient structure $\mathbb{R}/\underline{\square}_{(\equiv,\equiv)}$. It turns out that we only need one more axiom (schema), the vertical axiom AV shown below.

$$\langle X_1 | X_k = b.X_j + T, E_{-k}\rangle \\ =_\rho \\ \langle X_1 | X_k = a.X_j + b.X_j + T, E_{-k}\rangle \qquad \text{if } a\rho b$$

This axiom schema comfortably assumes each term in $\mathbb{R}/\underline{\square}_{(\equiv,\equiv)}$ to be a canonical LR expression (provable within the BPA_{LR} part) and therefore does not need to work in a context. This is particularly useful if $\underline{\square}_{(\rho,\rho^{-1})}$ was not a congruence. However, one of the conditions for the soundness of this axiom is not having empty observables in $\mathcal{O}$ which also ends up proving $\underline{\square}_{(\rho,\rho^{-1})}$ to be a congruence. For this reason we will prefer the second formulation which is more succinct.

In our preferred approach, we first translate BPA_{LR} to $BPA_{\rho LR}$ with terms of the two systems being identical. The translation of axioms of BPA_{LR} to $BPA_{\rho LR}$ involves the replacement of all $=$ by $=_\rho$, where $=_\rho$ on terms of $BPA_{\rho LR}$ corresponds to the relation $\underline{\square}_{(\rho,\rho^{-1})}$ on process graphs. In this manner we get the axioms $A1_\rho - A5_\rho$ and $R1_\rho, R2_\rho$ listed in figure 7. Note that the recursive equations E in a LR expression $\langle X|E\rangle$ remain of the form $\{X_i = T_i(\overline{X}) | i = 1,\ldots,n\}$ as the '$=$' in question here is a syntactical notation for specifying recursive equations and is not strong bisimilarity.

From the completeness of BPA_{LR}, we know that for all T, S in $Ter(BPA_{LR})$, $[T]\underline{\square}_{(\equiv,\equiv)}[S] \Longrightarrow BPA_{LR} \vdash T = S$. Taking the proof of $T = S$ in BPA_{LR} and replacing each application of the axioms of BPA_{LR} by the corresponding axioms of $BPA_{\rho LR}$, we end up constructing a proof of $T =_\rho S$ in $BPA_{\rho LR}$. That is, within $BPA_{\rho LR}$, we can prove the (ρ,ρ^{-1})-bisimilarity of every pair of strongly bisimilar processes. Using essentially the same arguments every closed term of $BPA_{\rho LR}$ can be proven to be (ρ,ρ^{-1})-bisimilar to a canonical LR expression within the framework of $BPA_{\rho LR}$.

In a similar manner, each instance of the *substitution* inference rule shown below is provable in $BPA_{\rho LR}$ by a translation of the corresponding proof [6] in BPA_{LR}.

$$\frac{T_k(\overline{X}) =_\rho T'_k(\overline{X})}{\langle X_1 | X_k = T_k(\overline{X}), E_{-k}\rangle =_\rho \langle X_1 | X_k = T'_k(\overline{X}), E_{-k}\rangle}$$

We introduce the axiom schema *Axiom-Vertical* AV_ρ (which is named so for reasons that will become clear in the proof of completeness) to

$b =_\rho a + b \quad \text{if } a\rho b$	AV_ρ
$x + y =_\rho y + x$	$A1_\rho$
$(x + y) + z =_\rho x + (y + z)$	$A2_\rho$
$x + x =_\rho x$	$A3_\rho$
$(x + y)z =_\rho xz + yz$	$A4_\rho$
$(xy)z =_\rho x(yz)$	$A5_\rho$
$\dfrac{p_i =_\rho \langle X_i \vert E\rangle,\ i = 1,\ldots,n}{p_1 =_\rho T_1(\overline{p})}$	$R1_\rho$
$\dfrac{p_i =_\rho T_i(\overline{p}),\ i = 1,\ldots,n}{p_1 =_\rho \langle X_1 \vert E\rangle}$ $\quad T_i(\overline{X})$ is guarded	$R2_\rho$

Figure 7: $\mathscr{A}_{(\rho,\rho^{-1})}$

obtain the axiomatization $\mathscr{A}_{(\rho,\rho^{-1})}$ in figure 7. The terms of $\mathscr{A}_{(\rho,\rho^{-1})}$ are the same as those of BPA_{LR}. The intuition of the axiom AV_ρ is similar to that of $A3_\rho$ which can prove that (with $x = b$) strong bisimilarity is insensitive to the addition of duplicate edges. AV_ρ says that $\underline{\Box}_{(\rho,\rho^{-1})}$ is insensitive to the addition of edges that are related by ρ to the existing ones.

In the case that ρ is also an equivalence relation, $a\rho b \Longrightarrow [b =_\rho a + b \wedge a =_\rho a + b] \Longrightarrow a =_\rho b$, which says that $=_\rho$ for such a ρ is indifferent to observables that lie in the same equivalence class with respect to ρ. Thus proving (ρ, ρ^{-1})-bisimilarity of graphs for such a ρ can be reduced to proving strong bisimilarity of graphs with labels from $\mathcal{O}/\rho$. This case not being so interesting, we shall only focus on the case where ρ is a preorder.

What should be easy to see is that all instances of AV as shown below, can be proven within the framework of this axiomatization.

$$\begin{array}{c} \langle X_1 \vert X_k = b.X_j + T, E_{-k}\rangle \\ =_\rho \\ \langle X_1 \vert X_k = a.X_j + b.X_j + T, E_{-k}\rangle \end{array} \qquad \textit{if } a\rho b$$

From $b =_\rho a + b \overset{A4_\rho}{\Longrightarrow} bX_j =_\rho aX_j + bX_j$, and using the congruence property of $=_\rho$, for any T we can prove that $b.X_j + T =_\rho a.X_j +$

$b.X_j + T$. From the substitution rule, with $T_j(\overline{X}) \equiv b.X_j + T$ and $T_j'(\overline{X}) \equiv a.X_j + b.X_j + T$, we get the required result.

With the result that each LR expression can be proven to be equivalent to a canonical LR expression upto (ρ, ρ^{-1}) bisimilarity within the scope of $BPA_{\rho LR}$, we could have alternatively chosen the above inference rule instead of the axiom AV_ρ. The above inference rule will also be useful in providing an intuitive understanding of the procedure of saturation. In figure 8, $g \equiv ||\langle X_1 | X_k = b.X_j + T, E_{-k}\rangle||$ and $g' \equiv ||\langle X_1 | X_k = a.X_j + b.X_j + T, E_{-k}\rangle||$ and the nodes p and p' correspond to X_k and X_j respectively. It is not hard to see that g and g' are in fact related by $\underline{\square}_{(\equiv, \rho^{-1})}$, and therefore also by $\underline{\square}_{(\rho, \rho^{-1})}$.

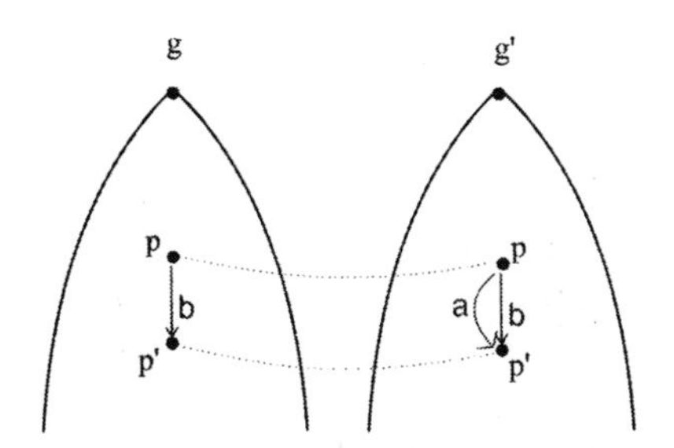

Figure 8: *Graphical intuition for AV*

The rest of the paper is devoted to the proof of soundness and completeness of the axiomatization. In section 7, we introduce the construction of derivatives of process graphs. The construction revolves around adding edges in the same way as the a edge is added in figure 8 to obtain g' from g.

It should be noted that the case of empty observables could lead to peculiar situations. For example, if a is an empty observable, then adding an a edge between p and p' has the effect of identifying p and p' and is not the addition of edges in any traditional sense. In fact, such addition corresponds to the addition of the term $a.X_j$ to X_k. However, the prefixing by an empty observable is undefined. This problem is summarized in definition 3 and part 3 of lemma 4. The presence of empty observables would force us to change the definition of non-recursive terms of BPA_{LR} to exclude empty observables, modify the mappings [ ] and [[ ]] to introduce empty loops at each node and restrict AV_ρ to work only for non empty observables in order to be sound. Not being able to add a particular observable leads to the failure of the completeness proof.

6 Soundness of $\mathscr{A}_{(\rho,\rho^{-1})}$

Theorem 17 **(Soundness)**

For all $T, S \in Ter(\mathscr{A}_{(\rho,\rho^{-1})})$ *in the absence of empty observables in* $\mathscr{O}$,

$$\mathscr{A}_{(\rho,\rho^{-1})} \vdash T =_{\rho} S \Longrightarrow [T] \sqsupseteq_{(\rho,\rho^{-1})} [S]$$

Proof In order to prove the soundness of the axiom AV_{ρ} we need to prove $[b] \sqsupseteq_{(\rho,\rho^{-1})} [a+b]$ for any $a, b \in \mathscr{O}$ such that $a \rho b$.

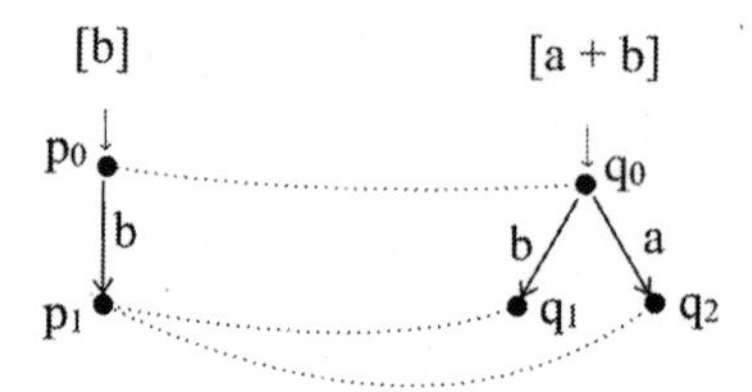

Figure 9: AV_{ρ}

From definition 12 we can see that the process graphs $[b]$ and $[a+b]$ are the ones given in figure 9. It is easily seen that $R : [b] \sqsupseteq_{(\rho,\rho^{-1})} [a+b]$ where $R = \{(p_0, q_0), (p_1, q_1), (p_1, q_2)\}$. Thus, AV_{ρ} is sound.

The rest of the proof follows along the lines of the soundness proof in [6] with the observation that $\sqsupseteq_{(\rho,\rho^{-1})}$ is a congruence and is preserved by the *prefix, tree, collaps* and *root-unwinding* operators employed in definition 12. ⊣

7 Saturation of Process Graphs

The method of proving completeness of $\mathscr{A}_{(\rho,\rho^{-1})}$ is inspired by the completeness proof of $BPA_{\tau LR}$ [6]. In [6], Bergstra and Klop define the operators Δ and E on process graphs which preserve observational congruence while reducing any two graphs related by observational congruence to strongly bisimilar graphs.

$$\begin{array}{lll} g & \approx^+ & h \\ \approx^+ & & \approx^+ \\ \Delta(g) & \approx^+ & \Delta(h) \\ \approx^+ & & \approx^+ \\ E(\Delta(g)) & \sim & E(\Delta(h)) \end{array}$$

Here, $\approx^+$ is observational congruence and $\sim$ is strong bisimilarity. The relation $g \approx^+ \Delta(g)$ is shown in a vertical fashion here. The proof of completeness of $BPA_{\tau LR}$ hinges upon such a diagram where $E(\Delta(g))$ and $E(\Delta(h))$ are strongly bisimilar if and only if g and h are observationally congruent.

Following a similar strategy of saturation, in this section we characterize certain relations between derivatives of graphs[‡] related by (ρ,σ)-bisimilarities. In particular, if $g \sqsubseteq_{(\rho,\rho^{-1})} h$ and ρ is a preorder, we are able to prove the existence of derivatives g' and h' such that $g \sqsubseteq_{(\equiv,\rho^{-1})} g'$, $h \sqsubseteq_{(\equiv,\rho^{-1})} h'$ and $g' \sqsubseteq_{(\equiv,\equiv)} h'$. Diagrammatically,

$$\begin{array}{lll} g & \sqsubseteq_{(\rho,\rho^{-1})} & h \\ \sqsubseteq_{(\equiv,\rho^{-1})} & & \sqsubseteq_{(\equiv,\rho^{-1})} \\ g' & \sqsubseteq_{(\equiv,\equiv)} & h' \end{array}$$

This result is critical for the proof of completeness of the axiomatization $\mathscr{A}_{(\rho,\rho^{-1})}$ and this entire section deals with proving the existence of such g' and h'. Roughly speaking, the transformation from g and h to g' and h' requires that all transitions in g and h that are matched by the relations ρ and σ get matched in g' and h' by $\equiv$ also. We start by constructing the derivatives $g[\pi_g]$ and $h[\pi_g]$ such that the transition $\pi_g \in trans(g)$ gets matched by $\equiv$ in $h[\pi_g]$. Also, any newly added transitions in $g[\pi_g]$ and $h[\pi_g]$ also get matched by $\equiv$ in $h[\pi_g]$ and $g[\pi_g]$ respectively. In this manner, the derivative pair $g[\pi_g], h[\pi_g]$ has fewer transitions that cannot be matched by $\equiv$ as compared to the pair g, h. This result is summed up in the partial saturation lemma 24. Performing this construction for all transitions of g and h we are able to get the required g' and h', resulting in the saturation lemma 27 and theorem 30.

From lemma 2.8, a (ρ,σ)-bisimilarity is an equivalence relation if and only if $\sigma = \rho^{-1}$ and ρ is a preorder. While we intend to tackle only equivalence relations and this condition is sufficient for all results proved in this section, we will continue to make only the minimal necessary assumptions in each lemma.

[‡]A process graph derived from another in some manner.

7.1 Partial Saturation

Let $R \subseteq nodes(g) \times nodes(h)$ be a binary relation between the nodes of process graphs g and h and let ρ and σ be binary relations on the set $\mathcal{O}$ of observables.

We write $Q(\pi_g)$ for the **h-image** of $\pi_g \in trans(g)$ and $P(\pi_h)$ for the **g-image** of $\pi_h \in trans(h)$, defining them as:

$$Q(p \xrightarrow{c} p') = \{q \xrightarrow{b} q' \in trans(h) \mid pRq \wedge c\rho b \wedge p'Rq'\}$$

$$P(q \xrightarrow{b} q') = \{p \xrightarrow{c} p' \in trans(g) \mid pRq \wedge c\sigma b \wedge p'Rq'\}$$

For sets of transitions $G \subseteq trans(g)$ and $H \subseteq trans(h)$, we write $\mathcal{Q}(G)$ and $\mathcal{P}(H)$ for the h-image of G and g-image of H, respectively.

$$\mathcal{Q}(G) = \bigcup_{p \xrightarrow{c} p' \in G} Q(p \xrightarrow{c} p')$$

$$\mathcal{P}(H) = \bigcup_{q \xrightarrow{b} q' \in H} P(q \xrightarrow{b} q')$$

We write $T_i(\pi_g)$ and $U_i(\pi_g)$ for the sets of **level i matches** of $\pi_g \in trans(g)$ in g and h respectively.

$$T_o(\pi_g) = \{\pi_g\}$$

$$T_{i+1}(\pi_g) = \mathcal{P}(U_i(\pi_g)) \qquad U_i(\pi_g) = \mathcal{Q}(T_i(\pi_g))$$

Note that $U_0(\pi_g) = \mathcal{Q}(T_0(\pi_g))$. $T(\pi_g)$ and $U(\pi_g)$, the sets of **closure matches** of π_g in g and h, can be defined as infinite unions of T_i and U_i. Intuitively, $T(p_o \xrightarrow{a} p_o')$ and $Q(p_o \xrightarrow{a} p_o')$ are the sets of transitions where an extra $\xrightarrow{a}$ edge must be added for the purpose of partial saturation with respect to $p_o \xrightarrow{a} p_o'$.

$$T(\pi_g) = \bigcup_{i \geq 0} T_i(\pi_g) \qquad \text{and} \qquad U(\pi_g) = \bigcup_{i \geq 0} U_i(\pi_G)$$

In figure 10, with $\pi = p_0 \xrightarrow{a} p_0'$, we depict a case where $T_0(\pi) = \{p_0 \xrightarrow{a} p_0'\}$, $U_0(\pi) = \{q_1 \xrightarrow{b_1} q_1'\}$, $T_1(\pi) = \{p_2 \xrightarrow{a_2} p_2', p_3 \xrightarrow{a_3} p_3'\}$, $U_1(\pi) = \{q_2 \xrightarrow{b_2} q_2'\}$ and $\forall i \geq 2 : T_i(\pi) = U_i(\pi) = \emptyset$. Thus, $T(\pi) =$

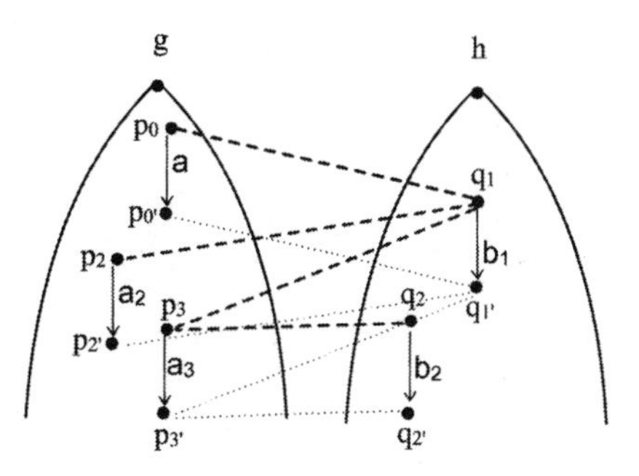

Figure 10: *Construction of $T(\pi)$ and $U(\pi)$*

$\{p_0 \xrightarrow{a} p_0',\ p_2 \xrightarrow{a_2} p_2',\ p_3 \xrightarrow{a_3} p_3'\}$ and $U(\pi) = \{q_1 \xrightarrow{b_1} q_1',\ q_2 \xrightarrow{b_2} p_2'\}$. Here R contains $(p_0, q_1), (p_0', q_1'), (p_2, q_1), (p_2', q_1'), (p_3, q_1), (p_3', q_1'), (p_3, q_2), (p_3', q_2')$. Also, $(a, b_1), (a_3, b_2) \in \rho$ and $(a_2, b_1), (a_3, b_1) \in \sigma$.

Definition 18 *Given a relation $R \subseteq nodes(g) \times nodes(h)$ on two process graphs $g, h \in \mathbb{G}$, a pair of relations on observables ρ, σ and a transition $\pi_g = p_o \xrightarrow{a} p_o' \in trans(g)$ we define the* ***partial saturation of** g **and** h **with respect to** π_g*, written as $g[\pi_g]$ and $h[\pi_g]$, as the process graphs that have the same set of nodes and root as g and h respectively, while the set of transitions are defined as follows*

1. $trans(g[\pi_g]) = trans(g) \cup \{p \xrightarrow{a} p' \mid p \xrightarrow{c} p' \in T(\pi_g), c \in \mathcal{O}\}$
2. $trans(h[\pi_g]) = trans(h) \cup \{q \xrightarrow{a} q' \mid q \xrightarrow{b} q' \in U(\pi_g), b \in \mathcal{O}\}$

A better intuition might be obtained by viewing this as a construction of $g[\pi_g]$ and $h[\pi_g]$ from g and h. We can refer to $g[\pi_g]$ as g' and $h[\pi_g]$ as h' as the transition in question $\pi_g = p_o \xrightarrow{a} p_o'$ is clear. Start with $g' := g$ and $h' := h$. The following steps correspond to adding the subgraph $a.p'$ at the node p if some transition $p \xrightarrow{c} p'$ is a closure match of π_g in g, and $a.q'$ at the node q if some transition $q \xrightarrow{b} q'$ is a closure match of π_g in h.

1. $\forall p \xrightarrow{c} p' \in T(p_o \xrightarrow{a} p_o')$, add $p \xrightarrow{a} p'$ in g'
2. $\forall q \xrightarrow{b} q' \in U(p_o \xrightarrow{a} p_o')$, add $q \xrightarrow{a} q'$ in h'

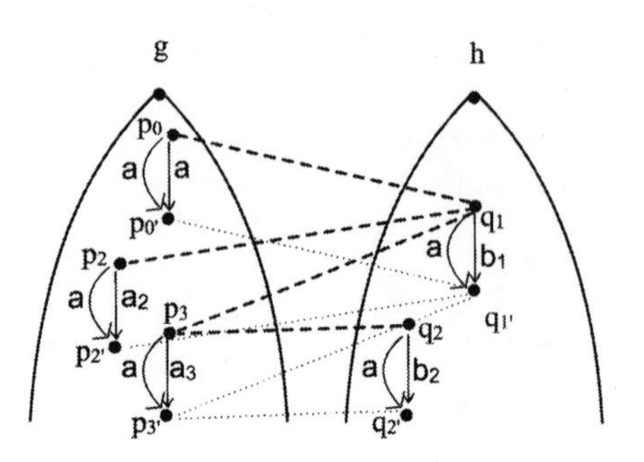

Figure 11: *Construction of* $g[p_0 \xrightarrow{a} p_0']$ *and* $h[p_0 \xrightarrow{a} p_0']$

Following the partial saturation process for the example of figure 10, we show the construction of $g[p_0 \xrightarrow{a} p_0']$ and $h[p_0 \xrightarrow{a} p_0']$ in figure 11. The axiom AV_ρ justifies each individual addition of this kind and we can break this construction into many single transition additions. However, in the presence of empty observables, it must be noted that it might not be possible to add an edge labelled by an arbitrary observable (see lemma 4 part 3). The inability to add edges at will, results in the failure of the various saturation results in this section.

Definition 19 *If* $v \in \mathcal{O}$ *and* $g \in \mathbb{G}$, *we define the* ***v-application of a transition*** $p \xrightarrow{w} p'$ ***to the graph*** g, *written* $g_v\langle p \xrightarrow{w} p'\rangle$, *as the graph obtained by adding the transition* $p \xrightarrow{v} p'$ *to* g.

We extend the definition to a set of transitions by the inductive definition $g_v\langle\{\pi\}\rangle = g_v\langle\pi\rangle$ *and* $g_v\langle G \cup \{\pi\}\rangle = g_v\langle\pi\rangle\langle G\rangle$.

Lemma 20 *If* ρ *is a preorder,* $h \in \mathbb{G}, a \in \mathcal{O}, q \xrightarrow{b} q' \in trans(h)$ *and* $a\rho b$, *then*

$$id_h : h \;\underline{\Box}_{(\equiv,\rho^{-1})}\; h_a\langle q \xrightarrow{b} q'\rangle$$

⊣

Lemma 21 *If* $g, h \in \mathbb{G}, R : g\underline{\Box}_{(\rho,\rho^{-1})}h$ *and* $\pi = p_o \xrightarrow{a} p_o' \in trans(g)$ *then*

1. $g_a\langle T(\pi)\rangle \equiv g[\pi]$
2. $h_a\langle U(\pi)\rangle \equiv h[\pi]$

⊣

Lemma 22 *The following properties of $\mathcal{Q}$ and $\mathcal{P}$ follow trivially from their definitions*

1. $\forall q \xrightarrow{b} q' \in \mathcal{Q}(T) : \exists p \xrightarrow{c} p' \in T : c\rho b$
2. $\forall p \xrightarrow{c} p' \in \mathcal{P}(U) : \exists q \xrightarrow{b} q' \in U : b\sigma^{-1}c$

$\dashv$

Lemma 23 *If $\rho \circ \sigma^{-1} \subseteq \rho$, $\rho \circ \rho \subseteq \rho$ (ρ is transitive) and $\equiv \subseteq \rho$ (ρ is reflexive) then $\forall i \geq 0$,*

1. $\forall\, p \xrightarrow{c} p' \in T_i(p_o \xrightarrow{a} p_o') : a\rho c$
2. $\forall\, q \xrightarrow{b} q' \in U_i(p_o \xrightarrow{a} p_o') : a\rho b$

$\dashv$

Lemma 23 says that, under the stated assumptions, the labels of all closure matches of a transition $p_o \xrightarrow{a} p_o' \in trans(g)$ are related to a through ρ. This can be proven by simultaneous induction on i.

Note that we will be dealing only with $\mathbb{R}$ in further sections although some results in this section are valid for $\mathbb{G}$. Therefore in sections 7.2 and 8, $trans(g)$ and $trans(h)$ are finite and hence $T(\pi_g) \subseteq trans(g)$ and $U(\pi_g) \subseteq trans(h)$ are also finite. Thus the partial saturations $g[\pi_g]$ and $h[\pi_g]$ are also finite process graphs.

7.2 Properties of Saturation

Lemma 24 (Partial Saturation)
For finite process graphs $g, h \in \mathbb{R}$ and a relation R such that $R : g \,\underline{\Box}_{(\rho,\sigma)}\, h$ where ρ, σ are such that $\equiv \subseteq \rho, \sigma$ (both are reflexive), $\rho \circ \rho \subseteq \rho$ (ρ is transitive) and $\rho \circ \sigma^{-1} \subseteq \rho$, the following diagram holds for all $\pi_g = p_o \xrightarrow{a} p_o' \in trans(g)$.

$$\begin{array}{ccc} g & \underline{\Box}_{(\rho,\sigma)} & h \\ \underline{\Box}_{(\equiv,\rho^{-1})} & & \underline{\Box}_{(\equiv,\rho^{-1})} \\ g[\pi_g] & \underline{\Box}_{(\rho,\sigma)} & h[\pi_g] \end{array}$$

Moreover,

1. $id_g : g \,\underline{\Box}_{(\equiv,\rho^{-1})}\, g[\pi_g]$

2. $id_h : h \sqsubseteq_{(\equiv,\rho^{-1})} h[\pi_g]$
3. $R : g[\pi_g] \sqsubseteq_{(\rho,\sigma)} h[\pi_g]$
4. (a) *For all* q *in* h' *such that* $(p_o, q) \in R$, $\exists q' : q \xrightarrow{a} q' \in h' \wedge (p'_o, q') \in R$
   (b) *For all transitions* $p \xrightarrow{a} p'$ *in* g' *that are newly added* [§] *and for all* q *in* h' *such that* $(p,q) \in R$, $\exists q' : q \xrightarrow{a} q' \in h' \wedge (p', q') \in R$
   (c) *For all transitions* $q \xrightarrow{a} q'$ *in* h' *that are newly added and for all* p *in* g' *such that* $(p,q) \in R$, $\exists p' : p \xrightarrow{a} p' \in g' \wedge (p', q') \in R$

Proof The proofs of parts 1 and 2 follow from lemma 20 and 21. If $U(\pi_g) = \{\pi_1, \ldots, \pi_n\}$ then from lemma 20 we get

$$h \sqsubseteq_{(\equiv,\rho^{-1})} h_a\langle\pi_1\rangle \sqsubseteq_{(\equiv,\rho^{-1})} \cdots \sqsubseteq_{(\equiv,\rho^{-1})} h_a\langle\pi_1\rangle \ldots \langle\pi_n\rangle$$

From the transitivity of $\sqsubseteq_{(\equiv,\rho^{-1})}$, the observation that by definition $h_a\langle\pi_1\rangle \ldots \langle\pi_n\rangle \equiv h_a\langle U(\pi)\rangle$, and from lemma 21, we may show $id_h : h \sqsubseteq_{(\equiv,\rho^{-1})} h[\pi_g]$. In a similar manner we prove part 1. Part 3 can be seen by noting $id_g^{-1} \circ R \circ id_h : g' \sqsubseteq_{(\rho,\sigma)} h'$ and $id_g^{-1} \circ R \circ id_h = R$.

To prove part 4.4c, consider the newly added transition $\pi_{h'} = q \xrightarrow{a} q'$ in $h' = h[\pi_g]$. Since $\pi_{h'}$ is newly added, it has been added by an a-application of some $q \xrightarrow{b} q' \in U(\pi_g)$ since $h[\pi_g] = h_a\langle U(\pi_g)\rangle$. Since $q \xrightarrow{b} q' \in U_i(\pi_g)$ for some i, we get $a\rho b$ by lemma 23. By the fact that $R : g \sqsubseteq_{(\rho,\sigma)} h$, for any p such that $(p,q) \in R$ we know that $\exists p', c[p \xrightarrow{c} p' \in g \wedge c\rho b \wedge (p', q') \in R]$. From the definition of T_{i+1}, we know that $p \xrightarrow{c} p' \in T_{i+1}(\pi_g) \subseteq T(\pi_g)$. The fact that g' is obtained from g by the a-application of $T(\pi_g)$ gives us $p \xrightarrow{a} p' \in g'$ as required. This proves part 4.4c. The part 4.4b is proven on symmetric lines.

For part 4.4a, notice that from $R : g \sqsubseteq_{(\rho,\sigma)} h$ and $(p_o, q) \in R$ we know that $\exists q', b[q \xrightarrow{b} q' \in g \wedge a\rho b \wedge (p'_o, q') \in R]$. On similar lines as the proof for part 4.4c, note that $q \xrightarrow{b} q' \in U_i(\pi_g)$ for some

[§]Newly added transitions in g' and h' are $trans(g') - trans(g)$ and $trans(h') - trans(h)$ respectively.

i. Thus the graph h' which is obtained by an a-application of $U(\pi_g)$ also contains the transition $q \xrightarrow{a} q'$ as required.

⊣

The construction of $g[p_0 \xrightarrow{a} p_0']$ and $h[p_0 \xrightarrow{a} p_0']$ is intended to make g and h closer in terms of bisimilarity. For the transition $p_0 \xrightarrow{a} p_0'$ in g, we add an edge $\xrightarrow{a}$ at the h matches of $p_0 \xrightarrow{a} p_0'$ to make sure that at least the transition $p_0 \xrightarrow{a} p_0'$ can be matched by the identity relation $\equiv$ on observables. This is captured by lemma 24.4a. These additions correspond to the additions due to U_0. For the example considered in figure 10 in section 7.1, we get the figure 12 upon the additions due to U_0.

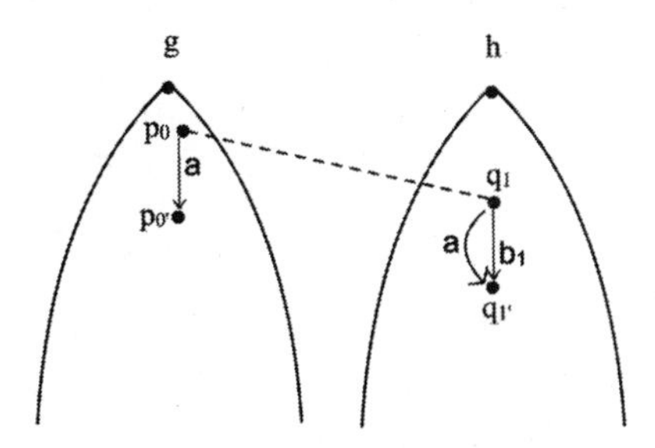

Figure 12: *Additions due to U_0*

This addition triggers off the need to add corresponding transitions (as given by T_1) in g to ensure that the newly added transitions in h also get matched in g by the identity on observables.

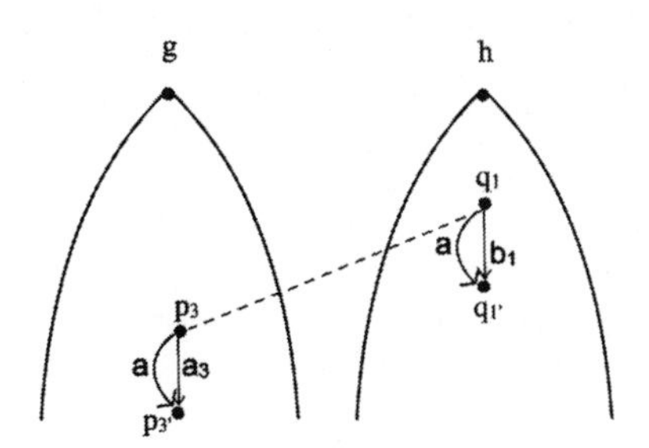

Figure 13: *An addition due to T_1*

In figure 13 we show one of the additions due to T_1. This addition, in turn, further triggers the need for additions due to U_1 as shown in figure 14.

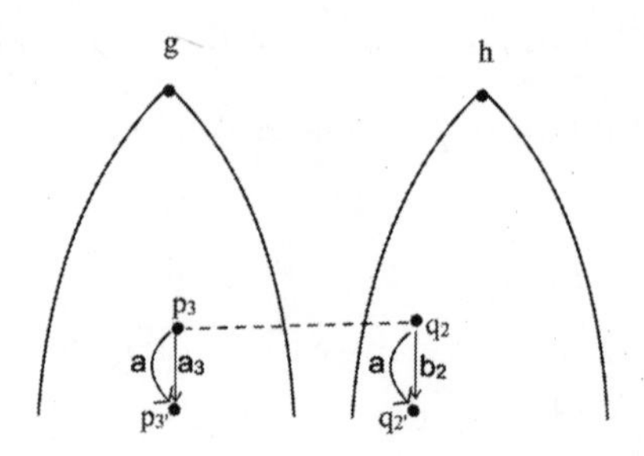

Figure 14: *An addition due to* U_1

Proceeding in this manner, in the end the transition $p_0 \xrightarrow{a} p_0'$ and all the newly added transitions end up being matched by $\equiv$ (identity on observables) in figure 11. This is captured by lemma 24.4b, 24.4c. Repeating the procedure for each transition $\pi \in g$ we end up with a g' and h' where each transition in g', old or new, is matched by transitions in h by $\equiv$. In this manner we can reduce $\sqsubseteq_{(\rho,\sigma)}$ to $\sqsubseteq_{(\equiv,\sigma)}$.

Definition 25 *Given a relation R, a pair of relations ρ,σ and process graphs $g,h \in \mathbb{R}$ such that $R : g \sqsubseteq_{(\rho,\sigma)} h$, we define the* ***partial saturation of*** g ***and*** h ***with respect to the set*** $G \subseteq trans(g)$*, written as $g[G]$ and $h[G]$, as follows.*

1. $g[\emptyset] = g, h[\emptyset] = h$
2. $g[G \cup \{\pi_g\}] = (g[G])[\pi_g], h[G \cup \{\pi_g\}] = (h[G])[\pi_g]$

Lemma 26 (Semi Saturation)
For finite process graphs $g,h \in \mathbb{R}$ and a relation R such that $R : g \sqsubseteq_{(\rho,\sigma)} h$ where ρ,σ are such that $\equiv \subseteq \rho,\sigma$ (both are reflexive), $\rho \circ \rho \subseteq \rho$ (ρ is transitive) and $\rho \circ \sigma^{-1} \subseteq \rho$, the following diagram holds.

$$\begin{array}{ccc} g & \sqsubseteq_{(\rho,\sigma)} & h \\ \sqsubseteq_{(\equiv,\rho^{-1})} & & \sqsubseteq_{(\equiv,\rho^{-1})} \\ g' & \sqsubseteq_{(\equiv,\sigma)} & h' \end{array}$$

where $g' = g[trans(g)]$ and $h' = h[trans(g)]$. The following hold as well.

1. $id_g : g \sqsubseteq_{(\equiv,\rho^{-1})} g'$
2. $id_h : h \sqsubseteq_{(\equiv,\rho^{-1})} h'$
3. $R : g' \sqsubseteq_{(\equiv,\sigma)} h'$

Proof The key to the proof of the semi saturation lemma is the "tiling" of the partial saturation diagram. From the partial saturation lemma 24, if $R : g[G_k] \sqsubseteq_{(\rho,\sigma)} h[G_k]$ for some $G_k \subseteq trans(g)$ then for $\pi_k \in trans(g)$ we get the following diagram

$$\begin{array}{ccc} g[G_k] & \sqsubseteq_{(\rho,\sigma)} & h[G_k] \\ \sqsubseteq_{(\equiv,\rho^{-1})} & & \sqsubseteq_{(\equiv,\rho^{-1})} \\ g[G_k][\pi_k] & \sqsubseteq_{(\equiv,\sigma)} & h[G_k][\pi_k] \end{array}$$

It helps to note that $trans(g) \subseteq trans(g[G_k])$ and therefore $\pi_k \in trans(g[G_k])$ as well and the above diagram comes directly from partial saturation lemma. Enumerating $trans(g) = \{\pi_1, \pi_2, \ldots, \pi_n\}$, we define $G_k = \{\pi_i \in trans(g) | i < k\}$. It should be obvious that $G_1 = \emptyset, G_{k+1} = G_k \cup \{\pi_k\}$ and $i < j \implies G_i \subset G_j$. Since $G_1 = \emptyset$, we get $g[G_1] \equiv g$ and $h[G_1] \equiv h$. Therefore, from the knowledge that $R : g \sqsubseteq_{(\rho,\sigma)} h$, the above diagram holds for $k = 1$. Note that $g[G_k][\pi_k] = g[G_k \cup \{\pi_k\}] = g[G_{k+1}]$ and similarly for h. By induction on k, we can show that the diagram holds $\forall k \geq 1$. By a composition of these diagrams, we can get the following diagram.

$$\begin{array}{ccc} g[G_1] & \sqsubseteq_{(\rho,\sigma)} & h[G_1] \\ \sqsubseteq_{(\equiv,\rho^{-1})} & & \sqsubseteq_{(\equiv,\rho^{-1})} \\ g[G_2] & \sqsubseteq_{(\rho,\sigma)} & h[G_2] \\ \sqsubseteq_{(\equiv,\rho^{-1})} & & \sqsubseteq_{(\equiv,\rho^{-1})} \\ \vdots & \vdots & \vdots \\ \sqsubseteq_{(\equiv,\rho^{-1})} & & \sqsubseteq_{(\equiv,\rho^{-1})} \\ g[G_{n+1}] & \sqsubseteq_{(\rho,\sigma)} & h[G_{n+1}] \end{array}$$

Note that since $G_{n+1} = trans(g)$, the last terms we obtain in this process are $g[trans(g)]$ and $h[trans(g)]$ as desired in the lemma. Using transitivity of $\sqsubseteq_{(\equiv,\rho^{-1})}$ we (nearly) get the diagram of semi saturation lemma. The other observation that g' and h' are related by $\sqsubseteq_{(\equiv,\sigma)}$ comes from using part 4 of the lemma 24 which implies that having each transition in g and all new ones in g' being matched by $\equiv$ in h', all transitions in g' end up matched by $\equiv$ in h'. For proving this, we first prove two subproofs.

1. Consider $\pi_{g'} = p \xrightarrow{a_k} p' \in g[G_{n+1}] - g[G_1]$ such that π'_g was added in the construction of $g[G_k][\pi_k]$ from $g[G_k]$ for

some k, with $\pi_k = p_k \xrightarrow{a_k} p'_k$. From lemma 24 part 4.4b, we know that $\forall (p,q) \in R\ \exists q' : q \xrightarrow{a_k} q' \in h[G_k][\pi_k] \wedge p'Rq'$. Since $trans(h[G_k][\pi_k]) \subseteq trans(h[G_{n+1}])$, $q \xrightarrow{a_k} q'$ is also a transition in $h[G_{n+1}]$.

2. If $\pi_g \in g[G_1]$ then $\pi_g = \pi_k = p_k \xrightarrow{a_k} p'_k$ for some k. In the construction of $h[G_k][\pi_k]$ from $h[G_k]$, we can use lemma 24 part 4.4a to see that $\forall (p_k,q) \in R\ \exists q' : q \xrightarrow{a_k} q' \in h[G_k][\pi_k] \wedge p'_kRq'$. Again, $q \xrightarrow{a_k} q'$ is also a transition in $h[G_{n+1}]$.

As a result of the above two subproofs, we know that for for all $(p,q) \in R$ and any $\pi = p \xrightarrow{c} p' \in g[G_{n+1}]$, there is a transition $q \xrightarrow{c} q'$ in $h[G_{n+1}]$ such that $(p',q') \in R$. This proves the first clause of the definition of $R : g' \underline{\Box}_{(\equiv,\rho^{-1})} h'$ where $g' = g[G_{n+1}]$ and $h' = h[G_{n+1}]$. The second part trivially follows from $R : g \underline{\Box}_{(\rho,\rho^{-1})} h$.

$\dashv$

By another application of the semi saturation lemma we can have every transition in h' also matched by $\equiv$ to some transition in g'. Thus we can reduce $\underline{\Box}_{(\equiv,\sigma)}$ further to $\underline{\Box}_{(\equiv,\equiv)}$, which is strong bisimilarity.

Lemma 27 (Saturation)
For finite process graphs $g, h \in \mathbb{R}$ and a relation R such that $R : g \underline{\Box}_{(\rho,\sigma)} h$, where ρ, σ are preorders and $\rho \circ \sigma^{-1} \subseteq \rho$, there exist derivatives g', h', g'', h'' such that the following diagram holds.

$$\begin{array}{ccc}
g & \underline{\Box}_{(\rho,\sigma)} & h \\
\underline{\Box}_{(\equiv,\rho^{-1})} & & \underline{\Box}_{(\equiv,\rho^{-1})} \\
g' & \underline{\Box}_{(\equiv,\sigma)} & h' \\
\underline{\Box}_{(\equiv,\sigma)} & & \underline{\Box}_{(\equiv,\sigma)} \\
g'' & \underline{\Box}_{(\equiv,\equiv)} & h''
\end{array}$$

Moreover, we have

1. $id_g : g\ \underline{\Box}_{(\equiv,\rho^{-1})} g'$ *and* $id_g : g'\ \underline{\Box}_{(\equiv,\sigma)} g''$
2. $id_h : h\ \underline{\Box}_{(\equiv,\rho^{-1})} h'$ *and* $id_h : h'\ \underline{\Box}_{(\equiv,\sigma)} h''$
3. $R : g'\ \underline{\Box}_{(\equiv,\sigma)} h'$ *and* $R : g''\ \underline{\Box}_{(\equiv,\equiv)} h''$

Proof The first part (g and h to g' and h') follows directly from lemma 26. We get $R : g' \underline{\square}_{(\equiv,\sigma)} h'$. By proposition 2.2, we get R^{-1} : $h' \underline{\square}_{(\sigma^{-1},\equiv)} g'$. Since $\equiv$ and σ^{-1} are both reflexive and transitive and we also have $\sigma^{-1} \circ \equiv^{-1} \subseteq \sigma^{-1}$, we use lemma 26 to get $R^{-1} : h'' \underline{\square}_{(\equiv,\equiv)} g''$ which is the same as $R : g'' \underline{\square}_{(\equiv,\equiv)} h''$. Parts 1, 2 and 3 follow easily as consequences of corresponding parts in lemma 26.

⊣

Corollary 28 *For finite process graphs $g, h \in \mathbb{R}$ and a relation R such that $R : g \underline{\square}_{(\rho,\sigma)} h$, where ρ, σ are preorders and $\sigma^{-1} \subseteq \rho$, there exist derivatives g'', h'' such that the following diagram holds.*

$$\begin{array}{ccc} g & \underline{\square}_{(\rho,\sigma)} & h \\ \underline{\square}_{(\equiv,\rho^{-1})} & & \underline{\square}_{(\equiv,\rho^{-1})} \\ g'' & \underline{\square}_{(\equiv,\equiv)} & h'' \end{array}$$

Also

1. $id_g : g \underline{\square}_{(\equiv,\rho^{-1})} g''$
2. $id_h : h \underline{\square}_{(\equiv,\rho^{-1})} h''$
3. $R : g'' \underline{\square}_{(\equiv,\equiv)} h''$

⊣

In the above corollary, we have shown how to obtain strongly bisimilar g'' and h'' from g and h by traversing down $\underline{\square}_{(\equiv,\rho^{-1})}$. This result is not restricted to only the cases where $\sigma^{-1} \subseteq \rho$. In the symmetrical case where $\rho^{-1} \subseteq \sigma$, we can also obtain strongly bisimilar graphs¶.

Corollary 29 *For finite process graphs $g, h \in \mathbb{R}$ and a relation R such that $R : g \underline{\square}_{(\rho,\sigma)} h$, where ρ, σ are preorders and $\rho^{-1} \subseteq \sigma$, there exist derivatives g'', h'' such that the following diagram holds.*

$$\begin{array}{ccc} g & \underline{\square}_{(\rho,\sigma)} & h \\ \underline{\square}_{(\equiv,\sigma)} & & \underline{\square}_{(\equiv,\sigma)} \\ g'' & \underline{\square}_{(\equiv,\equiv)} & h'' \end{array}$$

¶$\sigma^{-1} \subseteq \rho$ or $\rho^{-1} \subseteq \sigma$ is not an unusual condition. Simulation, strong bisimulation and observational equivalence have ρ and σ such that one of these conditions hold [5] .

Moreover,

1. $id_g : g \,\underline{\Box}_{(\equiv,\sigma)} g''$
2. $id_h : h \,\underline{\Box}_{(\equiv,\sigma)} h''$
3. $R : g'' \,\underline{\Box}_{(\equiv,\equiv)} h''$

$\dashv$

Theorem 30 *For finite process graphs $g, h \in \mathbb{R}$ and a relation R such that $R : g \underline{\Box}_{(\rho,\sigma)} h$, where ρ, σ are preorders and $\rho^{-1} = \sigma$, there exist derivatives g'', h'' such that the following diagram holds.*

$$\begin{array}{ccc} g & \underline{\Box}_{(\rho,\rho^{-1})} & h \\ \underline{\Box}_{(\equiv,\rho^{-1})} & & \underline{\Box}_{(\equiv,\rho^{-1})} \\ g'' & \underline{\Box}_{(\equiv,\equiv)} & h'' \end{array}$$

The following hold as well

1. $id_g : g \,\underline{\Box}_{(\equiv,\rho^{-1})} g''$
2. $id_h : h \,\underline{\Box}_{(\equiv,\rho^{-1})} h''$
3. $R : g'' \,\underline{\Box}_{(\equiv,\equiv)} h''$
4. $S : g \underline{\Box}_{(\rho,\rho^{-1})} h$ *where* $S = id_g \circ R \circ id_h^{-1}$

$\dashv$

Theorem 30.4 says that the diagram commutes. Using proposition 2.3, we can show that $id_g \circ R \circ id_h^{-1} : g \underline{\Box}_{(\equiv \circ \equiv \circ \rho, \rho^{-1} \circ \equiv \circ \equiv)} h$. So we have another relation $S = id_g \circ R \circ id_h^{-1}$ such that $S : g \underline{\Box}_{(\rho,\rho^{-1})} h$.

That the diagram commutes is critical for the completeness proof where we are able to prove $g \,\underline{\Box}_{(\rho,\rho^{-1})} g''$, $h \,\underline{\Box}_{(\rho,\rho^{-1})} h''$ and $h'' \,\underline{\Box}_{(\rho,\rho^{-1})} g''$ within the framework of the axiomatization and use the transitivity of $\underline{\Box}_{(\rho,\rho^{-1})}$ to prove $g \,\underline{\Box}_{(\rho,\rho^{-1})} h$ within the framework of the axiomatization. The reason we are unable to prove completeness for preorders is because when $\sigma \neq \rho^{-1}$ the diagram does not necessarily commute.

8 Completeness of $\mathscr{A}_{(\rho,\rho^{-1})}$

The proof of completeness of $\mathscr{A}_{(\rho,\rho^{-1})}$ hinges on reducing the two graphs to be compared g and h, to graphs g' and h' such that $g \underline{\Box}_{(\rho,\rho^{-1})} g' \wedge$

$h \underline{\square}_{(\rho,\rho^{-1})} h' \wedge g' \underline{\square}_{(\equiv,\equiv)} h' \Longleftrightarrow g \underline{\square}_{(\rho,\rho^{-1})} h$. That is, the graphs g' and h' turn out to be strongly bisimilar if and only if g and h were (ρ,ρ^{-1})-bisimilar. This relation between g, g', h, h' can also be shown diagrammatically in the following manner.

$$\begin{array}{ccc} g & \underline{\square}_{(\rho,\rho^{-1})} & h \\ \underline{\square}_{(\rho,\rho^{-1})} & & \underline{\square}_{(\rho,\rho^{-1})} \\ g' & \underline{\square}_{(\equiv,\equiv)} & h' \end{array}$$

The strong bisimilarity of g' and h' can be checked within the framework of BPA_{LR} as it is complete, while the relation between the pairs g, g' and h, h' can be proven using the axiom schema AV_ρ or AV.

Lemma 31 *For any graph $h \in \mathbb{R}$, transition $\pi = q \xrightarrow{b} q' \in trans(h)$ and $a \rho b$,*

$$\mathscr{A}_{(\rho,\rho^{-1})} \vdash ||h|| =_\rho ||h_a\langle q \xrightarrow{b} q'\rangle||$$

Proof Let $||h|| \equiv \langle X_1 | E\rangle$. Let X_k and X_j be the variables that correspond to the node q and q' respectively in the construction of $||h||$ (see lemma 15). Then T_k must be of the form $b.X_j + T$ for some term T. Thus $||h|| \equiv \langle X_1 | X_k = b.X_j + T, E_{-k}\rangle$. Consider the process graph $h_a\langle q \xrightarrow{b} q'\rangle$ which is derived from h by adding a single transition $q \xrightarrow{a} q'$ in the graph h, where $a \rho b$. Following the algorithm of $||\ ||$ given in lemma 15, it should be easy to see that $||h_a\langle q \xrightarrow{b} q'\rangle|| \equiv \langle X_1 | X_k = a.X_j + b.X_j + T, E_{-k}\rangle$. Thus in order to prove

$$\mathscr{A}_{(\rho,\rho^{-1})} \vdash ||h|| =_\rho ||h_a\langle q \xrightarrow{b} q'\rangle||$$

we need to prove $\begin{array}{lll} \mathscr{A}_{(\rho,\rho^{-1})} & \vdash & \langle X_1 | X_k = b.X_j + T, E_{-k}\rangle =_\rho \\ & & \langle X_1 | X_k = a.X_j + b.X_j + T, E_{-k}\rangle \end{array}$

Since $a \rho b$, the above is a direct application of AV (or application of AV_ρ, $A4_\rho$, congruence and substitution rule) and is hence provable.

Theorem 32 (Completeness)
If $g, h \in \mathbb{R}$ and $\rho \subseteq \mathscr{O} \times \mathscr{O}$ is a preorder where $\mathscr{O}$ has no empty observables, then

$$g \,\underline{\square}_{(\rho,\rho^{-1})} h \Longrightarrow \mathscr{A}_{(\rho,\rho^{-1})} \vdash ||g|| =_\rho ||h||$$

Proof If $g \sqsubseteq_{(\rho,\rho^{-1})} h$, there exists [5] a relation R such that $R : g \sqsubseteq_{(\rho,\rho^{-1})} h$. For $\pi_g = p_o \xrightarrow{a} p'_o \in trans(g)$ we can construct $T(\pi_g)$ and $U(\pi_g)$ as shown in section 7. Since it is finite, let $U(\pi_g) = \{\pi_1, \pi_2, \ldots \pi_n\}$ for some n. From lemma 23 we know that for any $q \xrightarrow{b} q' \in U(\pi_g)$, we have $a\rho b$. Hence by applying lemma 31 we have[||]

$$\mathcal{A}_{(\rho,\rho^{-1})} \vdash ||h|| =_\rho ||h_a\langle\pi_1\rangle||$$

$$\mathcal{A}_{(\rho,\rho^{-1})} \vdash ||h_a\langle\pi_1\rangle|| =_\rho ||h_a\langle\pi_1\rangle\langle\pi_2\rangle||$$

$$\mathcal{A}_{(\rho,\rho^{-1})} \vdash ||h_a\langle\pi_1\rangle\ldots\langle\pi_{k-1}\rangle|| =_\rho ||h_a\langle\pi_1\rangle\ldots\langle\pi_k\rangle||$$

Using the transitivity of $=_\rho$ (being an equational axiomatization $\mathcal{A}_{(\rho,\rho^{-1})}$ has the appropriate axioms for transitivity, reflexivity and symmetry but for brevity these are not mentioned) and the definition of application of a set of transitions (definition 19), we get

$$\mathcal{A}_{(\rho,\rho^{-1})} \vdash ||h|| =_\rho ||h_a\langle U_i(\pi_g)\rangle||$$

Similarly, let $T(\pi_g) = \{\chi_1, \chi_2, \ldots \chi_m\}$. From lemma 23 we know that for any $p \xrightarrow{c} p' \in T(\pi_g)$, we have $a\rho c$. Hence by lemma 31,

$$\mathcal{A}_{(\rho,\rho^{-1})} \vdash ||g|| =_\rho ||g_a\langle\chi_1\rangle||$$

$$\mathcal{A}_{(\rho,\rho^{-1})} \vdash ||g_a\langle\chi_1\rangle|| =_\rho ||g_a\langle\chi_1\rangle\langle\chi_2\rangle||$$

$$\mathcal{A}_{(\rho,\rho^{-1})} \vdash ||g_a\langle\chi_1\rangle\ldots\langle\chi_{m-1}\rangle|| =_\rho ||g_a\langle\chi_1\rangle\ldots\langle\chi_m\rangle||$$

Again, using the transitivity of $=_\rho$ we get

$$\mathcal{A}_{(\rho,\rho^{-1})} \vdash ||g|| =_\rho ||g_a\langle T_i(\pi_g)\rangle||$$

From lemma 21 we have $g_a\langle T_i(\pi_g)\rangle \equiv g[\pi_g]$ and $h_a\langle U_i(\pi_g)\rangle \equiv h[\pi_g]$ and thus,

$$\mathcal{A}_{(\rho,\rho^{-1})} \vdash ||h|| =_\rho ||h[\pi_g]||$$

[||]Note that $trans(h) \subseteq h_a\langle q \xrightarrow{b} q'\rangle$ and hence if a transition is in h, it is also in $h_a\langle q \xrightarrow{b} q'\rangle$. For instance the transition π_2 is also in $h_a\langle\pi_1\rangle$ and hence lemma 31 can be applied to it.

$$\mathscr{A}_{(\rho,\rho^{-1})} \vdash ||g|| =_\rho ||g[\pi_g]||$$

However, we are unable to prove the following even though from partial saturation lemma 24, we know $g[\pi_g] \,\underline{\square}_{(\rho,\rho^{-1})} h[\pi_g]$ to be true.

$$\mathscr{A}_{(\rho,\rho^{-1})} \vdash ||g[\pi_g]|| =_\rho ||h[\pi_g]||$$

Thus, the vertical relations between $g, g[\pi_g], h, h[\pi_g]$ in the diagram of partial saturation lemma (with the additional restriction $\sigma = \rho^{-1}$) can be proven within the context of $\mathscr{A}_{(\rho,\rho^{-1})}$ while the parts 3 and 4, which claim that $g[\pi_g]$ and $h[\pi_g]$ are related by $\underline{\square}_{(\rho,\rho^{-1})}$ and in fact are "closer in terms of strong bisimilarity", cannot be proven in this manner directly. To be accurate, we have only been able to prove $g\underline{\square}_{(\rho,\rho^{-1})}g'$ whereas the relation between g and g' in partial saturation lemma is $g\underline{\square}_{(\equiv,\rho^{-1})}g'$. However, for our purposes this will prove adequate.

Since the diagram of semi saturation lemma 26 is constructed by tiling up partial saturation lemma diagrams, each vertical relation in the semi saturation lemma diagram can also be proven within the context of $\mathscr{A}_{(\rho,\rho^{-1})}$ but we cannot prove $||g'|| =_\rho ||h'||$ although $g'\underline{\square}_{(\equiv,\rho^{-1})}h'$ is true. To be precise, we are able to prove $||g[G_k]|| =_\rho ||g[G_k][\pi_k]||$ within the framework for every $k \geq 1$, where G_k and π_k are as defined in the proof outline of semi saturation lemma. Using transitivity, we can prove $||g[G_1]|| =_\rho ||g[G_{n+1}]||$ and since $g[G_1] \equiv g, g[G_{n+1}] \equiv g'$, we get our desired result.

By similar reasoning, the vertical relations in the saturation lemma 27 diagram which are obtained by tiling two semi saturation lemma diagrams can also be proven within the context of the axiomatization. Once again we know $g''\underline{\square}_{(\equiv,\equiv)}h''$ to be true but cannot prove it in this manner. We now rely on BPA_{LR}, which is a complete axiomatization for $\underline{\square}_{(\equiv,\equiv)}$ and therefore the following can be proven in its framework

$$BPA_{LR} \vdash ||g''|| = ||h''||$$

By a translation of the above proof,

$$BPA_{\rho LR} \vdash ||g''|| =_\rho ||h''||$$

Since $BPA_{\rho LR} \subset \mathscr{A}_{(\rho,\rho^{-1})}$, we can also prove

$$\mathscr{A}_{(\rho,\rho^{-1})} \vdash ||g''|| =_\rho ||h''||$$

Now we use the fact that the diagram of theorem 30 commutes. Using symmetry of $=_\rho$, and putting together the relations between g, g'', h, h'' proven above, we get

$$\mathscr{A}_{(\rho,\rho^{-1})} \vdash \|g\| \;=_\rho\; \|g''\| \;=_\rho\; \|h''\| \;=_\rho\; \|h\|$$

By transitivity of $=_\rho$, we get

$$\mathscr{A}_{(\rho,\rho^{-1})} \vdash \|g\| =_\rho \|h\|$$

$\dashv$

We have proven $g \underline{\square}_{(\rho,\rho^{-1})} h$ in the framework of the axiomatization. Hence $\mathscr{A}_{(\rho,\rho^{-1})}$ is sound complete axiomatization of $\underline{\square}_{(\rho,\rho^{-1})}$ for finite process graphs, where ρ is an axiomatizable preorder on a set of observables that excludes all empty observables.

9 Conclusion

Under the assumption that the set of observables does not contain any empty observable, and relative to an axiomatization for the relation on observables ρ, we have been able to provide an axiomatization that is complete for all equivalence relations in the class of (ρ, σ)-bisimilarities. The axiomatization builds upon an existing axiomatization for strong bisimilarity [6], and after a simple translation of this, only requires one additional axiom AV_ρ. The intuition of this axiom is similar to that of $x = x + x$ which can prove that strong bisimilarity is insensitive to addition of a duplicate edge. AV_ρ says that $\underline{\square}_{(\rho,\rho^{-1})}$ is insensitive to the addition of all edges that are related by ρ to an existing edge.

We have also shown that under the absence of an empty observable, $\underline{\square}_{(\rho,\rho^{-1})}$ also turns out to be a congruence. In the presence of empty observables, a study of the nature of observables that are related to an empty action by ρ and σ might lead to a characterization of the smallest precongruence or congruence containing a given (ρ, σ) preorder or equivalence bisimilarity. Such observables are termed ρ-preemptive and σ-preemptive as they may occur without being matched by any observable (other than an empty action) of a related process which ends up preserving choice in contrast to the process that does a preemptive action.

The generalization presented in this paper cannot provide an axiomatization of observational equivalence because of the presence of empty observables in its characterization as $\underline{\square}_{(\hat{=}, \hat{=})}$, which was presented in [5]. The other problem is that while $\approx$ might be considered on finite process graphs, the use of $\mathcal{O} = Act^*$ in the definition of $\underline{\square}_{(\hat{=}, \hat{=})}$ converts the finite process graphs that have cycles into infinitely branching graphs for which our axiomatization has not been proven complete.

Another problem worth inquiring into is the inequational axiomatization of (ρ, σ) precongruences and preorders. Our method of saturation which is similar to the Δ-saturation method in [6] does not preserve the preorder in the sense that the diagram in lemma 27 does not commute unless $\sigma = \rho^{-1}$. A construction of derivatives that leads to a commuting diagram may be the key to finding a complete axiomatization for precongruences. An attempt to find such saturation methods for precongruences such as conformance [4] and efficiency precongruence [3] have led us to characterize some properties of the kernels of these relations. Presented in [12], these observations might also prove useful for an inequational extension of the axiomatization.

References

[1] LUCA ACETO. On relating concurrency and nondeterminism. In Stephen D. Brookes, Michael G. Main, Austin Melton, Michael W. Mislove, and David A. Schmidt, editors, *Mathematical Foundations of Programming Semantics*, Lecture Notes in Computer Science, pages 376–402. Springer-Verlag, Pittsburgh, PA, USA, 1991.

[2] S. ARUN-KUMAR. On bisimilarities induced by relations on actions. In *Proceedings 4th IEEE International Conference on Software Engineering and Formal Methods, Pune, India*. IEEE Computer Society Press, 2006.

[3] S. ARUN-KUMAR AND M. HENNESSY. An efficiency preorder for processes. *Acta Informatica*, 29:737–760, 1992.

[4] S. ARUN-KUMAR AND V. NATARAJAN. Conformance: A precongruence close to bisimilarity. In *STRICT, Berlin 1995*, volume number 526 in Workshops in Computing Series, pages 55–68. Springer-Verlag, 1995.

[5] S. ARUN-KUMAR, AMARINDER SINGH RANDHAWA, AND PRANAV SINGH. Parametrised bisimulations: Some characterisations. submitted.

[6] J.A. BERGSTRA AND J.W. KLOP. A complete inference system for regular processes with silent moves. In *F.R. Drake, J.K. Truss (Editors), Proceedings of Logic Colloquium 1986, North Holland, Amsterdam*, pages 21–81. 1988.

[7] FLAVIO CORRADINI, ROBERTO GORRIERI, AND MARCO ROCCETTI. Performance preorder and competitive equivalence. *Acta Informatica*, 34(11):805–835, 1997.

[8] ROBIN MILNER. A complete axiomatisation for observational congruence of finite-state behaviours. *Information and Computation*, 81(2).

[9] ROBIN MILNER. A complete inference system for a class of regular behaviors. *Journal of Computer and System Sciences*, 28:439–466, 1984.

[10] ROBIN MILNER. *Communication and Concurrency*. Prentice-Hall International, 1989.

[11] ARTO SALOMAA. Two complete axiom systems for the algebra of regular events. *J. ACM*, 13(1):158–169, 1966.

[12] PRANAV SINGH. Axiomatizations of bisimulation based relations. Master's thesis, Indian Institute of Technology, Delhi, Department of Computer Science and Engineering, 2005.

Path Logics with Synchronization

Wolfgang Thomas
RWTH Aachen University, Informatik 7, 52056 Aachen, Germany
thomas@informatik.rwth-aachen.de

Abstract

Over trees and partial orders, chain logic and path logic are systems of monadic second-order logic in which second-order quantification is applied to paths and to chains (i.e., subsets of paths), respectively; accordingly we speak of the path theory and the chain theory of a structure. We present some known and some new results on decidability of the path theory and chain theory of structures that are enhanced by features of synchronization between paths. We start with the infinite two-dimensional grid for which the finite-path theory is shown to be undecidable. Then we consider the infinite binary tree expanded by the binary "equal level predicate" E. We recall the (known) decidability of the chain theory of a regular tree with the predicate E and observe that this does not extend to algebraic trees. Finally, we study refined models in which the time axis (represented by the sequence of tree levels) or the tree levels themselves are supplied with additional structure.

Keywords: Monadic second order logic, path logics, regular trees

1 Introduction

In the verification of distributed systems with nonterminating behaviour, one has to consider all possible computation paths (or "system runs"), and synchronization is captured by a merge or a connection between certain points in different computation paths.

Usually the computation paths are considered as parts of a tree structure in which all paths are collected. For the model of infinite tree (where the branching may be finite or infinite) a powerful theory of verification is available for specifications that are formalizable in MSO-logic (monadic second-order logic). Starting from Rabin's Tree Theorem [9], many beautiful and strong decidability results have been obtained, in particular on numerous branching time logics (like CTL, CTL*, ECTL*).

Most branching time logics can be expressed already in proper fragments of MSO-logic in which set quantification only ranges over paths or chains. (By a chain we mean a subset of a path.) We call these systems path logic and chain logic, respectively. If we restrict even to quantifiers over finite paths or chains, we obtain finite-path logic and finite-chain logic, respectively. Accordingly we speak of the path theory, chain theory, finite-path theory, and finite-chain theory of a structure.

Starting from chain logic (or path logic), one can gain expressive power also in other ways than by proceeding to full MSO-logic. MSO-logic typically allows to formalize conditions on a global coloring of the tree under consideration, as this is required, for instance, in a statement that a tree automaton has a successful run over the tree.

A different decidable extension of chain logic arises by adding features that capture aspects of synchronization, for example by adjoining the binary "equal level predicate" (that connects vertices of the same tree level). If full MSO-logic over the binary tree is extended by this predicate one obtains an undecidable theory (see [12]).

In the present paper we pursue the idea of extending chain logic as well as the underlying models by constructs of synchronization. For several of such extensions (which are incomparable in expressive power to MSO-logic) we clarify the status of the model-checking problem.

We study two paradigmatic versions of "synchronization" or "merging of paths". For notational simplicity we confine ourselves just to structures with binary branching. So an element has two different successors, called its 0- and 1-successor.

The first type is given by a merge of two paths if their numbers of 0- and 1-successors coincide (regardless of their order). We obtain the infinite $(\mathbb{N} \times \mathbb{N})$-grid rather than the binary tree. It is well known that the MSO-theory of the grid is undecidable. We show that even the finite-path theory of the grid is undecidable.

As a second type of synchronization, we consider the expansion of the tree by the equal level predicate E mentioned above. It holds for two points of different paths if they occur at the "same time" if time progresses discretely along each path. We recall the result of [13] that the chain theory of a regular tree with the E-predicate is decidable. A simple argument will show that this result does not extend to algebraic trees (i.e., trees that are generated by deterministic pushdown automata). Finally, we address more refined types of models in which extra distinctions enter for the axis of time (by extra predicates on the sequence of levels) and for the indivual levels themselves. We analyze under which circumstances these extra distinctions lead to an undecidability of the chain theory.

The paper is structured as follows. In the subsequent section we fix the notation. Then we address path logic over the infinite grid, and afterwards turn to chain logic over regular and algebraic trees with the equal level predicate E. Finally, we treat the question of extending the logic by refining the time axis or the individual levels of a tree. In a concluding section we discuss some related issues and open questions.

2 Structures and Logics

Two types of structures are considered in this paper, the infinite two dimensional grid (in the labelled and unlabelled version) and the infinite binary tree.

We use a format of the infinite grid that follows the idea of an infinite matrix. The positions are pairs $(i,j) \in \mathbb{N} \times \mathbb{N}$, and we have two successor functions r (right) and d (down); so formally $d(i,j) = (i+1,j)$ and $r(i,j) = (i,j+1)$. We write $G_2 = (\mathbb{N} \times \mathbb{N}, d, r)$. We consider labelled grids with the label alphabets $\{0,1\}^n$, so that a labelling can be identified with an n-tuple $\overline{P} = (P_1, \ldots, P_n)$ of unary predicates where P_i contains those vertices whose n-bitvector label $(b_1 \ldots, b_n)$ satisfies $b_i = 1$.

The infinite binary tree is the structure $T_2 = (\{0,1\}^*, \cdot 0, \cdot 1)$ consisting of the finite bit words and equipped with the two successor functions $\cdot 0, \cdot 1$. Again, for the case of labelled trees we consider label alphabets $\{0,1\}^n$, and as above we identify a tree labelled in $\{0,1\}^n$ with a structure $T = (\{0,1\}^*, \cdot 0, \cdot 1, \overline{P})$, with a tuple $\overline{P} = (P_1, \ldots, P_n)$ of unary predicates. A tree $T = (\{0,1\}^*, \cdot 0, \cdot 1, \overline{P})$ is *regular* if for each $i = 1, \ldots, n$, the set P_i is regular. For notational simplicity we mostly confine ourselves to 0-1-labellings and correspondingly expansions of T_2 by a single predicate. So a tree $T = (\{0,1\}^*, \cdot 0, \cdot 1, P)$ is regular if for some finite automaton the state reached after processing w is accepting iff $w \in P$.

We also expand binary trees with the "equal level predicate" E defined by

$$E(u,v) \quad :\Leftrightarrow \quad |u| = |v|$$

We call a structure $T = (\{0,1\}^*, \cdot 0, \cdot 1, E, \overline{P})$, short $(T_2, E, \overline{P})$, a "tree with E".

For both kinds of structures (grids and trees) there is a natural notion of *path*, as a finite or infinite sequence $v_0, v_1, \ldots$ (respectively

$v_0, v_1, \ldots, v_k$) such that v_0 is the root and v_{i+1} is a successor of v_i. By root we mean the element $(0,0)$ of the grid, respectively ε in the binary tree, and "successor" refers to the respective functions d, r, $\cdot 0, \cdot 1$. A *chain* is a subset of some path; in terms of the partial order that is generated by the successor functions this means that chains are linearly ordered subsets of the universe.

Let us introduce the logics considered in this paper. MSO-logic (monadic second-order logic) has variables $x, y, z, \ldots$ for elements and variables $X, Y, Z, \ldots$ for subsets of the structure under consideration. In structures where the notions of path and of chain are meaningful (as is the case over the grid and the binary tree) we can restrict the second-order quantifications accordingly. If set quantifiers range only over paths, respectively chains, then we call the resulting system *path logic*, respectively *chain logic*. In the corresponding "weak" logics we restrict even further, namely to finite paths and chains, respectively. We call these systems finite-path logic and finite-chain logic.

If S is a structure, then the MSO-theory of S is the set of MSO-sentences (of appropriate signature) that are true in S. Similarly, referring to chain logic, path logic, and their "finite" restrictions, one defines the chain theory, path theory, finite-chain theory, and finite-path theory of S.

In some situations below it will be convenient to replace MSO-logic and chain logic with an expressively equivalent variant. Two modifications are assumed: First we convert our structures to relational ones; this means that we work with the successor relations S_d, S_r in place of the functions d, r and with the relations S_0, S_1 corresponding to the successor functions $\cdot 0, \cdot 1$. Second, we replace the first-order variables by second-order varibales ranging over singletons; this helps to avoid syntactic complications. So we identify MSO-logic with the logic that has the following atomic formulas:

- $X \subseteq Y$ ("X is a subset of Y"),
- Sing(X) ("X is a singleton"),
- $X \subseteq P$ for each unary predicate P in the signature,
- $X\ R\ Y$ for each binary relation R in the signature, including equality ("X, Y are singletons and we have xRy for their elements x, y")

3 Path Logic over the Grid

In the study of monadic logic, the infinite two-dimensional grid is prominent as a simple structure whose MSO-theory is undecidable.

The standard proof makes use of the possibility to describe global colorings (or labellings) of a structure in MSO-logic; even existential MSO-logic suffices for this purpose. A computation of a Turing machine M on its (left-bounded and right-infinite) tape can be represented by such a coloring of the grid, just by filling the rows of the grid successively with labellings that code the configurations of the computation. For a given Turing machine M it is easy to write down an existential MSO-sentence φ_M which describes a labelling that codes a halting computation of M when started on the empty tape. This shows that the (existential) MSO-theory of the grid G_2 is undecidable.

By another straightforward idea of coding we show that even the finite-path theory of G_2 is undecidable. We use a reduction to the termination problem of 2-counter machines (or 2-register machines). Such a machine M is given by a finite sequence

$$1 \text{ instr}_1; \ldots; k-1 \text{ instr}_{k-1}; k \text{ stop}$$

where each instruction instr_j is of the form

- $\text{Inc}(X_1)$, $\text{Inc}(X_2)$ (increment the value of X_1, respectively X_2 by 1), or
- $\text{Dec}(X_1)$, $\text{Dec}(X_2)$ (similarly for decrement by 1, with the convention that a decrement of 0 is 0), or
- If $X_i = 0$ goto ℓ_1 else to ℓ_2 (where $i = 1, 2$ and $1 \le \ell_1, \ell_2 \le k$, with the natural interpretation).

An M-configuration is a triple (ℓ, m, n), indicating that the ℓ-th instruction is to be executed and the values of X_1, X_2 are m, n, respectively. A terminating M-computation (for M as above) is a sequence $(\ell_0, m_0, n_0), \ldots, (\ell_r, m_r, n_r)$ of M-configurations where in each step the update is done according to the instructions in M and the last instruction is the stop-instruction (formally: $\ell_r = k$). The termination problem for 2-counter machines asks to decide, for any given 2-counter machine M, whether there exists a terminating M-computation that starts with $(1, 0, 0)$ (abbreviated as $M : (1, 0, 0) \to \text{stop}$). It is well-known that the termination problem for 2-counter machines is undecidable.

Theorem 1 *The finite-path theory of the infinite grid G_2 is undecidable.*

Proof

For any 2-register machine M we construct a finite-path formula φ_M such that $M : (1, 0, 0) \to \text{stop}$ iff $G_2 \models \varphi_M$.

The idea is to code a computation $(\ell_0, m_0, n_0), \ldots, (\ell_r, m_r, n_r)$ by three finite paths that basically proceed along the diagonal; a path signals value i by a deviation $r^i d^i$ from the diagonal to the right and then down by i steps. (The vertex reached after applying r^i before turning down is henceforth also called the "corner vertex"; in the figure below these corner vertices are indicated by bullets.) More precisely, we code a configuration (ℓ, m, n) by three path segments S_ℓ, S_m, S_n of length $2N+2$ each, where $N = \max(\ell, m, n)$:

$$S_\ell = d\,r\,r^\ell d^\ell (d\,r)^{N-\ell}, \; S_m = d\,r\,r^m d^m (d\,r)^{N-m}, \; S_n = d\,r\,r^n d^n (d\,r)^{N-n}$$

All three path segments start at the diagonal, say at vertex (j, j), and return to it at vertex $(j+N+2, j+N+2)$ (see Fig. 1 below for an illustration).

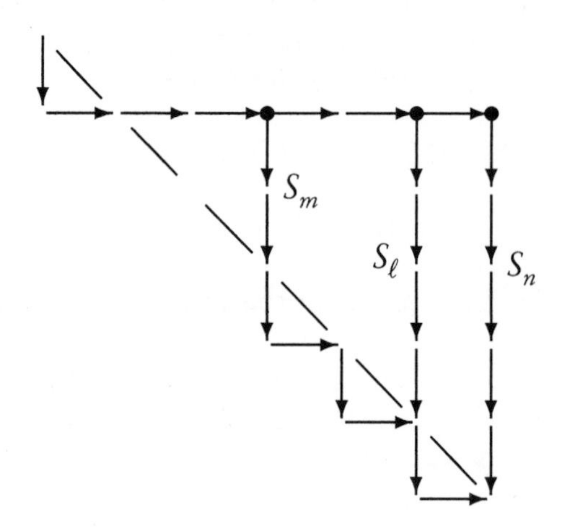

Figure 1: *Coding configuration* $(\ell, m, n) = (4, 2, 5)$.

The concatenation of the these triples of path segments for the configurations $(\ell_0, m_0, n_0), \ldots, (\ell_r, m_r, n_r)$ yields three finite paths P_0, P_1, P_2 which start at vertex $(0, 0)$ of the grid.

The desired formula φ_M expresses the existence of three finite paths P_0, P_1, P_2 that code a terminating computation of M. The formulation is a straightforward exercise once the following remarks are taken into account:

1. Vertex x is on the diagonal iff it can be reached from the origin $(0, 0)$ by a path $(d\,r)^*$; in other words: Each finite path P containing x and such that $r(d(y)) \in P$ implies $y \in P$ must contain the origin. This is expressible in finite-path logic; we use this idea also in analogous situations when saying that two points are "on the same diagonal".
2. The start of the code of a configuration is characterized by a path segment $d\,r\,r$ of P_0.

3. That two successive triples of path segments are in conformance with the instructions of M is expressibe by the following observations: The correct update of the instruction number is expressible by fixing the corresponding path segments S_ℓ explicitly. The correct update of the values m,n of the two M-variables X_1,X_2 is possible by checking whether the "corner vertices" of the successive path segments fit: The second corner vertex is on the same diagonal of S_i as the first corner vertex if the X_i-value stays the same, and it is located one vertex left, respectively right, of the diagonal through the first vertex if we have a decrement by 1, respectively increment by 1.
4. The condition that a computation is terminating is captured by the existence of a path segment $d\,r\,r^k d^k$ of P_0; this can be expressed in finite-path logic.

Trivially this undecidability result extends to path logic, finite-chain logic, and chain logic over the grid.

Another version of this result refers to satisfiability of path formulas in unlabelled finite grids (rather than satisfaction in the infinite grid). For this, one extends the formula φ_M of the proof above by the clause that the paths P_0,P_1,P_2 have to end by segments S_ℓ,S_m,S_n that code a terminating configuration. The modified formula φ'_M is satisfiable in a quadradic unlabelled finite grid iff M terminates when starting in configuration $(1,0,0)$. Hence we obbtain the following variant of Theorem 1:

Corollary 2 *Satisfiability of (finite-) path formulas over quadratic unlabelled finite grids is undecidable.*

4 Model-Checking over Trees with Equal Level Predicate

In this section we study the model-checking problem with respect to chain logic over structures (T_2,E,P) that are derived from the binary tree T_2 by two expansion steps: We add the equal level predicate E, capturing synchronization of vertices on the paths, and we add a unary predicate P corresponding to a labelling of the vertices by the values 0 and 1. In general, we can admit an n-tuple of predicates and thus labels from a larger alphabet $\{0,1\}^n$; we stay with the case $n=1$ for ease of notation. We consider chain logic over a structure (T_2,E,P). Let us recall a result from [13].

Theorem 3 *The chain theory of a regular binary tree with equal level predicate is decidable.*

Proof Let (T_2, E, P) be a regular tree with equal level predicate. We claim that there is a method to decide for any chain-sentence φ whether $(T_2, E, P) \models \varphi$.

For this, we refer to the variant of chain logic in which first-order variables are eliminated and the atomic formulas are of the form $X \subseteq Y$, $X \subseteq P$, and XEY (see Section 2). We shall provide a translation into the MSO-theory of $(\mathbb{N}, +1)$, also called Büchi's arithmetic S1S [1].

As a preparation we associate with each chain C in (T_2, E, P) a pair (α_C, β_C) of ω-words over $\{0,1\}$. The sequence α_C codes the leftmost path P_C of which C is a subset. (If C is infinite, then this path is unique; if C is finite, we agree on the leftmost branching after the last element of C to obtain uniqueness.) So $\alpha_C = d_0 d_1 d_2 \ldots$ is a sequence of "directions". The sequence β_C codes membership of elements in C along the path P_C: We have $\beta_C(i) = 1$ iff $d_0 \ldots d_{i-1} \in C$ (for $i = 0$ this means $\varepsilon \in C$). In order to code membership of P_C-vertices in the given regular set P, we also introduce a third sequence γ_C by setting $\gamma_C(i) = 1$ iff $d_0 \ldots d_{i-1} \in P$.

Let us identify the sequences $\alpha_C, \beta_C, \gamma_C$ with the corresponding sets of natural numbers. Using the correspondence between chains C and pairs (α_C, β_C) of 0-1-sequences and the coding of P along the path P_C by the sequence γ_C, we can rewrite any chain formula $\varphi(X_1, \ldots, X_n)$ speaking about the structure (T_2, E, P) as an S1S-formula $\varphi'(X_1, Y_1, Z_1 \ldots, X_n, Y_n, Z_n)$ such that

$$\begin{aligned}&(T_2, E, P) \models \varphi[C_1, \ldots, C_n] \\ &\text{iff} (\mathbb{N}, +1) \models \varphi'[\alpha_{C_1}, \beta_{C_1}, \gamma_{C_1}, \ldots, \alpha_{C_n}, \beta_{C_n}, \gamma_{C_n}]\end{aligned}$$

Now one observes that the reference to the γ_{C_i} is superfluous since they are (S1S-) definable from the α_{C_i}, using the fact that P is regular. Let us write $\varphi''(X_1, Y_1, \ldots, X_n, Y_n)$ for the resulting S1S-formula. Applying the transformation from φ to φ'' to sentences, we obtain a reduction of the chain theory of (T_2, E, P) to the MSO-theory of $(\mathbb{N}, +1)$ (which is decidable [1]).

A natural next step after showing that the chain theory of regular trees (T_2, E, P) is decidable would be to consider algebraic trees. A tree structure (T_2, E, P) is algebraic if P can be generated by a deterministic pushdown automaton with output: After processing the word w, the pushdown automaton gives output 1 iff $w \in P$. A simple example of an algebraic tree is given by the context-free set $P = \{0^i 1^i \mid i \geq 0\}$. It is well-known that the MSO-theory of an algebraic tree (T_2, P) without the E-predicate is decidable [3]. Here we are interested in the chain theory over algebraic trees expanded by the equal level predicate.

Theorem 4 *Let $P = \{0^i 1^i \mid i \geq 0\}$. The finite-chain theory of the algebraic tree (T_2, E, P) with equal level predicate is undecidable.*
Proof We note two simple facts:

First, the leftmost path of T_2 with the function $\cdot 0$ is a copy of the successor structure $(\mathbb{N}, +1)$.

Second, let $\varphi(x, y)$ be the formalization of the following condition as a finite-chain logic formula: "x, y are on the leftmost path, and given $x \in 0^*$, there is a vertex $z \in x \cdot 1^*$ which belongs to P, and the vertex $y \in 0^*$ is on the same level as z" (see Fig. 2).

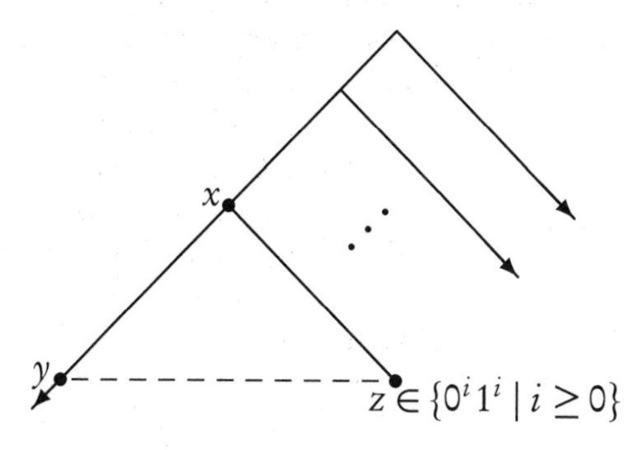

Figure 2: *Computing $y = 2x$.*

Clearly, the formula $\varphi(x, y)$ defines the double function $(x \mapsto 2x)$ on the copy of $(\mathbb{N}, +1)$ given by the leftmost path of T_2. More formally, we have

$$(T_2, E, P) \models \varphi[0^i, 0^j] \text{ iff } 2i = j$$

So we can interpret the weak MSO-theory of $(\mathbb{N}, +1, (x \mapsto 2x))$ in the finite-chain theory of (T_2, E, P). Since the weak MSO-theory of $(\mathbb{N}, +1, (x \mapsto 2x))$ is undecidable [10], this shows the claim.

5 Extra Structure

The previous result motivates to consider more modest kinds of extensions of regular tree models with E. In this section we analyze two kinds of such extensions. The first is given by extra (and of course non-regular) predicates on the time axis. Formally we represent this axis by the leftmost branch 0^ω of the tree which is identified with the set of natural numbers. For example, we may add the predicate to be a power of 2, given by the "time predicate" $Q = \{0^n \mid \exists i\ 2^i = n\}$ over the tree. We shall establish a close link between the MSO-theory of $(\mathbb{N}, +1, Q)$ and the chain theory of a structure (T_2, E, P, Q) where (T_2, E, P) is regular. (The results extend in the obvious way to the case where we consider an n-tuple $\overline{Q}$ of subsets $Q_i \subseteq 0^*$ instead of a single set $Q \subseteq 0^*$.) The second type of extension refers to additional expressive power over the individual levels of the tree structure.

5.1 Predicates on the time axis

First, let us remark that there are recursive predicates $Q \subseteq \mathbb{N}$ such that the MSO-theory of $(\mathbb{N}, +1, Q)$ – and *a fortiori* the chain theory of (T_2, E, Q) – is undecidable. To construct such a predicate Q over $\mathbb{N}$, consider a non-recursive, recursively enumerable set $R \subseteq \mathbb{N} \setminus \{0\}$ with effective enumeration $r_0, r_1, \ldots$. Define

$$Q := \{r_0, r_0 + r_1, r_0 + r_1 + r_2, \ldots\}$$

Then Q is recursive, and we have $n \in R$ iff $(\mathbb{N}, +1, Q)$ satisfies the sentence saying "there is $x \in Q$ such that the n-th successor of x is the next element in Q". This shows that even the first-order theory of $(\mathbb{N}, +1, Q)$ is undecidable.

Starting with work of Elgot and Rabin [4], many predicates Q have been found such that the MSO-theory of $(\mathbb{N}, +1, Q)$ is decidable. Prominent examples are the set of factorial numbers and, for each $k > 1$, the set of k-th powers and the set of powers of k. Further references on the subject are, for example, [2] and [11].

By a straightfoward adaptation of the proof of Theorem 3 we obtain:

Theorem 5 *Let $Q \subseteq \mathbb{N}$ and let $P \subseteq \{0,1\}^*$ be regular. The chain theory of (T_2, E, P) expanded by the time predicate Q is decidable iff the MSO-theory of $(\mathbb{N}, +1, Q)$ is decidable.*

Proof The direction from left to right is immediate since the MSO-theory of $(\mathbb{N}, +1, Q)$ is directly interpretable in the chain theory of (T_2, E, P, Q) (considering the leftmost branch of T_2).

For the other direction, we observe that the interpretation in the proof of Theorem 3 extends to the situation where the predicate Q is added: For each chain sentence φ in the signature with E, P, Q we can construct an MSO-sentence φ' over $(\mathbb{N}, +1, Q)$ such that

$$(T_2, E, P, Q) \models \varphi \text{ iff } (\mathbb{N}, +1, Q) \models \varphi'$$

By the assumption that MTh$(\mathbb{N}, +1, Q)$ is decidable the claim follows.

5.2 Quantifying over elements of levels

When a synchronization constraint refers to all existing computation paths at a certain moment, one might like to express properties of the corresponding level of the considered computation tree (T_2, E, P). Formally, we restrict to the set L of vertices at a given level and speak about a structure over the domain L. Since we are dealing with binary trees, where a natural left-to-right ordering of the paths exists, we allow the successor relation S_L and the ordering relation $<_L$ over L, and additionally unary predicates R that are chain-definable in the tree structure. More precisely, we consider a chain formula $\varphi(x)$ and introduce the corresponding predicate over L as

$$R_L^{\varphi} = \{v \in L \mid (T_2, E, P) \models \varphi[v]\}$$

Under which circumstances do we preserve the decidability of the chain theory of (T_2, E, P) when extending chain logic by features for expressing properties of level structures $(L, S_L, <_L, R_L^{\varphi}, \ldots)$?

Let us consider only two basic systems of this kind: "chain logic with E and FO$(S, <)$ on the levels", and "chain logic with E and MSO$(S, <)$ on the levels". In both systems we allow S and $<$ with their explained meaning (i.e., connecting vertices on a common level) and first-order, respectively monadic second-order quantifiers over elements belonging to a given level. Thus we include quantifiers $\exists y \in L_x$, $\exists X \subseteq L_x$ and the corresponding universal versions, meaning "there is a vertex y on the level of x", "there is a set X of vertices on the level of x", respectively. So, for example, in chain logic with E and MSO$(S, <)$ on the levels we can express that there is a level with an even number of vertices of a certain property φ. More generally, we can express the existence of certain partitions of the paths meeting a certain level. Such conditions occur, for example, in the study on logics of knowledge studied by Halpern and Vardi [6]. We show the following result, which indicates (in part (b)) severe limitations for decidability results.

Theorem 6

(a) The chain theory of a regular tree with E and FO$(S, <)$ on the levels is decidable.

(b) Even the finite-path theory of the unlabelled binary tree with E and MSO$(S, <)$ on the levels is undecidable.

Proof Part (a) is easy, since each FO$(S, <)$-formula restricted to a level of the tree model is directly expressible in chain logic with the equal level predicate E.

For part (b) we use an idea of [8] that allows to code a coloring of a binary tree up to (and exluding) level L by a coloring of level L itself. We simply transfer the color of vertex v (before level L) to the unique vertex $v' \in L$ which belongs to $v10^*$ (i.e. belongs to the leftmost path from the right successor of v). The map $v \mapsto v'$ is injective and definable in finite-path logic (given level L). Moreover, it is easy to see that the relations of being left or right successor in the tree are translated to definable relations over the level L under consideration.

Using this coding, an existential quantifier over finite sets in the binary tree is captured by an existential quantifier over the elements of an appropriate level of the tree. Thus, the weak MSO-theory of the binary tree with E is interpretable in the finite-path theory of the unlabelled binary tree with E and MSO($S,<$) on the levels. Since the weak MSO-theory of (T_2, E) is undecidable (see e.g. [12]), we obtain claim (b) of the Theorem.

6 Conclusion

In this paper, we analyzed path logic and chain logic over structures that arise from the binary tree by different versions of "merging paths". Concrete examples were the infinite grid and (labelled) infinite trees expanded by the "equal level predicate" that captures a synchronization of paths. We clarified the status of the model-checking problem of path logic and chain logic over these structures and certain natural extensions.

Many issues on path logics with features of synchronization are unresolved. A natural question is to extend the present results to the richer logics of "time granularities" studied by Franceschet, Montanari, Puppis et al. in [5, 7]. Also one might refine the results in the framework of branching time logics in order to get better complexity bounds (than the nonelementary bounds as inherent in the reduction to S1S). Furthermore, it should be analyzed whether stronger theories than S1S can be invoked for interpretation, in order to show decidability for more powerful mechanisms of synchronization. Another direction is the inclusion of aspects as they are used in timed systems (where computation paths are subject to constraints on durations).

7 Acknowledgment

Many thanks are due to Christof Löding for his remarks on an early version of this paper and to the anomymous referee for his suggestions.

References

[1] J. R. Büchi. On a decision method in restricted second order arithmetic, *Proc. 1960 International Congress on Logic, Methodology and Philosophy of Science*, E. Nagel at al., eds, Stanford University Press 1962, pp. 1-11.

[2] O. Carton, W. Thomas, The monadic theory of morphic infinite words and generalizations, *Information and Computation* 176 (2002), 51-76.

[3] B. Courcelle, I. Walukiewicz, Monadic second-order logic, graph coverings and unfoldings of transition systems, *Ann. Pure Appl. Logic* 92 (1998), 35-62.

[4] C.C. Elgot, M. O. Rabin, Decidability and undecidability of extensions of second (first) order theory of (generalized) successor, *J. Symb. Logic* 31 (1966), 169-181.

[5] M. Franceschet, A. Montanari, A. Peron, G. Sciavicco, Definability and decidability of binary predicates for time granularity, *J. Appl. Logic* 4 (2006), no. 2, 168-191.

[6] J. Y. Halpern, M. Y. Vardi, The complexity of reasoning about knowledge and time. I. Lower Bounds. *J. Comput. Syst. Sci.* 38 (1989), 195-237 (1989).

[7] A. Montanari, A. Peron, G. Puppis, On the relationships between theories of time granularity and the monadic second-order theory of one successor, *J. Appl. Non-Classical Logics* 16 (2006), 433-455.

[8] A. Potthoff, W. Thomas, Regular tree languages without unary symbols are star-free, in: *Proc. FCT 1993*, Springer LNCS 710 (993), 396-405.

[9] M.O. Rabin. Decidability of second-order theories and automata on infinite trees, *Trans. Amer. Math. Soc.* 141 (1969), 1-35.

[10] R.M. Robinson, Restricted set-theoretical definitions in arithmetic, *Proc. Amer. Math. Soc.* 9 (1958), 238-242.

[11] A. Rabinovich, W. Thomas. Decidable theories of the ordering of natural numbers with unary predicates, in: *Proc. 15th CSL 2006*. Springer LNCS 4207 (2006), 562-574.

[12] W. Thomas, Automata on infinite objects, in: *Handbook of Theoretical Computer Science, Vol. B* (J. v. Leeuwen, ed.), Elsevier, Amsterdam 1990, pp. 133-191.

[13] W. Thomas, Infinite trees and automaton definable relations over omega-words, *Theor. Comput. Sci. 103* (1992), 143-159.

Logic-Based Diagnosis for Distributed Systems*

Shaofa Yang[1]†, Loïc Hélouët[2] Thomas Gazagnaire[3]‡
[1] *UNU-IIST, Macao*
ysf@iist.unu.edu
[2] *INRIA Rennes, France*
loic.helouet @irisa.fr
[3] *Citrix Systems R&D Ltd., UK*
thomas.gazagnaire@citrix.com

Abstract

We address the problem of off-line fault diagnosis for distributed systems. It consists in finding explanations for a given partial observation of abnormal behaviour, using knowledge of system dynamics. For this, a diagnosis algorithm must decide whether there exists an execution that is compatible with our knowledge of the system and with the observation. We represent observations with restricted partial orders which model cause-effect relations among local states, and properties that hold at these states. We capture knowledge of system dynamics with a temporal logic which asserts the evolution of patterns of causal orders. We show that the corresponding diagnosis problem is undecidable. However, if we limit explanations to distributed behaviours in which each process causally influences every other process in a bounded manner, the restricted diagnosis problem becomes decidable.

Keywords: Diagnosis, partial orders, logic.

1 Introduction

For safety and economical reasons, diagnosis of faults is of paramount importance in domains such as telecommunication networks and embedded systems. Diagnosis can be performed off-line or online. In off-line diagnosis, the task is to infer missing information (unobserved events or states, faulty behaviors,...) from a partial observation of a

*work supported by the CREATE project of Region Bretagne.

†Work done while this author was at IRISA/INRIA Rennes, France supported by an INRIA post-doctoral fellowship.

‡Work done while this author was at ENS Cachan, antenne de Bretagne, France.

system, such as a log file. The observation has to be partial for two reasons. Firstly, some state information or actions may not be directly accessible. Secondly, due to the huge size of many distributed systems, it is simply not feasible to keep track of all state information. In online diagnosis, a system is continuously monitored and faults are supposed to be detected as soon as possible after their occurrence.

Traditionally fault diagnosis is performed by inductive reasoning, using expert heuristic rules between faults and observations. However, such expert knowledge is difficult to obtain and easily becomes obsolete when a system's configuration evolves. The so-called *model-based* approach brings more applicable solution to fault diagnosis. In this framework, one captures knowledge of system dynamics in some formal system model such as transition systems [17], Petri nets [2] or message sequence charts [10]. Faults are inferred from observation and the system model. One might be interested in determining if an unobserved fault has occurred, as in [17], or in finding all possible runs that may have led to the observation, as in [2]. Another objective ([8]) is to compute a summary of possible explanations, that is to annotate observations with information that help explaining what might have occurred. The additional information can be causal relations among observed events, known local properties of states or events in the observation, and events of interest not contained in the logged information.

One drawback of model-based diagnosis is that a complete model of the monitored system is not always available. Furthermore, it is difficult to update a system model when a system's configuration changes. Often, the only knowledge available for diagnosis consists of some partial *properties* of a system's behaviour. Considering this, we propose a logic-based approach to diagnosis. More precisely, we capture knowledge of a system's behaviour using formulae in some suitable temporal or modal logics. Temporal or modal logics enables one to specify properties of system's dynamics in a natural way. In many cases, updating properties of a system in the form of logic formulae can be done easily by changing a limited number of subformulae.

In this paper, we represent behaviours of distributed systems with restricted partial orders which define cause-effect relations among local states. These partial orders are called partially ordered computations. We propose a temporal logic over partially ordered computations and call it simply the *logic of partially ordered computations* (LPOC). The main feature of LPOC is to reason about evolution of patterns of causal orders. We use temporal operators similar to Computation Tree Logic. The design of LPOC is motivated by the fact that, for many distributed systems such as network protocols, properties about their executions are often available in the form of "whenever this pattern of causal ordering occurs, some other pattern will follow in future", where patterns are usually short sequences of message exchanges.

We study off-line diagnosis based on LPOC. The problem is to determine whether there exists an explanation for a given observation and a given LPOC formula Φ describing a system's dynamic properties. An explanation is a distributed behaviour which could have given rise to the observation and which satisfies the formula Φ. We show that this problem is undecidable in general. The undecidability result is mainly due to the undecidability of satisfiability problem of LPOC. We note that the satisfiability problem of several similar temporal logics in the literature are undecidable. These include m-LTL [15], a local temporal logic on Lamport diagrams, and template message sequence charts [9]. However, for a given K, if we limit explanations to so-called K-influencing distributed behaviours in which each process causally influences every other process in a bounded manner, then the restricted diagnosis problem is decidable (for the given K). Furthermore, one can effectively compute a compact summary of all K-influencing explanations.

In the next section, we introduce the syntax and semantics of the logic LPOC. Section 3 defines the diagnosis problem associated with LPOC and show that it is undecidable. Section 4 establishes the decidability of the restricted diagnosis problem where only K-influencing explanations are considered. We also analyze the complexity of the decision algorithm. Section 5 discusses related work and postulates some future directions. To reduce clutter, some proofs are omitted, but can be found in an extended version in [19].

2 Logic of Partially Ordered Computations

Through the rest of the paper, we fix a finite nonempty set $\mathscr{P}$ of process names, and $\mathscr{A}$ a finite nonempty set of atomic propositions. We let p, q range over $\mathscr{P}$.

Definition 1 *A* partially ordered computation *(or computation for short) over* $(\mathscr{P}, \mathscr{A})$ *is a tuple* $(S, \eta, \leq, V)$ *where:*

- S *is a finite set of* (local) states.
- $\eta : S \to \mathscr{P}$ *identifies the* location *of each state. For each* $p \in \mathscr{P}$, *we define* $S_p = \{s \in S \mid \eta(s) = p\}$.

- $\leq \subseteq S \times S$ *is a partial order, called the* causality *relation. Furthermore, for each* p, $\leq$ *restricted to* $S_p \times S_p$ *is a total order.*
- $V : S \rightarrow 2^{\mathscr{A}}$ *is a labeling function which assigns a set of atomic propositions to each state. We call* $V(s)$ *the* valuation *of* s.

Intuitively, a computation (also called Lamport diagram in the literature [15]) represents the causal ordering among local states in a distributed execution, in which states of each process are sequentially ordered. The valuation of local state s collects the atomic propositions that hold at s. Figure 1-a) shows a computation. States are designated by black dots with associated name $s_1, \ldots, s_6$. Processes P, Q, R are represented by vertical lines, and states located on a process line are ordered from top to bottom. Finally, valuations of states take value in $\{a, b, c\}$, and are represented between two brackets near the associated state. Note that we consider only *finite* computations, since we address only offline diagnosis in this paper. We will say that two computations $(S, \eta, \leq, V)$ and $(S', \eta', \leq', V')$ are *isomorphic* iff there is a bijection $f : S \rightarrow S'$ such that $\eta(s) = \eta'(f(s))$ for any $s \in S$, $s_1 \leq s_2$ iff $f(s_1) \leq' f(s_2)$ for any $s_1, s_2 \in S$, and $V(s) = V'(f(s))$ for any $s \in S$. We identify isomorphic computations and write $W \equiv W'$ if W and W' are isomorphic.

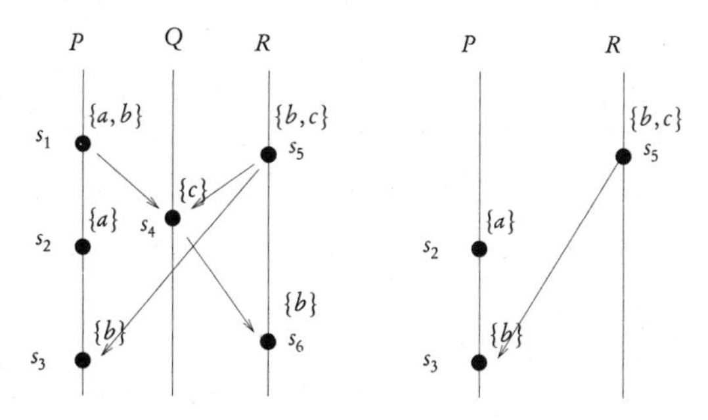

Figure 1: *a) A computation* W *b) the 1-view of* s_3 *in* W

Let $(S, \eta, \leq, V)$ be a computation, and $s, s' \in S$. As usual, we write $s < s'$ when $s \leq s'$ and $s \neq s'$. For each $p \in \mathscr{P}$, we define $\ll_p \subseteq S \times S$ as: $s \ll_p s'$ iff $s, s' \in S_p$, $s < s'$, and there does not exist $s'' \in S_p$ with $s < s'' < s'$. That is, $\ll_p$ is the "immediate" sequential ordering of states belonging to p. We let $\lessdot \subseteq S \times S$ be the least relation such that $\leq$ is the reflexive and transitive closure of $\lessdot$. For each $p, q \in \mathscr{P}$ with $p \neq q$, let us define $\ll_{pq} \subseteq S \times S$ as follows: $s \ll_{pq} s'$ iff $s \in S_p$, $s' \in S_q$, and $s \lessdot s'$. We also define $\ll = (\cup_{p \in \mathscr{P}} \ll_p) \bigcup (\cup_{p,q \in \mathscr{P}, p \neq q} \ll_{pq})$. We note that $<$ is in fact the transitive closure of $\ll$. If $s \ll s'$, we say

s' is a *(causal) successor* of s, and call s a *(causal) predecessor* of s'. We emphasize that $\ll$ is not equal to $\lessdot$. Indeed, $\lessdot$ does not necessarily capture the relations defined by the local ordering on processes. Consider for instance states s_5 and s_6 in Figure 1-a: we have $s_5 \ll s_6$, but not $s_5 \lessdot s_6$. A state s is *minimal* if it has no predecessor, and *maximal* if it has no successor. A *causal chain* is a sequence $s_1 s_2 \ldots s_n$ of states where $s_1 \ll s_2 \ll \ldots \ll s_n$.

In the sequel, we define a temporal logic of partial-order computations (called "LPOC" for short) to reason about distributed behaviors. It has two basic features. First, at a state s of a computation, atomic formulae assert that a "pattern" occurs in a bounded past or bounded future of s. Secondly, we consider a branching time framework with CTL-like operators and reason along sequences of causally ordered states.

Definition 2 *Let $(S, \eta, \leq, V)$ be a computation, and m be a natural number. The m-*view *of $s \in S$, denoted $\downarrow_m(s)$, is the collection of states s' in S such that there exists a causal chain of length at most m starting from s' and ending at s. More precisely, $\downarrow_m(s) = \{s' \mid \exists s_0, \ldots s_n \in S, n \leq m \text{ and } s' = s_0 \ll s_1 \ll \ldots \ll s_n = s\}$. Similarly, the m-*frontier *of s, denoted by $\uparrow_m(s)$, is the collection of states s' in S such that there exists a causal chain of length at most m starting from s and ending at s'.*

Figure 1-b) shows an example of m-view. Note that the 0-view and 0-frontier of a state s are both the singleton set $\{s\}$. Each state s has at most $|\mathcal{P}|$ successors, one belonging to each S_p. Thus, inductively, the m-view and m-frontier of s contains at most $\mathcal{N}_m = \sum_{i=0}^{m} |\mathcal{P}|^i = \frac{1-|\mathcal{P}|^{m+1}}{1-|\mathcal{P}|}$ states. In order to reason about the "pattern" of a computation, we also need a notion of *projection*.

Definition 3 *Let $W = (S, \eta, \leq, V)$ be a computation over $(\mathcal{P}, \mathcal{A})$, and let $A \subseteq \mathcal{A}$. The* projection *of W onto A is the computation $W' = (S', \eta', \leq', V')$ where $S' = \{s \in S \mid V(s) \cap A \neq \emptyset\}$, and η', $\leq'$, are the respective restrictions of $\eta, \leq$ to S' and $V'(s) = V(s) \cap A$ for every $s \in S'$.*

The atomic formulae of our logic will assert that the projection of the computation formed from the m-view or the m-frontier of a state is isomorphic to a given computation. We are now ready to define the logic LPOC.

Definition 4 *The set of LPOC formulae over a set of processes $\mathcal{P}$ and a set of atomic propositions $\mathcal{A}$, is denoted by LPOC$(\mathcal{P}, \mathcal{A})$, and is inductively defined as follows:*

- *For each $p \in \mathcal{P}$, the symbol loc_p is a formula in LPOC$(\mathcal{P}, \mathcal{A})$.*

- *Let m be a natural number, A be a subset of $\mathscr{A}$, and $T = (S, \eta, \leq, V)$ be a computation such that $V(s) \subseteq A$ for every $s \in S$. Then $\downarrow_{m,A}(T)$, $\uparrow_{m,A}(T)$ are formulae in $LPOC(\mathscr{P}, \mathscr{A})$.*
- *If φ, φ' are formulae in $LPOC(\mathscr{P}, \mathscr{A})$, then $\mathsf{EX}\varphi$, $\mathsf{EU}(\varphi, \varphi')$ are formulae in $LPOC(\mathscr{P}, \mathscr{A})$.*
- *If $\varphi, \varphi' \in LPOC(\mathscr{P}, \mathscr{A})$, then $\neg\varphi$ and $\varphi \vee \varphi'$ are formulae in $LPOC(\mathscr{P}, \mathscr{A})$.*

From now on, we shall refer to formulae in $LPOC(\mathscr{P}, \mathscr{A})$ simply as *formulae*. Their semantics is interpreted at local states of a computation. For a computation W and a given state s of W, we write $W, s \models \phi$ when W satisfies ϕ, which is defined inductively as follows:

- $W, s \models loc_p$ iff the location of s is p (i.e. $\eta(s) = p$),
- $W, s \models \downarrow_{m,A}(T)$ iff the projection of $\downarrow_m(s)$ onto A is isomorphic to T.
- $W, s \models \uparrow_{m,A}(T)$ iff the projection of $\uparrow_m(s)$ onto A is isomorphic to T.
- $W, s \models \mathsf{EX}\varphi$ iff there exists a state s' in W such that s' is a causal successor of s and $W, s' \models \varphi$.
- $W, s \models \mathsf{EU}(\varphi, \varphi')$ iff there exists a causal chain $s_1 s_2 \ldots s_n$ in W with $s = s_1$. Further, there exists an index i in $\{1, 2, \ldots, n\}$ with $W, s_i \models \varphi'$, and $W, s_j \models \varphi$ for every j in $\{1, 2, \ldots, i-1\}$.

The semantics for boolean combinations and negations of formulae is as usual. We assume the standard boolean operators. We define some derived temporal operators as follows: $\mathsf{EF}\varphi \equiv \mathsf{EU}(true, \varphi)$, $\mathsf{EG}\varphi \equiv \mathsf{EU}(\varphi, \varphi \wedge \neg\mathsf{EX}true)$, $\mathsf{AX}\varphi \equiv \neg\mathsf{EX}(\neg\varphi)$, $\mathsf{AF}\varphi \equiv \neg\mathsf{EG}(\neg\varphi)$, and $\mathsf{AG}\varphi \equiv \neg\mathsf{EF}\neg\varphi$. We can also assert the truth of an atomic proposition a at a state of a computation with $\varphi_a = \bigvee_{p \in \mathscr{P}} (loc_p \wedge \downarrow_{0,\{a\}}(T_{p,a}))$, where each $T_{p,a}$ is the computation containing a singleton state of location p and valuation $\{a\}$.

For a computation W and a LPOC formula φ, we say that W *satisfies* φ, written $W \models \varphi$, iff there exists some minimal state s_{min} of W such that $W, s_{min} \models \varphi$. We say that φ is *satisfiable* iff there exists a computation W such that $W \models \varphi$.

For application to diagnosis, it is useful to define the notion of a computations satisfying a collection of formulae, one for each process. Formally, for a computation $W = (S, \eta, \leq, V)$ and a $\mathscr{P}$-indexed family of formulae $\{\varphi_p\}_{p \in \mathscr{P}}$, we say W satisfies $\{\varphi_p\}_{p \in \mathscr{P}}$, written $W \models \{\varphi_p\}_{p \in \mathscr{P}}$ by abuse of notation, iff the following condition holds: for each $p \in \mathscr{P}$, $S_p \neq \emptyset$ and $W, s_p \models \varphi_p$ where s_p is the minimum state in S_p (i.e. $s_p \leq s$ for every $s \in S_p$). Note that $W \models \{\varphi_p\}_{p\in\mathcal{P}}$ iff $W \models \bigwedge_{p\in\mathcal{P}} \mathsf{EU}(\neg loc_p, loc_p \wedge \varphi_p)$.

We have chosen the existential until operator because it is essential in asserting properties such as "whenever some pattern T occurs, some other pattern T' will follow". More precisely, this demands that along every causal chain, whenever a pattern T occurs, pattern T' should occur later *and* no more pattern T can occur again before the point at which the pattern T' has occurred. This kind of properties are commonly needed in practical applications.

Let us define a simple example with LPOC. We define a formula meaning that whenever a connection phase described by a pattern T_{conn} occurs between two processes $Client$ and $Server$, then a data transfer described by a pattern T_{data} necessarily occurs later. This formula can be expressed by $AG\varphi$, where :
$\varphi = (loc_{Client} \wedge \uparrow_{2,A}(T_{conn})) \Longrightarrow$
$(EX(loc_{Client} \wedge EX(EU(loc_{Client}, loc_{Client} \wedge \uparrow_{2,A'}(T_{data}))))$, where the patterns T_{conn} and T_{data} are described in Figure 2,
$A = \{disc, noclient, client, connected\}$, and
$A' = \{DataSent, DataRecv, DataAck\}$.

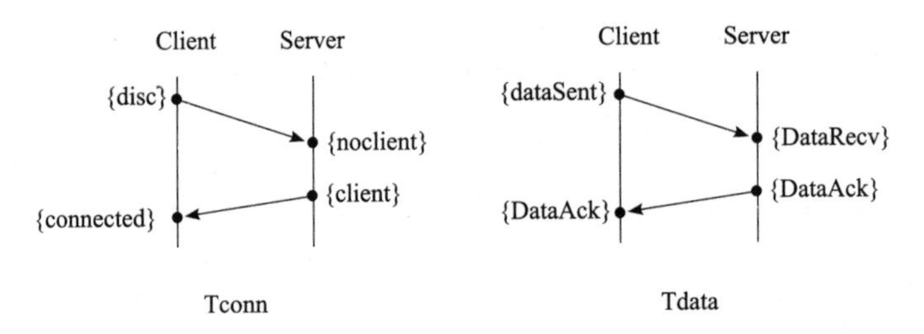

Figure 2: *Two patterns*

3 Diagnosis

An usual framework to perform diagnosis (see for instance [2, 10, 17]) in a distributed system is as follows. A central agent, called the *diagnoser*, collects information from some processes in the system. Each process is equipped with mechanisms that signal information to the diagnoser. This equipment can be implemented by means of code instrumentation, or by hardware mechanisms that raise alarms and sends them to the diagnoser. Of course, due to the size of runs of real systems, the diagnoser only collects a limited subset of what really occurs in the system. The collected information on observed states and causal dependencies is called an *observation*. From this observation and a model of the system, the diagnoser can then output an explanation of what have been observed. This generic framework is depicted in Figure 3: processes are represented by squares, connected by communication links. Black squares symbolize the local observation mechanisms, that send their observations to the diagnoser, symbolized by the central ellipse.

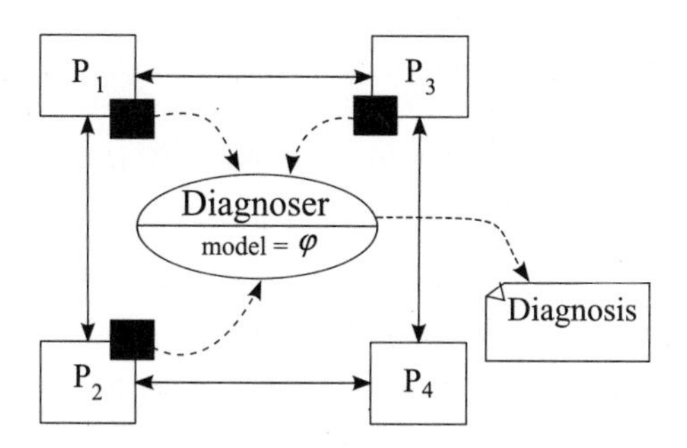

Figure 3: *A diagnosis framework*

Intuitively, an observation describes everything that is "recorded" during the execution of some prefix of a computation. We assume that the monitoring mechanisms transmit "recorded" observable atomic propositions to the central diagnosis system. Only records of the same process are guaranteed to be received by the central diagnosis system in the order they were sent. In addition to local observation on each process, observations contain some causal ordering among observed states located on different processes. This allows finest descriptions of observed behaviors, and to rule out some explanations that might be compatible with the shuffle of local observations during diagnosis. This causal ordering can be inferred if observations are tagged for instance with vectorial clocks [14, 6]. Nevertheless, the observation architecture and the way computations are logged is out of the scope of this work.

Our goal is to use LPOC for diagnosis. We shall represent knowledge of system dynamics by a $\mathcal{P}$-indexed family $\{\Phi_p\}_{p\in\mathcal{P}}$ of formulae. We also record observation of partial execution in the form of a computation. The objective of diagnosis is to find explanations, also in the form of computations, which could have led to the observation and satisfies the $\mathcal{P}$-indexed family of formulae. As there could be many explanations, it is also desirable to compute summarative information of the collection of explanations.

To this end, we fix $\mathcal{A}_{ob} \subseteq \mathcal{A}$ the subset of *observable* atomic propositions, and $\mathcal{A}_{ex} \subseteq \mathcal{A}$ the subset of *explanatory* atomic propositions. Typically, explanatory atomic propositions correspond to the faults or state information that cannot be directly accessed. On the other hand, observable atomic proposition indicate state information that can be directly "recorded", for instance, alarms and abnormal behavior. To formulate the diagnosis problem, we can now formalize the notions of *observation* and *explanation*.

Definition 5 *An* observation *is a computation* $(S, \eta, \leq, V)$ *such that for each state* s *in* S, $V(s) \subseteq \mathcal{A}_{ob}$. *Let* $O = (S_O, \eta_O, \leq_O, V_O)$ *be an observation and* $W = (S_W, \eta_W, \leq_W, V_W)$ *a computation. Then,* W *is an* explanation *for* O *iff there exists an injective mapping* $f : S_O \rightarrow S_W$ *such that:*

(i) $\forall s \in S_O$, $\eta_O(s) = \eta_W(f(s))$ *and* $V_O(s) = V_W(f(s)) \cap \mathcal{A}_{ob}$.
(ii) $\forall s, s' \in S_O$, *if* $s \leq_O s'$, *then* $f(s) \leq_W f(s')$.
(iii) For every s *in the image of* f, $V_W(s) \cap \mathcal{A}_{ob} \neq \emptyset$. *Furthermore, for any* $s' \in S_W$, *if* $\eta_W(s') = \eta_W(s)$, $s' \leq_W s$ *and* $V_W(s') \cap \mathcal{A}_{ob} \neq \emptyset$, *then* s' *is also in the image of* f.

Conditions i) to iii) come from the supposed faithful and non-lossy nature of the observation mechanism. More precisely, condition (i) means that the location mapping should be respected by the observation mechanism, and that in explanations, only states at which at least one observable atomic proposition holds may be "recorded", and hence appear in the observation. Intuitively, (the truth of) an observable atomic proposition corresponds typically to the *presence* of some alarm or observed abnormal behaviour. And the monitoring mechanism could only detect the *presence* of observable atomic propositions, but not their absence. We also suppose that the observation mechanisms do not produce atomic propositions that were not observed. Hence, if no observable proposition hold at a state s of an execution, then s is not recorded in the observation.

Condition (ii) asserts that the recorded causal orderings must originate from an actual dependence in the execution. The converse property (recording all causal dependencies) is *not* demanded, since some causal orderings in the explanation may not be "recorded". For generality, we do not impose any more specific condition on the recording of causality ordering. We remark however that our results could be extended easily to deal with more specific condition on recording of causalities, which may arise from particular application domains.

Condition (iii) states that there is no loss during recording of states: if a state of process p is recorded, then any causally preceding states s' of p must be recorded in case the valuation of s' contains observable atomic propositions. In other words, we assume that the monitoring mechanism is itself free from faults and thus do not miss any presence of observable abnormal signals. In some application domains such as telecommunication networks, the monitoring mechanism is also part of the observed system. In such situations, it may be sometimes necessary to consider also the potential faults incurred by the monitoring mechanism. However, in this paper, we will consider that the observation mechanisms is non-lossy.

Let us illustrate the notions of observation and explanation on the examples of Figure 4. In this example, there are two processes "Client" and "Server", and the observable atomic propositions consist of "Reset" and "Reboot" (shown in bold in Figure 4. Both W, W' are explanations for O in Figure 4. In W, the server glitches, reboots itself and requests the client to reset. Upon receiving the request from the server, the client then resets itself. In W', the server breaks down and reboots itself after repairing, while the client resets itself following detection of loss of connection ("LostConn") and failure to re-establish the connection.

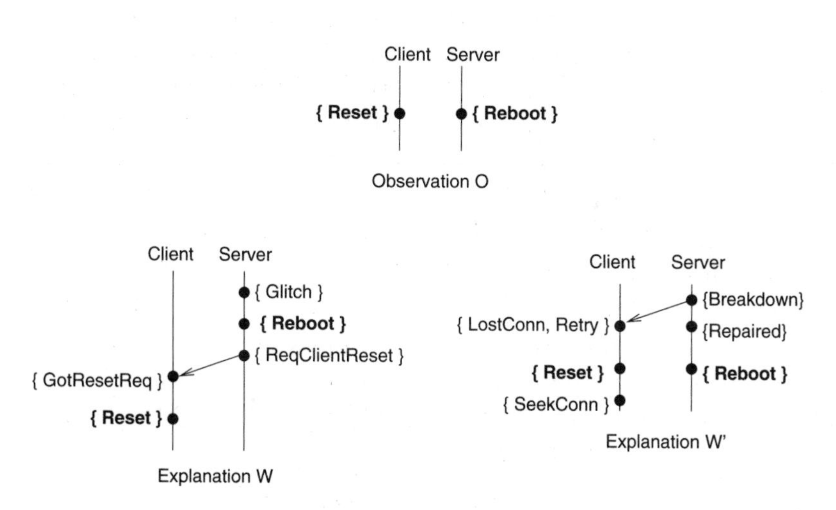

Figure 4: *Two computations W and W', and an observation O*

We emphasize that while an explanation W represents a full execution of a system, an observation induced by W, may have recorded information from a part of W. Thus, the image of f may be a proper subset of the states of W whose valuation contains observable atomic propositions. Similarly, some causal dependencies among observed states may not have been saved during the observation process. On the examples of Figure 4, we can notice that there is a causal dependency between the states of W labeled by $reboot$ and $reset$, but that this causality does not appear in the observation O. However, we suppose the observer is faithful, that is, it does not create wrong states or causalities. Note that for given O and W, if the mapping f from O to W exists, then it is necessarily unique. Thus, for each state $s \in S_O$, it makes sense to call $V_W(f(s))$ the *W-valuation of s*.

Definition 6 *Let O be an observation, and* $\{\Phi_p\}_{p\in\mathcal{P}}$ *be a* $\mathcal{P}$*-index family of formulae. W is a* $\{\Phi_p\}_{p\in\mathcal{P}}$-explanation *for O iff W is an explanation for O and* $W \models \{\Phi_p\}_{p\in\mathcal{P}}$.

Now we can define the diagnosis problem associated with LPOC.

Definition 7 *Let* $\mathcal{P}$ *be a set of processes,* $\mathcal{A}_{ob}$ *and* $\mathcal{A}_{ex}$ *be two sets of atomic propositions as before. The* LPOC-diagnosis *problem is defined as follows: given an observation* $O = (S_O, \eta_O, \leq_O, V_O)$ *and a* $\mathcal{P}$*-indexed family of formulae* $\{\Phi_p\}_{p\in\mathcal{P}}$ *over* $(\mathcal{P}, \mathcal{A}_{ob}, \mathcal{A}_{ex})$*, determine whether there exists a* $\{\Phi_p\}_{p\in\mathcal{P}}$*-explanation for O.*

In what follows, we will often omit the subscript $p \in \mathcal{P}$. The formulae $\{\Phi_p\}$ specify some knowledge about executions of the system, for instance some faulty behaviors. The objective of diagnosis is to figure out whether there exists an explanation for what has been observed, and hence detect if the actual behavior of the whole system was faulty or not. In case an explanation exists, we also want to obtain more detailed information about the possible truth values of propositions in $\mathcal{A}_{ex}$ at each observed state. We define the *(explanatory) summary* of O under $\{\Phi_p\}_{p\in\mathcal{P}}$ as the mapping $g : S_O \to 2^{\mathcal{A}_{ex}}$ such that for every $s \in S_O$, $g(s) \subseteq \mathcal{A}_{ex}$, and $a \in g(s)$ iff there exists an explanation W for O with $W \models \{\Phi_p\}_{p\in\mathcal{P}}$ and a is in the W-valuation of s. Thus, when the answer to the diagnosis problem is positive, we want further to compute the summary of O. Unfortunately, in the general case, the LPOC-diagnosis problem is undecidable (and so is the computation of summaries).

Theorem 8 *The LPOC-diagnosis problem is undecidable.*
Proof sketch: By a reduction from the Post Correspondence Problem (PCP), similar to that used in decision problems related to message sequence charts [9]. The complete proof of this theorem can be found in the extended version.

4 Diagnosis with K-Influencing Explanations

We have shown in previous section (Theorem 1) that the diagnosis problem is undecidable in general. There are two usual ways to overcome this problem. The first one is to consider a decidable fragment of the logic. Note however, that encoding a PCP becomes possible as soon as there is a way to describe sequences of properties located on a given process, and to define a mapping of states on different processes that respects the ordering. This is why in most cases very small fragment of partial order logics become undecidable when no restriction is imposed on the kind of model considered. Then, the question that naturally arises is whether we can identify a subclass of computations for which LPOC diagnosis is tractable. For this, we identify the subclass of K-influencing computations, and show that the LPOC diagnosis problem (and thus computing the summary) is decidable within this class.

Definition 9 *Let $\mathscr{P}$ be a set of processes, $\mathscr{A}_{ob}$ be a set of observable atomic propositions, $\mathscr{A}_{ex}$ be a set of atomic explanatory propositions. Let $W = (S, \eta, \leq, V)$ be a computation, and $p, q \in \mathscr{P}$ with $p \neq q$. The* causal degree *of p towards q in W is the maximum integer $n \in \mathbb{N}$ for which there exist $s_1, s_2, \ldots, s_n$ in S_p, and $s'_1, s'_2, \ldots, s'_n$ in S_q such that:*

(i) $s_1 < s_2 < \ldots < s_n$ and $s'_1 < s_2 < \ldots < s'_n$.
(ii) for $i = 1, 2, \ldots, n$, $s_i \ll s'_i$, that is, s_i is a predecessor of s'_i.
(iii) $s'_1 \not< s_n$.

For $K \in \mathbb{N}$, W is K-influencing *iff for any pair of processes p, q in $\mathscr{P}$ with $p \neq q$, the causal degree of p towards q is at most K.*

Intuitively, the causal degree of p towards q is the maximal number of events that precede some event on q that p can execute without having to wait for q. The general shape of $K-$influencing computations is illustrated in Figure 5. We now state the main result of this section.

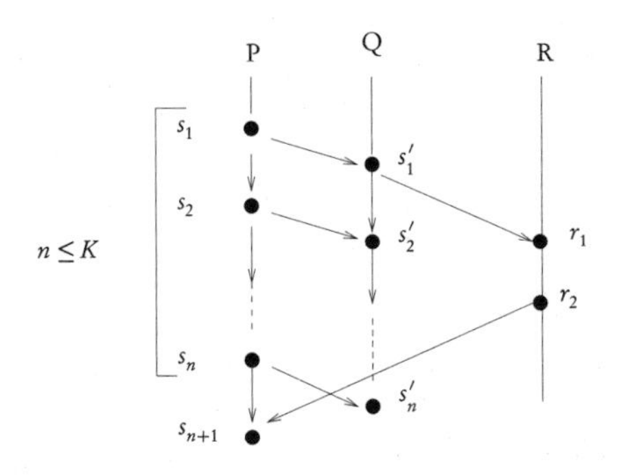

Figure 5: *$K-$influencing computations*

Theorem 10 *Given an observation O, a $\mathscr{P}$-indexed family $\{\Phi_p\}_{p\in\mathscr{P}}$ of LPOC formulae, and an integer $K \in \mathbb{N}$, one can effectively determine whether there exists a K–influencing computation W which is a $\{\Phi_p\}_{p\in\mathscr{P}}$-explanation for O.*

An important consequence of the above theorem is that one can effectively compute a summary of the K-influencing explanations of O. Let O be an observation and $\{\Phi_p\}_{p\in\mathscr{P}}$ a $\mathscr{P}$-indexed family of formulae. We can slightly adapt the definition of summaries in section 3 to K-influencing computations: the *K-summary* of O under $\{\Phi_p\}_{p\in\mathscr{P}}$ is the mapping $g : S_O \to 2^{\mathscr{A}_{ex}}$ such that for every s in S_O, $g(s) \subseteq \mathscr{A}_{ex}$, and $a \in g(s)$ iff there exists a K-influencing explanation W for O with $W \models \{\Phi_p\}_{p\in\mathscr{P}}$ and a is in the W-valuation of s.

Corollary 11 *Given an observation O, a $\mathscr{P}$-indexed family $\{\Phi_p\}_{p\in\mathscr{P}}$ of LPOC formulae, and an integer $K \in \mathbb{N}$, one can effectively compute the K-summary of O under $\{\Phi_p\}_{p\in\mathscr{P}}$.*

Through the rest of this section, we prove Theorem 10 and Corollary 11. We fix the integer K, the observation O and the formulae $\{\Phi_p\}_{p\in\mathscr{P}}$. Recall from section 2 that we can easily construct a single formula Φ such that for any computation W, $W \models \Phi$ iff $W \models \{\Phi_p\}_{p\in\mathscr{P}}$. In what follows, we fix Φ. We will assume that the computations used hereafter are nonempty. It will be clear from the proof that this involves no loss of generality. We let $\mathscr{W}_K$ denote the set of K-influencing computations.

The proof for Theorem 10 consists of two steps. Firstly, we show that K-influencing computations can be identified with Mazurkiewicz traces [5] over a suitable trace alphabet (Σ, I). This way, we can identify K-influencing computations with equivalence classes of finite sequences in $\Sigma^\star$. This encoding is in spirit the same as the of encoding of universally K-bounded message sequence charts with traces in [12]. Secondly, we construct three finite state automata Aut_K, Aut_Φ, Aut_O, running over linearizations of traces of (Σ, I). Aut_K checks if an input sequence represents a computation of $\mathscr{W}_K$. For a sequence σ representing a computation W_σ in $\mathscr{W}_K$, Aut_Φ accepts σ iff $W_\sigma \models \Phi$, and Aut_O accepts σ iff W_σ is an explanation for O. The crux is the construction of Aut_Φ. We shall give Aut_Φ in the form of a two-way alternating automaton, which can be transformed to a finite state automaton. The basic idea is similar to [7] and the usual translation from LTL to alternating automata [18]. The new technicality in our construction is in checking conformance with formulae of the form $\downarrow_{m,A}(T)$, $\uparrow_{m,A}(T)$. Then, there exists a sequence in $\Sigma^\star$ accepted by Aut_K, Aut_O, Aut_Φ iff $\mathscr{W}_K$ contains a computation W such that $W \models \Phi$

and W is an explanation for O. This then establishes Theorem 10. For Corollary 11, we will show that for any state s of O and any atomic proposition $a \in \mathcal{A}_{ex}$, one can find a finite state automaton $Aut_{s,a}$ which has the following property: if a sequence σ represents a computation W_σ in $\mathcal{W}_K$ such that $W_\sigma \models \Phi$, and W_σ is an explanation of O, $Aut_{s,a}$ accepts σ iff a is in the W_σ-valuation of O. As a result, one can then effectively compute the K-summary of O under $\{\Phi_p\}_{p \in \mathcal{P}}$.

Encoding K-influencing computations with Traces

We recall that a Mazurkiewicz trace [5] alphabet is a pair (Σ, I) where Σ is a finite alphabet, and $I \subseteq \Sigma \times \Sigma$ is an irreflexive and symmetric relation called the *independence relation*. The *dependence relation* D is given by $(\Sigma \times \Sigma) \setminus I$. A (finite) Σ-labelled poset is a pair $(E, \sqsubseteq, \lambda)$, where E is a finite set, $\sqsubseteq \subseteq E \times E$ a partial order, and $\lambda : E \to \Sigma$ a labeling function. As usual, we write $e \sqsubset e'$ if $e \sqsubseteq e'$ and $e \neq e'$. We let $\hat{\sqsubset} \subseteq E \times E$ denote the least relation whose reflexive and transitive closure is equal to $\sqsubseteq$. A (Mazurkiewicz) trace over (Σ, I) is a Σ-labelled poset $tr = (E, \sqsubseteq, \lambda)$ satisfying: (i) for any $e, e' \in E$, $\lambda(e) D \lambda(e')$ implies $e \sqsubseteq e'$ or $e' \sqsubseteq e$; (ii) for any $e, e' \in E$, $e \hat{\sqsubset} e'$ implies $\lambda(e) D \lambda(e')$. We define isomorphism of traces in the obvious way and write $tr = tr'$ if the traces tr, tr' are isomorphic.

Let $[K] = \{0, 1, \ldots, K-1\}$. We define the alphabets $\Gamma_{pre} = \{pre(p,i) \mid p \in \mathcal{P}, i \in [K]\}$, and $\Gamma_{suc} = \{suc(p,i) \mid p \in \mathcal{P}, i \in [K]\}$. Let $\Gamma = \mathcal{P} \times 2^{\Gamma_{pre}} \times 2^{\Gamma_{suc}} \times 2^{\mathcal{A}}$. We define Σ as the subset of Γ satisfying: $(p, \{pre(p_1,i_1), \ldots, pre(p_g,i_g)\}, \{suc(p'_1,i'_1), \ldots, suc(p'_h,i'_h)\}, A) \in \Sigma$ iff $p_1, \ldots, p_g$ are distinct members of $\mathcal{P} \setminus \{p\}$, and $p'_1, \ldots, p'_h$ are distinct members of $\mathcal{P} \setminus \{p\}$. We now define the dependence relation $D \subseteq \Sigma \times \Sigma$ as: $(p, PRE, SUC, A)\ D\ (p', PRE', SUC', A')$ iff one of the following conditions holds:

- $p = p'$.
- $p \neq p'$, and for some $i \in [K]$, $pre(p', i) \in PRE$ and $suc(p, i) \in SUC'$.
- $p \neq p'$, and for some $i \in [K]$, $suc(p', i) \in SUC$ and $pre(p, i) \in PRE'$.

We set the independence relation $I = \Sigma \times \Sigma - D$. It is trivial to verify that (Σ, I) is a trace alphabet. From now on, we fix the trace alphabet (Σ, I).

Let $W = (S, \eta, \leq, V)$ be a K-influencing computation. Let us define the Σ-labeling of W, denoted λ_W^Σ (or simply λ_W), as the following function from S to Σ: for $s \in S$, $\lambda_W(s) = (p, PRE, SUC, A)$ where:

- $p = \eta(s)$.
- $pre(q, i) \in \Gamma_{pre}$ is in PRE iff s has predecessor s' with $\eta(s') = q$ (such a s' is necessarily unique by the definition of predecessor) and $i = j \mod K$ where j is the number of states s'' in S_p satisfying $s'' < s$ and that s'' has a predecessor of location q.
- Further, $suc(\hat{q}, \hat{i}) \in \Gamma_{suc}$ is in SUC iff s has a successor $\hat{s}'$ with $\eta(\hat{s}') = \hat{q}$ (such a $\hat{s}'$ is also unique) and $\hat{i} = \hat{j} \mod K$ where $\hat{j}$ is the number of states $\hat{s}''$ in S_p satisfying $s'' < s$ and that s'' has a successor of location q.
- $A = V(s)$.

Note that PRE and SUC are not necessarily singletons, as a state may have one successor (reps. predecessor) on each process. Let us define $tr(W) = (S, \leq, \lambda_W)$. Remark that for $W \in \mathcal{W}_K$, $tr(W)$ is a trace over (Σ, I). Furthermore, if W' is a K-influencing computation, then $W = W'$ iff $tr(W) = tr(W')$. Figure 6 below shows an example of a 2-influencing computation with the associated labeling. For the sake of clarity, the subsets of atomic propositions that are true at each state are not explicitly given, but only described by $A_1, \ldots A_9$.

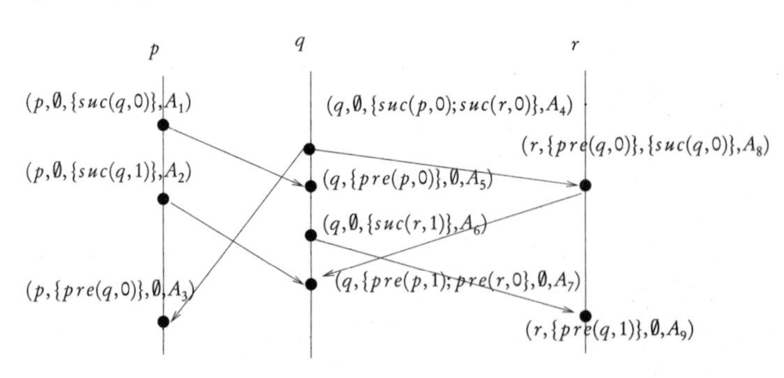

Figure 6: *A 2-influencing computation and its Σ-labeling*

A *linearization* of a trace $(E, \sqsubseteq, \lambda)$ is a sequence $\lambda(e_1) \ldots \lambda(e_n)$ where $e_1, \ldots, e_n$ are distinct members of E, $E = \{e_1, \ldots, e_n\}$, and for any $i, j \in \{1, \ldots, n\}$, $e_i \sqsubseteq e_j$ implies $i \leq j$. For any computation $W \in \mathcal{W}_K$, by a Σ-*linearization of* W, we refer to a linearization of $tr(W)$. We denote by $Lin^\Sigma(W)$ the set of Σ-linearizations of W. Let $Lin_K^\Sigma = \bigcup_{W \in \mathcal{W}_K} Lin^\Sigma(W)$. Note that computations in $\mathcal{W}_K$ can be uniquely constructed from sequences in Lin_K^Σ. For a non-null sequence σ in $\Sigma^\star$, let $last(\sigma)$ denote the last letter of σ. For σ, σ' in $\Sigma^\star$, we write $\sigma \precsim \sigma'$ iff σ is a prefix of σ'. We define the partial order $\sqsubseteq_{\Sigma^\star} \subseteq \Sigma^\star \times \Sigma^\star$ via: $\sigma \sqsubseteq_{\Sigma^\star} \sigma'$ iff σ, σ' are non-null, $\sigma \precsim \sigma'$ and there exist $\sigma_1, \sigma_2, \ldots, \sigma_h$, $h \leq |\Sigma|$, such that $\sigma \precsim \sigma_1 \precsim \sigma_2 \ldots \precsim \sigma_h \precsim \sigma'$ and $last(\sigma) D\, last(\sigma_1) D\, last(\sigma_2) \ldots last(\sigma_h) D\, last(\sigma')$. For every $\sigma \in Lin_K^\Sigma$, we define the computation $poc(\sigma) = (S_\sigma, \eta_\sigma, \leq_\sigma, V_\sigma)$, where:

- S_σ is the set of non-empty prefixes of σ.
- For $\tau \in S_\sigma$, $\eta_\sigma(\tau) = p$ iff $last(\tau) = (p, PRE, SUC, A)$ for some PRE, SUC, A.
- $\leq_\sigma$ is the restriction of $\sqsubseteq_{\Sigma^\star}$ to S_σ.
- For $\tau \in S_\sigma$, $V_\sigma(\tau) = A$ iff $last(\tau) = (p, PRE, SUC, A)$ for some p, PRE, SUC.

It is easy to verify that $poc(\sigma)$ is well-defined. Furthermore, for every $\sigma \in Lin_K^\Sigma$ and $W = poc(\sigma)$, we have $W \in \mathcal{W}_K$ and $\sigma \in Lin^\Sigma(W)$.

Automata construction.

We can build three finite state automata Aut_K, Aut_O, Aut_Φ which have the following properties:

- For $\sigma \in \Sigma^\star$, σ is accepted by Aut_K iff $\sigma \in Lin_K^\Sigma$.
- $\sigma \in Lin_K^\Sigma$ is accepted by Aut_O iff $poc(\sigma)$ is a K-influencing explanation for O.
- $\sigma \in Lin_K^\Sigma$ is accepted by Aut_Φ iff $poc(\sigma)$ satisfies Φ.

It follows that a sequence $\sigma \in \Sigma^\star$ is accepted by the product of Aut_K, Aut_O, Aut_Φ iff $poc(\sigma)$ is a K-influencing $\{\Phi_p\}$-explanation for O.

We do not detail the construction of Aut_K and Aut_O, that can be found in the extended version of this paper [19].

Proposition 12 *Let $\mathcal{P}$ be a set of processes, $\mathcal{A}$ be a set of atomic proposition, K be an integer, and Σ be Mazurkiewicz trace alphabet computed from $\mathcal{P}$, $\mathcal{A}$ and K. Then, there exists an automaton Aut_K of size $\mathcal{O}(|\Sigma|^{K \cdot |\mathcal{P}|^2})$ that recognizes linearizations of K-influencing computations over $\mathcal{P}$ with valuations in $\mathcal{A}$.*

We can reuse the construction of Aut_K to build Aut_O, the automaton that recognizes K-influencing explanations of O. At each state of this automaton, one must recall a state of Aut_K reached (ie, the current $K-$influencing linearization explored), the part of O that is embedded in this explanation, and some additional information about the causalities that may appear in the future.

Proposition 13 *Let O be an observation over a set of processes $\mathscr{P}$, with valuations in an alphabet $\mathscr{A}_{ob}$. Let K be an integer and $\mathscr{A} \supseteq \mathscr{A}_{ob}$ be a set of atomic propositions. The size of Aut_O is at most in $\mathcal{O}(|Aut_K| \cdot 2^{|O|} \cdot 2^{|\mathscr{P}|} \cdot (|O| \cdot |\mathscr{P}|)^2 \cdot (|\mathscr{P}| \cdot K+1)^K)$.*

Construction of Aut_Φ

We next give the description of Alt_Φ, a two-way alternating automaton [13] that recognizes linearizations of K–influencing computations satisfying Φ. This automaton can then be transformed into a standard finite state automaton Aut_Φ.

We introduce some new atomic formulae in order to simplify the structure of Φ. Recall that for a computation W and a state s of W, the m-view or the m-frontier of s contains at most $\mathscr{N}_m = \sum_{i=0}^{m} |\mathscr{P}|^i$ states. We introduce formulae of the form $\downarrow_m(T), \uparrow_m(T)$, where $m \in \mathbb{N}$ and T is a computation containing at most $\mathscr{N}_m$ states. Let $W = (S, \eta, \leq, V)$ be a computation and $s \in S$. Then $W, s \models \downarrow_m(T)$ iff the m-view of s is isomorphic to T. The semantics of $\uparrow_m(T)$ is given similarly. We note that a formula $\downarrow_{m,A}(T)$ is equivalent to $\bigvee_{T' \in \mathscr{T}} \downarrow_m(T')$, where $\mathscr{T}$ is the collection of computation T' such that T' contains at most $\mathscr{N}_m$ states and the projection of T' onto A is isomorphic to T. Similarly, we can write $\uparrow_{m,A}(T)$ as a disjunction of formulae $\uparrow_m(T)$ with an analogous semantics.

Let $W = (S, \eta, \leq, V)$ be a computation and $s \in S$. Let $s' \in S$ and $\tau = a_1 a_2 \ldots a_n$ be a non-null sequence in $\Sigma^\star$. If there exist $s_1, \ldots, s_n \in S$ such that $s' = s_n$, $s_n \ll s_{n-1} \ll \ldots s_1 \ll s$ and $\lambda_W(s_i) = a_i$ for $i = 1, \ldots, n$, then we say s' is a *τ-ancestor* of s. Recall that each state in W has at most $|\mathscr{P}|$ predecessors, one belonging to each S_p. Thus, we can in fact say s' is *the* τ-ancestor of s. We introduce formulae of the form $\downarrow(\tau, \tau')$ where τ, τ' are non-null sequences in $\Sigma^\star$. We define $W, s \models \downarrow(\tau, \tau')$ iff there exist states $\hat{s}, \hat{s}'$ such that $\hat{s}$ is the τ-ancestor of s, $\hat{s}'$ is the τ'-ancestor of s, and $\hat{s} \leq \hat{s}'$.

We argue that a formula $\downarrow_m(T)$ can be equivalently written as a boolean combination of formulae of the form $\downarrow(\tau, \tau')$. Assume without loss of generality of T contains a maximum state s_{max}. Thus, every state in T is the τ-ancestor of s_{max} for some τ of length at most the number of states of T. Hence, $\downarrow_m(T)$ is equivalent to asserting for each pair of states s, s' in T, whether $\downarrow(\tau, \tau')$, $\downarrow(\tau', \tau)$, or

$\neg\downarrow(\tau,\tau')\wedge\neg\downarrow(\tau',\tau)$, where s,s' are the respectively the τ-ancestor and τ'-ancestor of s_{max}.

Analogously, we define *τ-descendants* and introduce formulae of the form $\uparrow(\tau,\tau')$ where τ,τ' are non-null sequences in $\Sigma^\star$. It follows that a formula $\uparrow_m(T)$ is equivalent to a boolean combination of formulae of the form $\uparrow(\tau,\tau')$.

With the new formulae introduced above, we can assume without loss of generality that Φ is formed from $\neg,\wedge,\vee$ and the atomic formulae loc_p, $\downarrow(\tau,\tau')$, $\uparrow(\tau,\tau')$, $\mathsf{EX}\varphi$, $\mathsf{EU}(\varphi,\varphi')$. Furthermore, negations in Φ only apply to atomic formulae.

Now we are ready to describe the two-way alternating automaton Alt_Φ. The basic elements of Alt_Φ are similar as in usual translations of temporal logics to alternating automata (see e.g. [18]). The main difficulty is to deal with atomic formulae of the form $\downarrow(\tau,\tau')$, $\uparrow(\tau,\tau')$.

We informally recall some basics of two-way alternating automata and refer to [4, 13] for details. Let Alt be a two-way alternating automaton. An input word is delimited on the left by a left marker and on the right by a right marker. Initially, Alt is at the initial state with the head at the first letter of the input word. Upon reading the letter of the current head position, Alt can spawn several copies where each copy can move the head left or right and go to a new control state. Which combination of copies can be spawned are pre-determined by a transition relation. A run of Alt over an input word σ is a (finite) tree, where each branch terminates upon reaching the left or the right marker. And Alt accepts σ iff there exists a run over σ such that every leaf contains an accepting state.

For clarity, we describe only informally the operations of Alt_Φ. The exact construction of Alt_Φ can be found in the extended version. For illustration purpose, we fix an input word $\sigma=a_1a_2\ldots a_n$ in Lin_K^Σ. We write $\sigma,i\models\varphi$ iff $poc(\sigma),a_1\ldots a_i\models\varphi$.

Let $SF(\Phi)$ be the set of subformulae of Φ and their negations, where $\neg\neg\varphi$ is identified with φ. A state z of Alt_Φ consists of a formula φ in $SF(\Phi)$ and some alphabetic constraints. Such a state z must verify that φ holds at the current head position i and that the alphabetic constraints should be satisfied subsequently.

Note that $poc(\sigma)\models\Phi$ iff $\sigma,h\models\Phi$ where $a_1a_2\ldots a_h$ is a minimal state in $poc(\sigma)$. Thus, at the initial state, Alt_Φ searches for position h such that $a_j\ I\ a_h$ for $j=1,\ldots,h-1$, and upon reaching position h, it verifies that Φ holds at h.

It now suffices to explain how Alt_Φ verifies that a formula in $SF(\Phi)$ holds at the current head position. We proceed inductively from the atomic formulae of forms loc_p, $\downarrow(\tau,\tau')$, $\uparrow(\tau,\tau')$ and their negations, then to formulae of forms $\mathsf{EX}\varphi$, $\mathsf{EU}(\varphi,\varphi')$, and their negations. We then study conjunction and disjunction of formulae.

Firstly, Alt_Φ can easily check if loc_p or $\neg loc_p$ holds at the current head position, simply from the letter at the head position. Next we consider atomic formulae of the form $\downarrow(\tau,\tau')$, $\uparrow(\tau,\tau')$ and their negations. For the input sequence σ, we let $a_i = (p_i, PRE_i, SUC_i, A_i)$ for each i. Recall that $poc(\sigma) = (S_\sigma, \eta_\sigma, \leq_\sigma, V_\sigma)$ where S_σ is the set of prefixes of σ. For $s, s' \in S_\sigma$, we write $s \lll_\sigma s'$ iff s is a predecessor of s' in $poc(\sigma)$. Consider $g, h \in \{1,2,\ldots,n\}$, it is easy to see that $a_1\ldots a_g \lll_\sigma a_1\ldots a_h$ iff $g < h$ and one of the following conditions holds:

- $p_g = p_h$. And for each index i with $g < i < h$, $p_i \neq p_g$.
- $p_g \neq p_h$ and $a_g\ D\ a_h$. Further, there do *not* exist indices $i_1, i_2, \ldots, i_t$, $t \leq |\Sigma|$, such that $g < i_1 < i_2 < \ldots < i_t < h$, and $a_g\ D\ a_{i_1}\ D\ a_{i_2} \ldots a_{i_t}\ D\ a_h$.

For a formula $\downarrow(\tau,\tau')$, where $\tau = b_1 b_2 \ldots b_m$, $\tau' = b'_1 b'_2 \ldots b'_{m'}$, we note that $\sigma, \ell \models \downarrow(\tau,\tau')$ iff there exist indices $\ell_1, \ldots, \ell_m$, $\ell'_1, \ldots, \ell'_{m'}$, in $\{\ell+1, \ell+2, \ldots, n\}$ such that:

- $a_1 a_2 \ldots a_\ell \lll_\sigma \rho_1 \lll_\sigma \rho_2 \lll_\sigma \cdots \lll_\sigma \rho_m$, where $\rho_i = a_1 a_2 \ldots a_{\ell_i}$ for $i = 1,2,\ldots,m$. And $a_{\ell_i} = b_i$ for $i = 1,2,\ldots,m$. This asserts that the τ-ancestor of $a_1 a_2 \ldots a_l$ exists.
- $a_1 a_2 \ldots a_\ell \lll_\sigma \rho'_1 \lll_\sigma \rho'_2 \lll_\sigma \cdots \lll_\sigma \rho'_{m'}$, where $\rho'_i = a_1 a_2 \ldots a_{\ell'_i}$ for $i = 1,2,\ldots,m'$. And $a_{\ell'_i} = b'_i$ for $i = 1,2,\ldots,m'$. This asserts that the τ'-ancestor of $a_1 a_2 \ldots a_l$ exists.
- $a_1 a_2 \ldots a_{\ell_m} \leq_\sigma a_1 a_2 \ldots a_{\ell'_{m'}}$, that is, $a_1 a_2 \ldots a_{\ell_m} \sqsubseteq_{\Sigma^*} a_1 a_2 \ldots a_{\ell'_{m'}}$. This asserts that the τ-ancestor of $a_1 a_2 \ldots a_l$ causally precedes the τ'-ancestor of $a_1 a_2 \ldots a_l$.

Thus, to verify that a formula $\downarrow(\tau,\tau')$ or its negation holds at the current position, Alt_Φ moves to the left until it hits the left end marker and along the way checks the existence of indices $\ell_1, \ldots, \ell_m, \ell'_1, \ldots, \ell'_{m'}$ satisfying the above conditions. Analogously, it is clear how Alt_Φ can verify if a formula $\uparrow(\tau,\tau')$ or its negation holds at the current head position.

Finally, we note that formulae of forms $\mathsf{EX}\varphi$, $\mathsf{EU}(\varphi,\varphi')$ and their negation can be handled as in usual translations of temporal logics over traces to alternating automata (e.g. [7]). This is also the case for conjunction and disjunction of formulae. This completes the description of Alt_Φ.

It is not difficult to see that the number of states of Alt_Φ is of complexity $O(2^{|\Phi|} \cdot |\Sigma|^{|\Sigma| \cdot m})$, where m is the maximum length of τ, τ' for all atomic formulae of the form $\uparrow(\tau, \tau'), \downarrow(\tau, \tau')$. It follows from [13] that Alt_Φ can be transformed to a finite state automaton Aut_Φ with $2^{N \cdot 2^N}$ states where N is the number of states of Alt_Φ. Checking for the existence of an explanation then consists in checking the emptiness of the intersection of Aut_Φ and Aut_O built in proposition 13. The proof of Theorem 10 is now completed. ⊣

Proof of corollary 11

To prove Corollary 11, we first recall that if W is an explanation for O, then the injective mapping from the states of O to the states of W dictated in the definition of explanation is unique. Thus, it is easy to see that for any state s of O and any atomic proposition $a \in \mathscr{A}_{ex}$, one can construct a finite state automaton $Aut_{s,a}$ which has the following property: if σ a sequence σ representing a computation W_σ in $\mathscr{W}_K$ where $W_\sigma \models \Phi$ and W_σ is an explanation of O, $Aut_{s,a}$ accepts σ iff a is in the W_σ-valuation of s. $Aut_{s,a}$ can then be easily constructed from Aut_O by requiring that transitions that add s to the subset of observed states of O are labelled by letters with valuations that contain a. As a result, one can then effectively compute the K-summary of O under $\{\Phi_p\}_{p \in \mathscr{P}}$, by testing for each state s of O, each a in $\mathscr{A}_{ex}$, the non-emptiness of the product of Aut_Φ and $Aut_{s,a}$. ⊣

If the formula ϕ is such that all frontiers and views used are at most m-frontiers or m-views, then one can determine whether there exists a K-influencing explanation W for an observation O with complexity $\mathcal{O}(W_1 . 2^{W_2 . 2^{W_2}})$, where:

$$W_1 = |\Sigma|^{K . |\mathscr{P}|^2} \cdot 2^{|O|} \cdot 2^{|\mathscr{P}|} \cdot (|O| \cdot |\mathscr{P}|)^2 \cdot (|\mathscr{P}| \cdot K + 1)^K$$

$$W_2 = 2^{\left(|\phi| \cdot \mathscr{N}_m^2 \cdot 2^{\mathscr{A}^{ex}} \cdot \sum_{i=0}^{\mathscr{N}_m} f(i)\right)} \cdot |\Sigma|^{|\Sigma| \cdot m}$$

with $f(i) = i \cdot 2^{\frac{i^2}{4} + \frac{3i}{2} + ln(i)}$, and $\mathscr{N}_m = \frac{1 - |\mathscr{P}|^{m+1}}{1 - |\mathscr{P}|}$. From the definition of summaries, computing a summary for O can then be done in $\mathcal{O}(|O| \cdot |\mathscr{A}^{ex}| \cdot W_1 \cdot 2^{W_2 \cdot 2^{W_2}})$. The proof of these complexity results is not provided here, but can be found in the extended version of this paper [19].

5 Related Work and Conclusion

We have proposed a diagnosis framework based on a new partial order logic (LPOC) over partial orders (i.e. the truth of formulae is evaluated at local states). Unsuprisingly, satisfiability of LPOC formulae, and hence diagnosis are not decidable without restriction. To keep decidability of diagnosis, a restriction called K-influence is imposed on the models. As LPOC uses the existential until operator, for a given K, LPOC restricted to K-influencing computations is not definable in the first order logic over the Mazurkiewicz traces encoding K-influencing computations. However, it can be easily translated to MSO formulae. An interesting work would be to look for a fragment that is expressively complete for the first order logic over the traces encoding K-influencing computations.

Even with the restriction to K-influencing computations, diagnosis is very expensive (several exponential in the size of the formula and exponential in the size of the observed behavior). This high complexity could mean that diagnosis with LPOC is unfeasible. Note however that this complexity is in the worst cases. For instance, the exponential in the size of the observation comes from the maximal number of configurations in a partial order. In practice, for an observation with a bounded number of processes, the number of configurations can be much slower. The other costly part of the diagnosis problem comes from the translation from alternating automata to finite state automata. Again this is a worst case complexity. Note also that the translation of LPOC formulae into conjunction of formulae of the form $\uparrow(\tau,\tau')$, $\downarrow(\tau,\tau')$ is costly only when the pattern considered in the formulae are large. In general, basic patterns used in partial order languages such as message sequence charts are rather small, and we argue that this should also be the case with LPOC formulae. Some complexity gains can hence be expected by restricting the size and the number of partial order templates considered, but also the modalities of the formulae. Note however that most of the modalities chosen for LPOC seem important. The simple example of section 2 shows that the Until operator is essential to express properties of the form "when T1 occurs, T2 will occur later". One may also try to restrict the use of negation, that is LPOC formulae would only be conjunctions of positive assertions on the occurrence of patterns. Note however that the translation of an LPOC formula to a simplified formula on causal chains uses negation when two states of a pattern are not causally related. Hence, even in a restricted setting, negation of some properties will have to be checked. So, the small complexity gain that could occur may not justify the loss of expressiveness due to a restriction on negations.

In [16], D.Peled shows that model checking TLC^- formulae on High-level Message Sequence charts (HMSCs) is decidable. TLC^- is a subset of TLC that only contains next and until temporal operators, and describes the shape of causal chains in all the partial orders generated by a HMSC. TLC^- is clearly less expressive than LPOC.

The Propositional Dynamic Logic (PDL) for message passing systems proposed by [3], extends dynamic LTL for traces [11]. Model checking PDL properties over HMSCs is PSPACE complete. [15] proposes a local logic LD0 and several extensions over computations, with future and past modalities, and show that in the general case, satisfiability is undecidable. However, these logics become decidable when considering models of bounded size, or when computations can be organized as successive layers of finite message exchanges. LD0 only describes chains of causally related events occurring in the future or in the past of a local state, while the template matching in LPOC allows to describe a complete partial order in a bounded future or past of a local state. LPOC is then more discriminating than LD0, and if we restrict our models to Message Sequence Charts (a partial order where locality of events and messages are explicitly represented), it is also more expressive and discriminating than TLC^- and PDL.

Note also that for TLC^-, PDL, or $LD0$, partial orders are seen as models of formulae, but not as elements of the logic itself. The closest approach mixing logic and partial orders is called "Template Message Sequence Charts" [9]. A template MSC is an MSC that comports some “hole” and incomplete messages. Roughly speaking, models for a template MSC are obtained by filling the holes with new partial orders, and matching sendings and receptions of messages. The authors increase the power of template MSCs with pre/post condition operators. The models of these formulae are MSCs. This logic is very expressive, but satisfiability is undecidable when no bound is assumed on the set of models considered. However, a restricted fragment of the logic is proposed to model check existentially bounded Communicating Finite State Machines. Note however that models for template MSC formulae are MSCs, while models for LPOC formulae are arbitrary computations. Even if we only consider LPOC formulae over MSCs, LPOC and template MSCs remain uncomparable. On one hand, holes in template MSCs are not necessarily descriptions of what happens in the future or in the past of an event. By filling hole, one may add concurrent events, i.e. it is possible to say with template MSCs that whenever an action a occurs on process p, a concurrent action b occurs on process q. Clearly, this kind of formula can not be expressed with LPOC. On the other hand, some LPOC formulae that use the until operator do not find their equivalent in template MSC.

Note also that the works in [16],[3] and [9] rely on the existentially bounded nature of models to ensure decidability of model checking (that is, there is a bound b such that every MSC considered possess a linearization where the size of communication channel never exceeds b). This is not sufficient in our case to obtain decidability of diagnosis, as the PCP encoding of section 3 is existentially bounded. The K-influencing restriction is then closer to the universal bound on MSCs (the contents of communication channels in all linearizations of MSCs is bounded by some integer b) needed to model check HMSCs with global logics [1]. It might be interesting to see whether the layered computation restriction of [15] is sufficient to make diagnosis with LPOC formulae decidable.

References

[1] Rajeev Alur and Mihalis Yannakakis Model Checking of Message Sequence Charts. *CONCUR*, (1999) 114–129.

[2] Albert Benveniste, Eric Fabre, Claude Jard, and Stefan Haar. Diagnosis of asynchronous discrete event systems, a net unfolding approach. *IEEE Transactions on Automatic Control*, **48**(5),(2003) 714–727.

[3] Benedikt Bollig, Dietrich Kuske, and Ingmar Meinecke. Propositional Dynamic Logic for Message-Passing Systems. *FSTTCS*, (2007) 303–315.

[4] J.A. Brzozowski and E.L Leiss. On Equations for Regular Languages, Finite Automata, and Sequential Networks.*TCS*, **10**, (1980) 19–35.

[5] Volker Diekert and Gregor Rozenberg, editors. Book of Traces. *World Scientific*, Singapore, (1995).

[6] Colin Fidge. Logical time in distributed computing systems. *Computer*, **24**(8), (1991) 28–33.

[7] Paul Gastin and Madhavan Mukund. An Elementary Expressively Complete Temporal Logic for Mazurkiewicz Traces. *ICALP*, (2002) 938–949.

[8] Thomas Gazagnaire and Loïc Hélouët. Event Correlation with Boxed Pomsets. *FORTE*, (2007) 160–176.

[9] Blaise Genest, Markus Minea, Anca Muscholl, and Doron Peled. Specifying and Verifying Partial Order Properties Using Template MSCs. *FoSSaCS*, (2004) 195–210.

[10] Loïc Hélouët, Thomas Gazagnaire, and Blaise Genest. Diagnosis from Scenarios. *WODES*, (2006) 307–312.

[11] Jesper.G. Henriksen and P.S. Thiagarajan. Dynamic Linear Time Temporal Logic. *Ann. Pure Appl. Logic*, **96**(1-3), (1999) 187–207.

[12] Dietrich Kuske. Regular sets of infinite message sequence charts. *Information and Computation*, **187**(1), (2003) 80–109.

[13] Richard.E. Ladner, Richard.J. Lipton, and Larry.J. Stockmeyer. Alternating Pushdown and Stack Automata. *SIAM J. Comput.*, **13**(1), (1984) 135–155.

[14] Friedemann Mattern. On the relativistic structure of logical time in distributed systems. *Parallel and Distributed Algorithms*, (1989) 215–226.

[15] B. Meenakshi and Ramaswamy Ramanujam. Reasoning about Layered Message Passing Systems. *VMCAI*, (2003) 268–282.

[16] Doron Peled. Specification and Verification of Message Sequence Charts. *FORTE*, (2000) 139–154.

[17] M. SAMPATH, R. SENGUPTA, S. LAFORTUNE, K. SINNAMOHIDEEN, AND D.C TENEKETZIS. Failure diagnosis using discrete-event models. *IEEE Transactions on Control Systems Technology*, **4**(2), (1996) 105–124.
[18] MOSHE.Y VARDI. An Automata-Theoretic Approach to Linear Temporal Logic. *Banff Higher Order Workshop*, (1995) 238–266.
[19] SHAOFA YANG, LOÏC HÉLOUËT, AND THOMAS GAZAGNAIRE. Logic based diagnosis for distributed systems. *INRIA Technical report*, (2009).

Author Index